Forensic Speaker Recognition

Amy Neustein • Hemant A. Patil
Editors

Forensic Speaker Recognition

Law Enforcement and Counter-Terrorism

Springer

Editors
Amy Neustein
Linguistic Technology Systems
800 Palisade Avenue
Suite: 1809
Fort Lee, New Jersey 07024
USA
amy.neustein@verizon.net

Hemant A. Patil
Dhirubhai Ambani Institute of Information
and Communication Technology (DA-IICT)
Near Indroda Circle
Room 4103, Faculty Block 4
Gandhinagar, Gujarat 382007
India
hemant_patil@daiict.ac.in

ISBN 978-1-4614-0262-6 e-ISBN 978-1-4614-0263-3
DOI 10.1007/978-1-4614-0263-3
Springer New York Dordrecht Heidelberg London

Library of Congress Control Number: 2011939216

Printed on acid-free paper

Springer is part of Springer Science+Business Media (www.springer.com)

Foreword

Collection and storage of speech has become a very common phenomenon in recent times, thanks to the availability of the necessary electronic devices such as microphones and memory. All personal computers, cell phones and the like come equipped with these devices. With literally billions of people over the world having mobile phones, audio records are rapidly getting built up, sometimes without the knowledge of the user. In fact, many business and financial transactions are carried out over the phone without any authenticating documentation, thus creating a host of new legal problems. However, if this new mode of business is in the future likely to replace (with some degree of regularity) the conventional signed paperwork, we will need a robust authentication method for voice.

Yet, even with the increased use of voice technology, it seems highly unlikely at the moment that courts of law will accept the current speaker recognition technology as forensic evidence on par with signed documents, fingerprints, or DNA. The reason is that compared to a fingerprint or DNA or even a handwritten signature, voice has far greater variability. In addition, fatigue, common cold, emotions, among other factors can change the voice sample sometimes beyond recognition. In fact, in everyday life we as humans can sometimes incorrectly identify a speaker, so imagine how difficult it for a machine to consistently identify a speaker accurately. In spite of such limitations, which undoubtedly mitigate the evidentiary weight of speaker identification and verification findings that are presented to the court, speaker recognition can still play a significant role as a prime investigative tool in criminal prosecutions.

One of the goals in bringing this science to a higher level of performance must be to broaden the field of speaker recognition, or more aptly "voice" recognition, so that the scientific work of identifying a speaker would effectively incorporate the speaker's vocal tract characteristics into the identification process.

In addition to speaker recognition from good quality recordings, extensive research is needed to handle situations where signal to noise ratio is poor, or where many persons are speaking at the same time (e.g., multi-speaker speech in a meeting or a conference). In truth, recording levels can fluctuate over a large range, and some parts of speech may even take the form of whispers. The latter may significantly alter voice characteristics, resulting in little or no voicing, shift of lower formants, change in energy and duration characteristics, spectral slope, among other features.

One should not neglect the need for improved transcription either. Good transcription is important for improving the clarity of the record so that it can be listened to with ease. Swedish professor Anders Eriksson illuminates the readers to the importance of accurate transcription in his chapter titled "Aural/acoustic versus automatic methods in forensic phonetic case work". What we learn from this is that while undoubtedly making the voice sharp and clear for the purpose of transcription, some sounds, which perhaps are not voice-like at all, could be lost or distorted altogether, and these may be just the ones which hold vital clues for an important forensic investigation.

In the last analysis, forensic work requires a multi-disciplinary approach. We benefit from many years of research in developing digital signal processing techniques and in the understanding of the acoustics and aero-acoustics of production of speech (voice) signal. Moreover, as the signal processing does not have to be in real-time and can be done repeatedly, a larger variety of approaches, particularly non-linear methods such as Teager Energy Operator and computationally intensive methods—especially those involving searches of large databases—could find their place in the mainstream of forensic research.

The current volume brings together an excellent set of location markers, signposts and starting points for an interesting journey ahead.

Dhirubhai Ambani Institute of Information
and Communication Technology (DA-IICT)
Gandhinagar, India

Prof. S. C. Sahasrabudhe
(IEEE Fellow), Director

Preface

Forensic Speaker Recognition: Law Enforcement and Counter-Terrorism is an anthology of the research findings of thirty-five speaker recognition experts from around the world. The book provides a multidimensional look at the complex science involved in determining whether a suspect's voice truly matches forensic speech samples, collected by law enforcement and counter-terrorism agencies, that are associated with the commission of a terrorist act or other crime. Given the serious consequences for the suspect, who may go to jail or even (in the most extreme cases involving terrorism or murder) face the death penalty, a rigorous and reliable science must be used in finding a match between a suspect's voice and the speech samples collected in the forensic crime lab. The United States Supreme Court's ruling in *Daubert v. Merrell Dow Pharmaceuticals*, 509 U.S. 579 (1993) established a test for the legal admissibility of scientific evidence which requires that the theory and method upon which the evidence is based is testable, accepted, peer reviewed, and, where applicable, has a known equal error rate (EER). Similar standards for the validity and reliability of scientific evidence are used in other countries that have taken up the recommendations of the *National Academy of Sciences* (USA) and the *Law Commission* (UK).

Standards like these place a heavy burden on the expert who offers testimony in court. Each time forensic testimony is entered into evidence, the expert witness must prove *ab initio* that his science is reliable and valid in the case that is before the court before his or her testimony can properly qualify for admissibility. It follows that if forensic scientific methods are to be useful in legal contexts, they must hold up under judicial scrutiny, since in the end the question of admissibility will be decided by a trial judge.

Hence the consistent stress on bringing speaker authentication methods into line with the strict standards of legal admissibility is exactly what the reader will find in the work of this volume's diverse group of forensic speech scientists, whether they work side by side with investigators in crime labs, provide services to private companies that specialize in the design of speaker verification systems, or teach in university settings where they study (among other things) the effects of speech signal degradation on the quality of forensic speech samples. Forensic speaker recognition, as a probative science, must competently assist criminal investigators in minimizing both the occurrence of a "false positive"—in which the speech sample related to the commission of a crime or terrorist act is matched to the wrong suspect—or a "false negative," in which the real culprit's voice fails to match the crime lab's speech sample meant to fit him.

Although divided into eighteen chapters, addressing such varied topics as the challenges of forensic case work, handling speech signal degradation, analyzing features of speaker recognition to optimize voice verification system performance, and designing voice applications that meet the practical needs of law enforcement and counter-terrorism agencies, this book's material all sounds a common theme: how the rigors of forensic utility are demanding new levels of excellence in all aspects of speaker recognition. The book's contributors are among the most eminent scientists in speech engineering and signal processing; their work represents the best to be found at universities, research institutes for police science, law enforcement agencies and speech companies, in such diverse countries as Switzerland, Sweden, Italy, France, Japan, India and the United States.

Forensic Speaker Recognition opens with an historical and procedural overview of forensic speaker recognition as a science. Following this is a fascinating exposition by Professor Andrzej Drygajlo of the Swiss Federal Institute of Technology in Lausanne, whose chapter focuses on "the research advances in forensic automatic speaker recognition (FASR), including data-driven tools and related methodology that provide a coherent way of quantifying and presenting recorded voice as biometric evidence." Professor Drygajlo furnishes the reader with an in-depth discussion of the

> European Network of Forensic Science Institute's evaluation campaign through a fake (simulated) case, organized by the Netherlands Forensic Institute, as an example where an automatic method using the Gaussian mixture models (GMMs) and the Bayesian interpretation (BI) framework were implemented for the forensic speaker recognition task.

This first section, aptly titled "Forensic Case Work," is further enriched by the investigations of Swedish professor Anders Eriksson (of the University of Gothenburg) into the specific challenges of forensic case work. Drawing on a substantial number of investigations performed for the Swedish police, the author inspects in painstaking detail the differences between the aural/acoustic and the automatic methods in forensic case work, focusing on what works and what doesn't in real-life settings. The section concludes with a fascinating study of speaker profiling, based on the characteristics associated with speaker dialect. Manisha Kulshretha, a Haskins' Laboratory (Yale University) researcher, together with C. P. Singh of the Forensics Science Laboratory, Government of NCT of Delhi, and Professor R. M. Sharma of Punjab University, show that from a sample size of 210 speakers, acoustic features associated with lexical tone and sentence intonation, along with vowel quality and vowel duration, serve potentially to identify the speaker's particular dialect. Where dialect is an important element of identification, this method helps investigators to appreciably narrow the pool of potential suspects to those who reside in the region where that particular dialect is spoken.

The second section of the book, titled "Speech Signal Degradation: Managing Problematic Conditions Affecting Probative Speech Samples," devotes considerable attention to the stubborn problem of speech signal degradation that impedes the gathering of probative speech samples (that is, samples gathered for use in court processes) that are clear and audible. Since criminals, in their zeal to cover their tracks, often lower their voices even to a whisper, or make calls from public places where there is loud noise in the background (or use VoIP networks) the quality of the voice recording is often poor. Thus, much of the speech data available for forensic analysis are degraded by several factors such as background noise, transmission and channel impairments, microphone variability, multi-party conversations, whispered speech, and VoIP artifacts. As a result, speech scientists, as part of their efforts to manage the problematic conditions affecting the quality of probative speech samples, must carefully isolate and measure the effects of all such factors on speech signal degradation.

The authors presented in this section have met that challenge head-on. They bring to the discussion the results of years of careful study of degraded speech on the performance of an automatic speaker recognition (ASR) system, by concentrating on the following problems and, where available, their possible solutions:

1. speech under stress and the "Lombard Effect";
2. the wide range of artifacts of VoIP (speech codec, packet loss, packet reordering, network jitter, foreign-cross talk or echo) and the effect of such artifacts on the performance of an ASR system;
3. session variability ("mismatched" environments for collection of speech samples) and the use of the non-linear modeling techniques of Teager Energy Operator-based Cepstral Coefficients (TEOCC) and amplitude versus frequency modulation (AM-FM) to improve speaker recognition in mismatched environments;
4. noisy environments and the use of speaker-specific prosodic features to improve speaker recognition;
5. noisy backgrounds and the use of various noise reduction filters (Noise Reduction, Noise Gate, Notch Filter, Bandpass, and Butterworth Filter) in enhancing the speech signal for speaker identification; and
6. whispered speech and the use of an algorithm for whisper speech detection as part of a seamless neutral/whisper mismatched closed-set speaker recognition system.

This section has far too many contributors to name each one individually. They include University of Texas Professor John H. L. Hansen, University of Minnesota Professor Keshab K. Parhi, Raghunath S. Holambe, professor at SGGS Institute of Engineering and Technology, Nanded, India, and Jiju P. V., Senior Scientific Officer at the Forensic Science Laboratory, Government of NCT of Delhi, among other distinguished speech signal experts.

The third section, titled "Methods and Strategies: Analyzing Features of Speaker Recognition to Optimize Voice Verification System Performance in Legal Settings," presents the experimental research findings of some of the most innovative and forward-looking speech scientists who have isolated the important features of speaker recognition (some of which are appreciably less affected by signal degradation than others), and have carefully analyzed how such features may play an important role in improving forensic automatic speaker recognition.

The section begins with the experimental findings of Kanae Amino of the National Research Institute of Police Science in Japan (together with her research collaborators), showing that nasal sounds are effective for forensic speaker recognition despite the differences in speaker sets and recording channels. They show how "performance degradation caused by the channel difference, in this study of air- and bone-conduction … can be redressed by devising normalisation methods and acoustic parameters."

Next, T. V. Ananthapadmanabha, CEO of Voice and Speech Systems in Bangalore, describes his careful studies of the volume-velocity airflow through the glottis (or the glottal airflow). In so doing, he has explored the significance of speech source characteristics by utilizing rigorous analytical results from the aerodynamic and acoustic theory of voice production. Much of this work was inspired by the author's research collaboration with the late Professor Gunnar Fant at the Royal institute of Technology, Stockholm in the early 1980s. "A good understanding of the theory guides one in appropriate modeling and interpretation of voice source," he writes. In addition, Dr. Ananthapadmanabha contends that "habitually formed relative dynamic variations in voice source parameters are of greater significance in forensic speaker recognition."

The section is further enhanced by the analytic insights of Leena Mary, professor at Rajiv Gandhi Institute of Technology, Kottayam, India on the effectiveness of syllable-based prosodic features for speaker recognition. In her chapter, Professor Mary describes in painstaking detail a method for extracting prosodic features directly from the speech signal itself. "Applying this method," she tells us, speech is segmented into syllable-like regions using vowel onset points (VOP). The locations of VOPs (which entail Hilbert envelope of the linear prediction (LP) residual signal) serve as reference for extraction and representation of prosodic features."

Significantly, Professor Mary deliberately chose to analyze prosody—which reflects the learned/acquired speaking habits of a person and therefore contributes to speaker recognition—in as much as prosodic features are less affected by channel mismatch and noise, which are common causes of speech signal degradation in probative speech samples. Thus, prosodic features are particularly well suited to speaker forensics, a field that demands accurate identification of suspects and therefore a minimum of obstacles to robust speaker recognition, such as those posed by channel transmission problems.

The section is rounded off by the study findings of C. Chandra Sekhar, professor at the Indian Institute of Technology (IIT), Chennai, India, and his graduate student assistant, A. D. Dileep. The authors meticulously show that when the performance of Intermediate Matching Kernel (IMK)-based Support Vector Machines (SVMs) is compared to that of state-of-the-art GMM-based approaches to speaker identification (using the 2002 and 2003 NIST speaker recognition corpora in evaluation of different approaches to speaker identification), the IMK-based SVMs performed significantly better than the GMM-based approaches for speaker identification tasks. From this comparison, the authors draw the conclusion that because IMK-based SVMs are well suited to the basic challenges of providing reliable scores for intra-speaker variation of suspects and for inter-speaker variation within a potential population, they can play an important role in serving the needs of law enforcement and counter-terrorism agencies in performing forensic speaker recognition.

The final section of the book, titled "Applications to Law Enforcement and Counter-Terrorism," enlightens the reader about practical constraints in the use of forensic speaker recognition systems for the daily concerns of law enforcement and counter-terrorism agencies. The section begins with the research of V. Ramasubramanian, who serves as a senior member of Siemen's (Bangalore) technical staff, on automated telephony surveillance to detect if a person from a specific government watch-list is on the line at a given moment. As he points out:

> [S]uch an automatic solution is of considerable interest in the context of homeland security, where a potentially large number of wire tapped conversations may have to be processed in parallel, in different deployment scenarios and demographic conditions, and with typically large watch-lists, all of which make manual lawful interception unmanageable, tedious and perhaps even impossible.

His chapter begins with the "basic framework for watch-list based speaker-spotting, namely, open-set speaker identification, subsequently refined into a 'multi-target detection' framework." Dr. Ramasubramanian examines in detail "the main theoretical analysis available within the framework of multi-target identification, leading to performance predictions of such systems with respect to the watch-list size as the critical factor."

Taking an applications-oriented approach to forensic speaker recognition, he then outlines related speech topics—speaker change detection, speaker segmentation and speaker diarization—that can be useful in the design of automated telephony surveillance for border security and protecting critical infrastructure. These and other issues and concerns inhabit the broader context of homeland security. The author concludes with a summary of product level solutions currently available in the context of surveillance and homeland security applications, while acknowledging the realistic challenges and limitations faced by automated speaker-spotting systems.

Next, Patrick Perrot of the Forensic Research Institute of the French Gendarmerie and Gerard Chollet of Telecom-Paris take up the fascinating topic of criminals who disguise their voices to hide their actual identity, sometimes even impersonating someone else. Such disguises typically occur when criminals make telephone threats, malicious calls, extortion attempts and/or blackmail, or terrorist demands. The authors point out that while

> there are those cases when there are involuntary voice changes, as when there are alterations in voice characteristics due to poor transmission of telephonic communication … or even pathologies (both acute and chronic) that morph speech production … we limit this discussion to disguise which consists of a person who *deliberately* conceals his identity … as a means of misleading the human ear or even the automatic speaker recognition system.

Drs. Perrot and Chollet focus on specific voice characteristics to evaluate the recognition of a suspect's voice in the presence of voice disguise. Their analyses of voice transformation are based both on an acoustic approach, which they use to measure specific changes in speech, and on an automatic approach, which is employed to detect voice disguise. The acoustic analysis of specific features reveals that the effect of the disguise on voice characteristics is dependent upon the kind of disguise that is used, while in the automatic experiment the authors performed, they found that parallel fusion and SVM classifier provided the best results with a good level of discrimination.

The practical applications of these two French scientists' work can be seen in the fact that a major part of their research into voice disguise has been devoted to the study of voice disguise reversibility. Their studies of voice disguise reversibility have revealed that

> while it is not possible today to fully reverse a voice disguise in such a way that that the resulting waveform would sound completely natural to a listener (mainly due to limitations with the quality of converted voice synthesis), our study demonstrates, nevertheless, that a disguised voice could be reversed to a relatively "normal" voice as evaluated by current state of the art speaker verification systems.

Thus, the authors see a more robust speaker recognition on a reverse-disguised voice—that is, a voice that has already been converted back from its disguised form to normal speech—as a future practical application of their research, as well as evaluation of the performance of speech applications in such contexts.

The last two chapters of the book are authored by speech experts at Nuance Communications and Loquendo. Chuck Buffum, Nuance's Vice President, provides insightful lessons learned from commercial voice biometric deployments to forensic applications, giving the reader a better understanding of the evolution of speaker verification systems in forensic settings. Mr. Buffum points out that "commercial deployments of voice biometrics have predictably focused primarily on automating the correct acceptance of true users for telephony self-service. However, over the past few years, a trend has developed within the financial institutions to begin using voice biometric technology to look for duplicate enrollments or to investigate suspicious transaction activity," a trend that, he contends, "opens the discussion of bringing relevant techniques and experiences from commercial voice biometric deployments into the forensic voice biometric space."

Avery Glasser, consulting architect for the Italian-based company Loquendo, closely examines the practical needs of anyone wishing to implement investigatory voice biometric technology, and how best to bridge the gap between creators and implementers of this technology. The author points out that "there are critical problems that only voice biometrics can solve, but getting the solutions well positioned requires a deep understanding of the nature of government implementations that seems to escape the grasp of too many vendors. The chapter," according to Mr. Glasser's exordium, "will explore a number of critical use cases and provide perspective on how technology creators can position their solutions to meet those needs."

As the editors of this compendium we have endeavored to bring together notable forensic speaker recognition experts who, by virtue of their meticulous research and keen attentiveness to the needs of law enforcement and counter-terrorism agencies, have both individually and collectively brought forensic automatic speaker recognition (FASR) technology to a new plane. It is our hope that this science will continue to evolve so that the admissibility of speaker recognition evidence will no longer present a Sisyphean challenge to prosecutors who have come to depend on voice verification systems to make a convincing case to the court about the identity of a criminal suspect.

Fort Lee, NJ, USA — Amy Neustein, Ph.D.
Gandhinagar, India — Hemant A. Patil, Ph.D.

Contents

Contributors

Kanae Amino, Ph.D. National Research Institute of Police Science, 6-3-1 Kashiwanoha, Kashiwa-shi, Chiba 277-0882, Japan
e-mail: amino@nrips.go.jp

T. V. Ananthapadmanabha, Ph.D. Voice and Speech Systems, 53, "Girinivas", Temple Road, 13th Cross, Malleswaram, Bangalore 560003, India
e-mail: tva_vss@yahoo.com, tva.blr@gmail.com

Takayuki Arai, Ph.D. Department of Electrical and Electronics Engineering, Sophia University, 7-1 Kioi-cho, Chiyoda-ku, Tokyo 102-8554, Japan
e-mail: arai@sophia.ac.jp

Chuck Buffum, B.S. Nuance Communications, 1198 E. Arques Avenue, Sunnyvale, CA 94085, USA
e-mail: Chuck.Buffum@nuance.com

Saswat Chakrabarti, Ph.D. G.S. Sanyal School of Telecommunications, Indian Institute of Technology Kharagpur, Kharagpur 721302, West Bengal, India
e-mail: saswat@ece.iitkgp.ernet.in

Gérard Chollet, Ph.D. CNRS-LTCI, Telecom ParisTech, 46 rue Barrault, Paris 75013, France
e-mail: gerard.chollet@telecom-paristech.fr

Aaron E. Cohen, Ph.D. Leanics Corporation, 1313 5t St. SE, Mail Unit 70, Minneapolis, MN 55414, USA
e-mail: cohen082@umn.edu

Mangesh S. Deshpande, M.E. Department of Electronics and Telecommunication Engineering, SRES's College of Engineering, Kopargaon 423603, Maharashtra, India

A. D. Dileep, M. Tech. Department of Computer Science and Engineering, Indian Institute of Technology Madras, Chennai 600036, Tamilnadu, India

Andrzej Drygajlo, Ph.D. EPFL Speech Processing and Biometrics Group, UNIL School of Criminal Justice, Swiss Federal Institute of Technology Lausanne (EPFL), University of Lausanne (UNIL), Lausanne, Switzerland
e-mail: andrzej.drygajlo@epfl.ch

Anders Eriksson, Ph.D. Phonetics, Department of Philosophy, Linguistics and Theory of Science, University of Gothenburg, Box 200, 40530 Gothenburg, Sweden
e-mail: anders.eriksson@ling.gu.se

Xing Fan, B.S.A. Center for Robust Speech Systems (CRSS), Department of Electrical Engineering, Erik Jonsson School of Engineering and Computer Science, The University of Texas at Dallas, Richardson, TX 75080-3021, USA

Avery Glasser, B.L.S. Loquendo S.p.A., a Telecom Italia Group Company, Via Arrigo Olivetti, 6, 10148 Torino, Italy
e-mail: averyglasser@loquendo.com

John H. L. Hansen, Ph.D. Department of Electrical Engineering, Center for Robust Speech Systems (CRSS), Erik Jonsson School of Engineering and Computer Science, The University of Texas at Dallas, Richardson, TX 75080-3021, USA
e-mail: john.hansen@utdallas.edu

Raghunath S. Holambe, Ph.D. Department of Instrumentation Engineering, SGGS Institute of Engineering and Technology, Nanded 431606, Maharashtra, India
e-mail: rsholambe@sggs.ac.in

Toshiaki Kamada, B.E. National Research Institute of Police Science, 6-3-1 Kashiwanoha, Kashiwa-shi, Chiba 277-0882, Japan
e-mail: kamada@nrips.go.jp

Wooil Kim, Ph.D. Department of Electrical Engineering, Center for Robust Speech Systems (CRSS), Erik Jonsson School of Engineering and Computer Science, The University of Texas at Dallas, Richardson, TX 75080-3021, USA
e-mail: Wooli.Kim@utdallas.edu

Shashidhar G. Koolagudi, M. Tech. School of Information Technology, Indian Institute of Technology Kharagpur, Kharagpur 721302, West Bengal, India
e-mail: koolagudi@yahoo.com

Manisha Kulshreshtha, Ph. D. Haskins Laboratories, Yale University, 300 George St., Suite 900, New haven, CT 06511, USA

Vuppala Anil Kumar, M. Tech. G.S. Sanyal School of Telecommunications, Indian Institute of Technology Kharagpur, Kharagpur 721302, West Bengal, India
e-mail: anil.vuppala@gmail.com

Hisanori Makinae, Ph.D. National Research Institute of Police Science, 6-3-1 Kashiwanoha, Kashiwa-shi, Chiba 277-0882, Japan
e-mail: makinae@nrips.go.jp

Leena Mary, Ph.D. Rajiv Gandhi Institute of Technology, Kottayam 686501, Kerala, India
e-mail: leena.mary@rit.ac.in

Takashi Osanai, Ph.D. National Research Institute of Police Science, 6-3-1 Kashiwanoha, Kashiwa-shi, Chiba 277-0882, Japan
e-mail: osanai@nrips.go.jp

Keshab K. Parhi, Ph.D. Department of Electrical and Computer Engineering, University of Minnesota, Minneapolis, MN 55455, USA
e-mail: parhi@umn.edu

Hemant A. Patil, Ph.D. Dhirubhai Ambani Institute of Information and Communication Technology, DA-IICT, Gandhinagar, India
e-mail: hemant_patil@daiict.ac.in

Patrick Perrot, Ph.D. Gendarmerie Operational Unit, Gendarmerie Nationale, 3 rue de Dampierre, 17400 Saint Jean d'Angely, France
e-mail: patrick.perrot@gendarmerie.interieur.gouv.fr

Jiju P.V., M.Sc. Documents Division, Forensic Science Laboratory, Govt. of NCT of Delhi, Madhuban Chowk, Rohini, New Delhi 110085, India

V. Ramasubramanian, Ph.D. Siemens Corporate Research & Technologies—India, Bangalore 560100, India
e-mail: V.Ramasubramanian@siemens.com

K. Sreenivasa Rao, Ph.D. School of Information Technology, Indian Institute of Technology Kharagpur, Kharagpur 721302, West Bengal, India
e-mail: ksrao@iitkgp.ac.in

Ramu Reddy, B. Tech. School of Information Technology, Indian Institute of Technology Kharagpur, Kharagpur 721302, West Bengal, India
e-mail: ramu.csc@gmail.com

Abhijeet Sangwan, Ph.D. Department of Electrical Engineering, Center for Robust Speech Systems (CRSS), Erik Jonsson School of Engineering and Computer Science, The University of Texas at Dallas, Richardson, TX 75080-3021, USA
e-mail: Abhijeet.Sangwan@utdallas.edu

C. Chandra Sekhar, Ph.D. Speech and Vision Laboratory, Department of Computer Science and Engineering, Indian Institute of Technology Madras, Chennai 600036, Tamilnadu, India
e-mail: chandra@cse.iitm.ac.in

R. M. Sharma, Ph.D. Department of Forensic Science, Punjabi University, Patiala 147002, Punjab, India
e-mail: rmsforensics@gmail.com

C. P. Singh, Ph.D. Physics Division, Forensic Science Laboratory, Government of NCT of Delhi, Madhuban Chowk, Rohini, New Delhi 110085, India

Chi Zhang, M.S.E.E. Center for Robust Speech Systems (CRSS), Department of Electrical Engineering, Erik Jonsson School of Engineering and Computer Science, The University of Texas at Dallas, Richardson, Texas 75080-3021, USA

Part I
Forensic Case Work

Chapter 1
Historical and Procedural Overview of Forensic Speaker Recognition as a Science

Kanae Amino, Takashi Osanai, Toshiaki Kamada, Hisanori Makinae and Takayuki Arai

Abstract Forensic phonetics and acoustics are nowadays widely used regarding police and legal use of acoustic samples. Among many tasks included in this area, forensic speaker recognition is considered as one of the most complex problems. Forensic speaker recognition, sometimes called forensic speaker comparison, is a process for making judgments on whether or not two speech samples are from the same speaker. This chapter introduces the historical backgrounds of forensic speaker recognition including "voiceprint" controversy, human-based visual and auditory forensic speaker recognition, and automatic forensic speaker recognition. Procedural considerations in forensic speaker recognition processes and factors that affect recognition performances are also presented. Finally, we will give a summary of the progress and developments made in the forensic automatic speaker recognition.

1.1 Introduction

It is quite a common experience that we identify familiar people by their voices alone: family members, friends, colleagues, actors and actresses, radio and television personalities, and so forth [1–4]. Without actually seeing them speaking, you may know who it is immediately. This ability of human beings, to identify speakers by speech sounds, can be thought as a part of the language performance ability that humans possess, and may have evolved from our ancestors and primates. Perception and emission of the speakers' (callers') individualities are reported in nonhuman primates [5–7]. In primates' communities, and also in humans, it is important to know whether an invisible individual is an ally or an enemy, a subordinate or an executive.

Apart from making for smooth communication, perceptual speaker identification is important for forensic research. The brief history is given in the next section. As

K. Amino (✉)
National Research Institute of Police Science, 6-3-1 Kashiwanoha, Kashiwa-shi, Chiba 277-0882, Japan
e-mail: amino@nrips.go.jp

A. Neustein, H. A. Patil (eds.), *Forensic Speaker Recognition*,
DOI 10.1007/978-1-4614-0263-3_1,

you will see, its history is directly related to the developments of the recognition methodology. According to Bricker and Pruzansky [8], speaker recognition methods can be classified into three categories: recognition by machine, by reading spectrograms, and by listening.

Forensic speaker recognition, too, can occur in any of these three ways. In forensics, however, various factors are known to influence recognition and identification accuracy, in addition to the known variability of speech in general. Speech samples being compared may be recorded through different situations, a speaker may be disguising his/her voice, or utterances may be made under the influence of drugs or stress. Furthermore, in forensic situations, speaker recognition may be performed by either a professional or a layperson, a so-called earwitness.

This chapter briefly summarises the history of forensic speaker recognition and the factors affecting forensic speaker recognition by machine and by humans, as well as recognition by earwitnesses. We will also see the developments in automatic speaker recognition and some major approaches and techniques taken in recent systems.

1.2 Brief History of Forensic Speaker Recognition

The history of forensic speaker recognition is rather long. The first use of individual identification by speech sounds in criminal cases is said to be in the seventeenth century, in the trial of King Charles I of England [9]. The first scientific investigation was performed in 1937, in the court case of abduction of Charles A. Lindbergh's son that occurred in 1932 [10]. Today, there are many court cases where voice identification appeared as evidence.

After the invention of telephones and recording equipment followed by the invention of spectrography, speech science and technology made a dramatic progress. These changes also appeared in forensic speech sciences. When sonagraph, a brand name of the spectrograph produced by Kay Elemetrics, was created on the basis of Steinberg's idea [11], not only the inter-speaker similarities in diction, but also the intra-speaker similarities in speech patterns were focused [12]. The term "voiceprint" was first invented by the two researchers at Bell Laboratory, Grey and Kopp [13, cited in 14]. They reported that speaker identification by examining spectrograms could offer good possibilities. Kersta [15] re-examined this method and claimed that inter-speaker spectrographic differences were greater than intra-speaker differences. The term "voiceprint" became popular among the general public. It is still in vogue today and often televised in police dramas, although nobody actually uses it in scientific context. Consequently, there is a misunderstanding that forensic speaker recognition is reliable just as individual identification using fingerprint or genetic fingerprint (DNA) [16].

It was in 1966 that this visual method was first used in a court case in the United States [9]; nonetheless, some scientists and researchers were suspicious about the authenticity and reliability of the method especially when used in the court cases.

Thus, the visual method for speaker recognition has been examined over and over again until recently [15, 17–26]. Young and Campbell [17] showed that the participants could identify speakers at 78.4% accuracy by using voiceprints excerpted from the same contexts, but the performance was degraded by 40% when the materials taken from different contexts were used.

Stevens et al. [19] made an issue of whether visual method could excel primitive aural speaker recognition. They compared the two methods of speaker recognition, i.e., recognition by spectrography and by listening, and concluded that the aural method was more accurate than the visual method; the error rate for aural identification was 6% compared to 21% for visual identification. They also claimed that the participants were more confident of their judgments during the aural tests. Authenticity of speaker recognition by listening has also been repeatedly examined [*e.g.* 27–34].

Since 1970s, computer-based or automatic speaker recognition has become mainstreams of the research. Compared to visual and aural speaker recognition, automatic method has an advantage that it enables us quantitative analysis of the speech signals and an explicit presentation of the algorithm. We will closely see how automatic speaker recognition has developed, both in general and in forensics, later in this chapter.

Although the aural and spectrographic (human-based) methods are dependent on the subjective judgments of the examiner, it is widely approved by the practitioners in criminal investigation and court cases. Robustness against noise or voice disguise is higher in human-based methods compared to computer-based automatic methods. Therefore, aural identification is a suitable method, or at least can be a good aid for other methods, when voice disguise exists [22, 24]. On the other hand, computers are suitable for storing and processing huge amount of data. When handling a large size data, speaker recognition by machine is much more efficient [35]. This advantage also allows us to make enough verification experiments of a speaker recognition system.

In forensics, the speech data available for the analyses are often recorded with a poor quality through telephone, and the speakers are not always cooperative added to which some of them may disguise their voices [36]. Thus in forensics, we may reasonably take a semi-automatic approach, where spectrographic and/or aural perceptual speaker recognition is applied after lowering the number of the possible suspects by police investigation using machine recognition.

Whether we use human-based or semi-automatic or automatic methods, it is important for us to always remember that there are limitations in forensic speaker recognition, and identification and authentication using speech are not always perfect. In 2003, Bonastre et al. [37] sent us need-for-caution messages. They listed up the important factors that we should care about when interpreting speaker recognition results, and concluded, regardless of recognition methods, that "there is no scientific process that enables one to uniquely characterise a person's voice or to identify with absolute certainty an individual from his or her voice." In the next section, we will summarise those factors affecting forensic speaker recognition, in relation to the speech production and perception.

1.3 Factors that Affect Speaker Recognition Performances

Denes and Pinson [38] conceived human speech communication as “a chain of events linking the speaker’s brain with the listener’s brain.” Speech production starts with the linguistic stage in the speaker’s brain; the speaker arranges the thoughts into linguistic forms. Then as the result of controlling the speech organs, speech sounds are produced and transmitted through the air to the listener’s hearing organs. In the listener’s brain, the hearing mechanism is activated and the listener perceives and recognises the linguistic messages. As we all know, the primary significance of the speech sounds is to convey some message, and the speech chain is one of the most suitable frameworks to deal with various speech communication phenomena. We will therefore discuss speech production, the perception and recognition of the speaker identity, and the factors affecting them by following each step in the speech chain. We will regard here automatic speaker recognition as analogous to human speaker recognition with only difference that the listener is a computer.

1.3.1 Arrangement of the Linguistic Form and Speakers’ Attitudes

At the very first of the speech communication, the speaker decides what to say and arranges the thoughts into linguistic form. The contents of the message are transformed into sequences of the phonemes, the minimal units of sounds in a language, following the phonotactic rules of the language in question. O’Shaughnessy [35] points out that “what she or he says” can sometimes serve as a clue to the speaker identity. Intentions of the speaker may affect the speech production, too. In forensics, we often meet uncooperative speakers, who attempt not to be identified and disguise their voices.

Voice disguise has a considerable detrimental effect on speaker identification. Kuenzel [39] pointed out that most common voice disguises are as follows: changes in the phonation (falsetto, creaky voice, and whisper), faking a foreign accent, and pinching one’s nose. Zhang and Tan [40] studied ten types of common disguises and their effect on automatic speaker recognition performances. They showed that the most influential disguises were to mask one’s mouth, to change the phonation type (whisper), and to raise the pitch. Among other disguise types, to put something in the mouth, to change the speaking rate, and to pinch one’s nostrils slightly affected the recognition rate, but mimicry of a foreign accent did not have any effect. Reich and Duke [41] also tested various disguises and concluded that hypernasality degraded the performances the most. Orchard and Yarmey [42] investigated the effect of whispered speech and found that it produced a significant decrease in the identification rate. They also found that the difference was smaller when both the reference and test samples were whispered.

Another way to hide speaker identity is to mimic other dialects than one's own. Dialect mimicry certainly has an effect on correct speaker recognition rates [43, 44]. The linguistic knowledge stored in speaker's brain includes social and regional dialects, which may appear in articulatory level as speaker-specific characteristics. Some dialectal properties, such as laryngeal control and other fine phonetic differences that non-phoneticians do not notice [45], cannot be mimicked by the speakers of other dialects, and these properties are predicted to be powerful clues to the speaker identity in a forensic situation.

1.3.2 Physiological Properties and Articulatory Gestures

After the speaker constructs the linguistic form of an utterance, the speaker's brain sends impulses along the motor nerves in order to activate the articulatory organs. This stage is relevant to two phonetic levels: segmental and supra-segmental. Speech segments are the individual consonants and vowels in phoneme sequences, which go together to make up syllables and words. Supra-segments, termed sometimes as prosody, are the aspects of speech controlled at the initiation and phonation of the speech sounds. Supra-segmentals are relevant to any phenomena that occur in a broader range than segments.

At the stage of speech production, we have static and dynamic speaker individualities; the former includes speakers' physiological and anatomical properties, such as the length and thickness of the vocal folds, the length and shape of the vocal tract, and so forth; and the latter is exemplified by articulatory habits, sometimes including articulatory disorders such as lisp. Lisp is a functional speech disorder and occurs most frequently with the segment class called sibilants: /s, z, ʃ, t͡ʃ/. In frontal lisp, these sounds are pronounced as interdentals /θ, ð/, and in lateral lisp, with lateral airflow.

Although both static and dynamic speaker individualities are subject to the change over time, it is more often to be seen in speaker's physiological characteristics than in dynamic ones. In forensics, speech samples used for speaker recognition may be obtained at different points in time; the time elapse of a year or two is not unusual. There are few longitudinal studies on this topic. Among them, House and Stevens [46] investigated utterances spoken on two occasions separated by thirty years. They found that the speakers' productions were remarkably consistent over thirty years; especially the patterns of the changes in duration and mean fundamental frequency among vowels were stable. Hollien and Schwartz [47, 48] tested the effect of non-contemporary speech on auditory speaker identification. The identification rates dropped from 95% for contemporary speech samples to 70–85% for latencies from four weeks to six years. They also found that there was a drastic decrease down to 35% correct identification for twenty year latency.

1.3.3 Produced Speech Signals and Transmission

Once speech sounds are uttered, they travel through the air between speaker and listener. Transmission of the sound and its acoustic quality are also important to forensics. The most common problems facing forensic speech scientists are background noise and transmission properties. Background noise here means everything except the speech signals that we use in order to obtain identifying information about the speaker. The noise may be an unexpected bang, reverberation in the room, competing speakers, radio or TV playing in the background, or BGM (background music) of a shopping mall.

Practically, a large proportion of the speech data used in forensic speaker recognition is recorded over the telephone. Since most linguistic information conveyed by the speech sounds is concentrated below 8 kHz, the telephone line transmits only limited frequency components between 300 and 3,400 Hz. We can still communicate with each other over the telephone, however, frequency components of the speech signals themselves are present over a wide range, and the effect of the band elimination is often disadvantageous in forensics. Generally speaking, cellular phones and female speakers are known to make the situation worse. The situation may further be influenced by the types and models of the phones. It is important to know how and to what extent the bandwidth limitations affect speaker recognition. An experimental study will be introduced in later chapter, where we investigated the effect of channel differences [49].

Many studies [50–52] report that the frequencies of the vowel formants, especially the first formant of a high vowel, are shifted in telephone-transmitted speech. According to Kuenzel [50], they were shifted by approximately 6% for both male and female speech. We need to exercise caution when we use formant data for speaker recognition when the speech materials are recorded over the telephone.

Female speakers usually have higher pitch than male speakers. Female voice is said to have about 1.7 times higher fundamental frequencies on average than male voice [53]. This also affects the acoustic properties of female speech. For example, compared to male voice, female voice has: more glottal leakage, shorter pulses, lower sound pressure level, steeper spectral slope, and more noise fill in interformant regions [54]. Higher fundamental frequencies mean, at the same time, larger intervals between harmonics in the spectra. These harmonic spacings make it difficult to analyse female speech in forensic speaker recognition and other speech technologies.

1.3.4 Perception and Recognition of the Speech Sounds

Once the speech signals reach the listener's hearing organs, nerve impulses are produced and the speech sounds are perceived. Considerable nerve activities occur in the listener's brain and these activities are further modified. These modifications somehow bring about the listener's recognition. As mentioned above, it is important for a successful communication to express an appropriate attitude and speech modality according to the person you are talking to. In order to investigate how we perceive and recognise people only by their voices, a lot of studies have been done in cognitive psychology field. Forensics-oriented studies exist a lot, too. One of the topics that the researchers are eager to know is the limitations and decay of the human memory; that is, how long we can retain our memory on someone's voice, how accurate we can discriminate people only by their voices, and so forth.

1.3.4.1 Memory Retention and Languages

McGehee [27, 28] investigated auditory speaker recognition and the effect of the memory decay. She found that the recognition rate decreased as a function of time, from 80% after a lapse of one day down to 13% after five months. Later studies have supported her results, although the rates of the decay may vary from study to study. She also investigated the effect of the foreign accents [27]. Generally speaking, there is a decrease in speaker recognition performances when the speech materials with a foreign accent or in a foreign language are presented [27, 55–59]. Thompson [56] showed that the listeners were better at recognising a voice speaking their own native language than another language. Koester and Schiller [58] obtained similar outcomes; they also showed that listeners with no knowledge of a language performed worse on speaker recognition using a foreign speech, while listeners with some knowledge of it could perform as well as the native listeners.

1.3.4.2 Familiarity to the Speakers

Familiarity with the speakers also affects the identification performances. Van Lacker et al. [3, 4] showed that identification of familiar speakers and that of unknown speakers go through different processes, i.e., the former is rather like pattern recognition, while the latter is more like feature analysis. Hashimoto et al. [60] claimed that the contribution of the acoustic parameters, such as spectral information, the fundamental frequency, and tempo information, would differ according to whether the listeners are identifying familiar speakers or unknown speakers. Correlation between the subjective estimation of the familiarity to the speakers and the accuracy of speaker identification was reported in Schmidt-Nielsen and Stern [2].

1.3.4.3 Duration and Contents of the Materials

Although human speaker recognition is quite accurate, especially when the listeners know the speaker quite well, it is not always perfect [34, 36, 55, 61, 62]. One of the factors on which the identification accuracy is dependent is the duration and content of the speech materials presented to the listeners. Pollack et al. [29] showed that the identification rates of the familiar speakers would increase as a function of the stimulus duration, but it gets saturated at around 1.2 s. In Bricker and Pruzansky [30] and other studies [45, 62], the relationship among speaker identification accuracy, duration, and phonemic variation of the stimuli was examined. They all concluded that the identification rate increased with duration only if the longer stimuli contained more phonological variation.

As pointed out by Nygaard [63] there is an interaction between perception of linguistic contents and that of speaker identity, although these two kinds of information "are considered to be processed and identified separately" in the brain. One piece of evidence of the interaction is that the accuracy of speaker recognition depends on what linguistic contents are presented to the listeners. O'Shaughnessy and other researchers [35, 37] suggest identifying the phonemes that best indicate a speaker's individuality by conducting perceptual speaker identification experiments and using these phonemes to specify the speakers in machine recognition. As for the phonological contents of the stimuli, we will see in detail by showing our experimental results in later chapter [49].

1.3.4.4 Other Factors

Some other experimental factors are pointed out in previous studies. McGehee [27] found that male listeners responded more accurately than female listeners, whereas Thompson [56] found no such differences. Roebuck and Wilding [64] pointed out a "same-gender advantage," where listeners performed better when they were of the same gender as the speaker. This difference was also observed in Cook and Wilding [65].

1.3.5 Perception and Recognition in the Case of Earwitnesses

Most of the factors described above are also relevant to earwitness' perception, but some additional factors should be mentioned as peculiar issues. First, and most importantly, real life earwitnesses are in most cases not prepared for the criminal situation, and hear the voice under stress. Researchers investigating witness' perception must remember that we can never replicate a real situation. It is known that the performances of unprepared listeners are much worse than those who are prepared [31].

Memory and judgments of eye- and ear-witnesses are sometimes obscure; and they can easily be influenced by other factors. For example, the presence of a weapon suppresses memorisation of a situation (weapon focus effect) [66], and press reports and erroneous information mislead their judgments (post-event information effect and misinformation effect, respectively) [67]. Two other phenomena are reported as to the degradation of the performances: the effects of making statements (verbal overshadowing effect) [68–71] and face presentation (face overshadowing effect) [65, 72], although the latter effect was not observed in other study [73]. Some studies report a weak correlation between witness confidence and recognition accuracy [31, 42, 74]; but this, too, was not found in other studies [34, 60, 75].

Earwitness identification sometimes takes the form of earwitness lineups. This is true when an earwitness hears a criminal speaking but have not seen him or her. Later, speech samples of a suspect is recorded and embedded in a group of speech samples produced by other people, and presented to the earwitness. These sets of speech samples are called earwitness lineups or voice parades. As to the earwitness lineups, there are many important questions: the size of the lineups, the position of the target speaker, effect of the target absence, etc. Generally speaking, correct speaker identification rate decreases as lineup size increases. Bull and Clifford [75] made experiments on the size of the speaker lineups; they found that lineups with five or six foils may be the optimal size. They also showed that there is an effect of target position only when the target was put first in the lineup. Target absence also affects identification accuracy. Kerstholt et al. and other studies are in agreement that speaker identification is more accurate for target-present lineups than target-absent ones [60, 76, 77]. In addition, most of them found a significant tendency that listeners accept the speakers rather than reject them in target-absent lineups [60, 77].

1.4 Development of Automatic Speaker Recognition

1.4.1 Technology Developments

Most of the above-mentioned factors also affect automatic speaker recognition performances. Some factors such as acoustic quality of the speech materials and selection of the analysis parameters directly affect the performances of the system.

Apart from the recognition methods, speaker recognition can also be classified with respect to the procedural tasks, namely, matching and naming [8]. These two tasks can be associated with two applications in automatic speaker recognition: speaker verification and speaker identification, respectively. In Fig. 1.1 we depict these two applications. In speaker verification, a system judges whether the speaker is the same person as she or he claims to be; thus the response is either yes or no. On the other hand, judgments made in speaker identification are which of the pre-registered speech samples matches those of the applicants; and the response returned is someone's name. What these two processes have in common is that they both inspect the similarity between the input speech signals and the reference speech stored in the data bank, and make judgments based on the comparison between them. Automatic speaker recognition systems have developed by improving one or more of these processes: features to be extracted, normalisation and comparison methods, population of the speech data bank, and so forth.

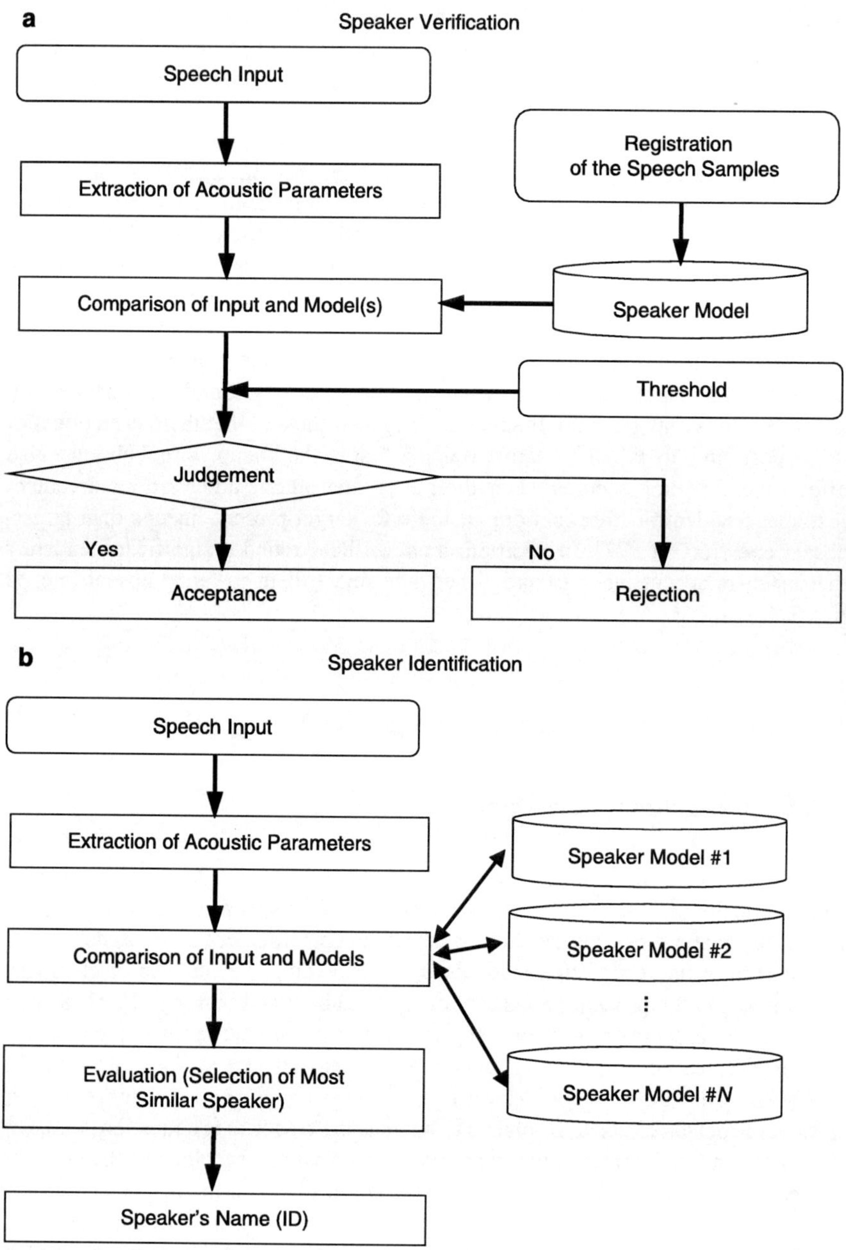

Fig. 1.1 Schematic representation for **a** speaker verification and **b** speaker identification. (Based on Furui [9])

In looking back over the chronological development of automatic speaker recognition, we notice that many of the 1970s research concentrated on searching for the effective acoustic features, and since mid-1980s the research focus shifted to the modeling of the speakers [78–100]. Also, the selected features for recognition have changed from the laryngeal source features such as fundamental frequency [78–88] to more filter-related features such as LPCC (Linear Prediction Cepstral Coefficients) and MFCC (Mel-Frequency Cepstral Coefficients) [89–100]. They reflect a speaker's vocal tract configurations, which is the predominant physiological individuality. In addition, modern systems use text-independent approaches rather than text-dependent template matching ones employed in "classic" studies. We also notice that recent systems take more probabilistic algorithms compared to previous ones.

In the past two decades, standardisation has been promoted and encouraged in speaker recognition, just as in other information technologies. Since 1996, NIST (National Institute of Standards and Technology) [101] has organised an SRE (Speaker Recognition Evaluation) every year or at least every other year [102]. It offers a speech corpus to work on, which is renewed each year, the performance criterion to achieve, and the time limit for challenge in accordance with the designated schedule. There are growing number of participants and published papers related to NIST evaluations. Doddington et al. [103] provides a detailed description on NIST-SRE, and Campbell et al. [16] gives a performance report on NIST-SRE systems.

1.4.2 Forensic Automatic Speaker Recognition Systems

Developments in automatic speaker recognition in the forensic area have followed a similar path with the speaker recognition technology in general. Modern forensic automatic speaker recognition (FASR) systems include MFCC as features, CMN (Cepstral Mean Normalisation), RASTA (RelAtive SpecTrAl) processing or analogous methods for channel compensation, and GMM (Gaussian Mixture Model) with UBM (Universal Background Model) normalisation for speaker modelling [104]. One important difference between commercial and forensic speaker recognition systems is that we must build and develop the system on legal and judicial perspectives [105], which means the evaluation part of the system is of greater weight.

In speaker verification system in general, the performance can be evaluated in terms of false rejection (FR, Type I error) rates and false acceptance (FA, Type II error) rates. These two types of errors are in tradeoff, and this tradeoff is used as a function of the decision threshold. Both types of errors (false rejection and false acceptance) must be specified in order to measure system performance, and in particular the relationship between Type I and Type II error. As such, the probability of FR versus FA, or the other way around, is depicted in the ROC (Receiver Operating Characteristic) curve or in the DET (Detection Error Tradeoff) curve [106]. In NIST-SRE, too, DET curve is recommended for evaluating the systems. This kind of evaluation method is suitable for commercial applications of speaker recogni-

Table 1.1 Selected recent literature on forensic automatic speaker verification

Lit.	Features	Speech materials	Pop.	Method
[108]	LPCC	Landline (JPN)	100 (M)	HMM
[109]	LPCC, delta-LPCC	Speech Corpus: Ahumada (SP)	25	GMM with BI
[110]	MFCC	Speech Corpus: Ahumada-Gaudi (SP)	249	GMM+CMN with BI
[111]	PLP Coef.	Landline (Swiss FR)	1000 (M)	GMM with BI
[112]	RASTA-PLP Coef.	Speech Corpora: Polyphone-IPSC02, NIST 2002 (EN)	12, 39	GMM with BI
[113]	MFCC	Speech Corpus: Ahumada III (SP)	61 (M)	GMM with UBM
[114]	FM, MFCC	Speech Corpus: NIST 2001 Cellular (EN)	174	GMM with BI
[115]	Vowel Formants	Speech Corpus: Pool Corpus (GER)	68 (M)	GMM with UBM
[116]	RASTA-PLP Coef.	Landline (GER)	182	GMM with UBM+MAP

tion. Besides, according to Bimbot et al. [107], some additional methods are proposed for forensic applications.

Nakasone and Beck [104] implemented "a confidence measure" of binary decisions in their automatic speaker verification system developed at the law enforcement agencies. In their system, the probabilistic certainty level of correctness is addressed for every verification decision (rejection/acceptance) based on statistics with known error rates generated from large sample populations. Another method is to use Bayesian approach through LR (Likelihood Ratio) of opposite hypothesis. Usually, LR is calculated as the ratio between the conditional probabilities of two competing hypotheses H0 and H1; where H0 stands for positive hypothesis (the suspected speaker is the source of the questioned recording), while H1 stands for the opposite hypothesis (the suspected speaker is not the source of the questioned recording) [110, 111].

Recent literature on FASR is summarised in Table 1.1 Most of them use above-mentioned filter-related features such as MFCC and vowel formants [110, 113–115]; some use PLP (Perceptual Linear Predictive) coefficients [111, 112, 116] proposed by Hermansky [117], or FM (Frequency Modulation) [114]. Also, some studies employ Bayesian Interpretation (BI) as the evaluation measure.

Selected chronology of the recent developments of FASR systems is now given in Table 1.2. Among these systems, IdentiVox developed by a research group in Madrid makes a practical implementation of the state-of-the-art method for FASR. It makes use of GMM as the basic technology, with options such as channel normalisation or UBM normalisation for speaker modelling. The system operates within a user-friendly platform (Win95/98/NT/2000/ME), and the user can configure, save, open, and print reports for each session, where all of the work is organised [122]. A commercial version of IdentiVox, Batvox, is marketed worldwide to law enforcement agencies in more than 22 countries by Agnitio S.L. [124]. It is reported that

Table 1.2 Established forensic automatic speaker recognition (FASR) systems. (Reviewed in [110] etc.)

System	Literature	Organisation	Country
SASIS	[118]	Rockwell International	U.S.A.
AUROS	[119]	Philips GmbH, BundesKriminalAmt (BKA)	Germany
C.A.V.I.S.	[120]	Los Angeles County Sheriff's Department	U.S.A.
IDEM	[121]	Fundazione Ugo Bordoni	Italy
SAUSI	[36]	University of Florida	U.S.A.
IdentiVox	[122]	Universidad Politecnica de Madrid	Spain
(N/A)	[123]	Swiss Federal Institute of Technology, Univ. Lausanne	Switzerland
SPES	[116]	BundesKriminalAmt (BKA)	Germany

by 2005 IdentiVox produced LRs which were considered sufficiently reliable for presentation in court. The number of the case reports has been grown from 30 in 2005 to 98 in 2008 [125].

1.5 Summary and Conclusion

In this chapter, we overviewed the history of forensic speaker recognition. We first summarized the backgrounds of the research developments in terms of the developments of the recognition methods. Then we focused on the factors that affect speaker recognition performances and that can be problematic in actual forensic cases within a speech-chain framework. The last part of the chapter concentrated on the development of forensic automatic speaker recognition, mentioning the trends in recent research and examples of the systems that implement the state-of-the-art methods.

One future direction of the research may be to solve standing problems in forensic automatic speaker recognition, such as coping with the mismatches in transmission channels and languages, and handling non-contemporaneous speech samples. We also need to resolve problems brought about by the factors described above; i.e., speakers' health conditions, voice disguise, disguise by using synthesised speech, and so forth. Also, fundamental research on the following topics must be deepened: the effects of noise and speech compression in digital recordings, female speech, effect of speaking style, and so on.

We will definitely need to have good speech corpora that meet forensic case simulations; namely, they must be large enough, recorded in realistic conditions, and desirably provided in each language. They may help us get an idea of how large and what sort of speaker population we must prepare in order to yield confidential statistics for the likelihood calculation.

Another direction of the research will be a collaborative work by those who apply the standard acoustic-phonetic, aural-visual, and speech technology (e.g., signal processing, artificial intelligence and natural language processing) approaches to speaker recognition. This may offer better solutions, for example, to establish the methodology for separating speaker individualities or their identifying characteristics from phonemic or phonological information conveyed by speech, or to search for effective speech portions for recognising speakers, or lastly to explore new methods for indicating the strength of evidence.

References

1. Nolan F (1983) The phonetic basis of speaker recognition. Cambridge studies in speech science and communiation. Cambridge University Press, Cambridge
2. Schmidt-Nielsen A, Stern KR (1985) Identification of known voices as a function of familiarity and narrow-band coding. J Acoust Soc Am 77:658–663
3. Van Lacker D, Kreiman J, Emmorey K (1985) Familiar voice recognition: patterns and parameters part 1: recognition of backward voices. J Phonetics 13:19–38
4. Van Lacker D, Kreiman J (1985) Familiar voice recognition: patterns and parameters part 2: recognition of rate-altered voices. J Phonetics 13:39–52
5. Cheney D, Seyfarth R (1980) Vocal recognition in free-ranging vervet monkeys. Anim Behav 28:362–367
6. Rendall D, Rodman PS, Emond RE (1996) Vocal recognition of individuals and kin in free-ranging rhesus monkeys. Anim Behav 51:1007–1015
7. Sugiura H (2001) Vocal exchange of coo calls in Japanese macaques. In: Matsuzawa T (ed) Primate origins of human cognition and behaviour. Springer, Tokyo, pp 135–154
8. Bricker P, Pruzansky S (1976) Speaker recognition. In: Lass N (ed) Contemporary issues in experimental phonetics. Academic Press, New York, pp 295–326
9. Furui S (1992) Acoustic and speech engineering (onkyo, onsei kougaku). Kindai Kagakusha Publishing Company, Tokyo
10. National Research Council (1979) On the theory and practice of voice identification. National Academy of Science, Washington, pp 3–13
11. Steinberg JC (1934) Application of sound measuring instruments to the study of phonetic problems. J Acoust Soc Am 6:16–24
12. Potter R (1945) Visible patterns of speech. Science 102:463–470
13. Grey CHG, Kopp GA (1944) Voiceprint identification. Bell Telephone Laboratory Annual Report, New York, pp 1–14
14. Tosi O, Oyer H, Lashbrook W, Pedrey C, Nicol J, Nash E (1972) Experiment on voice identification. J Acoust Soc Am 51:2030–2043
15. Kersta L (1962) Voiceprint identification. Nature 196:1253–1257
16. Campbell JP, Shen W, Campbell WM, Schwartz R, Bonastre JF, Matrouf D (2009) Forensic speaker recognition. IEEE Signal Process Mag 26:95–103
17. Young MA, Campbell RA (1967) Effects of context on talker identification. J Acoust Soc Am 42:1250–1254
18. Tosi O (1968) Speaker identification through acoustic spectrography. Proc Logoped Phoniatr, pp 138–145
19. Stevens KN, Williams CE, Carbonell JR, Woods B (1968) Speaker authentication and identification: a comparison of spectrographic and auditory presentations of speech material. J Acoust Soc Am 44:1596–1607
20. Bolt RH, Cooper FS, David EE Jr, Denes PB, Pickett JM, Stevens KN (1970) Speaker identification by speech spectrograms: a scientists' view of its reliability for legal purposes. J Acoust Soc Am 47:597–612
21. Bolt RH, Cooper FS, David EE Jr, Denes PB, Pickett JM, Stevens KN (1973) Speaker identification by speech spectrograpms: some further observations. J Acoust Soc Am 54:531–534

22. Koenig BE (1986) Spectrographic voice identification: a forensic survey. J Acoust Soc Am 79:2088–2090
23. Shipp T, Doherty TE, Hollien H (1987) Some fundamental considerations regarding voice identification. J Acoust Soc Am 82:687–688
24. Koenig BE, Ritenour DV Jr, Kohus BA, Kelly S (1987) Reply to 'Some fundamental considerations regarding voice identification'. J Acoust Soc Am 82:688–689
25. Lindh J (2004) Handling the voiceprint issue. Proc Fonetik, pp 72–75
26. Poza FT, Begault DR (2005) Voice identification and elimination using sural-spectrographic protocols. Proc AES Int'l Conf, pp 1–8
27. McGehee F (1937) The reliability of the identification of the human voice. J Gen Psychol 17:249–271
28. McGehee F (1944) An experimental study of voice recognition. J Gen Psychol 31:53–65
29. Pollack I, Pickett JM, Sumby WH (1954) On the identification of speaker by voice. J Acoust Soc Am 26:403–406
30. Bricker P, Pruzansky S (1966) Effects of stimulus content and duration on talker identification. J Acoust Soc Am 40:1441–1450
31. Clifford BR (1980) Voice identification by human listeners: on earwitness reliability. Law Human Behav 4:373–394
32. Papcun G, Kreiman J, Davis A (1989) Long-term memory for unfamiliar voices. J Acoust Soc Am 85:913–925
33. Yarmey AD, Matthys E (1992) Voice identification of an abductor. Appl Cogn Psychol 6:367–377
34. Yarmey AD, Yarmey AL, Yarmey M, Parliament L (2001) Commonsense beliefs and the identification of familiar voices. Appl Cogn Psychol 15:283–299
35. O'Shaughnessy D (2001) Speech communication—human and machine, 2nd edn. Addison-Wesley Publishing Company, New York
36. Hollien H (2002) Forensic voice identification. Academic Press, San Diego
37. Bonastre JF, Bimbot F, Boe LJ, Campbell JP, Reynolds DA, Magrin-Chagnolleau I (2003) Person authentication by voice: a need for caution. Proc Eurospeech, pp 1–4
38. Denes PB, Pinson EN (1993) The speech chain, 2nd edn. Worth Publishers, New York
39. Kuenzel H (2000) Effects of voice disguise on speaking fundamental frequency. Forensic Ling 7:149–179
40. Zhang C, Tan T (2007) Voice disguise and automatic speaker recognition. Forensic Sci Int 175:118–122
41. Reich AR, Duke JE (1979) Effects of selected vocal disguises upon speaker identification by listening. J Acoust Soc Am 66:1023–1028
42. Orchard TL, Yarmey AD (1995) The effects of whispers, voice-sample duration, and voice distinctiveness on criminal speaker identification. Appl Cogn Psychol 9:249–260
43. Sjoestroem M, Eriksson E, Zetterholm E, Sullivan KP (2006) A switch of dialect as disguise. Lund Univ. Linguistics and Phonetics Woking Papers, vol 52, pp 113–116
44. Markham D (1999) Listeners and disguised voices: the imitation and perception of dialect accent. J Speech Lang Law 6:289–299
45. Amino K, Arai T (2009) Dialectal characteristics of Osaka and Tokyo Japanese: analyses of phonologically identical words. Proc Interspeech, pp 2303–2306
46. House AS, Stevens KN (1993) Speech production: thirty years after. J Acoust Soc Am 94:1763
47. Hollien H, Schwartz R (2000) Aural-perceptual speaker identification: problems with noncontemporary samples. Forensic Linguist 7:199–211
48. Hollien H, Schwartz R (2001) Speaker identification utilizing noncontemporary speech. J Forensic Sci 46:63–67
49. Amino K, Osanai T, Kamada T, Makinae H, Arai T (2011) Effects of the phonological contents and transmission channels on forensic speaker recognition. In: Neustein A, Patil HA (eds) Advances in forensic speaker recognition. Springer

50. Kuenzel HJ (2001) Beware of the 'telephone effect': the influence of telephone transmission on the measurement of formant frequencies. Forensic Liguist 8:80–99
51. Byne C, Foulkes P (2004) The 'mobile phone effect' on vowel formants. J Speech Lang Law 11:1350–1771
52. Lawrence S, Nolan F, McDougall K (2008) Acoustic and perceptual effects of telephone transmission on vowel quality. J Speech Lang Law 15:161–192
53. Titze I (1989) Physiologic and acoustic differences between male and female voices. J Acoust Soc Am 85:1699–1707
54. Kent RD, Read C (2001) Acoustic analysis of speech, 2nd edn. Cengage Learning
55. Clarke FR, Becker RW (1969) Comparison of techniques for discriminating among talkers. J Speech Hear Res 12:747–761
56. Thompson CP (1987) A language effect in voice identification. Appl Cogn Psychol 1:121–131
57. Goggin J, Thompson CP, Strube G, Simental LR (1991) The role of language familiarity in voice identification. Mem Cognit 19:448–458
58. Koester O, Schiller NO (1997) Different influences of the native language of a listener on speaker recognition. Forensic Linguist 4:18–28
59. Philippon AC, Cherryman J, Bull R, Vrij A (2007) Earwitness identification performances: the effect of language, target, deliberate strategies and indirect measures. Appl Cogn Psychol 21:539–550
60. Hashimoto M, Kitagawa S, Higuchi N (1998) Quantitative analysis of acoustic features affecting speaker identification. J Acoust Soc Jpn 54:169–178
61. Hollien H, Majewski W, Doherty TE (1982) Perceptual identification of voices under normal, stress, and disguise speaking conditions. J Phonetics 10:139–148
62. Ladefoged P, Ladefoged J (1980) The ability of listeners to identify voices. UCLA Working Papers Phon 49:43–89
63. Nygaard L (2005) Perceptual integration of linguistic and nonlinguistic properties of speech. In: Pisoni DB, Remez RE (eds) The handbook of speech perception. Blackwell, Oxford, pp 390–413
64. Roebuck R, Wilding J (1993) Effects of vowel variety and sample length on identification of a speaker in a line-up. Appl Cogn Psychol 7:475–481
65. Cook S, Wilding J (1997) Earwitness testimony: never mind the variety, hear the length. Appl Cogn Psychol 11:95–111
66. Loftus EF, Loftus GR, Messo J (1987) Some facts about weapon focus. Law Human Behav 11:55–62
67. Loftus EF, Miller DG, Burns HJ (1978) Semantic integration of verbal information into a visual memory. J Exp Psychol Human Learn Mem 4:19–31
68. Schooler JW, Engstler-Schooler TY (1990) Verbal overshadowing of visual memories: some things are better left unsaid. Cogn Psychol 22:36–71
69. Chin JM, Schooler JW (2008) Why do words hurt? Content, process, and criterion shift accounts of verbal overshadowing. Eur J Cogn Psychol 20:396–413
70. Kitagami S (2001) Disruptive effect of verbal encoding on memory and cognition of nonverbal information. Kyoto Univ Dept Edu Bull Paper 47:403–413
71. Kasahara H, Ochi K (2008) Verbal overshadowing effect in earwitness perception. Proc Ann Conv Jpn Psychol Assoc 72:889
72. Cook S, Wilding J (2001) Earwitness testimony: effects of exposure and attention on the face overshadowing effect. Br J Psychol 92:617–629
73. Kasahara H, Ochi K (2006) Effect of face presence on memory for a voice. J Jpn Acad Facial Studies 6:71–76
74. Yarmey AD, Yarmey AL, Yarmey MJ (1994) Face and voice identifications in showups and lineups. Appl Cogn Psychol 8:453–464
75. Bull R, Clifford BR (1984) Earwitness voice recognition accuracy. In: Wells GL, Loftus EF (eds) Eyewitness testimony: psychological perspectives. Cambridge University Press, Cambridge, pp 92–123

76. Kerstholt JH, Jansen N, Van Amelsvoort AG, Broeders AP (2004) Earwitnesses: effects of speech duration, retention, internal and acoustic environment. Appl Cogn Psychol 18:327–336
77. Van Wallendael LR, Surace A, Parsons DH, Brown M (1994) Earwitness' voice recognition: factors affecting accuracy and impact on jurors. Appl Cogn Psychol 8:661–677
78. Pruzansky S (1963) Pattern-matching procedure for automatic talker recognition. J Acoust Soc Am 35:354–358
79. Li KP, Dammann JE, Chapman WD (1966) Experimental studies in speaker verification, using and adaptive system. J Acoust Soc Am 40:966–978
80. Glenn JW, Kleiner N (1967) Speaker identification based on nasal phonation. J Acoust Soc Am 43:368–372
81. Furui S, Itakura F, Saito S (1972) Talker recognition by the longtime averaged speech spectrum. IEICE Trans 55-A(1):549–556
82. Wolf JJ (1971) Efficient acoustic parameters for speaker recognition. J Acoust Soc Am 51:2044–2056
83. Atal BS (1972) Automatic speaker recognition based on pitch contours. J Acoust Soc Am 52:1687–1697
84. Furui S, Itakura F (1973) Talker recognition by statistical features of speech sounds. Electron Commun Jap 56-A:62–71
85. Atal BS (1974) Effectiveness of linear prediction characteristics of the speech wave for automatic speaker identification and verification. J Acoust Soc Am 55:1304–1312
86. Sambur MR (1975) Selection of acoustic features for speaker identification. IEEE Trans Acoust Speech Sig Process 23:176–182
87. Hollien H, Majewski W (1977) Speaker identification by long-term spectra under normal and distorted speech conditions. J Acoust Soc Am 62:975–980
88. Matsumoto H, Nimura T (1978) Text-independent speaker identification based on piecewise canonical discriminant analysis. Proc Int Conf Acoust Speech Sig Process, 3:291–294
89. Markel JD, Davis SB (1979) Text-independent speaker recognition from a large linguistically unconstrained time spaced data base. IEEE Trans Acoust Speech Sig Process 27:74–82
90. Furui S (1981) Cepstral analysis technique for automatic speaker verification. IEEE Trans Acoust Speech Sig Process 29:254–272
91. Li KP, Wrench EH (1983) Text-independent speaker recognition with short utterances. Proc Int Conf Acoust Speech Sig Process, 8:555–558
92. Soong F, Rosenberg A, Rabiner L, Juang BH (1985) A vector quantization approach to speaker recognition. Proc Int Conf Acoust Speech Sig Process, 387–390
93. Rosenberg A, Soong F (1986) Evaluation of a vector quantisation talker recognition system in text independent and text dependent modes. Proc Int Conf Acoust Speech Sig Process, 11:873–876
94. Shirai K, Mano K, Ishige D (1987) Speaker identification based on frequency distribution of vector-quantised spectra. IEICE Trans 70-D:1181–1188
95. Rosenberg A, Lee CH, Soong F (1990) Sub-word unit talker verification using Hidden Markov Models. Proc Int Conf Acoust Speech Sig Process, 1:269–272
96. Higgins A, Bahler L, Porter J (1991) Speaker verification using randomized phrase prompting. Digit Signal Process 1:89–106
97. Tishby NZ (1991) On the application of mixture AR Hidden Markov Models to text-independent speaker recognition. IEEE Trans Acoust Speech Sig Process 39:563–570
98. Reynolds AD, Carlson B (1995) Text-dependent speaker verification using decoupled and integrated speaker and speech recognizers. Proc Eurospeech, pp 647–650
99. Reynolds AD, Rose R (1995) Robust text-independent speaker identification using Gaussian mixture speaker models. IEEE Trans Speech Audi Process 3:72–83
100. Che C, Lin Q (1995) Speaker recognition using HMM with experiments on the YOHO database. Proc Eurospeech, pp 625–628
101. NIST webpage. http://www.nist.gov/index.html
102. NIST-SRE. http://www.itl.nist.gov/iad/mig//tests/sre/

103. Doddington GR, Przybocki MA, Martin AF, Reynolds DA (2000) The NIST speaker recognition evaluation—overview, methodology, systems, results, perspective. Speech Commun 31:225–254
104. Nakasone H, Beck SD (2001) Forensic automatic speaker recognition. Proc A Speaker Odyssey—the speaker recognition workshop, pp 139–142
105. Drygajlo A (2007) Forensic automatic speaker recognition. IEEE Signal Process Mag 24:132–135
106. Martin A, Doddington G, Kamm T, Ordowski M, Przybocki M (1997) The DET curve in assessment of detection task performance. Proc Eurospeech, pp 1895–1898
107. Bimbot F, Bonastre JF, Fredouille C, Gravier G, Magrin-Chagnolleau I, Meignier S, Merlin T, Ortega-Garcia J, Petrovska-Delacretaz D, Reynolds DA (2004) A tutorial on text-independent speaker verification. EURASIP J Appl Signal Process 4:430–451
108. Noda H, Darada K, Kawaguchi E, Sawai H (1998) A context-dependent approach for speaker verification using sequential decision. Proc Int Conf Spoken Lang Process
109. Ortega-Garcia J, Cruz-Llanas S, Gonzalez-Rodriguez J (1998) Quantitative influence of speech variability factors for automatic speaker verification in forensic tasks. Proc Int Conf Spoken Lang Process
110. Gonzalez-Rodriguez J, Ortega-Garcia J, Lucena-Molina JJ (2001) On the application of the Bayesian approach to real forensic conditions with GMM-based systems. Proc a speaker odyssey—the speaker recognition workshop, pp 135–138
111. Meuwly D, Drygajlo A (2001) Forensic speaker recognition based on a Bayesian framework and Gaussian Mixture Modelling (GMM). Proc a speaker odyssey—the speaker recognition workshop, pp 145–150
112. Alexander A, Botti F, Drygajlo A (2004) Handling mismatch in corpus-based forensic speaker recognition. Proc odyssey04 the speaker and language recognition workshop, pp 69–74
113. Ramos D, Gonzalez-Rodriguez J, Gonzalez-Dominguez J, Lucena-Molina JJ (2008) Addressing database mismatch in forensic speaker recognition with Ahumada III: A public real-casework database in Spanish Proc Interspeech, pp 1493–1496
114. Thiruvaran T, Ambikairajah E, Epps J (2008) FM features for automatic forensic speaker recognition. Proc Interspeech, pp 1497–1500
115. Becker T, Jessen M, Grigoras C (2008) Forensic speaker verification using formant features and Gaussian Mixture Models. Proc Interspeech, pp 1505–1508
116. Becker T, Jessen M, Alsbach S, Bross F, Meier T (2010) SPES: The BKA forensic automatic voice comparison system. Proc Odyssey—the Speaker and Language Recognition Workshop, pp 58–62
117. Hermansky H (1989) Perceptual linear predictive (PLP) analysis of speech. J Acoust Soc Am 87:1738–1752
118. Paul JE, Rabinowitz AS, Riganati JP, Richardson JM (1975) Semi-automatic speaker identification system (SASIS)—analytical studies. Final Report C74–11841501, Rockwell International
119. Bunge E (1977) Speaker recognition by computer. Philips Tech. Review 37(8):207–219
120. Nakasone H, Melvin C (1989) C.A.V.I.S.: (Computer assisted voice identification system). Final Report 85-IJ-CX-0024. National Institute of Justice
121. Falcone M, De Sairo N (1994) A PC speaker identification system for forensic use: IDEM. Proc ESCA workshop on automatic speaker recognition, identification and verification, pp 169–172
122. Gonzalez-Rodriguez J, Ortega-Garcia J, Lucena-Molina JJ (2001) IdentiVox: a PC-Windows tool for text-independent speaker recognition in forensic environments. Prob Forensic Sci 47:246–253
123. Drygajlo A, Meuwly D, Alexander A (2003) Statistical methods and Bayesian interpretation of evidence in forensic automatic speaker recognition. Proc Eurospeech, pp 689–692
124. Agnitio, Sociedad Limitada. http://www.agnitio.es/index.php
125. Morrison GS (2009) Forensic voice comparison and the paradigm shift. Sci Justice 49:298–308

Chapter 2
Automatic Speaker Recognition for Forensic Case Assessment and Interpretation

Andrzej Drygajlo

Abstract Forensic speaker recognition (FSR) is the process of determining if a specific individual (suspected speaker) is the source of a questioned voice recording (trace). The forensic expert's role is to testify to the worth of the voice evidence by using, if possible, a quantitative measure of this worth. It is up to the judge and/or the jury to use this information as an aid to their deliberations and decision. This chapter aims at presenting research advances in forensic automatic speaker recognition (FASR), including data-driven tools and related methodology, that provide a coherent way of quantifying and presenting recorded voice as biometric evidence, as well as the assessment of its strength (likelihood ratio) in the Bayesian interpretation framework, compatible with interpretations in other forensic disciplines. Step-by-step guidelines for the calculation of the biometric evidence and its strength under operating conditions of the casework are provided in this chapter. It also reports on the European Network of Forensic Science Institutes (ENFSI) evaluation campaign through a fake (simulated) case, organized by the Netherlands Forensic Institute (NFI), as an example, where an automatic method using the Gaussian mixture models (GMMs) and the Bayesian interpretation (BI) framework were implemented for the forensic speaker recognition task.

2.1 Introduction

Speaker recognition is the general term used to include all of the many different tasks of discriminating one person from another based on the sound of their voices. Forensics means the use of science or technology in the investigation and establishment of facts or evidence in the court of law. The role of forensic science is the provision of information (factual or opinion) to help answer questions of importance to investigators and to courts of law. Forensic speaker recognition (FSR) is the process of determining if a specific individual (suspected speaker) is the source of a questioned voice recording (trace). This process involves the comparison of record-

A. Drygajlo (✉)
EPFL Speech Processing and Biometrics Group,
UNIL School of Criminal Justice, Swiss Federal Institute of Technology Lausanne (EPFL),
University of Lausanne (UNIL), Lausanne, Switzerland
e-mail: andrzej.drygajlo@epfl.ch

A. Neustein, H. A. Patil (eds.), *Forensic Speaker Recognition*,
DOI 10.1007/978-1-4614-0263-3_2, © Springer Science+Business Media, LLC 2012

ings of an unknown voice (questioned recording) with one or more recordings of a known voice (voice of the suspected speaker) [17, 50].

There are several types of forensic speaker recognition [50, 51]. When the recognition employs any trained skill or any technologically-supported procedure, the term technical forensic speaker recognition is often used. In contrast to this, so-called naïve forensic speaker recognition refers to the application of everyday abilities of people to recognize familiar voices.

The approaches commonly used for technical forensic speaker recognition include the aural-perceptual, auditory-instrumental, and automatic methods [51]. Aural-perceptual methods, based on human auditory perception, rely on the careful listening of recordings by trained phoneticians, where the perceived differences in the speech samples are used to estimate the extent of similarity between voices [44, 45]. The use of aural-spectrographic speaker recognition can be considered as another method in this approach. The exclusively visual comparison of spectrograms in what has been called the "voiceprint" approach has come under considerable criticism in the recent years [9, 35]. The auditory-instrumental methods involve the acoustic measurements of various features such as the average fundamental frequency, articulation rate, formant centre-frequencies, etc., and comparisons of their statistical characteristics [51].

Forensic automatic speaker recognition (FASR) is an established term used when automatic speaker recognition methods are adapted to forensic applications. In automatic speaker recognition, the deterministic or statistical models of acoustic features of the speaker's voice and the acoustic features of questioned recordings are compared [17].

2.1.1 *Forensic Automatic Speaker Recognition (FASR)*

Biometrics is the science of establishing identity of individuals based on their biological and behavioral characteristics [32]. FASR offers data-driven biometric methodology for quantitative interpretation of recorded speech as evidence. Despite the variety of characteristics, the biometric processing chain that measures biometric differences between people have essentially the same architecture and many factors are common across several biometric modalities. This generic processing chain of biometric recognition, in particular automatic speaker recognition, starts from signal sensing, passes through features extraction and their modeling and ends at the stage of features against model comparison and interpretation of similarity scores (Fig. 2.1). Biometrics based FASR, presented in this chapter, is a relatively recent application of digital speech signal processing and pattern recognition for judicial purposes and particularly law enforcement.

Results of FASR based case assessment and interpretation may be of pivotal importance at any stage of the course of justice, be it the very first police investigation or a court trial. In the police *investigative mode*, abduction, is at the root of investigation [31]. Abductive reasoning follows a process of generating likely explanations, testing these with new observations and eliminating or re-ranking the expla-

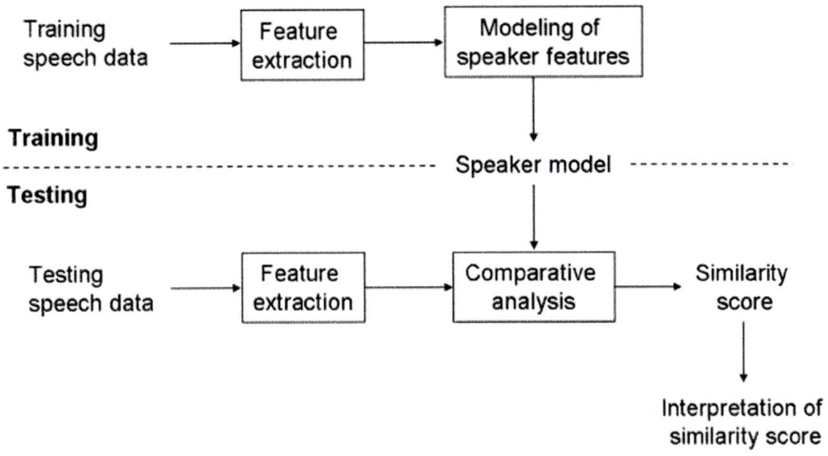

Fig. 2.1 Generic processing chain of automatic speaker recognition

nations. In this way, the investigator arrives at the best explanation of the observations, continually refining that view as further observations are made. In the forensic *evaluative mode* for a court trial, an opinion of evidential weight, based upon case specific propositions (hypotheses) and clear conditioning information (framework of circumstances) should be provided for use as evidence in court [31]. If there are two, mutually exclusive, competing propositions, exhaustive in the framework of circumstances of the case, then the odds form of Bayes' theorem can be used. The evaluative opinion of the forensic expert should be based around an assessment of a likelihood ratio of the observations given specific individual propositions (hypotheses) for the scientific findings. In the sequel of this chapter, the FASR application is limited to the evaluative mode of forensic case assessment and interpretation.

2.1.2 Overview of European Research on FASR in Case Assessment and Interpretation

The first published proposal that the likelihood-ratio framework be adopted for forensic voice comparison appears to have been made by Lewis [34]. This clearly had very little effect on the research community because it was not shown how such an idea could be implemented in practice. There was then more than decade-long time period before the idea appeared in publication again, this time showing an implementation. In April 1998 Meuwly, El-Maliki, and Drygajlo presented the paper entitled "Forensic Speaker Recognition Using Gaussian Mixture Models and a Bayesian Framework" at the COST-250 Workshop on Speaker Recognition by Man and by Machine: Directions for Forensic Applications [40]. They described the rationale for the use of the likelihood-ratio framework for forensic voice comparison, and described the design and results of tests of a Gaussian-Mixture-Model (GMM) system which calculated likelihood ratios. A substantial forensic opinion argument which has had a greater impact on the research community was made by Champod and Meuwly, initially at the Workshop on Speaker Recognition and its Commercial and Forensic Applications (RLA2C) in April 1998, with a subsequent journal article published in 2000 [15]. This paper drew on the existing literature on the evaluation and interpretation of forensic evidence in fields such as DNA to make a lucid argument for its adoption in forensic voice comparison. Meuwly and Drygajlo also described the application of the likelihood-ratio framework to forensic voice comparison at the Congrès Français d'Acoustique in September 2000 [38]. At the International Speech Communication Association (ISCA) "A Speaker Odyssey, The Speaker Recognition Workshop" in June 2001, papers describing forensic automatic speaker recognition systems, using likelihood ratio and GMMs, were presented by Meuwly and Drygajlo, as well as by González-Rodríguez, Ortega-García, and Lucena-Molina [24, 39]. At ISCA's Interspeech conference in 2003 Drygajlo organized the first special session on forensic speaker recognition in the history of this conference [20, 23, 42]. Then, at Interspeech 2005 a tutorial on forensic automatic speaker recognition was presented by Drygajlo, and at Interspeech 2008 a keynote address was given by González-Rodríguez in which the likelihood-ratio framework was a central focus.

At two successive Interpol Forensic Science Symposia, in 2001 and 2004, Broeders presented reviews of developments in forensic voice comparison from 1998 to 2001 and 2001 to 2004 respectively [11, 12]. In both reports he discussed the need for forensic voice comparison evidence to be evaluated using the likelihood-ratio framework, and noted that a number of automatic systems could output likelihood ratios. At the Interpol Forensic Science Symposium in 2007, in the review on forensic audio and visual evidence, the following opinion was expressed by Jessen: "At least since 2004 forensic automatic speaker recognition has outgrown the initial developmental stages and is now a mature speech technological discipline in which there is solid knowledge about the ranges of recognition rates that can be obtained with this method. There also seems to be broad agreement as to which essential components in the three stages feature extraction, feature modelling and the calculation of distances a speaker recognition system must have as well as how the evidence is evaluated in a Bayesian approach to forensic decision making" [7, 8].

Important journal articles describing the likelihood-ratio framework and its use for the calculation of data-based likelihood ratios in forensic automatic speaker recognition were published by the European research groups in the middle of the last decade [4, 10, 17, 25, 26], and some important Ph.D. theses in the domain were completed by Meuwly in 2000 [36], Alexander in 2005 [2] and Ramos in 2007 [46]. A special chapter entitled "Forensic Evidence of Voice" by Drygajlo was introduced in the Encyclopedia of Biometrics in 2009 [19].

The 20 years, between 1984 and 2004, of pioneering research work allowed for carrying out a collaborative exercise in the Expert Working Group for Forensic Speech and Audio Analysis (FSAAWG) within the ENFSI (European Network of Forensic Science Institutes) by the Netherlands Forensic Institute (NFI), which has shown that there is increasing interest in using the automatic and auditory-instrumental, approaches to forensic voice comparison within the framework of Bayesian interpretation of forensic evidence [14, 18]. This chapter reports only on the automatic approach used for that collaborative exercise.

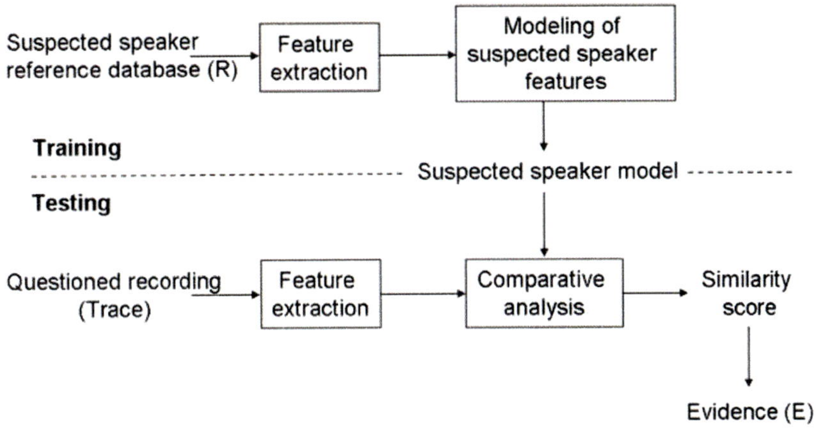

Fig. 2.2 Processing chain for calculating biometric evidence *E*

2.2 Voice as Biometric Evidence

The ongoing paradigm shift [41, 43, 52] in the evaluation and presentation of evidence in the forensic sciences which deal with the comparison of the quantifiable properties of samples of known and questioned origin is a shift towards requiring that evidence be evaluated and presented in a logically correct manner and that the reliability of the results be demonstrable. This approach needs biometric methods for recognition of individuals based on their biological and behavioural characteristics, as a common practice [16, 32, 53].

When using forensic automatic speaker recognition (FASR) the goal is to identify whether an unknown voice of a questioned recording (trace) came from a suspected speaker (source). Consequently, the *biometric evidence consists of the quantified degree of similarity between speaker-dependent features extracted from the trace and speaker-dependent features extracted from recorded speech of a suspect, represented by his or her model* [19, 20, 50] (Fig. 2.2).

To compute the evidence, the processing chain (Fig. 2.2) based on the generic biometric processing chain of automatic speaker recognition may be employed [20]. However, the calculated value of evidence does not allow the forensic expert alone to make an inference on the identity of the speaker.

As no ultimate set of speaker specific features is present or detected in speech, the recognition process remains in essence a statistical-probabilistic process based on models of speakers and collected data, which depend on a large number of design decisions. Information available from the acoustic features and their evidentiary value depend on the speech organs and language used [44]. The various speech organs have to be flexible to carry out their primary functions such as eating and breathing as well as their secondary function of speaking. The number and flexibility of the speech organs result in a high number of "degrees of freedom" when producing speech. These "degrees of freedom" may be manipulated at will or may be subject to variation due to external factors such as stress, fatigue, health, and so on. The result of this plasticity of the vocal organs is that no two utterances from the same individual are ever identical in a physical sense. Moreover, no two individuals sound identical because their vocal tract shapes, larynx sizes, and other parts of their voice production organs are different. In addition to this, the linguistic mechanism (language) driving the vocal mechanism is itself far from invariant. Each speaker has his or her characteristic manner of speaking, including the use of a particular accent, rhythm, intonation style, emphasis, choice of vocabulary and so on. Speaker recognition thus involves a situation where neither the physical basis of a person's speech (the vocal organs) nor the language driving it, are constant.

2.2.1 Features

The feature extraction module (Fig. 2.2) transforms the raw speech data into feature vectors in which speaker-specific properties are emphasized and statistical redundancies suppressed [33]. In the training mode (feature extraction and modeling modules), a suspected speaker model is created (trained) using the feature vectors. In the comparative analysis (testing) mode, the feature vectors extracted from the tested utterance (questioned recording) are compared against the suspected speaker model to give a similarity score of the evidence (E).

From the viewpoint of their physical interpretation, features commonly used for automatic speaker recognition are based on the various speech production and perception models. The speech signal continuously changes due to articulatory movements, and therefore, the signal must be broken down in short frames. Short-term spectral features, as the name suggests, are computed from short frames of about 20–30 ms in duration. Within this interval, the signal is assumed to remain stationary. These acoustic feature vectors are usually descriptors of the short-term spectral envelope which is an acoustic correlate of the resonance properties of the supralaryngeal vocal tract [6]. Thus some form of spectral envelope based features is used in most speaker recognition systems even if they are dependent on external recording conditions, e.g., Mel-Frequency Cepstral Coefficients (MFCCs) or RelAtive SpecTrAl Perceptual Linear Prediction (RASTA-PLP) coefficients [22, 28, 29].

2.2.2 Speaker Models

Automatic speaker recognition systems can be text-dependent and text-independent. By using acoustic feature vectors extracted from a given speaker's training utterance, a speaker model is trained and stored into the recognition system as a reference. In text-dependent systems [27], suited for cooperative users, the model is utterance-specific and it includes the temporal dependencies between the feature vectors. In text-independent systems, there are no constraints on the words which the speakers are allowed to use [33]. Thus, the reference (what are spoken in training) and the test (what are uttered for testing) utterances may have completely different linguistic content, and the recognition system must take this phonetic mismatch into account. In forensic applications, a text-independent automatic speaker recognition system is preferable to a text-dependent one, since the speakers can be considered non-cooperative as they do not specifically wish to be recognized. Text-independent recognition is the much more challenging of the two tasks. In general, phonetic variability represents one adverse factor to accuracy in text-independent speaker recognition.

Classical speaker models can be deterministic or statistical [13], also known as nonparametric and parametric models, respectively. In deterministic models, training and test feature vectors are directly compared with each other with the assumption that either one is an imperfect replica of the other. The amount of distortion between them represents their degree of dissimilarity. Dynamic time warping (DTW) and vector quantization (VQ) are representative examples of deterministic models for text-dependent and text-independent recognition, respectively [22].

In statistical models, each speaker is modeled as a probabilistic source with an unknown but fixed probability density function. The training phase is to estimate the parameters of the probability density function from a training sample. Comparison is usually done by evaluating the likelihood of the test utterance with respect to the model. The hidden Markov model (HMM) and the Gaussian mixture model (GMM) are the most popular statistical models for text-dependent and text-independent recognition, respectively [48].

In summary, a speaker is characterized by a speaker model such as DTW, VQ, HMM or GMM. At comparison analysis (testing), an unknown voice is first represented by a collection of feature vectors, and then evaluated against the speaker models [33].

Thus, the most persistent real-world challenge in this field is the variability of speech. There is within-speaker (within-source) variability, between-speakers (between-sources) variability and differences in recording session conditions for training and testing. Consequently, using any of the feature extraction techniques and any of the speaker models (deterministic or statistical), forensic speaker recognition methods should provide a statistical-probabilistic evaluation, which attempts to give the court an indication of the strength of the evidence, given the estimated within-source variability and the between-sources variability [20, 51], and this evaluation should be compatible with other interpretations in other forensic disciplines [21, 26, 36, 37]. The Bayesian interpretation framework, using a likelihood ratio concept, offers such interoperability. At a high level of abstraction, Bayesian data analysis is extremely simple, following the same, basic recipe: via Bayes' Theorem, we use the data to update prior beliefs about unknowns [30]. There is much to be said on the implementation of this procedure in any specific application, e.g., forensic speaker recognition, and these details are the subject of the present chapter.

2.3 Bayesian Interpretation of Biometric Evidence to Satisfy Evidentiary Requirements

To address the variability of speech, a probabilistic model [1], Bayesian inference [15] and data-driven approaches [20] appear to be adequate. In FASR statistical techniques the distribution of various features extracted from a suspect's speech is compared with the distribution of the same features in a reference population with respect to the questioned recording. The goal is to infer the identity of a source [1], since it cannot be known with certainty.

The inference of identity can be seen as a reduction process, from an initial population to unity [37]. Recently, an investigation concerning the inference of identity in forensic speaker recognition has shown the inadequacy of the speaker verification and speaker identification (in closed set and in open set) techniques for forensic applications [15].

Speaker verification and identification are the two main automatic techniques of speech recognition used in commercial applications. When they are used for forensic speaker recognition they imply a final discrimination decision based on a threshold. Speaker verification is the task of deciding, given a sample of speech, whether a specified speaker is the source of it. Speaker identification is the task of deciding, given a sample of speech, who among many speakers is the source of it. Therefore, these techniques are clearly inadequate for forensic purposes, because they force the forensic expert to make decisions which are devolved upon the court. Consequently, the state-of-the-art speaker recognition algorithms using dynamic time warping (DTW) and hidden Markov models (HMMs) for text-dependent tasks, and vector quantization (VQ), Gaussian mixture models (GMMs), ergodic HMMs and others for text-independent tasks have to be adapted to the Bayesian interpretation framework which represents an adequate solution for the interpretation of the evidence in the judicial process [1, 49].

The court is faced with decision-making under uncertainty. In a case involving FASR it wants to know how likely it is that the speech samples of questioned recording have come from the suspected speaker. The answer to this question can be given using the Bayes' theorem and a data-driven approach to interpret the evidence [20, 49, 50].

The odds form of Bayes' theorem shows how new data (questioned recording) can be combined with prior background knowledge (prior odds) to give posterior odds for a judicial outcome (Eq. 1.1). This allows the forensic expert to revise the odds measure of uncertainty based on new information, by calculating the likelihood ratio of the evidence given the pair of competing hypotheses (propositions), e.g.: H_0-the suspected speaker is the source of the questioned recording, H_1-the speaker at the origin of the questioned recording is not the suspected speaker.

$$\underbrace{\frac{p(H_0|E)}{p(H_1|E)}}_{\substack{\text{posterior knowledge} \\ \textit{posterior odds} \\ \text{(province of the court)}}} = \underbrace{\frac{p(E|H_0)}{p(E|H_1)}}_{\substack{\text{new data} \\ \textit{likelihood ratio} \\ \text{(province of the expert)}}} \cdot \underbrace{\frac{p(H_0)}{p(H_1)}}_{\substack{\text{prior knowledge} \\ \textit{prior odds} \\ \text{(province of the court)}}} \tag{1.1}$$

This reasoning method, based on the odds form of the Bayes' theorem, allows evaluating the likelihood ratio of the evidence that leads to the statement of the degree of support for one hypothesis against the other. As a result, the suspect's voice can be recognized as the recorded voice of the trace, to the extent that the evidence supports the hypothesis that the questioned and the suspect's recorded voices were generated by the same person (source) rather than the hypothesis that they were not the same person.

The value of a likelihood ratio depends critically on the choices one makes for describing the hypotheses. The hypotheses proposed in this section are not the only ones possible [3]. The numerator of the likelihood ratio can be considered a similarity term, and the denominator a typicality term [51]. In calculating the strength of evidence, the forensic scientist must consider not only the degree of similarity of the evidence with respect to the suspect, but also its degree of typicality with respect to the relevant population [41, 50].

The ultimate question relies on the evaluation of the probative strength of the voice evidence provided by an automatic speaker recognition method [25].

In developing an opinion based on Bayesian interpretation of evidence, the forensic expert has to utilize some form of inference process (from observations to the source). This evaluative opinion of evidential weight based upon the estimation of a likelihood ratio, should be based upon case specific propositions (hypotheses) and clear conditioning information (framework of circumstances) that is provided for evidential use in court [31].

2.4 Calculating the Strength of Biometric Evidence

The strength of the forensic evidence of voice is the result of the interpretation of the evidence, expressed in terms of the likelihood ratio given two alternative hypotheses. The principal processing chain for the interpretation of the evidence is presented in Figs. 2.3, 2.4 and 2.5 [20].

The methodological approach based on a Bayesian interpretation (BI) framework, presented in this chapter, is independent of the automatic speaker recognition method chosen, but the practical solution presented here as an example uses text-independent speaker recognition system based on Gaussian mixture model (GMM) [39].

The GMM method is not only used to calculate the evidence by comparing the questioned recording (trace) to the GMM of the suspected speaker (source), but it is also used to produce data necessary to model the within-source variability of the suspected speaker (Fig. 2.3) and the between-sources variability of the potential population of relevant speakers (Fig. 2.4), given the questioned recording. The Bayesian interpretation of the evidence consists of calculating the likelihood ratio using the probability density functions (pdfs) of the within-source and between-sources similarity scores and the single score E representing the value of evidence (Fig. 2.5).

The information provided by the analysis of the questioned recording (trace) leads to specify the initial reference population of relevant speakers (potential popu-

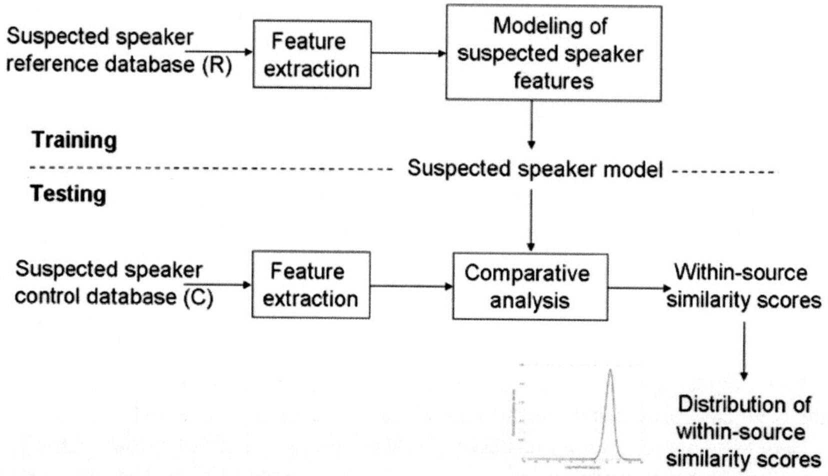

Fig. 2.3 Processing chain for calculating within-source similarity scores and their distribution

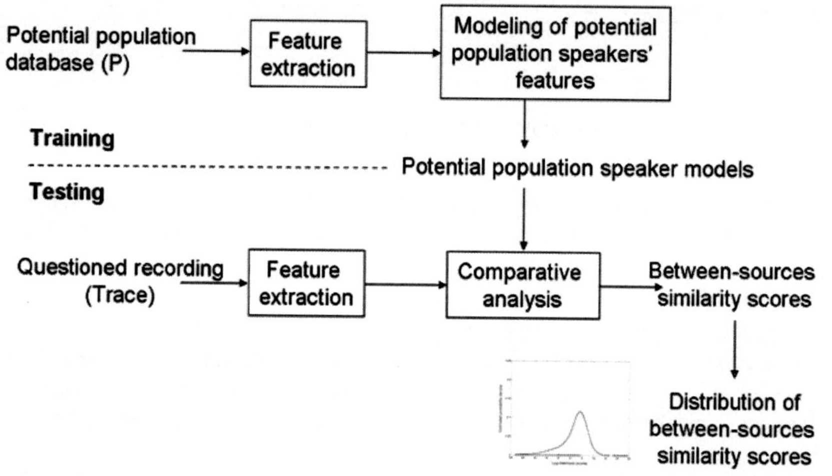

Fig. 2.4 Processing chain for calculating between-sources similarity scores and their distribution

lation) having voices similar to the trace, and, combined with the police investigation, to focus on and select a suspected speaker. The methodology presented needs three databases for the calculation and the interpretation of the evidence: the potential (relevant) population database (P), the suspected speaker reference database (R) and the suspected speaker control database (C) [39].

The potential population database (P) is a database for modeling the variability of the speech of all the potential relevant sources, using the automatic speaker recognition method. It allows evaluating the between-sources variability given the

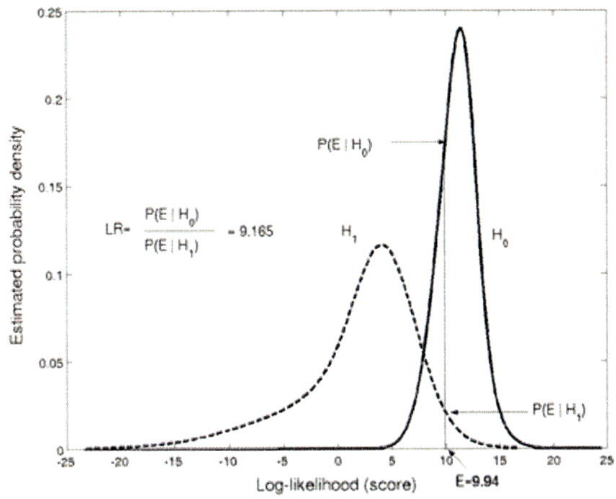

Fig. 2.5 The likelihood ratio (*LR*) estimation given the value of the evidence *E* and the probability density functions (pdfs) of the within-source and between-sources similarity scores

questioned recording, which means the distribution of the similarity scores that can be obtained, when the questioned recording is compared to the speaker models (GMMs) of the potential population database. The calculated between-sources variability pdf is then used to estimate the denominator of the likelihood ratio $p(E|H_1)$. Ideally, the technical characteristics of the recordings (e.g. signal acquisition and transmission) should be chosen according to the characteristics analyzed in the trace.

The suspected speaker reference database (R) is recorded with the suspected speaker to model his/her speech with the automatic speaker recognition method. In this case, speech utterances should be produced in the same way as those of the P database. The suspected speaker model obtained is used to calculate the value of the evidence, by comparing the questioned recording to the model.

The suspected speaker control database (C) is recorded with the suspected speaker to evaluate her/his within-source variability, when the utterances of this database are compared to the suspected speaker model (GMM). This calculated within-source variability pdf is then used to estimate the numerator of the likelihood ratio $p(E|H_0)$. The recording of the C database should be constituted of utterances as far as possible equivalent to the trace, according to the technical characteristics, as well as to the quantity and style of speech.

The basic method proposed has been exhaustively tested in mock forensic cases corresponding to real caseworks [2, 14, 36]. In an example presented in Fig. 2.5, the strength of evidence, expressed in terms of likelihood ratio gives $LR=9.165$ for the evidence value $E=9.94$, in this case. This means that it is 9.165 times more likely to observe the score E given the hypothesis H_0 than H_1. The important point to be made here is that the estimate of the *LR* is only as good as the modeling techniques and databases used to derive it. In the example, the kernel density estimation technique was used to estimate pdfs from the data representing similarity scores [1, 2].

The likelihood ratio (short form—LR) summarizes the statement of the forensic expert in the casework.

2.5 ENFSI-NFI Speaker Recognition Evaluation Through a Fake (Simulated) Case

When the Expert Working Group for Forensic Speech and Audio Analysis within the European Network of Forensic Science Institutes (ENFSI) was formed in 1997, one of its main goals was to gain insight into the different methods that are employed in the field of speaker recognition within these institutes. In 2004, a collaborative evaluation exercise was constructed at the Netherlands Forensic Institute (NFI) with English material that was recorded especially for this purpose [14]. Twelve reports were returned by the start of 2005, together with the results of all measurements that had been done and a completed questionnaire asking about the experience of the expert, the time spent on the collaborative exercise, the software that was used, etc. In this paper, the collaborative evaluation exercise is described, and a summary of the results using automatic speaker recognition method is presented based on the case report [5].

2.5.1 Formulation of the Case Key Issue

Twelve audio recordings were provided, by the Netherlands Forensic Institute (NFI) as part of a fake case evaluation, consisting of two reference recordings $R1$ and $R2$, and ten questioned recordings $Q1$–$Q10$. The ten questioned recordings consisted of conversations between two speakers, i.e., each containing the speech of a known speaker and an unknown speaker. The two reference recordings consisted of two conversations between a known speaker and a suspected speaker.

The aim of the analysis was to determine whether the speech of the unknown speaker in each of the questioned recordings was produced by the suspected speaker in the reference recordings. The case key issue was phrased as follows: "The question in this case is whether the speaker referred to as 'NN-male' in the questioned material is the same person as the suspect, the speaker referred to as 'Peter' in the (uncontested) reference material.

2.5.2 The Case Recordings

2.5.2.1 Original Format

1 CD-ROM with 12 recordings in 16 kHz, 16-bit Linear PCM wave files were provided. According to the accompanying documentation, these recordings were recorded directly from a phone line onto a Digital Audio Tape (DAT), at 44 kHz and then down-sampled to 16 kHz and later transferred to a computer using Cool Edit Pro 2.0.2. Detailed transcriptions of the recordings with the corresponding dates, time and telephone numbers were also provided.

2.5.2.2 Preprocessing of the Case Recordings

Preprocessing, consisting of segmentation of the audio into the speech of individual speakers and removal of non-speech regions, was performed in order to prepare the recordings for the databases creation. It was ascertained by NFI, that all of the recordings provided were performed with a fixed telephone network and that there was no mobile (GSM) channel effect in the recording conditions of all the recordings in the case. Because of this, no attempt at compensating for mismatched conditions was made.

The preprocessing chain constituted of the three following steps:

- *Acquisition and Down-sampling:* Acquisition was unnecessary as the files were already in digital format. However, in order to maintain consistency with the other databases used for comparison, it was necessary to down-sample the audio files to 8 kHz, 16-bit Linear PCM files using Cool Edit Pro 2.0.
- *Segmentation:* The questioned recordings and the reference recordings were in the form of conversations between two speakers. In order to compare the speech of individual speakers it was necessary to segment each of the conversations. This segmentation was performed aurally, with the help of the transcripts provided. Zones of overlap, laughs, and other anomalies were discarded.
- *Removal of Non-Speech Regions:* The recordings were passed through a voice activity detector (VAD) [47], which separates speech and non-speech regions, using instantaneous signal to noise ratio (SNR). The non-speech regions of the recording contain information about the conditions of ambient noise present in the recording and no speaker-dependent information. Removal of these non-speech regions better allows for speaker specific characteristics to be considered, when modeling voice.

2.5.3 Databases

The methodology of Bayesian interpretation of voice as biometric evidence, presented in previous sections, was used as a means for calculating the value of evidence and its strength. This methodology requires, in addition to the questioned recordings (Q), the use of three databases: a suspected speaker reference database (R), a suspected speaker control database (C) and a potential population database (P). The set of recordings obtained, along with their durations, is presented in Table 2.1.

The two recordings R1 and R2, called by NFI reference recordings of the suspected speaker were divided into reference recordings of database R and control recordings of database C. The files R01 Peter.wav and R02 Peter.wav were further segmented into:

- two reference files R01_Peter_Ref1.wav (2 m 17 s) and R02_Peter_Ref1.wav (2 m 19 s) (R database)
- seven control recordings R01_Peter_C01.wav, R01_Peter_C02.wav, R01_Peter_C03.wav, R01_Peter_C04.wav, R02_Peter_C01.wav, R02_Peter_C02.wav and R02_Peter_C03.wav of 20 s each (C database)

Table 2.1 Individual speakers segments and their durations

No.	Source original recording	Speaker segmented recordings analyzed	Length of segmented recording (s)
1	Q1.wav	Q01_Eric.wav	169.46
2	Q1.wav	Q01_NN_Male.wav	172.28
3	Q2.wav	Q02_Eric.wav	20.73
4	Q2.wav	Q02_NN_Male.wav	11.51
5	Q3.wav	Q03_Eric.wav	91.38
6	Q3.wav	Q03_NN_Male.wav	57.59
7	Q4.wav	Q04_Eric.wav	298.23
8	Q4.wav	Q04_NN_Male.wav	279.03
9	Q5.wav	Q05_Eric.wav	25.59
10	Q5.wav	Q05_NN_Male.wav	15.86
11	Q6.wav	Q06_Eric.wav	132.09
12	Q6.wav	Q06_NN_Male.wav	88.57
13	Q7.wav	Q07_Eric.wav	10.23
14	Q7.wav	Q07_NN_Male.wav	6.39
15	Q8.wav	Q08_Eric.wav	26.62
16	Q8.wav	Q08_NN_Male.wav	15.86
17	Q9.wav	Q09_Eric.wav	32.76
18	Q9.wav	Q09_NN_Male.wav	16.89
19	Q10.wav	Q10_Eric.wav	33.53
20	Q10.wav	Q10_NN_Male.wav	18.68
21	R1.wav	R01_Jos.wav	109.01
22	R1.wav	R01_Peter.wav	432.29
23	R2.wav	R02_Jos.wav	44.79
24	R2.wav	R02_Peter.wav	197.62

The potential population database (P) used is the PolyCOST 250 database. We have used 73 speakers from this database. This database was chosen among the available databases because it was found to be best suited to the case, especially in the language (English spoken by European speakers) and technical conditions (fixed European telephone network) under which the reference recordings of the suspect were made.

2.5.4 Calculation of the Value of Evidence and Its Strength

The following processing was then applied to each questioned recording Qn and the R, C and P databases:

- Feature extraction and creation of models of the speakers' voices: Extraction of 12 RASTA-PLP features for each analysis frame and creation of a statistical model by means of a 64 component Gaussian mixture model (GMM)
- Calculation of the evidence score E: Comparison between the questioned recording Qn and GMM of the suspected speaker (created using database R)

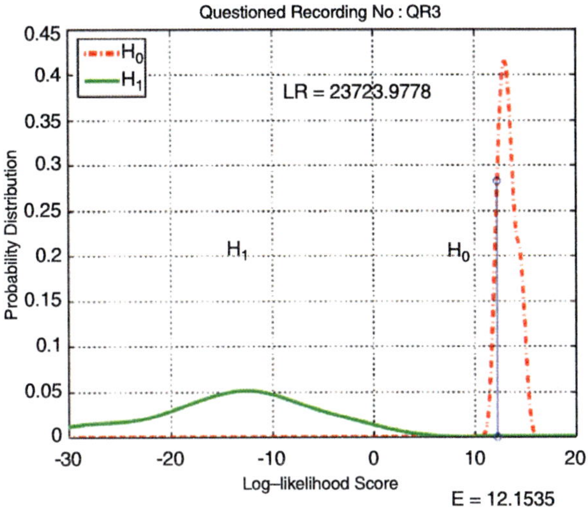

Fig. 2.6 The likelihood ratio (*LR*) estimation for Case 3 (Questioned recording Q03_NN_Male.wav)

- Within-source similarity scores: Comparison between GMM of the features of the reference recording (R) and the features of the control recordings of the suspected speaker (C)
- Between-sources similarity scores: Comparison between the features of the questioned recording Qn and the GMMs of the voices of the speakers from the database representing the potential population (P)
- Calculation of the strength of evidence: Calculation of the likelihood ratio (*LR*) by evaluating the relative likelihood ($p(E|H_0)/p(E|H_1)$) of observing the evidence score (E) given the hypothesis that the source of the questioned recording is the suspect (H_0) and the likelihood of observing the evidence score given hypothesis that someone else in the potential population was its source (H_1). Kernel density estimation was used to calculate the probability densities of distribution of scores for each of the hypotheses.

Each of the ten questioned recordings (Q1, Q2, …, Q10) is considered as a separate case (Case 1, Case 2, …, Case 10).

2.5.4.1 Example: Case 3

For Case 3 we consider the question: Is Peter in the reference recordings (R1 and R2) the same speaker as the unknown speaker in the recording Q3?

In Fig. 2.6 the distribution of scores for H_0 obtained when comparing the features of the suspected speaker control recordings (C database) of the suspected speaker, Peter, with the two statistical models of his speech (created using files from the R database) is represented by the red dotted line. The distribution of scores for H_1 obtained by comparing the segment of the questioned recording Q3, corresponding to the unknown speaker (Q03_NN_Male), with the Gaussian mix-

Table 2.2 Likelihood ratios and conclusions for all ten cases

Questioned Recording	Biometric Evidence (*E*)	Likelihood Ratio (*LR*)	Ground Truth	Conclusion Statement
Q1	10.86	6.56	different	inconclusive
Q2	11.20	163.41	same	inconclusive
Q3	12.15	23723.98	same	correct
Q4	12.68	21720.97	same	correct
Q5	13.51	11631.8	same	correct
Q6	11.63	329.0	same	correct
Q7	12.48	38407.33	same	rejected
Q8	10.68	0.660	different	inconclusive
Q9	12.92	3033.47	same	correct
Q10	7.19	4.36×10^{-23}	different	inconclusive

ture models of the speakers of the potential population database (P) is represented by the green solid line. The average score (*E*), represented by the point on the log-likelihood score axis in Fig. 2.6, obtained by comparing the questioned recording with the Gaussian mixture models of the suspected speaker, Peter's speech is 12.15. A likelihood ratio of 23,723.98, obtained in Fig. 2.5, means that it is 23,723.98 times more likely to observe this score (*E*) given the hypothesis H_0 (the suspect is the source of the questioned recording) than given the hypothesis H_1 (that another speaker from the relevant population is the source of the questioned recording).

We also observe that this score of *E*, is statistically significant (at a 5% statistical significance level) in the distribution of scores corresponding to hypothesis H_0.

2.5.4.2 Example: All Cases

A summary of the results of the automatic speaker recognition for all ten cases is presented in Table 2.2. For each case we consider the question "Is the speaker, in the reference recordings R1 and R2, the same speaker as the unknown speaker in the questioned recording Qn?".

The conclusions with respect to each of the ten questioned recordings Qn have in each case been placed on the scale of conclusions that the expert uses. In Table 2.2, they are designated as correct or incorrect (the strength of the biometric evidence (*E*) is given by *LR*), inconclusive, or rejected. The results for Q3, Q4, Q5, Q6 and Q9 are correct. The latter category (rejected) includes the results for recordings that are judged to be too short (Q7), and the category inconclusive includes results that are not statistically significant (Q1, Q2, Q8, Q10). This means that the statistical significance analysis does not allow us to progress the case in any direction. The conclusions of the remaining participants of the ENFSI-NFI speaker recognition evaluation through a fake case are presented in [14] for comparison.

2.6 Summary

We discussed some important aspects of forensic speaker recognition, focusing on the necessary statistical-probabilistic framework for both quantifying and interpreting recorded voice as biometric evidence. Methodological guidelines for the calculation of the evidence and its strength under operating conditions of the casework were presented. As an example, an automatic method using the Gaussian mixture models (GMMs) and the Bayesian interpretation (BI) framework were implemented for the forensic speaker recognition task. The BI method represents neither speaker verification, nor speaker identification. These two recognition techniques cannot be used for the task, since categorical, absolute and deterministic conclusions about the identity of source of evidential traces are logically untenable because of the inductive nature of the process of the inference of identity. The method, using a likelihood ratio to indicate the strength of the biometric evidence of the questioned recording (trace), measures how this recording scores for the suspected speaker model compared to relevant non-suspect speaker models.

This chapter also reports on the first ENFSI evaluation campaign through a fake case, organized by the Netherlands Forensic Institute (NFI), as an example, where the proposed automatic method was applied. The aim of the case assessment and interpretation was to determine whether the recordings of unknown speakers, in the ten questioned recordings, were produced by the suspect (suspected speaker) present in the reference recordings. Note that the given conclusions take into consideration the likelihood ratios as well as other factors such as the length and content of the recordings and the statistical significance of the results.These factors may influence the statement of the conclusions coming from the likelihood ratio only. In such a case, the forensic expert can change the statement to inconclusive, e.g., if the statistical significance level is too low.

Statistical evaluation of voice as biometric evidence, and particularly probabilistic Bayesian methods such as calculation of likelihood ratios based on automatic speaker recognition methods, have been criticized, but they are the only demonstrably rational means of quantifying and evaluating the value of voice evidence available at the moment.

References

1. Aitken C, Taroni F (2004) Statistics and the evaluation of evidence for forensic scientists. Wiley, Chichester
2. Alexander A (2005) Forensic automatic speaker recognition using Bayesian interpretation and statistical compensation for mismatched conditions. Ph.D. thesis, EPFL
3. Alexander A, Drygajlo A (2004) Scoring and direct methods for the interpretation of evidence in forensic speaker recognition. In: 8th international conference on spoken language processing (ICSLP 2004), Jeju, Korea, pp 2397–2400
4. Alexander A, Dessimoz D, Botti F, Drygajlo A (2005) Aural and automatic forensic speaker recognition in mismatched conditions. Int J Speech Lang Law 12(2):214–234

5. Alexander A, Drygajlo A, Botti F (2005) NFI: speaker recognition evaluation through a fake case. Case Report, EPFL-UNIL, Lausanne
6. Arcienega M, Alexander A, Zimmermann P, Drygajlo A (2005) A Bayesian network approach combining pitch and spectral envelope features to reduce channel mismatch in speaker verification and forensic speaker recognition. InterSpeech, Lisbon
7. Association of Forensic Science Providers (2009) Standards for the formulation of evaluative forensic science opinion. Sci Justice 49:161–164
8. Bijhold J, Ruifrok A, Jessen M, Geradts Z, Ehrhardt S, Alberink I (2007) Forensic audio and visual evidence 2004–2007: a review. 15th INTERPOL forensic science symposium, Lyon, France
9. Bolt RH et al (1979) On the theory and practice of voice identification. National Academy of Sciences, Washington
10. Botti F, Alexander A, Drygajlo A (2004) On compensation of mismatched recording conditions in the Bayesian approach for forensic automatic speaker recognition. Forensic Sci Int 146S:S101–S106
11. Broeders A (2001) Forensic speech and audio analysis, Forensic Linguistics 1998 to 2001: a review. 13th interpol forensic science symposium Lyon, INTERPOL, France, pp D2-53–D2-54
12. Broeders A (2004) Forensic speech and audio analysis, Forensic Linguistics 2001 to 2004: a review. 14th interpol forensic science symposium Lyon, INTERPOL, France, pp 171–188
13. Campbell J (1997) Speaker recognition: a tutorial. Proc IEEE 85(9):1437–1462
14. Cambier-Langeveld T (2007) Current methods in forensic speaker identification: results of a collaborative exercise. Int J Speech Lang Law 14(2):223–243
15. Champod C, Meuwly D (2000) The inference of identity in forensic speaker identification. Speech Commun 31(2–3):193–203
16. Dessimoz D, Champod C (2008) Linkages between Biometrics and Forensic Science. In: Jain A, Flynn P, Ross A eds Handbook of biometrics. Springer, New York, pp 425–459
17. Drygajlo A (2007) Forensic automatic speaker recognition. IEEE Signal Process Mag 24(2):132–135
18. Drygajlo A (2009) Statistical evaluation of biometric evidence in forensic automatic speaker recognition. In: Geradts ZJ, Franke KY, Veenman CJ eds Computational forensics. Springer, Berlin, pp 1–12
19. Drygajlo A (2009) Forensic Evidence of Voice. In: Li SZ ed Encyclopedia of biometrics. Springer, New York, pp 1388–1395
20. Drygajlo A, Meuwly D, Alexander A (2003) Statistical methods and Bayesian interpretation of evidence in forensic automatic speaker recognition. 8th European conference on speech communication and technology (Eurospeech 2003), Geneva, Switzerland, pp 689–692
21. Evett I (1986) A Bayesian approach to the problem of interpreting glass evidence in forensic science casework. J Forensic Sci Soc 26(1):3–18
22. Furui S (1997) Recent advances in speaker recognition. Pattern Recognit Lett 18(9):859–872
23. Gfroerer S (2003) Auditory-instrumental forensic speaker recognition. 8th European Conference on Speech Communication and Technology (Eurospeech 2003). Geneva, Switzerland, pp 705–708
24. González-Rodríguez J, Ortega-García J, Lucena-Molina JJ (2001) On the application of the Baysian framework to real forensic conditions with GMM-based systems. A Speaker Odyssey, The Speaker Recognition Workshop, Crete, Greece
25. Gonzalez-Rodriguez J, Drygajlo A, Ramos-Castro D, Garcia-Gomar M, Ortega-Garcia J (2006) Robust estimation, interpretation and assessment of likelihood ratios in forensic speaker recognition. Comput Speech Lang 20(2–3):331–355
26. Gonzalez-Rodriguez J, Rose P, Ramos D, Toledano DT, Ortega-Garcia J (2007) Emulating DNA: rigorous quantification of evidential weight in transparent and testable forensic speaker recognition. IEEE Trans Audio Speech Lang Process 15(7):2104–2115
27. Hébert M (2008) Text-dependent speaker recognition. In: Benesty J, Sondhi M, Huang Y eds Springer handbook of speech processing. Springer, Heidelberg, pp 743–762

28. Hermansky H (1994) RASTA processing of speech. IEEE Trans Speech Audio Process 2(4):78–589
29. Huang X, Acero A, Hon H-W (2001) Spoken Language Processing. Prentice Hall PTR, Upper Saddle River
30. Jackman S (2009) Bayesian analysis for the social sciences. Wiley, Chichester
31. Jackson G, Jones S, Booth G, Champod C, Evett I (2006) The nature of forensic science opinion—a possible framework to guide thinking and practice in investigations and in court proceedings. Sci Justice 46:33–44
32. Jain AK, Flynn P, Ross AA, eds (2008) Handbook of Biometrics. Springer, New York
33. Kinnunen T, Li H (2010) An overview of text-independent speaker recognition: from features to supervectors. Speech Commun 52:12–40
34. Lewis SR (1984) Philosophy of speaker identification, police applications of speech and tape recording analysis. Proc Inst Acoust 6(1):69–77
35. Meuwly D (2000) Voice analysis. In: Siegel J, Knupfer G, Saukko P eds Encyclopedia of forensic sciences, Academic Press, London, pp 1413–1421
36. Meuwly D (2001) Reconnaissance de locuteurs en sciences forensiques: l'apport d'une approche automatique. PhD dissertation, University of Lausanne, Lausanne, Switzerland
37. Meuwly D (2006) Forensic individualisation from biometric data. Sci Justice 46(4):205–213
38. Meuwly D, Drygajlo A (2000) Reconnaissance automatique de locuteurs en sciences forensiques: Modélisation de la variabilité intralocuteur et interlocuteur. 5ème Congres Français d'Acoustique, Lausanne, pp 522–525
39. Meuwly D, Drygajlo A (2001) Forensic speaker recognition based on a Bayesian framework and Gaussian mixture modelling (GMM). A Speaker Odyssey, The Speaker Recognition Workshop, Crete, pp 145–150
40. Meuwly D, El-Maliki M, Drygajlo A (1998) Forensic speaker recognition using Gaussian mixture models and a Bayesian framework. COST-250 workshop on speaker recognition by man and by machine: directions for forensic applications, Ankara, Turkey, pp 52–55
41. Morrison G (2009) Forensic voice comparison and the paradigm shift. Sci Justice 49:298–308
42. Nakasone H (2003) Automated speaker recognition in real world conditions: controlling the uncontrollable. European conference on speech communication and technology (Eurospeech 2003), Geneva, Switzerland, pp 697–700
43. National Research Council (2009) Strengthening forensic science in the United States: a path forward. National Academies Press, Washington
44. Nolan F (1983, reissued 2009) The phonetic bases of speaker recognition. Cambridge University Press, Cambridge
45. Nolan F (2001) Speaker identification evidence: its forms, limitations, and roles. Conference on Law and Language: Prospect and Retrospect, Levi (Finnish Lapland), pp 1–19
46. Ramos Castro D (2007) Forensic Evaluation of the Evidence using Automatic Speaker Recognition Systems. Ph.D. thesis, Universidad Autonoma de Madrid, Madrid, Spain
47. Renevey P, Drygajlo A (2001) Entropy based voice activity detection in very noisy conditions. 7th European Conference on Speech Communication and Technology (Eurospeech 2001), Aalborg, Denmark, pp 1887–1890
48. Reynolds D, Quatieri T, Dunn R (2000) Speaker verification using adapted Gaussian mixture models. Digital Signal Process 10(1):19–41
49. Robertson B, Vignaux G (1995) Interpreting evidence. Evaluating forensic science in the courtroom. Wiley, Chichester
50. Rose P (2002) Forensic speaker identification. Taylor and Francis, London
51. Rose P (2006) Technical forensic speaker recognition: evaluation, types and testing of evidence. Comput Speech Lang 20(2–3):159–191
52. Saks MJ, Koehler JJ (2005) The coming paradigm shift in forensic identification science. Science 309:892 895
53. Wayman J et al (eds) (2005) Biometric systems: technology, design and performance evaluation. Springer, New York

Chapter 3
Aural/Acoustic vs. Automatic Methods in Forensic Phonetic Case Work

Anders Eriksson

Abstract In this chapter focus will be on speech analysis in a forensic context. Both so called aural/acoustic approaches and automatic methods will be considered and their application in a forensic context described. Forensic casework introduces many challenges not found in the laboratory settings where the applied methods were originally developed. Forensic phonetic case work may involve many different types of tasks like voice comparison, speaker profiling, phonetic transcription and enhancement of poor quality recordings but in the present chapter only voice comparison will be described in any detail. The purpose of forensic speech science is to produce evidence that can be used in court. This introduces other types of challenges like the choice of presentation format. Pros and cons applying the traditional approach using verbal scales vs. using a likelihood ratio approach will be considered and put in the context of a current debate within forensics in general about the presentation of evidence.

3.1 Introduction

Until recently, most forensic case work involving speech has been carried out using what we may call aural/acoustic methods. During the past decade, however, automatic methods have been successively gaining ground. We have no statistics showing the precise proportion of case work done using automatic methods vs. aural/acoustic methods. We may say, however, that the majority of case work is still done using aural/acoustic methods only. A combination of both types is used by the police crime labs in Germany and France and by independent analysts in other places. In this chapter we will present pros and cons of both methods based, among other things, on our experience from a substantial amount of case work for the Swedish police. The focus will be on application and usefulness in actual forensic speech science case work rather than the scientific research that lies behind it.

A. Eriksson (✉)
Phonetics, Department of Philosophy, Linguistics and Theory of Science, University of Gothenburg, Box 200, 40530 Gothenburg, Sweden
e-mail: anders.eriksson@ling.gu.se

A. Neustein, H. A. Patil (eds.), *Forensic Speaker Recognition*,
DOI 10.1007/978-1-4614-0263-3_3, © Springer Science+Business Media, LLC 2012

This chapter will primarily be concerned with forensic phonetic case work involving what is often referred to as speaker identification, although we will see that appropriateness of this term may be questioned. But forensic speech science case work may also involve other tasks that should not be forgotten. One such task is transcription of recordings. This may sound as an easy task that anyone can do without any specific forensic or linguistic/phonetic training. But this is very far from the truth. In forensic transcription it is often crucial to know exactly what is said. It is also absolutely necessary not to allow any guessing. If the transcriber cannot say with absolute certainty what words, expressions, figures are used this must be clearly stated. Untrained transcribers have a strong tendency to write down not what is actually said, including mispronunciations, speech errors, hesitations, repetitions etc., but what they "know" based on their linguistic knowledge as native speakers that the speaker "must have said" although it may in some cases be absolutely impossible to say with certainty if that was actually what was said. There are many reasons why transcription may be difficult and require a lot of training and experience to get it right. Poor recoding quality is one such factor. Other factors are overlapping speech, slur, holding the telephone away from the mouth, background noise, poor transmission of telephone calls etc. Let me cite just two examples where accurate transcription played a crucial role. The first case involved an illegal arms deal between criminal gangs. A conversation during the arms pick-up was recoded from the bugged mobile phone used by one of the criminals. But the telephone call had nothing to do with the deal itself. It just so happened that the speaker tried to speak to someone over the phone while at the same time instructing someone else who was physically present about where to get the automatic rifles. While doing so he moved the mobile phone away from the mouth making the reception over the phone extremely poor. It was nevertheless possible to extract enough of the content to say with certainty that what went on was connected with an arms deal. Another case involved suspected illegal trading on the stock market. In his case the sound quality of the bugged mobile phone calls was fairly good but here the speakers frequently used code words which made it absolutely essential to transcribe the exact words used even though they in many cases made no logical sense to the uninformed listener. I mention these cases just to make it clear that forensic speech science can be a lot more than just determining whether speech sample A and speech sample B come from the same speaker.

I also want to say a few words about cleaning poor quality sound files. As we all know from films and TV crime series, this is as easy as a, b, c. A few clicks on the buttons on the computer screen and the previously unintelligible speech in the sound file comes out as clear as crystal. In real life it is not quite like that. What most people, including many police officers, do not know is that cleaning in almost all cases involves the removal of potentially useful information. This may make the recording less annoying to listen to, but may at the same time remove information that in uncontrollable ways distorts acoustically based analyses. In certain cases, where the goal is to better hear and understand what is said, acoustic cleaning may be helpful. Cleaning is most successful (i.e. removes potentially important information only minimally) if the noise is strictly periodic and has a fairly narrow bandwidth. One such case involved a robbery of a taxi driver. During the robbery, which was recorded over the surveillance system installed in the taxi, there was a very annoying, narrow-band beep repeated with a regular frequency. Such sounds are fairly easy to remove without removing much else. Doing so made the conversation considerably easier to transcribe. This example illustrates that cleaning may indeed be useful but also that rather special conditions apply when this is the best solution. In most normal cases cleaning attempts make little or no difference or may even be harmful.

3.2 A Brief History of Forensic Speaker Recognition

The rest of the chapter will be about speaker recognition or comparison as we prefer to see it. But let me begin by giving a brief historic background of speaker recognition in a forensic context in order to give you a better understanding of where we are now and why we are just were we happen to be. For obvious reasons speaker recognition in the old days could only be aural. It was what we today would refer to as earwitness evidence. By that we mean, technically speaking, that the evidence is based solely on memory of the voice by the witness. The first case on record where earwitness testimony in a modern sense played a crucial role in a trial dates back to 1660. The case involved a certain William Hulet who was accused of regicide in the execution of King Charles I. A witness testified that he had heard the executioner, whose face was obscured, and that he knew that it was Hulet "by his speech". The jury found Hulet guilty of high treason and he was sentenced to death. Earwitness testimony should properly be regarded as a field of forensic voice comparison and there is a rich research literature on the topic. But since the present volume is devoted to the analysis of recorded material we will have to leave earwitness evidence for some other occasion.

The advent of recording machines opened new avenues for forensic speech science. A particular milestone in that development was the invention of new analysis techniques most notably the spectrograph. The basic ideas were published in the thirties [56] and the first commercially available machine (the *Sonagraph*) was manufactured in the early forties. Its usefulness for phonetic analysis is undisputable, but the possibility of performing speaker identification based on visual inspection of the spectrograms, a technique that came to be known by the name of *voiceprinting*, was grossly overestimated. Possibly the illusion was created by the superficial similarities between the patterns in a spectrogram and fingerprints. Anyhow for many years to come, voiceprinting, that is speaker recognition based on visual voice comparison of spectrograms, played a significant role as an investigative tool although it was not generally accepted as evidence in court. It also has to be said that it was not recognized as a reliable method by the leading phoneticians of the time and the accuracy claimed by its supporters could never be verified in scientific studies. If we look at this development from a theory of science perspective we may perhaps call it a paradigm shift away from aural-only identification towards acoustically based methods. Today voiceprints are only used by some private investigators and interestingly also as an investigative tool by the FBI but its role in forensic speech science and case work is otherwise negligible. A comparable development took place in the Soviet Union and there is some evidence that automatic (or at least semiautomatic) methods were also tried. Very little about this is known today but there is an interesting account of such attempts at semiautomatic speaker recognition in the novel *The First Circle* by Solzhenitsyn.

3.3 Current Methods

As was mentioned in the introduction, the predominantly used approach in forensic phonetic case work today and since two or three decades back is the aural/acoustic approach. This approach is used by most forensic phonetic analysts in Europe and also elsewhere. Among the members of the only professional organization of forensic speech scientists, the *International Association for Forensic Phonetics and Acoustics* (IAFPA), the majority of members work within this framework. But as was also mentioned in the introduction, automatic methods are gaining ground. Both approaches will be briefly outlined in the following two sections. More details will follow in subsequent sections.

3.3.1 Aural/Acoustic Analysis Methods

Aural/acoustic methods mean a combination of linguistic judgements made by the examiner listening to the speech material combined with the use of acoustic measurements traditionally used by phoneticians to describe speech. In the former category we find such factors as speech errors, pathological speech problems, idiosyncrasies, dialect, foreign accent etc. In the latter we typically find measurement of formants, fundamental frequency mean and standard deviation, speaking rate etc.

We may schematically divide an aural/acoustic analysis process into the following steps:

1. Careful listening to the speech material
   This is to provide a first assessment of general factors like the duration of useful material and sound quality but also to identify traits that seem to be worth exploring in more depth.
2. Transcription of the material
   I am not aware to what extent transcription is routinely used by others but in our case work in Gothenburg we have found it very useful. First of all the entire

material becomes searchable in a convenient fashion. Secondly, the transcription may be used in order to find and extract units in an automatic way for various types of analysis, for example formant analysis. The way to do this can be realized in many different but equivalent ways, but the way we do it is by using scripts in the acoustic analysis package *Praat*.

3. Linguistic analysis
   Valuable information may be found by examining the speech samples from a linguistic point of view analysing dialect, sociolect, accent, grammar, etc. This is the part that is the most difficult to imagine how it might be replaced by automatic methods.
4. Acoustic analysis
   The acoustic analysis always incorporates an analysis of the most common parameters like formants, fundamental frequency, intonation patterns etc., but many more parameters may be used. This will be explained in some detail in Sect. 3.4.
5. Summarizing the findings in a report
   This part is, in my view, the Achilles heel of the aural/acoustic approach as it is most commonly applied. If we were to follow the guidelines suggested by (p. 10, [42]) the analysis should be performed in two stages—feature extraction and feature comparison. Steps 1–4 may represent the feature extraction part. From the brief outline above it may be seen that these steps are fairly straightforward. But the last step, feature comparison, is more problematic. Nolan suggests that each compared speaker should be represented by a point in a multi-dimensional space using the values from the analysis of each parameter as co-ordinate values. Based on these representations some distance metric should be applied in order to estimate the inter speaker distances. It is just that this procedure is hardly ever followed and if we were to do so, there are no commonly agreed standards on how to do it.

An alternative way of summarizing the results would be by applying the Likelihood Ratio framework, but this is at present fairly uncommon. One reason is that if we are to use all or many of the parameters described in Sect. 3.4 there is simply not enough data available on the distribution of parameter values in the relevant population. Some forensic speech analysts therefore suggest that the Likelihood Ratio approach is not a realistic approach at the present moment. An alternative approach has therefore been suggested in a UK position statement [19]. How this approach differs will be discussed in Sect. 3.8.3.

The aural/acoustic approach is largely experience based and analyses may therefore differ between individual analysts. This does not necessarily mean that the conclusions are inaccurate, but it is not easy to see how accuracy could be measured and compared. This is in my view the principle weakness of the approach.

Most references in Sect. 3.4, where strengths and weaknesses of various parameters used in forensic case work will be discussed, refer to studies or case work carried out in the aural/acoustic tradition reflecting the predominance so far of this approach.

For the interested reader who wants an in-depth account of the aural/acoustic approach there are numerous text books that may be consulted. The following may be mentioned as a guide: [3, 23, 42, 49].

3.3.2 Automatic Methods

Strictly speaking the word 'automatic' as in 'Automatic Speaker Recognition' (ASR) only means that we leave the task to be carried out to a computer (or some other type of machine) rather than do it manually. Analysis of all the various parameters described in Sect. 3.4 could therefore in principle be carried out using fully automatic or semiautomatic methods in order to save time if for nothing else.

As things stand today, however, most automatic speaker recognition systems should more properly be called automatic voice recognition systems or voice comparison systems as we prefer to call them since 'recognition' makes reference to individuals rather than speech samples which is what the systems work with. The use of the word 'speaker' is misleading in the same way. What is recognized is not, strictly speaking, a speaker interpreted as an individual. That would require knowledge of all kinds of other evidence (e.g. fingerprints, DNA, eyewitness reports and so on) which is not available to the forensic speech analyst. The reason why 'speaker' has come to be used is that there is also 'speech recognition' which is something completely different.

Most existing ASR systems do not compare speech but voices. This is an important distinction to keep in mind. Various techniques used in ASR are described in other chapters of the book so it suffices just to remind the reader that what is analysed in the most widely used ASR systems is the voice, or more precisely what we may refer to as the *timbre* of the voice. The Mel Frequency Cepstral Coefficients (MFCCs) that are used in those systems are well suited to describe timbre and moreover modelled on studies of the human perceptual system. So, if it is voice (in this particular sense) comparison we are after, this approach is just right. But as we have seen, speech so is much more than voice. A given speaker may be described in many other ways like speaking style, dialect, speaking rate, patterns of intonation, word stress etc., but also more speaker specific traits like speech errors, idiosyncratic choice or words, mispronunciations etc. There are ongoing attempts to incorporate some such factors in ASR but to our knowledge no such system is used today in actual forensic case work. In our work we make a clear distinction between voice and speech. Our understanding of 'voice' is inspired by the *Modulation Theory of Speech* which will be described later. It differs a bit from how voice is usually seen but timbre is an important factor.

Recognizing an individual, in our case a speaker, would require a method by which it is possible to single out a single individual in an open set. This is a particularly important aspect to keep in mind in connection with forensic speech science. In some areas we may perhaps approach open set recognition. With respect to DNA and fingerprints it would, at least in principle, be possible to construct a complete population database. Given such a database and sufficiently reliable analysis procedures, it would perhaps be possible to pick out a single individual matching a given sample.

Would something similar be possible for speech even in principle? Well, in theory it would be possible to construct a population database for speech, but it has to be remembered then that such a database, in contrast to a DNA or fingerprint database, would soon be outdated. Whereas DNA or fingerprints do not change over time for a given individual (except in the case of hand injury destroying the fingerprints) human voice and speech changes over time. Voice changes are for example caused by ageing, disease, smoking habits, and injury. Also, voice characteristics may change even in a short time perspective for example due to a cold or the emotional state of the speaker, under the influence of drugs, or environmental conditions such as background noise.

So, the bottom line here is that we should perhaps reserve the expression 'automatic speaker recognition' for the Science Fiction domain and focus on what we can do, namely speaker comparison, at present by and large limited to voice comparison. Automatic Voice Comparison (AVC) would be a more appropriate name for it.

A known problem for automatic systems is so called mismatch conditions, for example differences in sound quality between speech samples due to transmission channel differences. If the questioned speech samples are recorded over the telephone and the reference database consists of direct recorded speech then the performance of the system is worse than if the reference data are also telephone speech. Now simple cases of variation like that can to some extent be taken care of, but the in forensic phonetic case work variation in channel quality is much wider. What if the questioned sample is recorded with a poor quality microphone three meters away from a shouting bank robber in a room with severe reverberation, a case which is typical rather than uncommon? In real life case work there are countless examples of mismatch conditions of this type. The very severe types of mismatch conditions that we see in actual case work are a serious problem for automatic as well as aural/acoustic analysis.

In our own work at the University of Gothenburg, we use an ASR system for voice comparison. The system we use is the French ALIZE SpkDet packages which are developed and released as open source software under a so-called LGPL licence [4–5]. Voice comparison results are then combined with traditional aural/acoustic analysis and the combined results expressed using the verbal 9-point scale described in a later section. A step forward would be to combine the results in a Bayesian Likelihood Ratio framework, but at present it is not clear how this may be implemented.

As far as we are aware, the most commonly used or tested system by police labs is the Spanish *Batvox*[1] system by *Agnitio*. It has been tested by the police in Sweden and several other countries. We have no information, however, to what extent it is being used in actual case work. Batvox has consistently performed well in evalua-

[1] http://www.agnitio.es/ingles/batvox.php.

tions and we have no doubt it is a very qualified system. What is of some concern, however, is its reputed user-friendliness. This might sound like a contradiction, but isn't really. If we refer to the *Daubert* ruling (as cited in p. 10, [41]) saying that an expert's testimony should be based on "scientific knowledge" then a system which is user-friendly to the extent that it can be used by anyone trained to operate the system, but without requiring any particular insights into the principles and workings of the system itself, is a potential threat to the principle of scientific knowledge of the expert. A forensic speech expert should know the tools used inside out and basically only regard them as a way of saving time compared to doing it all by pencil and paper. I have absolutely no evidence that the Batvox system or any other similar system *is* misused, I am just cautioning against a potential danger that we should be aware of.

3.4 A Look at Some of the Most Important Parameters Used in Forensic Speech Science

In the following no sharp distinction will be made between analyses performed within the aural/acoustic paradigm and those using automatic methods. Although it is true that most automatic methods so far have been based on voice analysis, most notably using a Mel Frequency Cepstral Coefficient (MFCC) approach, there is no reason why automatic or semiautomatic methods should not be possible to extend to other types of analyses. And the aural/acoustic approach may, of course, also include voice analysis. Several studies referred to below have been performed within a Likelihood Ration framework. This becomes important when presenting the evidence in court. Here the Likelihood Ratio based approach offers a standardised way of presenting the results and all types of analyses may in principle be presented in identical ways. The largely experience based aural/acoustic approach does not easily lend itself to any similar degree of standardized presentation. This issue will be discussed in the section on how to present the results in court together with an account of new trends in forensic science in a legal context in general and what some claim is the beginning of a paradigm shift in the forensic speech science domain [38, 39].

One problem with many of the commonly used acoustic measures is that they are not particularly robust with respect to such factors as speaking style, recording quality, channel bandwidth etc. In the following a brief description will be given of the problems encountered in the context of speaker comparison when using the different parameters and measures but also something about how the problems may be dealt with.

The following account will allow a rather prominent place for a description of the use of two traditional parameters—fundamental frequency and formants. As will become apparent, intra-speaker variation and many other factors make these two parameters rather dull instruments for speaker comparison. Analysis based primarily on formants and fundamental frequency is therefore slowly being replaced by more sophisticated approaches but since formants and fundamental frequency have played such a prominent role in the historical development of forensic speech analysis and are still widely used in forensic case work, they merit a reasonably prominent place in a summary of forensic speech analysis. Describing them in some detail will also serve as an introduction that paves the way for a description of more sophisticated approaches.

3.4.1 Fundamental Frequency (F0)

Fundamental frequency is the frequency of vibration of the vocal cords in phonated speech. Its usefulness for speaker comparison in forensic case work has been an issue for a long time. As descriptors of individual differences in fundamental frequency, long-term distribution measures such as arithmetical mean and standard deviation are often suggested [49]. To some extent means and standard deviations depend on the duration of the speech sample but there is no general agreement on what minimum duration is required to yield representative results. Horii [24] suggests that recordings should exceed 14 s. In other studies, ranges from 60 s [42] up to 2 min [3] have been suggested as a minimum. Rose [49], reports that F0 measurements "/…/ for seven Chinese dialect speakers stabilised very much earlier than after 60 seconds", as suggested by [42], implying that duration requirements may be language specific. Braun [7] points out that the measures are also dependent on physiological factors such as age, smoking habits, decease, intoxication and psychological factors such as the emotional state of the speaker, sleepiness or vocal effort, but maintains that 15–20 s should be sufficient 'if the communicative behaviour may be considered "normal"'. She also mentions the influence of ambient noise on speaking style. In spite of all these sources of variation, fundamental frequency has nevertheless been shown to be a useful forensic phonetic parameter in several investigations [2, 3, 22, 29, 42, 49]. Voice disguise is another factor that may affect fundamental frequency. A rather common type of disguise is to raise or lower the fundamental frequency. In a study by [31] where 100 subjects were asked to read a text using a voice disguise of their own choosing, a very common choice was to either raise or lower fundamental frequency. There was a tendency for speakers with lower than average fundamental frequency to lower it even further and for those with a normally high level to raise it. For obvious reasons such manipulations will render fundamental frequency comparisons rather meaningless in forensic case work.

Positive skewing of the F0 distribution for a speaker is typical (Jassem et al. 1973). This means that the values will not be symmetrically distributed around the mean. Using the traditional measures for describing fundamental frequency level, such as mean or median values, may therefore yield misleading results. Positive skewing occurs primarily because there is much more room for fundamental frequency variation upwards that downwards[2]. If fundamental frequency is measured in semitones skewness is reduced, but not eliminated.

In order to make the F0 measure insensitive to some of the above mentioned types of variation alternative ways of representing fundamental frequency have been searched for. We use what we call the baseline as a representation of fundamental frequency since it better represents a speaker specific fundamental frequency level. A further advantage is that this measure is considerably more robust than the mean or median. In a study by [35] it was shown to be almost completely insensitive to variation in channel quality and emotional expression. The concept of baseline will be explained in Sect. 3.6.

All the above factors concern intra-speaker variation, but in forensic case work the usefulness of a parameter also depends on inter-speaker variation. Even if we are able to determine the characteristic frequency for a given speaker quite precisely, the predictive power of fundamental frequency may still be low if the fundamental frequency of the speaker falls in a region were most other speakers cluster. The F-ratio has been proposed as way of expressing the intra- vs. inter-speaker variation [5, 25, 28]. The likelihood ratio, which will be explained later, provides a framework for combining data on intra- and inter-speaker variation. Given that we have access to sufficiently large and appropriate databases this would give us a fair estimate of the usefulness of fundamental frequency data in a given case.

3.4.2 *Formants*

Formants are resonance frequencies in the vocal tract. They are determined by the shapes and volumes of the different cavities of the tract. From this description alone one may draw the conclusion that the formant frequencies will change continuously as we vary jaw opening, tongue position and lip shape during speech. There are, however, certain regularities in the formant frequencies. During normal speaking conditions they are fairly invariant for a given speaker, speech rate and vowel. Also, the vowel schwa, which is produced with the speech organs in a relaxed neutral position, may be correlated with vocal tract length. The schwa vowel may also be used as a measure of the timbre of the voice under neutral conditions. Traunmüller [58] in his *Modulation Theory of Speech* has suggested that this quality may be seen as a carrier wave which, when we speak, is modulated to convey the linguistic message but also to express various paralinguistic qualities. This will be further explained in Sect. 3.6.

Even though formant frequencies vary with the articulation of speech sounds, they also depend on the size and proportions of the speech organs of an individual

[2] A typical male speaker reading a sentence in a neutral tone may have a mean of 110 Hz, a minimum F0 of 85 and a maximum value of 160 Hz. In a livelier reading the minimum stays about the same, whereas the maximum may exceed 200 Hz.

speaker. Given that we were able to separate the contribution of the "neutral" vocal tract size and proportions from the modulations caused by articulation, formant frequencies should in principle be possible to use as characteristics of a given speaker. There are, however, many normally occurring conditions which may alter the formant frequencies making such a separation very difficult. Speaking rate is one such factor. Imaizumi and Kiritani (1989) have shown that the two highest of the formants normally considered (F3 and F4) "vary significantly with rate".

Formant frequencies have also been shown to change as a function of vocal effort [61]. In loud or shouted speech, which by the way is the speaking style we find in most recordings of bank robberies, the first formant (F1) has been shown to increase [14]. Reference recordings in bank robbery cases are, however, often taken from interrogations with the suspect speaking in a relaxed tone of voice, which makes acoustic comparison problematic. To some extent we may predict the change in formant frequencies as a function of vocal effort but far too little is known about how vocal effort affects formant frequencies in general to make any more precise predictions.

Channel bandwidth is another factor known to affect formant frequencies. This is the case for telephone recorded speech samples. The effect has been examined in several studies involving aural voice recognition. The first studies of this kind used telephone quality simulations like pass-band filters [53]. In other studies (e.g. Rathborn et al. 1981) landline phone recorded speech samples were used. Studies of the effect of telephone sound quality on aural voice recognition have, however, in general shown no significant effect of telephone quality on recognition accuracy. In acoustic analyses, however, sound quality effects have been observed. A marked effect on the first formant was observed in a study by [34] for example. In a study with the somewhat alarmist title "Beware of the telephone effect", [32] reports a strong effect on primarily the first formant (*F1*) in telephone mediated speech samples. In many cases the questioned speech samples are bugged telephone calls and the reference samples direct recorded police interrogations. In such cases, Künzel argues in rather strong terms, formant analyses should not be used. Künzel's views are not universally endorsed but we may say that there is general agreement that formant analysis should be treated with caution where a mixture of direct and telephone recorded samples are involved.

Today most bugged telephone calls in connection with crimes are made using mobile phones. Reports from forensic investigators in Sweden, the UK and Germany indicate that a substantial and increasing number of cases involve mobile phone recorded speech [43]. A detailed technical overview of the various ways in which mobile phone transmission affects sound quality may be found in [21]. Mobile phone transmission introduces a number of other sound quality degrading effects in addition to bandwidth limitations and these effects have been shown to affect formant analysis. Byrnc and Foulkes [8] showed that mobile phone transmission had a significant effect on the first formant ("29 per cent higher than in the direct condition") and in the study by [21] the three first formants were all affected and under certain transmission conditions rather dramatically so.

Some types of disguise may also affect formant frequencies. Papcun [44] looked at imitation performed by both professional and amateur mimics and found that mimics succeeded to approach the formant values of the targets, at least to some degree. Similar results have been obtained in other studies [15, 17]. It has to be said, however, that imitations of a given target are rare in actual case work but changing ones voice in similar ways may also have the effect of changing the formant values away from those of the speaker's normal voice.

To sum up we may say that although formant positions depend, among other things, on the physical properties of the vocal tract of a speaker and are thus to some extent typical of a given speaker, there are so many other factors that may influence formant positions that one may seriously question the value of formant analysis for forensic speaker comparison.

3.4.3 Rhythm and Other Patterns in the Time Domain

Various measures conventionally connected with speech rhythm have been suggested also for speaker comparison but also other measures that reflect motor patterns connected with speech articulation have been suggested. Not enough is yet known so it is difficult to predict how successful these attempts will be for speaker comparison but what little we already know is reason for a certain degree of optimism. If we begin at a rather fundamental level we may observe that highly automatic motor patterns have been shown to vary from individual to individual but tend to be stable and invariant once they have been firmly established. This has for example been demonstrated to be the case for typing [57] and gait [6, 65]. If such motor patterns are sufficiently stable and inter-individual differences sufficiently large it should be possible to use them in person recognition. This has also been demonstrated for gait (see for example the two references cited above). Speech articulation is another example of such highly automatic motor activities and if articulatory patterns are also fairly characteristic of a given speaker then they may be useful for speaker comparison. A study of impersonation by [17] showed, for example, that although the professional impersonator was quite successful at approaching the target values in terms of mean fundamental frequency, over all timing, global speaking rate, and to some extent also formant frequencies, his timing at the segmental level was always closer to his own personal timing patterns than to those of the target voice. Three renditions of the same texts were compared—the target version, the imitation, and the impersonator's rendition in his own natural speaking style. The three versions were transcribed at the segmental level marking the segment onsets, producing three timing sequences (target, imitation, and natural) for each utterance. The timing deviations from the target, segment by segment, for the imitations and the natural renditions were plotted for whole utterances, selected phrases and words. In all cases were there minimal differences between the two versions produced by the impersonator but rather substantial deviations from the targets. In other words, although the impersonator was quite successful at approaching the targets for the other tested parameters he seemed unable to change his own personal timing patterns in the direction of the target to any appreciable degree. Another example along the same lines are the studies of formant dynamics by McDougall [36, 37] suggesting that individual timing differences in selected articulatory movements may be used for speaker comparison. Several attempts have been made to describe the rhythmic character of languages [20, 48]. Various duration based measures used for language classification like the Pairwise Variability Index (PVI) and %V (percent of the total duration that consists of vowels) have been tested in experiments aiming at speaker comparison [10, 64, 66]. In a study by [11], 10 German speakers were recorded reading a short text 5 times with speech rates varying between normal and "as fast as possible". Test variables were %V and PVI (mentioned above) and ΔC (standard deviation of consonantal interval duration). All test variables remained stable within a given speaker across speaking conditions as did inter-speaker differences. In an experiment involving disguise [12], one native speaker of English read 29 sentences in his own normal voice and using disguise in the form of dialect imitation. In the disguise condition the speaker also increased fundamental frequency mean and standard deviation and voice breathiness. In spite of these manipulations, a number of tested temporal measures "revealed no differences between the normal and disguised conditions". It is too early to say how useful these measures will turn out to be in real life case work, but the results so far are very promising.

3.4.4 Speaking Rate

Speaking rate is the number of speech units produced per minute or per second. The most common measure is *words per minute* (wpm). *Syllables per second* (syll/s) is a more fine grained measure and should be preferred. In addition there is the question of how to handle pauses. When we talk about *speech rate* pauses are included. In most cases it makes sense, however, to separate the rate when actually speaking from pauses. In this case we speak of *articulation rate*. Typical articulation rates for normal speech are 5–7 syllables per second. Syllable rate is measured syllable onset to syllable onset or vowel onset to vowel onset. In reasonably good quality speech syllable onsets may be detected automatically, but as speech quality deteriorates so does the precision in syllable onset detection. In poor quality recordings there is often no other alternative than manual annotation. We may say that speech rate reflects speaking style and that articulation rate reflects motor timing much in the sense described in the previous section. Based on what we know about the stability of motor timing we would predict that articulation rate might also be reasonably speaker specific showing moderate intra-speaker variability making it an interesting parameter to investigate for forensic purposes. And this assumption has indeed been confirmed in at least a handful of studies. Künzel [30] studied various aspects of speaking tempo, varying the speaking style from read to spontaneous speech. Articulation rate turned out to be "almost entirely unaffected" by speaking style and "remarkably constant within speakers". The discrimination power for articulation rate (taking both intra- and inter-speaker variation into account) also turned out to be quite good. Speech rate which also reflects speaking style showed, as might be expected, a higher degree of intra-speaker variation and a considerably lower discrimination power.

3.4.5 Voice Quality

Voice quality may also be a useful parameter in speaker comparison. Until now most research on voice quality has been carried out in connection with studies of speech pathology. It may well be the case, however, that some of the measures used in this field may also be useful for speaker comparison, but this possibility has been tested to a rather minor extent. Two of most commonly used acoustic voice quality descriptors in speech pathology are *jitter* and *shimmer*. *Jitter* is the period-to-period variation in fundamental frequency and *shimmer* the corresponding variation in amplitude. Some studies do exist that have tested the usefulness of jitter and shimmer for speaker comparison. Wagner [63] used measures of jitter to successfully distinguish between a group of speakers with pathological voices and speakers with normal voices. Farrús and Hernando [18] used both jitter and shimmer as additional factors in an automatic speaker verification system. Jitter and shimmer used on their own did not produce "good enough" *sic.* results but adding them to a system based on prosodic and spectral parameters improved the results of the system.

Other possible, but so far untested, voice descriptors include glottal pulse shape and glottal source spectrum. Voice creak should also be mentioned. Voice creak is in general terms a perceptual category which may have several different causes. When the transglottal pressure decreases below a certain level, typically towards the end of a phrase, this often results in an abrupt halving of the fundamental frequency. It may also result in what is called diplophonia which is a regular pulse-to-pulse variation in amplitude with every second pulse being considerably weaker. The degree of creak varies between speakers but if the inter speaker variation is great enough to be forensically useful has not been tested. Most other descriptors are primarily auditory based and therefore suffer from inevitable subjectivity. Some of those are *leaky voice* (glottal folds do not close completely letting a stream of air pass through), *harsh voice* (tense vocal folds), *whispery voice* (incomplete voicing of normally voiced segments), *hoarse voice* (irregular closing) and *falsetto* (increased fundamental frequency to a point resulting in sharply falling glottal spectrum). None of this has been tested with a view to forensic application as far as I am aware.

3.4.6 Linguistic Factors

Last but not least, the importance of linguistic factors must be pointed out. Such factors are dialect, accent, idiosyncratic expressions, unusual pronunciation errors etc. In actual forensic case work it is not uncommon for such factors to carry a rather decisive weight.

3.5 Disguise

Recordings of disguised voices are not very common in forensic case work but they do occur. As mentioned above, changing fundamental frequency is one rather common type of disguise but there are others. In the study by [31] mentioned above changing the fundamental frequency was one of the chosen types of disguise. Künzel further reports that the most common other types of disguise found in cases analysed at the crime lab of the German federal police (BKA) were falsetto, pertinent creaky voice, whispering, faking a foreign accent and pinching one's nose. Although all these types of disguise are of a fairly simple nature they may nevertheless make speaker comparison considerably more difficult. A more detailed account of various types of disguise and their effect on speaker identification may be found in [16].

Harsh voice is sometimes used as a form of disguise. It is quite easy to do and may severely diminish the chances for successful speaker comparison. Automatic methods in particular suffer greatly. We had a kidnapping case where the perpetrator used harsh voice for his ransom calls delivered over a mobile phone. The reference was two mobile phone calls the suspect made to a female acquaintance. The automatic analysis failed to produce any conclusive results and although there was quite a bit of similarity in terms of certain expressions and speaking style that matched the known background of the suspect and made us fairly convinced that the speaker in the ransom calls was the same as the speaker in the reference calls, it was not possible to draw any conclusions that would have a chance of standing up in court. This is an example of how even a very simple form of disguise may effectively destroy the possibilities of successful forensic speaker comparison.

Speaker comparison of disguised speech may be performed in four steps, (1) Identifying the type of disguise, (2) Applying known models of the acoustic consequences of the disguise in question, (3) Eliminating as far as possible the distortions caused by the disguise from the speech sample and (4) Using the now 'cleaned' speech sample for speaker comparison. There is ongoing research addressing the problem along these lines [45–46] but it is still too early to say how far it will take us.

3.6 The Modulation Theory of Speech

In our work we use two concepts inspired by the *Modulation Theory of Speech* [58]—a definition of 'voice' inspired by the notion of carrier in the modulation theory and a measure of fundamental frequency we call the 'baseline'. A brief description of the modulation theory and why it is relevant in an account of forensic speaker comparison is therefore in its place.

As you may notice you do not find any mention of "voice source" in the Table 3.1. That is because *voice source* does not play any role as an independent parameter in this theory. The partly corresponding parameter is instead the *carrier* which is a fundamental concept. The concept of carrier is inspired by, but not identical to, the concept of carrier wave in signal theory. In the modulation theory all types of information in speech communication is transmitted as modulations of the carrier, hence the name "Modulation Theory". What then is the carrier? In the 1994 paper where Traunmüller first presented his theory the carrier is described in the following way: "we should think of the carrier as having the properties characteristic of a 'neutral' vowel, approximately [ə]".

The relevance of the concept of carrier in the present context is that it seems as a very reasonable way of thinking about what 'voice' should mean; a schwa vowel produced with a relaxed vocal effort and fundamental frequency. The nice thing about it is that one way of representing the voice of a speaker acoustically would then be an MFCC representation of the carrier. In our case work we make a distinction between voice and speech and our definition of voice is the carrier. We do not use the word carrier in our reports, however, since the term is not well known but refer to it as 'voice'. Now, an MFCC analysis is not limited to the analysis of neutrally produced schwa vowels but include a great variation of speech sounds; that is sounds represented by modulations of the carrier in the modulation theory sense. How to handle this in voice comparison while maintaining the idea of carrier comparison will be explained in 3.6.1.

Another idea inspired by the modulation theory is the idea of a neutral, speaker specific fundamental frequency level. In [58] this is described as "a stable point about 1.5 SD below the mean value of F0". This level has been empirically determined by studying speaker behaviour. For example this level has been shown to behave like an eigenvector for fundamental frequency when speakers vary the live-

Table 3.1 A schematic representation of the various speech qualities recognized in the *Modulation Theory* of speech. (The table is an adaptation from a corresponding table in [59])

Quality	Information	Phenomena
Linguistic		
Conventional, social	Dialect, accent, speech style, …	Words, speech sounds, prosodic patterns, …
Expressive		
Psychosocial within speaker variation	Emotion, attitude, …	Phonation type, register, vocal effort, speech rate, …
Organic		
Anatomical between speaker variation	Age, sex, pathology, …	Larynx size, vocal tract length, …
Perspectival		
Spatial, transmittal	Place, orientation, channel, …	Distance, channel distortion, …

liness in their speech. We refer to this level as the fundamental frequency baseline and it may be thought of as the frequency a speaker returns to after fundamental frequency excursions used for prosodic or other purposes. In a study by [60] this level was estimated as 1.43 standard deviations below the mean. This can be translated as approximately 7% up from the lowest frequencies used by the speaker and going much lower will normally result in creak which speakers tend to avoid. There is, in principle, no corresponding upper limit, however, resulting in a distribution bias towards higher frequencies.

3.6.1 More About Voice vs. Carrier

In the modulation theory it is assumed that a complete separation may be made between the carrier and the information (both linguistic and paralinguistic) that is transmitted by modulating the carrier. As was mentioned, a neutrally produced schwa vowel would be a close approximation of the carrier. That said it should be clear that a neutrally produced /a/ comes less close to the carrier. From this we may draw the conclusion that the MFCC:s used in the automatic voice comparison systems are "contaminated" by the message, in particular the phonology of the language (the set of contrastive speech sounds used by the language in question). Does this mean that for automatic voice comparison to produce reliable result one must make sure to use a reference database which closely matches the particular language, dialect, speech style etc. used in the known and questioned samples to be analysed? This is basically an empirical question that has to be examined in experiments comparing different combinations of test material and reference databases. If the MFCC:s represented the carrier and only the carrier such tests would not be needed, at least not as a function of language. The carrier only depends on the physiological properties of the speaker (size and proportions of the speech organs). It seems reasonable to suggest, however, that these properties are independent of language. If we had a perfectly working automatic carrier comparison system, then if the known and questioned samples were spoken in German it would not matter if the reference database was Spanish or German. With the MFCC approach it does matter but precisely how much has not been studied to the extent that would be needed to give a precise answer. Informal test using German speech samples tested against a Spanish database indicate, however, that matching the language is not as crucial as one might think (Künzel, p.c.). Dialect within a given language should influence the results even less. If we assume that there are no important anatomical differences between speakers of different Spanish dialects in Spain, and that the inventory of meaningful speech sounds is about the same in the dialects then it should not make much of a difference if we use one dialect or another for the reference database. This hypothesis has in fact also been tested. In a study presented by Moreno et al. at the IAFPA annual meeting in 2006, it was shown that that the differences between three tested dialects of Spanish were not great enough to make any appreciable difference for the recognition results. In the study, using the MFCC based Batvox system from Agnitio, the "suspects" were speakers of Andalusian and the reference databases that were used were Andalusian, Castilian and Galician. The test results were by and large the same regardless of reference dialect. The conclusion we may draw from this, albeit with some caution given that this is just one study, is that although the MFCC:s represent not just the carrier but also the phonology of a given dialect, the *differences* between voices obtained in the automatic voice comparison come close to the assumed differences in carriers between the speakers. It therefore makes some sense to think or the *inter-speaker differences* obtained in automatic voice comparison as *carrier differences* while keeping in mind that the MFCC:s themselves are not the carrier.

3.7 A Summary So Far

In a book on phonetically based speaker recognition by [42] he suggested that the following criteria should apply to parameters used in speaker recognition.

1. High between speaker variability.
2. Low within speaker variability.
3. Resistance to attempted disguise or mimicry.
4. Availability.
5. Robustness in transmission.
6. Measurability.

As we have seen in the survey above, the commonly used parameters do not in general comply fully with these criteria. Criteria 4 and 6, availability and measurability, is usually not a problem but all the parameters reviewed above have shortcomings with respect to the other criteria. A way to express the interdependence of criteria 1 and 2 is to calculate the Likelihood Ratio (See Appendix B). This requires access to suitable databases, but representative databases are seldom available. This is a severe shortcoming. There are, however, attempts at the level of basic research that show how this may work. Using a small database of elicited speech data from 13 male speakers of Standard Japanese, [25] used formant data from the database recordings in a recognition test comprising of 90 same-speaker pairs and 180 different-speaker pairs using the Likelihood Ratio approach. Five of the 180 different-speaker pairs were wrongly classified as the same speaker and 9 of the 90 same-speaker pairs were classified as different speakers. Others have performed similar experiments. Alderman [5] used formant values for *F1*, *F2* and *F3* extracted from recordings of eleven male speakers of Australian English. Vowels were elicited in a /h_d/ context to be comparable to data in the reference database (the so called Bernard data). Same-speaker pairs were correctly identified in 72.2% of the cases and different-speaker pairs correctly identified in 99.6% of the cases. Descriptions of other experiments along similar lines may be found in [52] and [50].

In Sect. 3.4 it was seen that using single measures to describe a parameter, say the mean value to describe fundamental frequency, does not lead to very precise speaker discrimination. What we may learn from other scientific fields, for example genetics, is that combining several measures representing different aspects of the same parameter may substantially increase the predictive power. The effect of using such a combination of measures is nicely illustrated in [26] and [27]. In the first of the two studies [26], she tested the discriminative power of fundamental frequency long term mean and Standard Deviation. The reference data were from a corpus of recorded spontaneous speech by 90 male native Japanese speakers. Test data were casual speech by 12 native Japanese speakers recorded for the experiment. The discriminative power of the tested measures turned out to be low. Kinoshita sums up the results the following way: "These LRs are practically unity; meaning that the evidence produced using this method has no informative value with respect to the given speaker's identity". In the second study [27] not only long term *F0* mean and *SD* but also *kurtosis*, *skew*, *modal F0* and the probability density of modal F0 were included. All measures were found to be useful to some degree for speaker discrimination, but the real advantage came from combining them all. What we may learn from these two experiments is that the use of a single descriptor of a given parameter may suggest that the parameter is a weak or even useless parameter in speaker comparison while combining many aspects relevant for the description of the given parameter may make a radical change.

Robustness in transmission is another factor to be considered according to [42]. One should perhaps generalize that to "robustness" in general. Transmission in the technical sense of channel quality, like GSM transmission, may indeed be a problem, but there are many more threats to robustness. Some of them, like speech liveliness, vocal effort and disguise have been mentioned above in connection with describing the analysis of different parameters. We may also mention background noise and reverberation which is a common problem. Combining many aspects of a given parameter or using a Likelihood Ratio approach does not solve these problems. If the robustness of the used factors is poor then the combined results will also be poor and an LR approach does not change that. It is therefore important to find ways of representing the various components that are as robust as possible. One such measure, mentioned above, is the base line as a measure of fundamental frequency suggested by [35] which is far more robust against some of the more common sources of error like speaking style and recording quality than the mean or median commonly used. A very promising approach from the point of view of robustness is also the use of measures in the time domain described above. These measures should be particularly robust against channel quality degradation. Measuring durations is often possible even in recordings of such poor quality that formant measurement or even *F0* measurement is more or less meaningless. We need to see more studies where timing measures are used, however, before any more far reaching conclusions may be drawn.

3.8 Presenting Evidence in Court?

In this section I will discuss how results may be presented in court. As will be seen there is no single agreed upon way of doing this and we seem to be in the middle of a transition period, some say moving towards a paradigm shift [54], where several competing ways of presenting the results co-exist. One might expect that the only existing professional organization of forensic speech scientists, the IAFPA mentioned earlier, should have guidelines instructing members how to present analysis results. But this is at present not the case. This situation is not unique but rather a mirror image of a lively international debate over the presentation of forensic evidence in general.

3.8.1 Presenting Evidence in the Aural/Acoustic Tradition

The presently most common way of presenting the results of a forensic phonetic investigation is to use a verbal scale. The following nine-point scale is the one recommended by the Swedish police [62] and used by us in most case work.

+4	The results support the hypothesis with near certainty
+3	The results strongly support the hypothesis
+2	The results support the hypothesis
0	The results are inconclusive
−1	The results contradict the hypothesis to some degree
−2	The results contradict the hypothesis
−3	The results strongly contradict the hypothesis
−4	The results contradict the hypothesis with near certainty

The “hypothesis” is usually that the analysed speech samples are from the same speaker. Verbal assessments of the same or a similar type are used by the Finnish, French and German police (p.c.), and most likely in many other places. It goes without saying that such a scale invites subjectivity, in particular when weighing and combining the results from analyses of several parameters. This does not necessarily mean that the analyses are unreliable. It is likely that several qualified and experienced analysts working with the same material would come to the same conclusions in most cases. When we have performed analyses in co-operation with others there has been little disagreement about the conclusions.

3.8.2 The Likelihood Ratio Framework: A Possible Paradigm Shift

There has, however, been criticism of the relative subjectivity of the present practice and a move towards a Likelihood Ratio based approach [38–39, 51]. A broader criticism of forensic science in general has been raised both in the US [41] and the UK [33]. Forensic speech science is not covered in the two reports but most of what is said is obviously relevant also with respect to speech science. The points of criticism include: Disparities in the Forensic Science Community, Lack of mandatory Standardization, Certification, and Accreditation, Interpretation, and Measures of performance (pp. 5–8, [41]) and further that "The simple reality is that the interpretation of forensic evidence is not always based on scientific studies to determine its validity" (p. 9). Not surprisingly, DNA evidence is seen as a model.

> With the exception of nuclear DNA analysis, however, no forensic method has been rigorously shown to have the capacity to consistently, and with a high degree of certainty, demonstrate a connection between evidence and a specific individual or source. (p. 7, [41])

We may see the trend towards an increased use of the Likelihood Ratio approach in forensic speech science as a reaction to the subjectively based reporting described above and there is an obvious attempt to raise the status of forensic speech evidence to a DNA like level if possible. This is all very well, but we must at the same time recognize the substantial differences between the two. A more detailed comparison with DNA is misleading for several reasons. A DNA sequence is discrete (after removing measurement errors) whereas speech variables are for the most part scale variables. Speech may also be expressed by a large number of different (and at least partly independent) variables like voice quality, fundamental frequency, formant structure, timing etc. DNA is stable and invariant over the life time of an individual, speech is not. Even if we consider only one parameter like *F0*, we have seen that it varies not just over the life time of an individual but sometimes even on a minute by minute basis. Also, whereas the Likelihood Ratio approach may be applied to all parameters that may be described using scale variables, there are other aspects of speech for which there are no obvious descriptions of this type. We may think of speaking style, dialect, foreign accent, speech errors etc. In principle we may think of estimates of the number of speakers of a given dialect etc., but that too introduces a number of problems; dialectality may also be seen as a gradual variable for example. In the *Daubert* ruling (See Appendix A) the requirement for expert testimony is that it should be based on "scientific knowledge" and "evidentiary reliability", but the Supreme Court also describes the *Daubert* standard as "flexible". We may perhaps interpret this as a licence to use DNA-like methods as far as this is possible, but may also allow ourselves to supplement this type of evidence with scientifically based judgement that must not be readily expressible in quantitative terms but has been shown to possess some degree of "evidentiary reliability". While we should indeed strive towards more stringent descriptions, we must also be aware of what is and is not realistic. These last observations are more or less what motivated the UK position statement described in the next section.

An advantage with the verbal scale quoted above is that it is easy to understand and also that it is commonly used, the disadvantage is that it is subjective. The Likelihood Ratio is the exact opposite, it is to a very minor extent subjective, but it is far from obvious how the results should be presented in a way that may be easily understood by non-specialists. Does a Likelihood Ratio of 2374 really mean anything to a jury or a judge? Strictly following the Bayesian framework the number means that whatever your prior evidence led you to believe it will now have to be revised by saying that it is now 2374 times more likely that the combined evidence weighs against the suspect. And if the judge or the jury are sitting there with Bayes' formula in front of them eagerly waiting to fill in these numbers, then fine. But this is a highly unlikely scenario. They are more likely not even to have heard about Bayes. Concerns like these have been raised also by statisticians. For an instructive example based on a real life case see [13]!

Attempts have been made to translate LRs to verbal expressions ([9]; Gonzalez-Rodriguez et al. 2001).

1–10	Limited support
10–100	Moderate support
100–1,000	Strong support
Over 1,000	Very strong support

The suggestion makes some sense, especially since we may see the verbal scale only as a suggested interpretation of what the reported LRs mean. There is, however, at present no general agreement on what is the best way to present LRs in court.

3.8.3 The UK Position Statement

The UK Position Statement [19] grew out of awareness that the traditional way of presenting the evidence invites an interpretation known as the *prosecutor's fallacy*. If the results of a voice comparison is presented as "it is *highly likely* that the samples come from the same speaker" but without considering the possibility that other speech samples, not analysed, could have been equally similar, the presentation is unfairly biased against the defendant. As we have seen the likelihood ratio approach takes care of this problem so why not just switch to this approach then? The main reason mentioned by the proponents of the statement is the present lack of suitable reference databases which makes numerical assessments impossible for most parameters. The proposed solution is what is basically a verbal version of the likelihood approach. The key concepts here are *consistency* and *distinctiveness* (which roughly parallels *similarity* and *typicality*). Consistency is "a decision … concerning whether the known and questioned samples are compatible, or consistent, with having been produced by the same speaker". The outcome can be: Consistent, not consistent or no decision. Distinctiveness is an evaluation of to what extent the same features are shared by other people in the population. Distinctiveness can be: Exceptionally distinctive, highly distinctive, distinctive, moderately distinctive or not distinctive.

This approach is currently accepted by most analysts in the UK, but I have personal information that they too are considering moving towards a likelihood ratio approach. Likelihood ratios are also discussed by police labs in Finland, Germany and France (p.c.) but no decisions have been taken so far one way or the other.

3.9 Summary and Conclusions

It should hopefully be clear from what has been said in this chapter that speaker comparison for forensic purposes is quite a lot more complex than what many people may think. The complexity arises primarily from the fact that those traits in speech which may be used for speaker comparison are also used for speech communication and are therefore not constant. An example that was mentioned is formant frequencies. While it is true that they are partly a function of the physical size and shape of the vocal tract of an individual, they are also constantly changing due to the changes in tongue shape and position, jaw opening, and lip opening used to produce speech. This is in stark contrast to DNA or fingerprints for example which remain constant over the life time of an individual.

The task is now to find what, if anything is reasonably invariant in the speech of a given individual and to find out if such traits show sufficient inter-speaker variation to be useful for speaker comparison. What we know from the various attempts so far seems to suggest that no single trait possesses these properties but that we must look for combinations of factors characterizing a given parameter. One example mentioned [27] was the combination of *mean, standard deviation, kurtosis, skew, modal F0* and the *probability density of modal F0* for the fundamental frequency parameter.

In our case work we have found it useful to make a distinction between voice and speech. Voice similarity is not the same as speaker identity. It is quite possible for a voice comparison to yield very high Likelihood Ratios in an automatic voice comparisons analysis while the analyses of other factors indicate that the samples come from different speakers. A somewhat drastic but illustrative example would be a test of two voice samples that results in a LR of 1000. In good old television series style you might be tempted to cry "We have a match", but what if it turns out that the two speakers speak different languages? In doing actual case work it is therefore useful to keep mind that voice and speech are not the same thing.

Another important take home message is that we should abandon the idea of *identification*. Reliable recognition in the sense of identification of a specific individual from speech samples is not just difficult but virtually impossible. And this is true, some would say, not just for speech based evidence but for all types of forensic evidence including DNA (For an excellent overview see [54]!).

Access to relevant reference databases is presently a practical problem for most types of analyses. Work is in progress in many places to remedy this shortage by creating new databases, but for now and several years to come we will have to base typicality estimates on less reliable data.

Finally some concern was mentioned over what it means to be a forensic expert witness. With more automatic, user friendly methods available there is a risk that should not be underestimated that unqualified analysts will rely completely on the machines they operate without really knowing what goes on behind the screen. We have no evidence that this is a real problem today but we should be aware of a development in that direction as a potential threat to the scientific validity of forensic case work.

Appendix A: The Daubert Ruling

The US Supreme Court *Daubert* ruling[3] in 1993 came in the aftermath of a controversy over the admissibility of scientific evidence in connection with the *Daubert v. Merrill Dow Pharmaceuticals* trial. Prior to the *Daubert* ruling most courts relied on the *Frye* ruling[4] from 1923, which held that scientific evidence should be based on principles generally accepted within the relevant scientific community. We may say that the question of admissibility rested on the degree of consensus in the scientific community in a given case. Without going into detail we may observe that there are obvious pros and cons concerning the *Frye* ruling. On the one hand, consensus in the scientific community gives the evidence a high degree of scientific credibility. On the other hand we also know that scientific progress almost always begins with minority views.

The *Daubert* ruling, replacing the *Frye* ruling, presented a different view replacing it with a set of criteria meant as a general guide [47]:

1. Is the evidence based on a testable theory or technique;
2. has the theory or technique been peer reviewed;
3. in the case of a particular technique's does it have a known error rate and standards controlling the techniques operation;
4. is the underlying science generally accepted?

The court cautioned, however, that the list should not be regarded as "a definitive checklist or test".

At first sight this seems very reasonable, stressing scientific validity and the competence of the expert witness, but without requiring consensus in the scientific community. The *Daubert* ruling has not been without critics, however, and it is easy to see where the weak point lies. The final word concerning admissibility is shifted from the scientific community to the trial judge who in most cases is likely to possess little or no competence to evaluate scientific validity and reliability. Two Supreme Court Justices (Stevens and Rehnquist) voiced concern over such a development warning against judges taking on the role of "amateur scientists".

This also seems to have become the case to a considerable extent. Many judges have invented their own admissibility standards that well exceed what is required in the scientific community, the result of which is that it has become very difficult for plaintiffs in cases involving health effects of toxic waste or side effects of drugs even to have their cases tried before a jury. The cases are often dismissed by a summary judgement and most often it seems in favour of the defendant (For a summary

[3] *Daubert v. Merrell Dow Pharmaceuticals, Inc*, 509 US 579 (1993).

[4] *Frye v. United States*, 293 F. 1013 (D.C. Cir. 1923).

see: [47]. The effect in criminal cases seems to have been less severe, at least until now.

If we can develop techniques and methods that comply with the four guiding criteria formulated by the *Supreme Court* we may, however, have a chance to bring back the power over admissibility to the scientific community, at least to some extent. It also has to be said that the question of admissibility does not exist in all countries and where it does it may differ from the rules that apply in the US, but the same kind of reasoning concerning validity and reliability reappears in the recommendations by the *National Research Council* and the *UK Law Commission* and is likely to be a topic of great significance for the future development of forensic phonetic expert testimonies.

Appendix B: The Likelihood Ratio Framework

I have referred to Likelihood Ratios (LR) on several occasions in the chapter. For some readers this is a very familiar concept, for others less so. The very brief account below should be sufficient to understand its place in the discussion for readers who are not already sufficiently familiar with the subject.

The Likelihood Ratio expresses the quotient between the strength of the evidence that the known and questioned samples have the same origin and the alternative hypothesis that they do not. Another way of expressing it is that it represents the quotient between a *similarity* score and a *typicality* score.

The technical way of expressing the LR is the following equation:

$$LR = \frac{p(E|Hso)}{p(E|Hdo)}$$

In the equation, the numerator expresses the probability of the evidence (E) under the hypothesis that the known and questioned samples have the same origin. The denominator expresses the strength of the evidence under the alternative hypothesis, namely that the samples have different origins. If the two probabilities are the same, i.e. the LR is equal to 1, then obviously no definite conclusion may be drawn. If the quotient is greater than 1, then the evidence speaks in favour of the "same origin" hypothesis; the greater the number, the stronger the evidence. For LRs less than 1, the opposite obviously holds.

A few words also need to be said about the so called Bayesian framework. This approach is based on Bayes' theorem:

$$\frac{p(Hso|E)}{p(Hdo|E)} = \frac{p(E|Hso)}{p(E|Hdo)} \times \frac{p(Hso)}{p(Hdo)}$$

In the equation the term to the left of the equality sign is referred to as *posterior odds*, the first term on the right hand side is the LR and the second term is the so called *prior odds*. The posterior odds is the combined weight of the all the evidence. One way of verbally defining *Prior odds* would be to say that it represents the trier of fact's belief about the two competing hypotheses prior to the "new" (hence 'prior') evidence the strength of which is expressed by LR. The posterior odds will as a consequence have to me modified by a factor expressed by the LR. It goes without saying that the prior odds should be based on some other type of solid evidence and not just beliefs in general. The term 'prior' should, however, not be interpreted as any requirement of an ordering in time of various types of evidence. It is just a way of expressing how beliefs may be modified as more and more evidence is presented. This said, it should be obvious that this is something for the judge, jury, lawyers to think about but not the expert witness presenting the evidence. There may for example be fingerprints, DNA, eyewitness reports etc. which the court needs to consider, but which should be completely disregarded by the expert when performing the forensic speech analysis. It therefore makes sense to call the approach used by the expert witness the *Likelihood Ratio framework* rather than the *Bayesian framework*. This is also the suggestion by several forensic speech scientists [40]. There is, however, one type of prior odds that the forensic speech science expert may have to consider and that has to do with the choice of a relevant reference database. There is generally complete agreement about the sex of the speaker, often also about approximate age and dialect. In such cases the speech analyst may limit the reference database accordingly.

References

1. Alderman TB (2005) Forensic speaker identification—a likelihood ratio-based approach using vowel formants. LINCOM Europa, München
2. Atal BS (1972) Automatic speaker recognition based on pitch contours. J Acoust Soc Am 52(6):1687–1697
3. Baldwin J, French P (1990) Forensic phonetics. Pinter, London
4. Bonastre J-F, Scheffer N, Matrouf D, Fredouille C, Larcher A, Preti A et al (2008). ALIZE/SpkDet: a state-of-the-art open source software for speaker recognition. In: Paper presented at the Odyssey 2008: the speaker and language recognition workshop, Stellenbosch, South Africa, 21–24 January 2008
5. Bonastre J-F, Wils F, Meignier S (2005) ALIZE, a free toolkit for speaker recognition. In: Paper presented at the ICASP 2005, Philadelphia, PA, USA
6. Bouchrika I, Nixon MS (2008) Gait recognition by dynamic cues. In: Paper presented at the 19th IEEE international conference on pattern recognition, Tampa, FL, USA
7. Braun A (1995). Fundamental frequency – how speaker specific is it? In: Braun A, Köster J-P (eds) Studies in forensic phonetics. WVT Wissenschaftlicher, Trier, pp 9–23
8. Byrne C, Foulkes P (2004) The 'mobile phone effect' on vowel formants. Int J Speech Lang Law 11(1):83–102
9. Champod C, Evett IW (2000) Commentary on A. P. A. Broeders (1999) 'Some observations on the use of probability scales in forensic identification', Forensic Linguistics, 6(2):228–41. Int J Speech Lang Law 7(2):238–243
10. Dellwo V (2010) The automatic extraction of time-domain based speaker idiosyncratic features. In: Paper presented at the IAFPA 2010, Trier, Germany, 18–21 July 2010

11. Dellwo V, Koreman J (2008) How speaker idiosyncratic is measurable speech rhythm? In: Paper presented at the IAFPA 2008, Lausanne, Switzerland, 20–23 July 2008
12. Dellwo V, Ramyead S, Dankovicova J (2009) The influence of voice disguise on temporal characteristics of speech. In: Paper presented at the IAFPA 2009, Cambridge, UK, 3–5 August 2009
13. Donnelly P (2005) Appealing statistics. Significance 2(1):46–48
14. Elliott J (2000) Comparing the acoustic properties of normal and shouted speech: a study in forensic phonetics. In: Paper presented at the 8th Australian International Conference on Speech Science and Technology, Canberra, Australia
15. Endres W, Bambach W, Flösser G (1971) Voice spectrograms as a function of age, voice disguise, and voice imitation. J Acoust Soc Am 49:1842–1848
16. Eriksson A (2010) The disguised voice: imitating accents or speech styles and impersonating individuals. In: Llamas C, Watt D (eds) Language and identities. Edinburgh University Press, Edinburgh, pp 86–96
17. Eriksson A, Wretling P (1997) How flexible is the human voice?—A case study of mimicry. In: Paper presented at the EuroSpeech '97, Rhodes, Greece
18. Farrús M, Hernando J (2009) Using jitter and shimmer in speaker verification. IET Signal Process Special Issue Biometr Recogn 3(4):247–257
19. French JP, Harrison P (2007) Position statement concerning use of impressionistic likelihood terms in forensic speaker comparison cases. Int J Speech Lang Law 14(1):137–144
20. Grabe E, Low EL (2002) Durational variability in speech and the rhythm class hypothesis. In: Gussenhoven C, Warner N (eds) Papers in laboratory phonology, vol 7. Mouton de Gruyter, Berlin, pp 515–546
21. Guillemin BJ, Watson C (2008) Impact of the GSM mobile phone network on the speech signal: some preliminary findings. Int J Speech Lang Law 15(2):193–218
22. Hollien H (1990) The acoustics of crime. Plenum Press, New York
23. Hollien H (2002) Forensic voice identification. Academic Press, San Diego
24. Horii Y (1975) Some statistical characteristics of voice fundamental frequency. J Speech Hear Res 18(1):192–201
25. Kinoshita Y (2001) Testing realistic forensic speaker identification in Japanese: a likelihood ratio-based approach using formants. Unpublished Ph.D thesis, The Australian National University, Canberra
26. Kinoshita Y (2005) Does Lindley's LR estimation formula work for speech data? Investigation using long-term f0. Int J Speech Lang Law 12(2):235–254
27. Kinoshita Y, Ishihara S, Rose P (2009) Exploring the discriminatory potential of F0 distribution parameters in traditional forensic speaker recognition. Int J Speech Lang Law 16(1)
28. Kirkland J (2003) Forensic speaker identification using Australian English fucken: a Bayesian likelihood ratio-based auditory and acoustic phonetic investigation. Unpublished Honours Thesis, Australian National University
29. Künzel HJ (1987) Sprechererkennung: Grundzüge forensicher Sprachverarbeitung. Kriminalistik, Heidelberg
30. Künzel HJ (1997) Some general phonetic and forensic aspects of speaking tempo. Forensic Linguist 4(1), 48–83
31. Künzel HJ (2000) Effects of voice disguise on speaking fundamental frequency. Forensic Linguist 7:149–179
32. Künzel HJ (2001) Beware of the 'telephone effect': the influence of telephone transmission on the measurement of formant frequencies. Forensic Linguist 8(1):80–99
33. Law Commission (2009) The admissibility of expert evidence in criminal proceedings in England and Wales: a new approach to the determination of evidentiary reliability
34. Lawrence S, Nolan F, McDougall K (2008) Acoustic and perceptual effects of telephone transmission on vowel quality. Int J Speech Lang Law 15(2):161–192
35. Lindh J, Eriksson A (2007) Robustness of long time measures of fundamental frequency. In: Proceedings of Interspeech 2007, Antwerp, Belgium, pp 2025–2028

36. McDougall K (2004) Speaker-specific formant dynamics: an experiment on Australian English /aI/. Int J Speech Lang Law: Forensic Linguist 11(1):103–130
37. McDougall K (2006) Dynamic features of speech and the characterization of speakers. Int J Speech Lang Law 13:89–126
38. Morrison GS (2009a) Forensic voice comparison and the paradigm shift. Sci Justice 49(4):298–308
39. Morrison GS (2009b) The place of forensic voice comparison in the ongoing paradigm shift. In: Proceedings of the 2nd international conference on evidence law and forensic science conference, vol 1, pp 20–34
40. Morrison GS (2010) Forensic voice comparison. In: Freckelton I, Selby H (eds) Expert Evidence. Thomson Reuters, Sydney (Ch 99)
41. National Research Council (2009) Strengthening forensic science in the United States: a path forward. The National Academies Press, Washington D.C.
42. Nolan F (1983) The phonetic bases of speaker recognition. Cambridge University Press, Cambridge
43. Öhman L, Eriksson A, Granhag PA (2010) Mobile phone quality vs. direct quality: how the presentation format affects earwitness identification accuracy. Eur J Psychol Appl Legal Context 2(2):161–182
44. Papcun G (1988) What do mimics do when they imitate a voice? J Acoust Soc Am 84(S114)
45. Perrot P, Aversano G, Chollet G (2007) Voice disguise and automatic detection: review and perspectives. In: Stylianou Y, Faundez-Zanuy M, Esposito A (eds) Progress in nonlinear speech processing. Springer, Berlin, pp 101–117
46. Perrot P, Preteux C, Vasseur S, Chollet G (2007) Detection and recognition of voice disguise. In: Paper presented at the IAFPA 2007, Plymouth, UK
47. Project on Scientific Knowledge and Public Policy (2003) Daubert—the most influential Supreme Court ruling you've never heard of
48. Ramus F, Nespor M, Mehler J (1999) Correlates of linguistic rhythm in the speech signal. Cognition 73(3):265–292
49. Rose P (2002) Forensic speaker identification. Taylor & Francis, New York
50. Rose P (2006) Technical forensic speaker recognition: evaluation, types and testing of evidence. Comput Speech Lang 20(2–3):159–191
51. Rose P, Morrison GS (2009) A response to the UK position statement on forensic speaker comparison. Int J Speech Lang Law 16(1):139–163
52. Rose P, Osanai T, Kinoshita Y (2003) Strength of forensic speaker identification evidence: multispeaker formant- and cepstrum-based segmental discrimination with a Bayesian likelihood ratio as threshold. Forensic Linguist 10:179–202
53. Rothman HB (1979) Further analysis of talkers with similar sounding voices. In: Hollien H, Hollien PA (eds) Current issues in the phonetic sciences, vol 9. John Benjamins, Amsterdam, pp 837–846
54. Saks MJ, Koehler JJ (2008) The individualization fallacy in forensic science evidence. Vanderbilt Law Rev 61(1):199–219
55. Solzhenitsyn AI (1968) The first circle (trans: Whitney TP). Northwestern University Press, Evanston
56. Steinberg JC (1934) Application of sound measuring instruments to the study of phonetic problems. J Acoust Soc Am 6(1):16–24
57. Terzuolo CA, Viviani P (1980) Determinants and characteristics of motor patterns used for typing. Neuroscience 5(6):1085–1103
58. Traunmüller H (1994) Conventional, biological and environmental factors in speech communication: a modulation theory. Phonetica 51:170–183
59. Traunmüller H (2000) Evidence for demodulation in speech perception. In: Proceedings of ICSLP 2000, Beijing, China, pp 790–793
60. Traunmüller H, Eriksson A (1995) The perceptual evaluation of F0-excursions in speech as evidenced in liveliness estimations. J Acoust Soc Am 97(3):1905–1915

61. Traunmüller H, Eriksson A (2000) Acoustic effects of variation in vocal effort by men, women and children. J Acoust Soc Am 107(6):3438–3451
62. Utlåtandeskalan (2008) http://www.skl.polisen.se/Global/www%20och%20Intrapolis/Informationsmaterial/SKL/Utlatandeskalan.pdf
63. Wagner I (1995) A new jitter-algorithm to quantify hoarseness: an exploratory study. Forensic Linguist 2(1):18–27
64. Wiget L, White L, Schuppler B, Grenon I, Rauch O, Mattys SL (2010) How stable are acoustic metrics of contrastive speech rhythm? J Acoust Soc Am 127(3):1559–1569
65. Yoo J-H, Hwang D, Moon K-Y, Nixon MS (2008) Automated human recognition by gait using neural network. In: Paper presented at the IPTA 2008, first workshops on image processing theory, tools and applications
66. Yoon T-J (2010) Capturing inter-speaker invariance using statistical measures of rhythm. In: Paper presented at the proceedings of speech prosody 2010, Chicago, IL, USA, 10–14 May 2010

Part II
Speech Signal Degradation: Managing Problematic Conditions Affecting Probative Speech Samples

Chapter 4
Speaker Profiling: The Study of Acoustic Characteristics Based on Phonetic Features of Hindi Dialects for Forensic Speaker Identification

Manisha Kulshreshtha, C. P. Singh and R. M. Sharma

Abstract Personal identification has been proved to be one of the important aspects of forensic science in which biometric features have been evaluated for information related to individuals. Biometric techniques of identification such as handwriting characteristic, speaker identification, fingerprint and DNA Profiling etc. are significant in forensic science for solving specific type of crime cases. Another aspect of identification of individuals through voice is speaker profiling, in which information about speakers are extracted from the recorded speech material. One of the speaker profile characteristics is the dialectal accent feature that can establish the speaker's identity through his dialect. Khariboli, Bundeli, Kannauji, Haryanvi, Chattisgarhi, Marwari and Bhojpuri dialects are chosen from different parts of the Hindi speaking belt for this study. Khariboli is considered as the basic language due the close approximation to standard Hindi. Also it is the dialect that forms the basis of the modern standard Hindi. The prepared texts are transliterated using same script (Devnagri) but different vocabularies are used within the dialect group. Speakers have been selected from a uniform area of the relevant regional dialect. 15 male and 15 female speakers were selected from each dialectal region keeping the selection criteria in mind, giving a total of 210 speakers. Vowel quality and quantity of dialect speakers has been measured with the help of formant frequencies and vowel length and compared with Khariboli speakers. Intonation and tone has also been observed and compared. Bhojpuri, Chattisgarhi, Kannauji, Marwari, Khariboli, Bundeli and Haryanvi dialects are found unique for characterization in terms of vowel quality and vowel duration when compared with Khariboli. Acoustic features associated with lexical tone and sentence intonation are also found unique and useful to a dialect identification for speaker profiling. Our results illustrate that vowel quality, quantity, intonation and tone of a speaker as compared to Kahriboli (standard Hindi) could be the potential features for identification of dialect accent.

R. M. Sharma (✉)
Department of Forensic Science, Punjabi University, Patiala 147002, Punjab, India
e-mail: rmsforensics@gmail.com

A. Neustein, H. A. Patil (eds.), *Forensic Speaker Recognition*,
DOI 10.1007/978-1-4614-0263-3_4, © Springer Science+Business Media, LLC 2012

4.1 Introduction

In the present scenario, investigators are encountering cases, where speech of a person is important physical evidence, if recorded and proved to be the only available evidence to establish the link of a suspect with the criminal activity. Reason being, modes of voice communication are changing drastically in today's world. There are mobile phone-sets with in-build multi-media facility and other sophisticated digital recorders having small size memory chips, which are easy to handle and affordable by general public. Taking advantage of such hi-tech communication devices, criminals are using telephone or mobile phone communication while committing various offences like, kidnapping, bomb threat calls, extortion, terrorist activities, hoax calls, bribery and many other offences. On the other hand, the victim of such offences generates recorded conversations as evidence through mobile phone-sets and digital recorders. In such situations, forensic science helps the investigators in identifying the individuals involved in criminal activities and in differentiating between the guilty and innocence so that the guilty is punished and innocent is prevented from the unjust punishment. There are various methods in the hands of forensic experts to establish the identity of alibi depending upon the nature and type of collected evidences. Identification through voice is one of the well-established methods used by the investigating agencies; in case, the recorded conversation is available as physical evidence. Identification of a speaker on the basis of the voice characteristics is performed when the characteristic features of recently recorded questioned speech samples of an unknown person are compared with the features of specimen speech samples of the suspect(s). The method is technically termed as Forensic Speaker Identification. In some of the cases, where the suspected culprits are not available or they refuse to give specimen sample to investigating agencies (as one cannot force an individual to give his voice sample) and it is difficult to prove the involvement of an individual in a criminal activity or speaker identification is not possible in the absence of required control samples, speaker profiling provides clue for the investigating agencies in order to pin point the actual criminal and to narrow down the field of investigation. Speaker Profiling includes the extraction of individual information about the anonymous perpetrator from incriminated speech material like gender, age, dialect, features of respiration, phonation, articulation and manner of speaking etc. Therefore, a person belonging to a regional dialectal group can be identified with accent features reflected in his or her speech. Accent feature of a person also depends upon the educational background, mother tongue of the person and the regional language of area where he or she is living for last many years. In India, most of the people are bilingual, especially those living in the border areas. While speaking the non-native language, it is always affected by the native accent of that person.

Hindi as official language is widely spoken among the population of India. Some popular dialects of Hindi namely, Khariboli, Bhojpuri, Bundeli, Chattisgarhi, Kanauji, Marwari and Haryanvi have been selected and experiments have been performed to study the acoustic characteristics of native accent that are reflected in

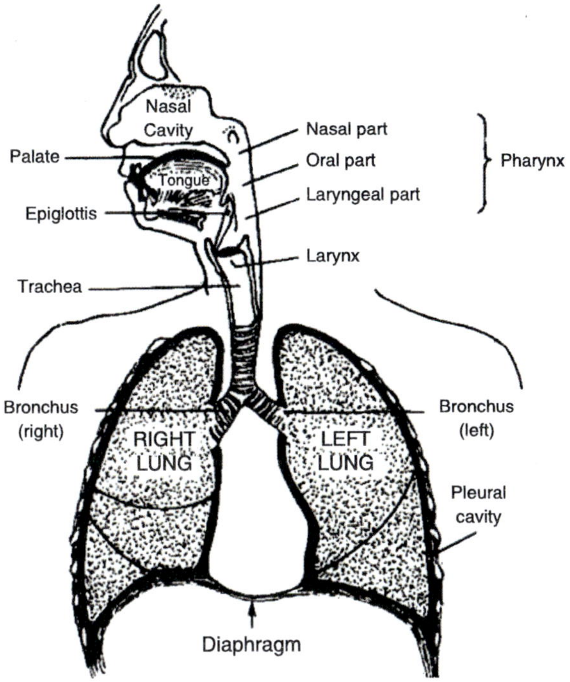

Fig. 4.1 Organs of speech. [27]

non-native language on the basis of phonetic features, considering Khariboli (standard Hindi) as base language. This study is aimed to find the acoustic description of the accent features with primary purpose of speaker profiling. Study also includes finding the effect of selecting clue materials from accented and non-accented texts for comparison by spectrographic methods. In order to understand this effect, vowel quality and quantity is to be studied for dialects chosen with reference to that of Khariboli [1, 56].

4.2 Speech Production Mechanism

Vocal organs are those active parts of a human body, which are directly or indirectly involve in the production of speech and also termed as organs of speech. The main organs of speech are lungs, chest muscles, trachea, larynx, pharynx, lips, teeth, tongue, roof of mouth and vocal folds. The organs all together are termed as vocal tract as shown in Fig. 4.1. When a speaker intends to produce a speech sound, the brain is the first organ that sends set of signals to the responsible organs involved in the process of speech production [1]. Diaphragm that divides the abdominal cavity and thoracic cavity moves upward due to the action of abdominal muscle. The process expands and contracts the lungs so that the air from outside is

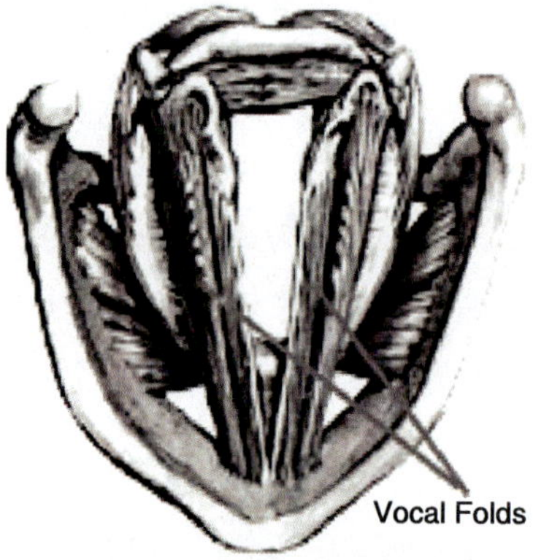

Fig. 4.2 Vocal folds structure. [46]

drawn in and pushed out alternatively. The most important function of lungs that is relevant to speech production is respiration and it is responsible for the movement of air. In the production process, the air is pushed out of the lungs and passes through the windpipe, i.e., trachea. At the top of windpipe there is a cartilaginous structure, called larynx and at the upper- end there is a passage between the two horizontal folds of elastic muscular tissues (vocal folds) called glottis. Vocal folds are pair of lips like structure that can close or open at the glottis and can be brought together by the use of laryngeal muscles [28]. Larynx is used to control the steady flow of air into and out of the lungs. Rising in the pressure on the underside of the shelf takes place when the edges of the vocal cords are held together. When this pressure reaches to a certain level, it overcomes the resistance offered by the obstruction and the vocal cords open. These structures are made up of ligaments and muscle fibers that have certain degree of elasticity, which make them to return to their initial position and lower the air pressure. Figure 4.2 shows the structure of vocal folds. Again the pressure rises and the cycle of opening and closing is repeated. Mass, length and tension of the vocal folds are the physical variables that can be controlled by the speaker to produce the sound of the required frequency range. Normally, men use lower fundamental frequency, women uses middle range and the children use the highest fundamental frequency of sound. The main physiological function of the vocal folds is to close the lungs at the time of eating or drinking so that no solid particle of food or liquid enters the lungs through the windpipe. Due to their tissue structure, the vocal folds are capable of vibrating with different frequencies, when the air passes through and the vibration is called voice. After passing through the glottis and reaching the pharynx, the outgoing airstreams (expressive air-stream) can escape either through the nasal cavity (nostrils) or through the oral cavity (mouth). When the air escapes through mouth by bringing the velum into close contact with pharyngeal wall, can close the nasal cavity. Such a type of closure of the nasal passage is called the velic closure. When there is velic closure, the air can escape only through mouth. The nasal cavity can also be kept open when the air passes through the mouth allowing part of the air to pass through nose also. Normal breathing occurs when air passes through the nasal cavity. The whole process is known as the respiration mechanism, an important part of the speech production mechanism [46].

4.3 Supra-Segmental Features

Vowels and consonants are considered as small segments of the speech, which together form a syllable and make the utterance. Specific features that are superimposed on the utterance of the speech are known as supra-segmental features. Common supra-segmental features are the stress, tone and duration in the syllable or word for a continuous speech sequence. Sometimes even harmony and nasalization are also included under this category. Supra-segmental or prosodic features are often used in the context of speech to make it more meaningful and effective. Without supra-segmental features superimposed on the segmental features, a continuous speech can also convey meaning but often loses the effectiveness of the message being conveyed. The most widely discussed phonological supra-segmental is the syllable. A syllable is considered as a phonological unit, generally consists of three phonetic parts namely, the onset, the peak and the coda. In a monosyllabic word /bat/, /b/ is the onset, /a/ is the peak and, /t/ is the coda. A syllable is further discriminated as closed syllable comprising a vowel (V) as a nuclei, preceded and followed by a Consonants (C) as onset and coda, i.e., CVC or VCV structure and open syllable, in which either onset or coda is absent (CV or VC). Features like stress, tone and duration are always present in almost all the utterances of a language. Thus, all the utterances can be characterized by different degrees of these features.

4.3.1 Stress

Stress is one of the supra-segmental features of utterances. It applies not to individual vowels and consonants but to whole syllable. A greater amount of energy is required to pronounce a stressed syllable than that of required while pronouncing the unstressed syllable because more air is pushed out from the lungs by extra contraction of the abdominal muscles in addition to the laryngeal muscles, which results in the additional rise in pitch. Increase in the amount of air pushed out also increases the loudness of the sound produced.

There is a misconception of expressing stress in terms of loudness; whereas, loudness is simply the amount of acoustic energy involved in the process. There are many more factors affecting the stress like length, change of pitch and quality of the vowel. These features all together can express the variation of stress over utterances in a language. In a situation, where words involve more than one syllable, a language specific stress pattern is expected. Disyllabic words are distinguished as stressed and unstressed whereas; a polysyllabic word is distinguished as primary stressed or unstressed and secondary stressed or unstressed.

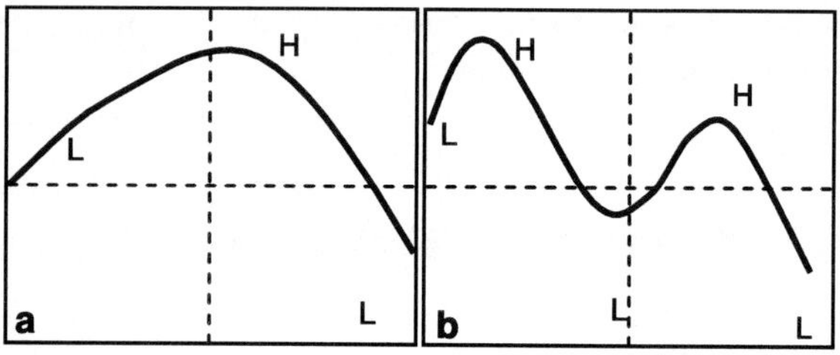

Fig. 4.3 The pattern of pitch movement in two different languages

4.3.2 Intonation

Intonation can be defined acoustically as the gradual variation of fundamental frequency with respect to time. If the variation is along the sentence or phrase, it is termed as sentence or phrasal intonation whereas along the word or word-segment, it is called lexical intonation or tone and the languages with such features are known as tonal languages. More analytically, intonation is something, which is super-imposed upon the meaning of words uttered and carries important semantics function in many languages. It is important to note that such semantic functions through intonations can be brought out in the speech continuum by rhythmic modulations.

4.3.3 Tone

Tone may be rising, level or falling tone depending on the increase, steady and decreasing of pitch respectively. Tone or intonation of a speaker can be represented as low (L), high (H) or the combination of these two. Figure 4.3 shows the intonation pattern in case of two different accents of two languages. Rules of representation, which are common to all the languages are known as conventional rules whereas, dialect or language specific rules are termed as prescription rules.

4.3.4 Length

Languages can also be discriminated on the basis of the length of the segment. Length is basically concerned with the duration of the vowel used in the segment. There are languages in which words having same structure but different length convey different meaning. For example in Hindi /gana/ is the word used for 'song' whereas, /ga: na/ is the word with long vowel conveying the meaning 'to sing'. Symbol /:/ after the vowel represents the long vowel.

4.4 Variability in Speech

The recorded speech obtained through telephonic transmission channels is affected by noise associated with the environment from where the talker is speaking. The resonance and the reverberation of the room from where the talker is speaking through telephone may cause damping or amplification by some bands of frequencies. Ideally speaking, the unknown and known samples are to be recorded with in the same room in order to minimize these types of resonance affects on the process of speaker identification and elimination. In actual cases, the environment in which the criminal is using the telephone is not known. However it is the opinion [56] that effect of this resonance is not significant in most of the cases. Another factor affecting the recorded speech is due to the factors of resonance or response curve of the telephone line utilized. The response curve of the telephone line may amplify or attenuate some band of frequency within the range of frequency, normally from 150 to 4000 Hz. The pickup device and method of connecting the receiving telephone to the recorder and also the tape recorder itself can also introduce distortion the unknown and known samples. It is mentioned [56] that in every case the response curve of each of the transmitting and recording elements in use during the recording of both known and unknown voice should be obtained to have information of this type of information. On the other hand speech sounds are generally posed with variability during the production of speech. Generally speaking, variability in speech can be of two types, i.e., Intra speaker and Inter speaker variability. Intra speaker variability, which exists within the same speaker, is due to many factors. Some of these factors are emotions, rate of utterance, mode of Speech, disease, mood of the speaker and the emphasis given to word at a particular moment. One of the important factors which are not having and quantitative measurements is the temporal variation within the speaker. The inter speaker variation, which exist among the different speakers are arises mainly due to anatomical differences in the vocal organs and from learned differences in the use of speech mechanism.

4.5 Principles of Voice Identification

The principle underlined the technique of voice identification for use in the personal identification is due to the fact and assumptions discussed herewith. The identification of a person for speech sound is based on the fact that the same words uttered by different speakers are quite apparent to any listener. The variability known as inter-speaker variability and arises mainly due to anatomical differences in vocal tract and from learner differences. The difference in the anatomical and physiological mechanism and in the use of these speech organs in turns produce frequency mixtures, specific to the individual, Inter speaker variability is always greater or different than the intra speaker variability regardless of the parameters involved in this variability. In addition to anatomical differences every person experimentally develop individual or unique process of learning to speech contributing to built a unique speech spectra for each individual when uttering a given word. The disguising of voice by a person may produce significant variations in his speech samples. However, an expert can always detect these variations in disguised voice. Disguising voice is not a problem as experts are trained to disregard or discard the distorted or disguised speech samples.

4.6 Methods of Voice Identification

4.6.1 Auditory Analysis

Phonetic is the study of the phonemes related to a language or dialect and their idiosyncratic pronunciation by a speaker. The technique is based on the process called critical listening. Under this technique, a particular speaker is to be identified using phonetic sequences and the phonetic events undertaken while speaking accented speech as well as regional dialect. Speech of person is characterized using this technique with some phonetic features like stylistic impression, phonation, nasality, dynamic of loudness and flow of speech. Based on these features the linguistic characteristics of a person in specimen are compared with that of the person in questioned.

4.6.2 Spectrographic Method

When the voice of a person is subjected to spectrographic analysis, the sound energy is converted into electrical energy, which operates the stylus of the spectrograph (the wave analyzer) and the stylus in turn creates a trace in the form of a graph on the paper on the drum. The graph is known as a spectrogram. Horizontal dimensions represent the time interval of the speech signal. The vertical dimensions indicate the frequency and the darkness indicates the intensity of the loudness of the voice. The visual display of the disputed and the sample utterance (of the same word or text) are compared visually. The process of using the method is simple. The suspects are made to utter relevant word(s) and the voice is recorded on a device similar to one used in recording the disputed utterance. The recorded voices are then turned into their visual spectrograms through the sound spectrograph and the spectrograms are compared and evaluated. Spectrographic analysis depends upon the use of filters and filters are examples of resonant systems. Such systems may be sharply tuned and have little damping, or they may be more highly damped and not so sharply tuned. Based on the filters used in the analysis, the spectrogram is known as wideband or narrow band spectrograms as shown in Fig. 4.4a and b.

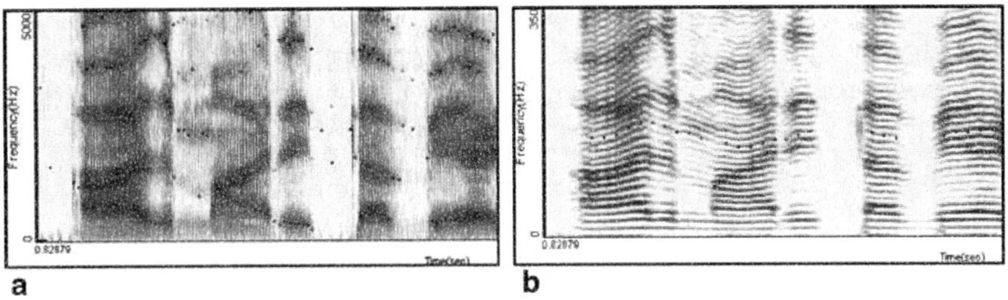

Fig. 4.4 **a** Wide-band spectrogram. **b** Narrow-band spectrogram

4.6.3 Tests and Errors in Voice Identification

According to the composition of unknown and know voice samples, tests of voice identification or elimination can be classified into three group; discrimination tests, open tests and close tests. In the discrimination tests the examiner is provided with one unknown voice sample and one known voice sample. Two types of errors can be produced in the discrimination tests: (a) false elimination, when the examiner decides that both samples belong to different talkers, but they are actually from the same talkers and (b) false identification, when the examiner decides both samples belong to the same talkers, when they actually do not. In the open tests, the examiner is given one unknown voice sample and several known voice samples. He is told that the unknown sample may or may not be found among the known samples. This type of test can yield three types of errors. The first is false elimination, when the unknown voice sample is among the known samples. But the examiner decides that it is not. The second and third types are errors of false identification that can originate from two possibilities (a) when one of the known samples is the same as the unknown one, and (b) none of the known samples is same as the unknown one then the examiner decides that one of them is same as the unknown. In the closed tests of voice identification, the examiner is given one unknown voice sample and several known voice samples but he is told that the unknown voice sample is also included in the known voice samples. Consequently, here only one type of error may be produced, i.e., an error of false identification in which the examiner selects the wrong one.

4.7 Hindi Language

Language is an arbitrary system by which one can exchange his or her thoughts and ideas with others. It has a set of specified symbols (graphemes) assigned to specific phones and termed as script of a language. The script contains the words or vocabulary used within a linguistic community to express their views as well as to convey the meanings according to the prescribed rules. Different communities are using different sets of sounds for communication, thus, have different languages. Versions of a language that sound different but are mutually intelligible are called dialects of the language. Generally, dialects of a language do not have any written script. Speaker, while speaking different dialect uses same basic language but different vocabularies and also pronounces words differently. Due to this reason, speech of a person belonging to a dialectal group always has its specific dialectal accent. In fact, native accent is always affected while speaking non-native language and shows the potential in the investigation of crime as accented speech carries linguistic information regarding the regional dialect of an individual.

Hindi is considered to be one of the Middle Indo-Aryan languages. There is no specific time mentioned in the literature about the birth of Hindi, but 1,000 AD is commonly accepted. Hindi sometimes is also referred as Hindavi or Hindustani and also Khariboli. Approximately 600 million people across the globe speak Hindi either as first or as second language. Based on a survey conducted in 1997, 66% of all Indians can speak Hindi and 77% of the Indians regard Hindi as 'one language' across the nation. Literary Hindi has an approximately 300 years old, well attested rich literary and grammatical tradition. As on today, Hindi is placed as world's third most spoken language and is spreading all over the world. In the era of technological advancements and the 'global village', Hindi assumes much importance as it is spoken by a large number of people across the globe. The use of word Hindi in the context of Hindi language was seen for the first time in the epic JAFARNAMA in 1424 AD, in the initial days of its development as a language. The apbransh of Prakrit and Pali led to the rise of different regional dialect of Hindi. These dialects paved the path for Khariboli, a form in which Hindi is recognized today. Over nearly a thousand years of Muslim influence, when Muslim rulers controlled much of northern India during the Mughal Empire, many Persian and Arabic words were borrowed [34].

4.8 Dialects of Hindi

Hindi is often divided into Western Hindi and Eastern Hindi, which are further divided into its various dialects. Around 200 of regional dialects are identified within Hindi language itself. Many linguists consider only western and eastern dialect as the proper dialects of Hindi and rest are considered as sub language or separate language. There is a report [55] noted that the classification of the dialect under various branches and their classification as a dialect of Hindi or as an independent language depend upon the perception of the linguist. Accordingly dialects of Hindi are categorized into five groups; some of the Hindi dialects are listed below.

4.8.1 Western Hindi

- **Khariboli:** It is also called as Sarhind or Kauravi. It is originally spoken in the western Uttar Pradesh (the districts of Saharanpur, Muzaffarnagar, Merrut, Gaziabad, Bijnor, Rampur, Moradabad, and district of Dehardun in Uttaranchal) and Delhi region. It is the dialect that forms the basis of modern standard Hindi.

- **Braj Bhasha:** This dialect is mainly spoken in the districts of Mathura, Aligarh, Dhaulpur, Mainpuri, Etah, Badaun and Bareilly of south-central Uttar Pradesh. It has a rich poetic literal tradition, especially linked with the Hindu divinity Krishna.
- **Haryanvi:** It is spoken in the state of Haryana and it is heavily influenced by Punjabi language.
- **Bundeli:** It is a dialect mainly concentrated in MP and is found spoken in the districts of Jhansi, Jalaun and Hamirpur in Uttar Pradesh and Gwalior, Bhopal, Sagar, Narsinghpur, Seoni, Hoshangabad, etc. in Madhya Pradesh.
- **Kannauji:** This is a dialect of the districts of Etawah, Farrukhabad, Shahjahanpur, Kanpur, Hardoi and Pilibhit in Uttar Pradesh.

4.8.2 Eastern Hindi

- **Awadhi:** Spoken in central and parts of eastern Uttar Pradesh, in the districts of Allahabad, Fatehpur, Mirzapur, Unnao, Raebareli, Sitapur, Faizabad, Gonda, Basti, Bahraich, Sultanpur, Pratapgarh and Barabanki. The famous Hindu scripture Ramcharitamnas was written by Tulsidas in this dialect.
- **Bagheli:** Dialect of the districts of Rewa, Nagod, Shadol, Satna and Maihar etc. in Madhya Pradesh.
- **Chattisgarhi:** Spoken mostly in districts of Bastar, Bilaspur, Dantewada, Durg, Raigarh and Raipur in Chattisgarh.

4.8.3 Rajasthani

This is mostly spoken in the state of Rajasthan and consists of several dialects:

- **Marwari:** It is a dialect known as Western Rajasthani, spoken in the Indian state of Rajasthan, but is also found in the neighboring state of Gujarat and in Eastern Pakistan. Marwari is still spoken widely in and around the districts of Jodhpur region.
- **Jaipuri:** It is also known as Eastern Rajasthani and is spoken in Dausa, Jaipur, Jhunjhunu and Sikar district of Rajasthan.
- **Mewati:** This dialect is also known as Northern Rajasthani and is spoken in Bharatpur and Dholpur districts of Rajasthan.
- **Malwi:** This dialect comes under Southern Rajasthani. According to the survey conducted in 2001, only 4 sub-dialects were found, i.e., Ujjaini, which is spoken in Ujjain, Indore, Dewas and Sehore districts, Rajawadi, spoken in Ratlam,

Mandsaur, Neemuch districts, Umadwadi spoken in Rajgarh district and Sondhwadi, which is spoken in Jhalawar district of Rajasthan.

4.8.4 Pahari

Pahari *is* a general term for various dialects spoken in the Indian part of the central Himalayan range.

- **Eastern Pahari:** This dialect includes Nepali, which is now considered as a separate language. It is an Indo-Aryan language spoken in Nepal, Bhutan, and some parts of India and Myanmar (Burma). Also it is the official language of Nepal. This dialect is also known as Gorkhali or Gurkhali, "the language of the Gurkhas", and Parbatiya, "the language of the mountains."
- **Central Pahari:** It includes Garhwali and Kumauni as sub-dialects of the newly created state of Uttaranchal. Garhwali itself has many regional dialects spoken in different parts of the state, like Jaunsari and Jadhi. Kumaoni is a local dialect of Kumaon Division of Uttaranchal, a region in the Indian Himalayas.
- **Western Pahari:** It includes a number of dialects including Bhadrawahi, Bhattiyali, Bilaspuri, Chambeali, Churahi, Dogri-Kangri, Gaddi, Hinduri, Jaunsari, Harijan Kinnauri, Mandeali, Kullu Pahari, Mahasu Pahari, Pahari-Potwari, Pangwali, and Sirmauri, which are the sub-dialects of western pahari spoken in various districts of Himachal Pradesh.

4.8.5 Bihari

- **Maithali:** It is a dialect of the east champaran, Muzaffur, Munger, Bhagalpur, Darbhanga, Purnia and North Santhal and Pargana districts of Bihar.
- **Magahi or Magadhi:** The Dialect is spoken in the districts of Gaya, Patna, Munger and Bhagalpur of Bihar state and Palamu, Hazaribagh and Ranchi districts of Jharkhand state.
- **Bhojpuri:** It is a dialect spoken in Gorakhpur, Deoria, Mirzapur, Varanasi, Jaunpur, Ghazipur, Ballia districts of eastern Uttar Pradesh, Chappra, Siwan, Gopalganj and Bhojpur districts of Western Bihar and a small part of Palamu and Ranchi of Jharkhand.

Depending upon the perception, linguist also includes various dialects under Hindi, such as Nimari, Maiswari, Vajjika, Angika, etc. For the purpose of experiments in this chapter, some popular dialects of Hindi, namely, Khariboli, Bundeli, Kannauji and Haryanvi among the western Hindi belt, Chattisgarhi from eastern Hindi dialetal group, Marwari from Rajasthani group and Bhojpuri from Bihari group of Hindi belt have been selected. Figure 4.5 represents the distribution of Hindi and its dialects along the northern belt.

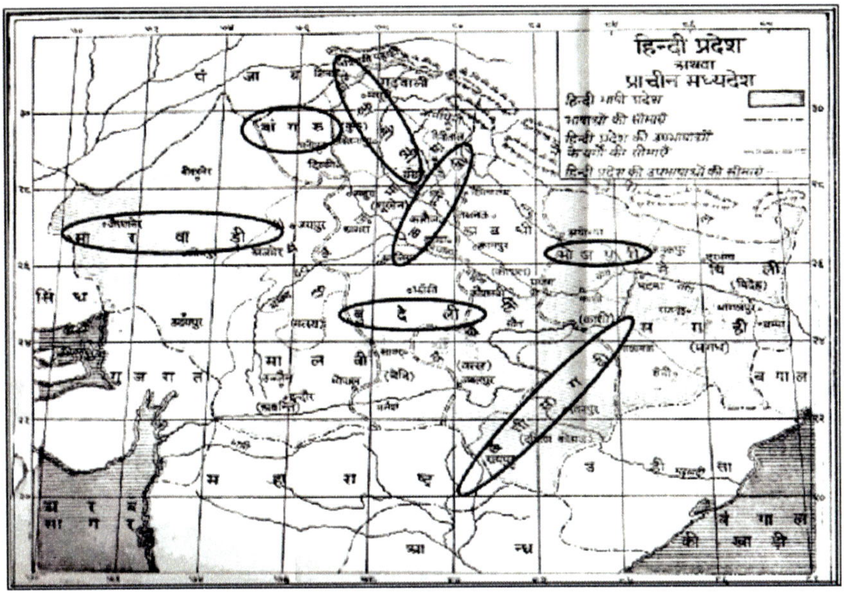

Fig. 4.5 Map showing the extent of Hindi in the northern belt of India. [24]

4.9 Phonology of Hindi Language

Phonology is a subfield of linguistics, which studies the sound system of language or languages. An important part of phonology is to study distinctive sounds within a language. There are 11 vowels and 35 consonants frequently used in Hindi Speech with few sounds borrowed from Persian and Arabic languages but now considered as a part of Hindi language. These sounds can be further classified according to the place and manner of the articulation. Vowels of Hindi are almost similar to that of other languages but there are certain differences and special features in Hindi consonants, which are common to other languages.

4.9.1 Vowels of Hindi

Vowels of Hindi are classified into 11 different vowel sounds as mentioned in various literatures. In addition, there are extra sounds, which sometimes are included as a part of vowel sound system of Hindi language. Vowel sound system of Hindi language is described in Table 4.1. The short-open-mid-unrounded vowel /ɛ/ (as e in get) does not have any symbol or diacritic in Hindi script. It occurs only as conditioned allophone of schwa. Thus, the pronunciation of the vowel अ occurs in two forms. When this vowel is followed by word-middle /h/, or it surrounds word-middle /h/, or is followed by word ending /h/, it changes allophonically to short

Table 4.1 Description of the vowels sounds in Hindi language

Alphabet	Pronunciation	IAST equiv.	Description
अ	/ə/	a	Short or long Schwa: as the *a* in above or ago
आ	/ɑ:/	ā	Long Open back unrounded vowel: as the *a* in father
इ	/i/	i	Short close front unrounded vowel: as *i* in bit
ई	/i:/	ī	Long close front unrounded vowel: as *i* in machine
उ	/u/	u	Short close back rounded vowel: as *u* in put
ऊ	/u:/	ū	Long close back rounded vowel: as *oo* in school
ए	/e:/	e	Long close-mid front unrounded vowel: as *a* in game (not a diphthong)
ऐ	/æ:/	ai	Long near-open front unrounded vowel: as *a* in cat
ओ	/o:/	o	Long close-mid back rounded vowel: as *o* in tone (not a diphthong)
औ	/ɔ:/	au	Open-mid back rounded vowel: as au in caught

/ɛ/. In all other cases it is the mid central vowel schwa. Thus, the following words 'sehar', 'rehna' and 'keh' are pronounced as /ʃɛhɛr/, /rɛhnɑ:/ and /kɛh/ and not as /ʃəhər/, /rəhənɑ:/ and /kəh/.

The short-open-back-rounded vowel /ɒ/ (as o in hot) does not exist in Hindi at all, other than for English loanwords. In orthography, a new symbol has been invented for it, i.e., ऑ and included in Hindi phonology. There are some additional vowels traditionally listed in the Hindi alphabet. They are, ऋ (a vowel-like syllabic retroflex approximant), pronounced in modern Hindi as /ri/, used only in Sanskrit loan words. All vowels in Hindi, short or long, can be nasalized except ऑ. In Sanskrit and in some dialects of Hindi (as well as in a few words in Standard Hindi), the vowel ऐ is pronounced as a diphthong /əi̯/ or /ai̯/ rather than /æ:/. Similarly, the vowel औ is pronounced as the diphthong /əu̯/ or /au̯/ rather than /ɔ:/. Other than these, Hindi does not have true diphthongs—two vowels might occur sequentially but then they are pronounced as two syllables (a glide might come in between while speaking). The schwa vowel (/ə/) is pronounced very short otherwise it will be very difficult to pronounce few words in the absence of schwa. Such a situation arises when there is a consonantal cluster at the end of the word. Thus, for phonological purposes, a word-ending grapheme without a halant or any other vowel-diacritic must be treated as consonant ending. The schwa in Hindi is usually dropped (syncopated) in Khariboli even at certain instances in word-middle positions, where the orthography would otherwise dictate so. e.g., रुकना (to stay) is normally pronounced as /ruknɑ:/, while according to the orthography, it should have been /rukəna:/ [55].

4.9.2 Consonants of Hindi

There are 38 distinct consonants reported in Hindi phonology. Due to the regional variability of Hindi language it is very difficult to tell the exact number of sounds to be included in the category of Hindi consonants. The consonants system consists of 25 occlusive, where the airstreams from the mouth is completely blocked and 8 sonorant and fricatives. It also includes 7 sounds borrowed from Persian and Arabic but is now considered as Hindi sounds (Table 4.2).

Few sounds, borrowed from the other languages like Persian and Arabic are written with a dot (bindu or nukta) (Table 4.3). Many native Hindi speakers, especially those who come from rural backgrounds and do not speak really good Khariboli or Urdu, pronounce these sounds as the nearest equivalents in Hindi.

ड़ /ɽə/ and ढ़ /ɽʱə/ are not of Persian/Arabic origin, but they are allophonic variants of simple voiced retroflex stops [55].

4.10 Speech Data Collection

Specific texts have been prepared for collection of speech material, in Khariboli as base language including the words or phrases commonly found in telephonic interception of criminal act like bomb hoax and intimidation, kidnapping for ransom, match fixing, call girls rackets, extortion and bribery etc (see Appendix 1). This prepared text is transliterated in different dialects chosen for the study considering the accent as well as the uses. In order to minimize the intra dialectal variation due to regions, speakers have been selected from a uniform area of the regional dialect as well as with in a closed age group (20–25) with at least higher secondary education and also can speak its pure regional dialect. Condition is that the speaker should not have any influence of other native language on its dialect. Considering the criteria mentioned above, 15 male and 15 female speakers were selected from each dialect. Information related to the speakers and their dialectal backgrounds have been noted. Out of various districts in which a dialect is being spoken, only one place has been chosen to collect the samples of the speakers. Name of the regional dialect and the place from where the speech samples of the speakers have been collected are mentioned below in Table 4.4.

4.11 Recording of Speech Samples

4.11.1 Microphones

A microphone is a transducer that converts sound energy into electrical energy. Sound information exists as patterns of air pressure and the microphone changes this infor-

Table 4.2 Consonants of Hindi

Devnagri letters	Transliteration	IPA
क	k	k
ख	kh	k^h
ग	g	g
घ	gh	$g^ɦ$
ङ	ṅ	ŋ
च	c	tʃ
छ	ch	$tʃ^h$
ज	j	dʒ
झ	jh	$dʒ^ɦ$
ञ	gn	ɲ
ट	ṭ	ʈ
ठ	ṭh	$ʈ^h$
ड	ḍ	ɖ / ɽ
ढ	ḍh	$ɖ^ɦ$ / $ɽ^ɦ$
ण	ṇ	ɳ
त	t	t
थ	th	t^h
द	d	d
ध	dh	$d^ɦ$
न	n	n
प	p	p
फ	ph	p^h
ब	b	b
भ	bh	$b^ɦ$
म	m	m
य	y	j
र	r	ɾ
ल	l	l
ळ	ḷ	ɭ
व	v	v
श	ś	ɕ
ष	ṣ	ʂ
स	s	s
ह	h	h

mation into patterns of electric current. There are varieties of mechanical techniques that can be used to build a microphone. The two most commonly used methods are the magneto-dynamic and variable condenser. Majority of microphones used in recording of sounds are either capacitor (electrostatic) or dynamic (electromagnetic)

Table 4.3 Consonants of Hindi (Borrowed from Persian and Arabic)

Devnagri Letters	IPA Equivalent
क़	/qə/ voiceless uvular plosive
फ़	/fə/ voiceless labiodental
ख़	/xə/ voiceless velar fricative
ग़	/ʁə/ voiced velar fricative
ज़	/zə/ voiced alveolar fricative
ड़	/ɽə/ unaspirated retroflex flap
ढ़	/ɽʱə/ aspirated retroflex flap

Table 4.4 Selected regional dialect and their place of recording

Name of the regional dialect	Place of recording
Bundeli	Jhansi (Uttar Pradesh)
Khariboli	Anwalkhera (Uttar Pradesh)
Marwari	Alwar (Rajasthan)
Bhojpuri	Deoria (Gorakhpur)
Chattisgarhi	Charoda (Raipur)
Kanauji	Kayanpur (Kanpur)
Haryanvi	Hisaar (Haryana)

models, which employ a moving diaphragm to capture the sound, but make use of a different electrical principle for converting the mechanical energy into an electrical signal. The efficiency of this conversion is very important because the amount of acoustic energy produced by voices and musical instruments is very small. The wide availability of electorate condenser microphones has greatly simplified the problem of obtaining high quality recordings. Electret microphones respond directly to the sound pressure of the speech signal. Directional electorate microphones can be obtained that respond differentially to sounds coming from one direction. This can be an advantage if one is recording the samples in a noisy environment.

The choice of the microphone depends on the goal of a particular purpose. In the present study, the microphone used is a dynamic microphone of Philips (Model DM295) with 1.8 mV/Pa sensitivity, 600 Ω impedance, and frequency range of 100–12,000 Hz. Using the above-mentioned specifications of microphone speech samples has been recorded directly on computer using inbuilt multimedia card. Speech samples of speakers (dialect speakers) are recorded in three repetitions, in Khariboli as well as in their regional dialect.

4.11.2 *Digitization of Speech Samples*

In the present scenario, most of the equipments are computer based and work in digital domain. For the purpose of analysis in digital domain, all analog signals need to be converted into digital signals. An electronic circuits that covert continuous analog signal into discrete digital numbers is called Analog-to Digital converter (abbreviated ADC).

4.11.3 Sampling of Speech Exemplars

The recorded utterances of the speakers chosen for the study have been subjected to preliminary auditory analysis for selection of appropriate speech data from the raw data. The utterances are chosen in which the accent features of the speakers are well reflected and are found suitable on the basis of speech quality or clarity of the recorded sample. Speech exemplars of 15 male speakers and 15 female speakers are chosen in each dialectal group.

4.12 Instrumentation

4.12.1 Sound Spectrograph

Alexander Melville bell in 1867 developed a visual representation of the spoken words later named as 'visible speech' at Bell telephone laboratory. In 1940s Potter, Kopp and Green developed a new method of speech sound analysis using speech spectrograph. Dr. Ralph Potter introduced an electromechanically acoustic spectrograph in 1941. In 1962 Lawrence Kersta, an engineer and a staff member of the bell laboratories re-examined the voiceprint method at the request of law enforcement group and introduced the instrument named sound spectrograph as a potential tool for Forensic Speaker Identification. Basic function of this device was to convert the speech into visual representation of its frequency and intensity components. A sound spectrograph has four parts, magnetic recorder, electronic filters, a rotating drum carrying paper around, on which the spectrogram is recorded and an electrically operated stylus. The traditional analog version of the sound spectrograph records the input signal on a magnetic medium that goes round the outside edge of the thin drum. Magnetic image is formed on the thin recording disc by the recording head just like a conventional tape recorder when the sound spectrograph switch is on recording mode.

The voice or sound spectrograph is of three types:

- Analog Spectrograph: In Analog spectrograph speech from the microphone is fed into a band pass filter. Harmonic of the voice whose frequency falls within the range of that filter gives the output with the amplitude proportional to its strength and produce a three dimensional record with the stylus on a paper, showing the change in frequency and amplitude w.r.t. time.
- Digital spectrograph: A Digital spectrograph consists of special circuits embedded in the microprocessor systems to produce the spectrogram instanteously with the speech. Voice identification Inc. USA has produced a real time digital

Fig. 4.6 Computerized Speech Laboratory system

spectrograph, which produces video display of spectrograms. It can determine the duration of the speech segments; calculate fundamental frequency and formant ranges etc.

- Hybrid spectrograph: Hybrid spectrograph is the combination of two spectrograms mentioned earlier.

The sound spectrograph analysis requires only a limited amount of utterance at a time, which could be recorded in one revolution of the drum. It converts the speech signal into a visual spectrum in the form of traces of the graph. The dimensions and the intensity of the traces are dependent upon the utterance being analyzed. Now a day, the use of computers in conjunction with the spectrograph has increased the volume of the recording.

4.12.2 *Computerized Speech Laboratory (CSL)*

Computerized Speech Lab (CSL) for windows is a hardware and software system for the acquisition, acoustic analysis, display and playback of speech signals. It records, edits and quickly analyses the speech signal. It is possible to carry out the detailed studies of the utterances through segmentation of the recordings.

CSL is suitable for any acoustic signal characterized by changing spectra over time. It is windows based programs, which require a computer operating under Windows 95 and Windows 98. Operations of CSL include acquisition, storing speech to disk memory, graphical and numerical display of speech parameters, audio output, signal editing. A variety of analysis, namely, spectrographic analysis, pitch contour analysis, LPC Analysis, Cepstrum analysis, FFT and Energy contour analysis etc can be performed through this instrument. It gives results easily and quickly in comparison with the old speech spectrograph and can handle large speech data at a time. Speech-exemplars chosen during the sampling process have been analyzed using Computerized Speech Laboratory model 4300 B as shown in Fig. 4.6.

Table 4.5 Syllables selected for the study

Selected Word (C_1VC_2)	Vowel	Description of the vowel
/kʌl/	/ʌ/	Open-mid-back-un-rounded
/bat/	/a/	Open-back-un-rounded
/tʰik/	/i/	Close-front-un-rounded
/kon/	/o/	Open-mid-back-rounded
/tum/	/u/	Close-back-rounded

4.13 Analysis of Speech Exemplars

4.13.1 *Auditory Phonetic Analysis*

Recorded speech samples have been analyzed on the basis of perceptual characteristics. The chosen sentence and word segment suitable for analysis have been studied for sentence and lexical intonation of speakers belonging to the regional dialects. A comparative study of these features is also performed as compared to the Prescription Model of Prosody among the Khariboli speakers.

The intonation pattern is observed along the selected sentence commonly uttered by all the dialect speakers, which is given below:

/kʌl//muʤɛ//bʌhu̪t//bukʰɑr//tʰɑ/

Similarly the lexical intonation has been observed along the phrase **/kɑm/** for all the regional dialect speakers.

4.13.2 *Study of Vowel Quality and Quantity*

The vowel quality is a term by which one can identify the difference between two vowel sounds. Whereas, vowel quantity is referred or the duration or length of a vowel. In order to study vowel quality and quantity C_1VC_2 word segments of five different vowels, namely, /ʌ/, /a/, /i/, /o/ and /u/ are chosen from the recorded utterances of all the dialects. These vowels are respectively represented in chapter as vowel 1, vowel 2, vowel 3, vowel 4 and vowel 5 while illustrating in the graphs. The chosen C_1VC_2 syllables from Khariboli as well as from other dialects are given in Table 4.5.

Now, to observe the vowel quality of a subject while pronouncing one of the selected vowels, the formant frequencies F1 (first formant frequency), F2 (second

Table 4.6 Values of F1, F2 and duration of Khariboli male speakers in vowel /o/

Speakers	Ist formant frequency (F1) (Hz)	IInd formant frequency (F2) (Hz)	Duration of syllabic nuclei (ms)
KhM-1	551	995	0.12
KhM-2	565	982	0.12
KhM-3	590	982	0.12
KhM-4	511	928	0.13
KhM-5	632	1062	0.09
KhM-6	551	928	0.18
KhM-7	618	982	0.14
KhM-8	618	1,062	0.12
KhM-9	551	995	0.14
KhM-10	591	1,008	0.12
KhM-11	551	995	0.12
KhM-12	565	982	0.12
KhM-13	590	982	0.12
KhM-14	511	928	0.13
KhM-15	632	1,062	0.09
Average	*575*	*992*	*0.13*

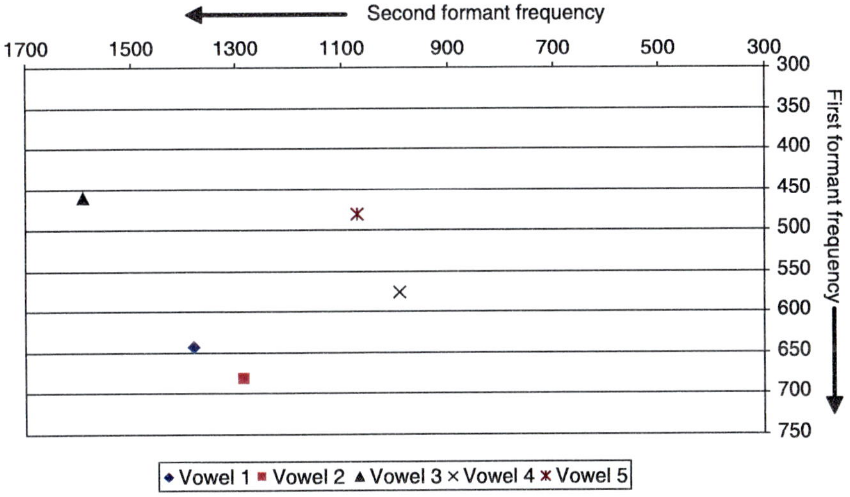

Fig. 4.7 Vowel quadrilateral of average F1 and F2 for Khariboli male

formant frequency) and duration of syllabic nuclei have been measured at appropriate location of the vowel formants. Table 4.6 is an example of these measurements for all male subjects of Khariboli while producing rounded vowel /o/. Similar measurements have been done for all five vowels in Khariboli speakers (male and female) as well as in other regional dialect speakers. These measured values of F1(y axis) and F2 (X axis) are plotted in vowel quadrilateral to get the vowel quality.

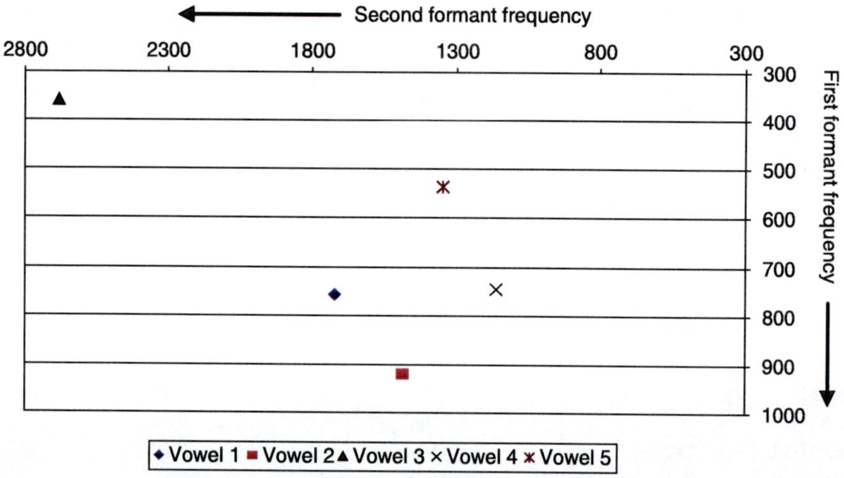

Fig 4.8 Vowel quadrilateral of average F1 and F2 for Khariboli females

Figure 4.7 shows the vowel quality of Khariboli male speakers for all selected vowels of Hindi. Similarly, Fig. 4.8 shows the vowel quality of Khariboli female speakers. Vowel quality of speakers belonging to other regional dialects has also measured and plotted in the similar manner and compared with respect to that of Khariboli speakers. Because it is not possible to display all the graphs and measured values in this chapter, authors have chosen to display the graph and measurements of the base language, i.e., Khariboli. Similarly the mean vowel duration for dialect speakers is plotted as a with the mean vowel duration of Khariboli speakers in case of all five vowels. Figure 4.9a and b shows the comparative vowel duration of Bhojpuri and Chattisgarhi dialect with Khariboli respectively for vowel 1, vowel 2, vowel 3, vowel 4 and vowel 5. The graph clearly show that the vowel quantity of Bhojpuri and Chattisgarhi dialect is distinctive than that of Khariboli speakers.

4.13.3 Prosody Analysis

Prosody analysis has been performed on the selected sentence and word from the text based on the phonetic significance and clarity of the speech. Sentence intonation and lexical tone has been observed and represented in term of High [H] raising or Low [L] falling pitch pattern. The intonation pattern and the tonal characteristic of a Khariboli subject are shown in Fig. 4.10. Figure 4.10a represent the intonation pattern reflected by one of the Khariboli speaker along the sentence /kʌl//muʤɛ// bʌhut̪//bukʰɑr/ /tʰɑ/. Based on the data, intonation pattern shown by majority of male and female speakers along the sentence is identified to be the intonation pattern in Khariboli dialectal group and named as prescription model of prosody. Figure 4.10b and c shows the pitch variation along the word and hence represent the

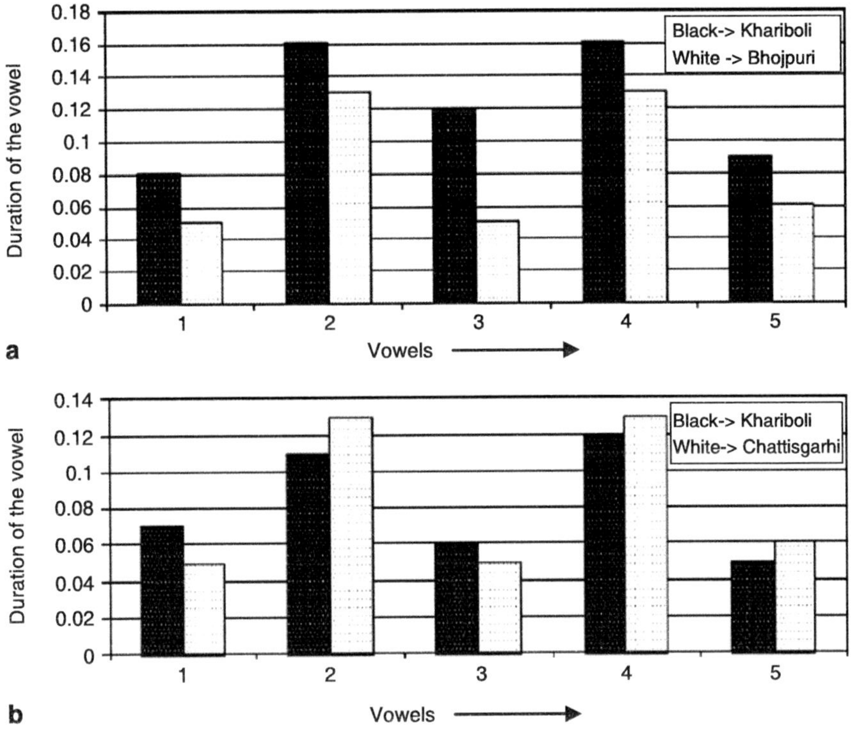

Fig. 4.9 Vowel duration of Bhojpuri (**a**) and Chattisgarhi (**b**) speaker w.r.t. Kahriboli

tone of the Khariboli male and female speakers respectively. On the basis of the analysis for tonal characteristics majority of Kahriboli speakers use a falling tone. Falling tone is considered as Prescription Model of lexical tone in Khariboli accent. Similar observations have been tabulated for each subjects of each dialect for males as well as for female. Though, due to space limitation in the chapter, it is not possible to explain and display the intonation and tone for all the dialect groups. The specification of CSL instrument at the time of analysis and measurement was as follows, sampling rate of 25 kHz, which is down sampled at 12.5 kHz and wide band analytical filter of 183.11 Hz, has been used with the quantization at 16 bits.

4.14 Results of the Study

The dialectal accent features of Khariboli, Bhojpuri, Chattisgarhi, Marwari, Kannauji, Bundeli and Haryanvi have similarity and dissimilarities in the intonation pattern, tones, vowel quality and quantity. In terms of intonation pattern, Bhojpuri, Chattisgarhi, Marwari, Kannauji, Bundeli and Haryanvi dialect speakers have iden-

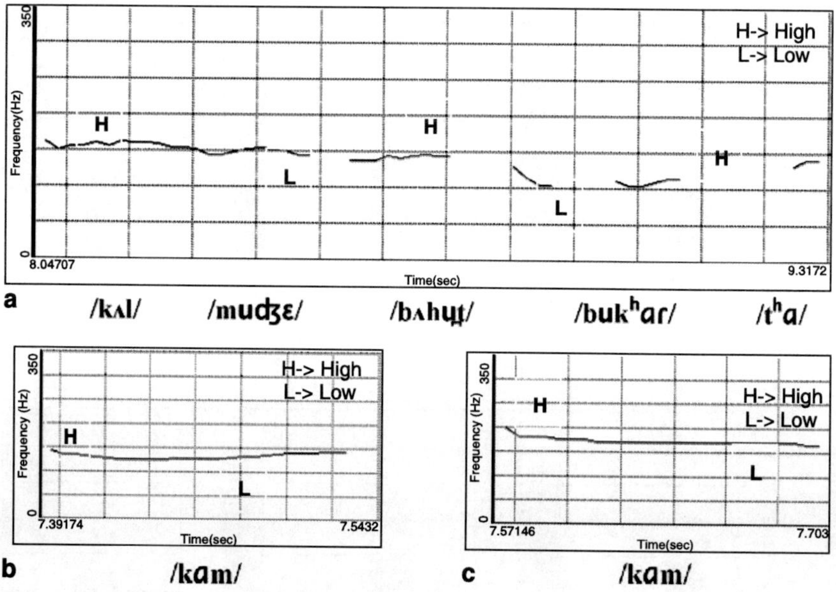

Fig. 4.10 Intonation (**a**) and tonal pattern (**b**) and (**c**) of Khariboli speaker

tified Prescription Model initiated in a similar manner whereas, Khariboli dialect speakers initiated in a different manner. At the end of the utterance, Khariboli, Bhojpuri, Chattisgarhi, Marwari, Bundeli and Haryanvi speakers have a similar intonation pattern, whereas, Kannauji speakers have distinct pattern of intonation. In the middle portion of the utterance, Chattisgarhi, Marwari and Haryanvi dialect speakers have similar intonation, Kannauji and Bhojpuri speakers have similar intonation and Khariboli speakers have similar intonation pattern to that of Bundeli speakers. In terms of Prescription Model for lexical prosody or the tonal characteristics, Chattisgarhi, Marwari, Kannauji, Bundeli and Haryanvi dialects speakers have tone different from that of Khariboli speakers. Tonal features of the Bhojpuri dialect could not be ascertain. It is suggested in a study [3] conducted on native Cantonese speakers that the tone contrast effect may be language specific. In this study tonal features are found to be dialect specific though the there are similarities either at the end or at the initial of the sentence uttered. On the basis of the observations and measured values of F1 and F2 the vowel Quality for all the subjects belonging to a particular regional dialect has been plotted as a graph and the distribution of all the five vowels has been studied with in the regional group for all the subjects.

Vowel quality of majority of Bhojpuri dialect speakers is different from Khariboli speakers in the case of chosen vowels as the native speakers of Bhojpuri dialect use a bit closer form of the vowels than the vowels used by Khariboli speakers. Vowel quality of Chattisgarhi dialect speakers is different from that of Bhojpuri in case of vowel /ʌ/, which is used in an opened form and vowel /i/ is pronounced very close to the cardinal point of the vowel by most of the Chattisgarhi speakers.

Majority of Marwari speakers have a distinguishable vowel quality in case of vowel /u/ form that of Bhojpuri and Chattisgarhi speakers. Marwari speakers have a vowel quality of vowel /u/ more towards the back region of the vowel quadrilateral. Kannauji speakers have a distinctive feature in vowel /o/, which is used in a more open manner than the other dialects and dialect speakers have a tendency to shift the vowels quality toward the central region of the vowel quadrilateral. Vowel quality of Bundeli is similar to that of Bhojpuri except vowel /u/, which is used in open manner by the Bundeli speakers. Haryanvi speakers pronounce vowel /o/ in a more open form, which is similar to that of the feature of Kannauji dialect but vowel /i/ is used very close to the cardinal point of the vowel in case of Haryanvi dialect as well.

Average vowel quantity of Bhojpuri and Bundeli speakers is found to be more or less similar to that of Khariboli speakers. Vowel quantity of Kannauji speakers is similar to the average vowel length of the Marwari speakers in comparison with the Khariboli speakers. The identified Prescription Model for sentence prosody and lexical prosody of the speakers belonging to the chosen dialects are found to be dialect specific and carries the accent information of the individual. The findings are supported by the Study conducted by Cummins and reported that lexical and sentence prosodic features for identification of accented feature of a dialectal group or language and lexical tone and phrasal intonation is an important marker of accent and stress in every language [1.a.i.21]. In terms of acoustic information, which are related to prescription models of prosody, vowel quality and vowel quantity, Non-native speakers can be distinguish from native speakers. The finding is in agreement with the study conducted on non-native speakers of Hindi by native speakers of Punjabi, Dogri and Kashmiri [51].

4.15 Summary and Conclusion

The study conducted on acoustic characteristics of vowel quality and quantity of dialectal speakers reveals the following facts. The vowel quality of each dialect is distinguishable while comparing from Khariboli and this distinction is unique to the dialectal accent. Though there are similarities of vowel quality among the regional dialect speakers. However, by considering the overall vowel quality of the dialect as a whole, vowel quality is proved to useful feature for profiling of speakers. These features based on the vowel quality can also be collaborated with features of prosody as discussed above.

As far as vowel quantity is concerned dialectal speakers of Bhojpuri, Chattisgarhi, Kannauji, Marwari, Bundeli and Haryanvi dialects use long vowel /ʌ/ as compare to Khariboli. Dialectal speakers of Bhojpuri, Chattisgarhi, Bundeli, Kannauji and Marwari dialects use longer vowel /a/ as compare to Khariboli whereas, Haryanvi dialect speakers use short vowel. Dialectal speakers of Chattisgarhi, Kannauji, Haryanvi and Marwari dialects use longer vowel /i/ as compare to Khariboli whereas, Bhojpuri and Bundeli dialect speakers use short vowel. Dialectal speakers of Bhojpuri, Chattisgarhi, Kannauji, Bundeli and Marwari dialects use long vowel

/o/ as compare to Khariboli whereas, Haryanvi dialect speakers use short vowel. Dialectal speakers of Chattisgarhi and Haryanvi dialects use long vowel /u/ as compare to Khariboli whereas, Bhojpuri, Kannauji, Bundeli and Marwari dialect speakers use short vowels.

Specific acoustic features of Bhojpuri, Chattisgarhi, Kannauji, Marwari, Khariboli, Bundeli and Haryanvi based on the Prescription Model of prosody are found unique for characterization of speakers belonging to the regional dialectal group. Acoustic features associated with lexical and sentence intonation are found unique to a dialectal groups and found useful for speaker profiling. Front vowels' quality expressed in terms of first and second formants are found more significant as profile characteristic than that of back vowels. An overall distribution of vowel quality on vowel quadrilateral is important in the characterization of dialectal accent based speaker profiling from the utterances of an unknown speaker. Though there is possibility of change scenario in accent feature due to exposure to other non-native dialects, accent features are likely to remain as profile characteristic for quite some time. Some of the speakers' vowel quality and quantity is deviating from the vowel quality and quantity of the accented dialect. The observations imply that some of the regional dialect speakers use vowel quality and quantity other than the vowel quality and quantity of their own dialect. While selecting the clue words from accented utterances to be compared with the unaccented conversation, care should be taken in order to choose same vowel quality and quantity, particularly the vowel quality. In actual crime case examination for Speaker identification if the auditory impression is of accented utterances either in questioned or specimen speech exemplars. It is recommended to study the vowel quality and quantity of the utterances in the questioned sample as well as in the specimen sample. In doing so, as many as possible syllabic nuclei of same vowels is required to be selected and preliminary study of variant of vowels within questioned as well as within specimen is required to be conducted in order to understand the variant of vowel quality due to the production of accented and unaccented utterances by the accused or by the suspect(s).

As the population distribution of regional dialect speakers of Kannauji is in close proximity to the population distribution of Khariboli speakers, there are similarities in features among the native speakers of Khariboli and Kannauji dialect. Though Haryanvi dialect speakers' distribution is in the close proximity to the Khariboli, there are larger distinctive features among the native speakers of Haryanvi and Khariboli as the Haryanvi dialect of Hindi is heavily influenced by the Panjabi on the north west of Haryanvi region. Likewise, the Bundeli and Chattisgarhi dialect speakers are showing some similarities in the initial and the end of the sentence intonation. However, differences occur in the middle of the sentence. The clear distinction between Bundeli and Chattisgarhi can be made in terms of the vowel quality, where Bundeli speakers are showing a closer form of the vowels than the Chattisgarhi speakers. The study reveals that most of the female native speakers of the regional dialect are showing a constraint to the process of language change as far as accented delivery is concerned. The Prescription Model identification is better from among female speakers of the dialect than that from the male speakers. Most of the dialect speakers of Hindi are found to be different in terms of tonal characteristics as compare to Khariboli dialect. In this study, majority of the speakers of the chosen dialects use tonal features different from the tonal features used by the majority of the Khariboli speakers.

Though there is influence of regional dialects or deviated accented features among the Khariboli speakers, identification of Prescription Model with larger number of representative data, i.e., $(80 \pm 10)\%$ is possible. Likewise, Haryanvi speakers can also be identified with the larger representative data while identifying the Prescription Model. Other dialects of Hindi, namely, Bhojpuri, Chattisgarhi, Kannauji, Marwari, Khariboli, and Bundeli are observed to be influenced by non-native accent and lesser number of representative data for identification of Prescription Model is used specially among the male speakers. The observations imply that there is a constant influence of other regional dialect on the native accented features of these dialects of Hindi. The change in the dialectal accent is more affected among the male speakers than the female speakers of a dialect. The findings of the study is encouraging in the sense that Prescription Model of prosody, vowel quality and quantity data is recommended to collect from rest of the dialect of Hindi in order to profile the speakers based on dialectal accent. Such database is one of the essential information databases for forensic labs dealing with Speaker Identification task.

APPENDIX -1

Text - 1

आज मेरी तबीयत ठीक है : कल मुझे बहुत बुखार था । मेरे ऑफिस में बहुत काम है । इस बार भी काम की वजय से मेरी तबीयत खराब हो गयी । मैंने डाक्टर की बात मान कर दवा खा ली है । डाक्टर के हिसाब से बुखार जान भी ले सकता है । मैं ऑफिस हमेशा समय से पहुंच जाता हूं । 10:30 के बाद भी लोगों का आना जाना लगा रहता है । मेरा बड़ा भाई भी मेरे साथ ही काम करता है । अभी उसने 10 लाख का मकान खरीदा है । जीने के लिए पैसा बहुत जरूरी है । हम सब सिर्फ पैसे के लिए इतना काम करता है । बिना पैसा के दुनिया में जीना मुश्किल हो जाएगा ।

Text - 2

आज का काम कल पर नहीं छोड़ना चाहिए । व्यायाम करना सेहत के लिए ठीक है । रक्षा बन्धन में बहने अपने भाई की लम्बी उम्र के लिए प्रार्थना करती हैं । हम अपने माता पिता से बहुत प्यार करते हैं । Meeting में समय पर पहुंच जाना । मैं हर बात एक बार एक बार कहुंगी । मुझे खाना बनाना अच्छा लगता है । अपनी जान सबको प्यारी होती है । लाख पैसा कमा लो पर इन्सान साथ ले के कुछ नहीं जाता । कौन कहां जाएगा किसे पता ।

<u>**Text-3**</u>

1. तुम कितना इंतजाम कर सकते हो ।

2. कौन बोल रहा है ।

3. कब तक हो जायेगा ।

4. मुझे माल time पर चाहिए ।

5. मैं तुम्हे फिर फोन करूगा ।

6. जैसा मैं कहूं बैजा करना ।

References

1. Abberton E, Fourcin AJ (1978) Intonation and speaker identification. Lang Speech 21(4):305–318
2. Ahmed R, Agrawal SS (1969) Significant features in the perception of Hindi consonants. J Acoust Soc Am 45(3):758–763
3. Alexander LF, Valter C (2001) Lexical tone contrast effects related to Linguistic experience. J Acoust Soc Am 109(5):2475
4. Allard J, Wayland R, Wong S (2000) Acoustic characteristics of English fricatives. J Acoust Soc Am 108(3):1252–1263
5. Atal BS (1974) Effectiveness of linear prediction characteristics of speech wave for automatic speaker identification and verification. J Acoust Soc Am 55(6):1304–1312
6. Atkinson JE (1976) Inter and intra speaker variability in fundamental voice frequency. J Acoust Soc Am 60(2):440–445
7. Black J, Lashbrook W, Nash E, Oyer H, Pedry C, Tosi O, Truby H (1973) Reply to speaker identification by speech spectrograms: some further observations. J Acoust Soc Am 54:535–537
8. Bolt R, Cooper F, David E, Denes P, Picket J, Stevens K (1973) Speaker identification by speech spectrograms: some further observations. J Acoust Soc Am 54:531–534
9. Bolt RH, Cooper FS, David EE, Denes, PB, Picket JM, Stevens KN (1970) Speaker identification by speech spectrograms: a scientist view of its reliability for legal purpose. J Acoust Soc Am 47:597–612
10. Braun B, Kochanski G, Grabe E, Rosner BS (2006) Evidence for attractors in English intonation. J Acoust Soc Am 119(6):40006–4015
11. Bricker P, Pruzansky S (1966) Effect of stimulus content and duration on talker identification. J Acoust Soc Am 40(6-II):1441–1449
12. Caroline RW, James DH (2006) The influence of Gujrati and Tamil L1s on Indian English: a preliminary study. Word English 25(1):91–104
13. Chao YR (1933) Tone and intonation in Chinese. Bull Inst Hist Philol 4:121–134
14. Cheung RS, Eisenstein BA (1978) Feature selection via dynamic programming for text independent speaker identification. IEEE Trans Acoust Speech Signal Process ASSP-26(5):396–403
15. Chiba T, Kajiyama M (1941) The vowel: its nature and structure. Kaiseikan, Tokyo

16. Clarke FR, Bricker RW (1969) Comparison techniques for discriminating among talkers. J Speech Hear Res 12:747–761
17. Coleman R (1973) Speaker identification in the absence of inter subject differences in glottal source characteristics. J Acoust Soc Am 53:1741–1743
18. Connell B (2000) The perception of lexical tone in Mambila. Lang Speech 43(2):163–182
19. Cooper WE, Serensen JM (1981) Fundamental frequency in sentence production. Springer, New York
20. Crystal D (1969) Prosodic systems and intonation in English. Cambridge Studies in Linguistics. Cambridge University Press, Cambridge
21. Cummins F, Gers F, Schmidhuber J (1999) Comparing prosody across many languages. IDSIA Technical Report, IDSIA-07
22. Dara C, Marc DP (2006) The interaction of linguistic and affective prosody in a tone language. J Acoust Soc Am 119(5):3303–3304
23. Das SK (1969) A method of decision making in pattern recognition. IEEE Trans 18:329–333
24. Dhirendra Verma (1996) Hindi bhasha aur lipi. Hindustani Academy, Allahabad
25. Dik JH, Joost VG (1991) The frequency scale of speech intonation. J Acoust Soc Am 90(1):97–102
26. Donald JS, Hemeyer T (1972) Identification of place of consonant articulation from vowel formant transitions. J Acoust Soc Am 51(2):652–658
27. Fry DB (1970) Prosodic phenomenon. In: Malmberg B (ed) Manual of phonetics. North Holland, Amsterdam
28. Fry DB (1979) The physics of speech. Cambridge University Press, Cambridge
29. Fujimura O (1962) Analysis of nasal consonants. J Acoust Soc Am 34:1865–1875
30. Fujimura O (1971) Sweep tone measurements of vocal characteristics. J Acoust Soc Am 49(2):541–558
31. Glenn JW, Norbert K (1968) Speaker identification based on nasal phonation. J Acoust Soc Am 43(2):368–372
32. Gray CH, Kopp GA (1944) Voice print identification. Bell Telephone Laboratories Report, New Jersey, pp 13–14
33. Green N (1972) Automatic speaker recognition using pitch measurements in conversational speech Joint Speech Research Unit, Report No 1000
34. Grierson GA (1928) Linguistic Survey of India, and Shapiro, MC (2001) Facts about the world's languages, Hindi
35. Hazen B (1973) Effects of differing phonetic context on spectrographic speaker identification. J Acoust Soc Am 54(3):650–658
36. Hillenbrand J, Getty LA, Clark MJ, Wheeler K (1995) Acoustic characteristics of American English vowels. J Acoust Soc Am 97:3099–3111
37. James DH (1996) Pitch range and focus in Hindi. J Acoust Soc Am 99(4):2493–2500
38. Kenneth NS (1966) Acoustical description of syllabic nuclei: an interpretation in terms of a dynamic model of articulation. J Acoust Soc Am 40(1):123–131
39. Kersta LG (1962) Voice print identification infallibility. J Acoust Soc Am 34(12):1978–1978
40. Koenig EB (1986) Spectrographic voice identification: a forensic survey. J Acoust Soc Am 79(6):2088–2090
41. Ladefoged P (2001) A course in phonetics, 4th ed. Harcourt, Fort Worth, p 177
42. Ladefoged P (1962) Elements of acoustic phonetics. University of Chicago Press, Chicago
43. Lori HH, Andrew J, Lotto K, Kluender R (2001) Influence of fundamental frequency on stop-consonant voicing perception: a case of learned co-variation or auditory enhancement. J Acoust Soc Am 109(2):764–774
44. Nolan F (1983) The phonetic bases of speaker recognition. Cambridge University Press, Cambridge
45. Ohala JJ (1978) Production of tone. In: Fromkin VA (ed) Tone: a linguistic survey. Academic Press, New York, pp 5–39
46. Ohala JJ (1983) Cross-language use of pitch: an ethological view. Phonetica 40:1–18

47. Peterson GE, Lehiste I (1960) Duration of syllable nuclei in English. J Acoust Soc Am 32:693–703
48. Rose P (2002) Forensic speaker identification. Forensic science series. Taylor and Francis, London
49. Ruth, HM, Elaine RS (2003) Hemisphere differences in prosody production: a new look. Int Soc Phonetic Sci 87:9–17
50. Ryo K, Youngon C (1999) Effects of native language on the perception of American English /R/ and /L/: a comparison between Korean and Japanese. ICPh 99, San Francisco, pp 1429–1432
51. Singh CP, Singh SR (1998) Voice spectrographic study of class characteristics of Hindi utterances of Punjabi, Dogri and Kashmiri speakers. J Indian Acad Forensic Sci 37(1, 2):40–45
52. Smrkovski L (1975) Study of speaker identification by aural and visual examination of non contemporary speech samples. J Official Anal Chemist 59:927–937
53. Stevens KN, Williams CE, Carbonell, JR, Woods B (1972) Speaker authentication and identification: a comparison of spectrographic and auditory presentations of speech material. J Acoust Soc Am 51:2030–2043
54. Stevens SS, Volkman J (1940) The relation of pitch to frequency: a revised scale. Am J Psychol 53:329–353
55. Tiwari B (1966/2004) हिन्दी भाषा (Hindī Bhāshā), Kitāb Mahal, Allahabad, ISBN 81-225-0017-X
56. Tosi O, Oyer M, Lashbrock W, Pedey C, Nical J, Nash E (1972) Experiment on voice identification. J Acoust Soc Am 51:2030–2043
57. Web page /http://www.phonam.de/sprecher_e.html/
58. Wolf JJ (1972) Efficient acoustic parameters for speaker recognition. J Acoust Soc Am 51(6):2044–2057
59. Woo N (1969) Prosody and phonology. PhD dissertation, MIT
60. Yi Xu, Wallace A (2004) Multiple effects of consonant manner of articulation and intonation type on F0 in English (A). J Acoust Soc Am 115(5):2397
61. Young M, Campbell R (1967) Effect of contexts on talker identification. J Acoust Soc Am 42:1250–1254

Chapter 5
Speech Under Stress and Lombard Effect: Impact and Solutions for Forensic Speaker Recognition

John H. L. Hansen, Abhijeet Sangwan and Wooil Kim

Abstract In the field of voice forensics, the ability to perform effective speaker recognition from input audio streams is an important task. However, in many situations, individuals will change the manner in which they produce their speech due to the environment (i.e., Lombard Effect), their speaker state (i.e., emotion, cognitive stress), and secondary tasks (i.e., task stress at hand, both physical and/or cognitive). Automatic recognition schemes for both speech and speaker ID are impacted by the variability introduced in these conditions. Extensive research in the field of speech under stress has been performed for speech recognition, primarily for low-vocabulary isolated-word recognition. However, limited formal research has been performed for speaker ID/verification primarily due to the lack of effective corpora in the field. This chapter addresses speech under stress including Lombard effect for the purposes of speaker recognition. Domains where stress/variability occur (Lombard Effect, Physical Stress, Cognitive Stress) will first be considered. Next, to perform effective speaker recognition it is necessary to detect if a subject is under stress, which is a useful trait in and of itself for voice forensics and biometrics, and therefore we consider prior research on the detection of speech under stress. Next, the impact of stress on speaker recognition is considered, and finally we address ways to improve speaker recognition in these domains (TEO features, alternative sensors, classification schemes, etc.). While speech under stress has been considered, the domain of speaker recognition represents an emerging research aspect which deserves further investigations.

5.1 Introduction

In the field of voice forensics, the ability to identify an individual using automatic speaker recognition techniques can be an effective tool in removing or including an individual within a subject pool. For automatic speaker recognition algorithms,

---

J. H. L. Hansen (✉)
Department of Electrical Engineering, Center for Robust Speech Systems (CRSS), Erik Jonsson School of Engineering and Computer Science, The University of Texas at Dallas, Richardson, TX 75080-3021, USA
e-mail: john.hansen@utdallas.edu

A. Neustein, H. A. Patil (eds.), *Forensic Speaker Recognition*,
DOI 10.1007/978-1-4614-0263-3_5,

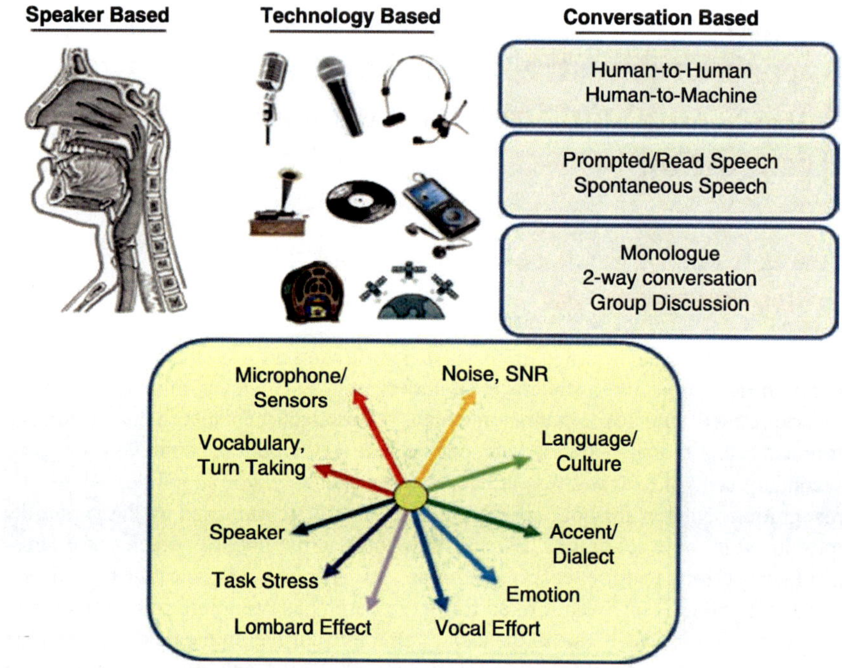

Fig. 5.1 Range of speech, speaker, technology and environmental factors that can influence speech for speaker recognition within audio forensics

great care is necessary to ensure there is limited mismatch between training and testing conditions in the audio stream. Though human listeners can often detect potential mismatch between training and test material (at least at the qualitative level) they do not always agree on the level, type, extent, or cause of the mismatch. In the primary speaker recognition evaluation (SRE) competitions organized by the U.S. National Institute of Standards (NIST-SRE), which are held biannually, a number of mismatch conditions have been considered (NIST SRE [1]). From 2006 to 2010, these have focused primarily on noise-free audio data for which the primary mismatch is (i) microphone (one of approximately 14 mics), (ii) handset (standard landline telephone, cordless telephone, cellphone), and (iii) language (for NIST SRE-2008). Little, if any, work has been focused on speakers under non-neutral conditions (i.e., speech under stress, Lombard effect (changes in speech due to the environment, such as background noise), or cognitive/emotional factors). The range of mismatch due to speaker-based variability, speech from human-human/ human-machine variations, and technology or environmental factors is quite extensive. Figure 5.1 illustrates a broad perspective across these domains. The focus of this chapter is to consider mismatch due to stress and Lombard effect for speaker recognition (The primary author's other co-written chapter, appearing in the first section of this book, considers speech production differences due to vocal effort including whisper speaking style).

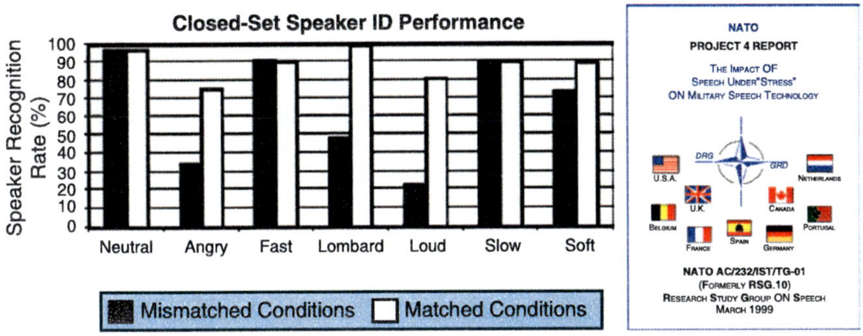

Fig. 5.2 Closed-set speaker recognition results based on neutral trained models and speech under stress/emotion/Lombard effect SUSAS data. [9]

As shown in Fig. 5.1, the range of speaker variability, technology introduced mismatch, and variations in conversation-based engagement causes changes to the audio stream that impacts speaker recognition performance. For *Speaker Based* mismatch, these include (i) speech under stress (physical, psychological/cognitive, speaking style including emotion, etc.), (ii) vocal effort (whisper to shouted), (iii) emotion, (iv) Lombard effect (speech produced in the presence of noise), (v) accent/dialect (idiosyncratic differences within a given language). For *Technology Based* mismatch, these include (i) microphone, (ii) handset/phone, (iii) communications/channel effects, (iv) audio recording/storage/compression effects, (v) environmental, including room noise and echo. Finally, *Conversation Based* mismatch includes (i) various forms of human-to-human communications, (ii) human-machine interactions, (iii) read versus spontaneous speech, (iv) single person spontaneous monologue, (v) 2-way person-to-person communication, either face to face or via telephone system, (vi) multiple speaker scenarios including group discussions, debates, etc. Given these constraints, speech that is captured in a controlled noise-free setting and used for training a speaker recognition model will not perform as well when test materials are drawn from these mismatched conditions.

As noted, historically, there has been limited work on speaker recognition under stress due primarily to the lack of formal audio corpora necessary to investigate this important domain. Early work on speech under stress, including speaking styles, Lombard effect, emotion (anger), and task stress was performed by Hansen [2–5]. This early work resulted in the SUSAS "*S*peech *U*nder *S*imulated and *A*ctual *S*tress" database, which was later released through the LDC [6, 7]. A comprehensive write-up and analysis was also considered by Hansen and Bou-Ghazale [8]. Further discussions regarding research based on this corpus will follow in this chapter. The primary reason for discussing this here is that as part of the NATO RSG.10 research study on speech under stress [9], an experiment was completed which highlighted the loss of close-set speaker recognition performance (reported in 1999) using SUSAS speaker data. The results from [9] are included in Fig. 5.2, for a GMM based speaker recognition system trained with neutral speech, and tested with: neutral, angry, fast, slow, loud, soft, and Lombard effect speech (The results presented here are without cepstral-mean normalization, but similar results were also obtained with CMN engaged). The results show that closed-set speaker ID performance for Lombard effect, loud, and angry speech conditions were severely impacted when using neutral trained speaker models. Performance does improve when speaker models are trained with matched stress speech conditions; however, achieving such performance would require a front-end stress state detector to first direct the proper speaker trained model. In subsequent sections here, we will consider (i) Sect. 5.2: Speech Under Stress corpora, (ii) Sect. 5.3: Detection of Speech Under Stress, (iii) Sect. 5.4: Speaker Recognition under Stress, and (iv) Sect. 5.5: Speech Technology and Advancements for Stress/Emotion Detection and Speaker Identity.

5.2 Stress and Emotion Databases

In the past, speech databases of stress/emotion were largely collected in controlled lab conditions with few speakers and in mostly simulated emotion-states [8, 10, 11]. Additionally, the focus was on a few extreme emotional states, and rarely involved speech data in an interactive setup such as spontaneous conversations. As a result, the full range of human stress/emotions were not captured effectively as the collection paradigm often did not allow for more natural human interactions. Furthermore, previous data also tended to consist of isolated words and/or phrases. On one hand, the focused data collection in the early days of corpus development allowed for insightful comparisons across speakers as they spoke the same words across different stress conditions; and a number of novel algorithms for stress/emotion detection and classification were proposed. However, it was found that systems trained on staged stereotypical data were of limited use in real-world settings. Hence, there was a critical need to collect corpus materials that reflected the full-range of human stress/emotions.

More recently, there has been increased efforts in creating larger and more diverse corpora [12–14]. In general, these collection efforts have focused on capturing a full-range of emotions while providing rich annotations. A number of these collections are multimodal, namely, UT-Drive [15–17], SMARTKOM [18], Humaine [13], and others. Additionally, data collection has moved out of the laboratory setting to more naturalistic and real-world conditions. For example, the UT-Drive collection has been performed exclusively inside a car while driving to incorporate the many interactive task stress encountered while driving. However, real-world collection of stress/emotion data poses challenges for corpus development, especially along moral/ethical lines. A number of collection paradigms require that the subject be placed under stressful conditions. However, corpus designers have to be conscious of exposing their subjects to physical harm and moral/ethical dilemmas.

Some of the major aspects of data collection and corpus design when collecting stress/emotion data are summarized below. These aspects determine the scope of the data and the type of systems that can be built with the data.

Simulated vs. Natural Stress/Emotion: Simulated emotions are easier to extract from subjects and likewise easy to annotate and rate as well. However, past research has shown that there are large differences between emotion in controlled lab settings and the real-world [19]. Natural emotions are subtle and hard for human evaluators to reliably identify and rate.

Ground Truth: Establishing ground truth for type as well as the level of stress/emotion being experienced by subjects is fundamentally hard for even people to judge. Whenever possible, simultaneous collection through alternative modalities can help to mitigate this issue to a certain extent. For example, heart-rate measurements can be a reasonable indicator of physical stress. A number of corpora use the visual channel as it captures facial expressions, postures, and gestures which can assist human judges to rate stress/emotions more consistently and reliably.

Session Variability: Collecting multiple sessions with speakers allows the corpus to capture session-to-session variability of emotion/stress. This enables researchers to conduct more realistic evaluations, and build more practical and robust systems.

The remainder of this section discusses several corpora that reflect the diversity of material being collected in the area of stress/emotion [6–8, 12–20].

5.2.1 SUSAS: Speech Under Simulated and Actual Stress, Lombard Effect

SUSAS speech database [6–8] was collected for research, analysis and development of new algorithms for speech recognition in noise and stress. The SUSAS database refers to Speech Under Simulated and Actual Stress, and is intended to be employed in the study of how speech production and recognition varies when speaking during stressed conditions. The database consists of five main stress domains covering a wide variety of stresses and emotions. The data contains over 16,000 utterances from a total of 32 speakers (13 female, 19 male), with ages ranging from 22 to 76. The five stress domains include: (i) talking styles (slow, fast, soft, loud, angry, clear, question), (ii) single workload tracking task or speech produced in noise (Lombard effect), (iii) dual tracking computer response task, (iv) actual subject motion-fear tasks (G-force, Lombard effect, noise, fear), (v) psychiatric analysis data (speech under depression, fear, anxiety). The collected speech consists of 35 aircraft communication words. The corpus also consists of speakers producing speech in neutral as well as in stressful scenarios (i.e., pilots flying Apache helicopters in emergency situations, and subjects riding on amusement park roller-coasters).

5.2.2 UT-Scope: Speech Under Physical, Cognitive, and Lombard Effect

The UT-Scope database consists of speech data collected under physical, cognitive, and environmental noise stress [12, 21]. Additionally, speech data was also collected from all subjects under neutral conditions. The UT-Scope database consists of sentences from the well-known TIMIT corpus. The speech data under environmental stress was collected while speakers were exposed to highway, large-crowd, and pink noises. Here, the noise presentation levels varied from 65 to 90 dB-SPL to induce Lombard effect. A total of 59 native as well as non-native speakers of American English participated in the collection for speech under environmental stress.

The cognitive stress/physical task stress portion of UT-Scope has 118 total sessions, with 77 unique native speakers of American English. For each task, the subject was prompted to say the same 35 sentences. The physical task stress [22] was induced by using an elliptical stair stepper exercise machine. To measure the amount of physical task stress induced in the speakers, heart rate (HR) in beats per minute (BPM) was also recorded for the neutral, cognitive stress, and physical task stress portions of the database.

5.2.3 HUMAINE: Emotion in Natural Settings

The HUMAINE corpus focuses on collecting emotions in natural settings [13]. A large proportion of the collection is audio-visual, where the visual channel in the corpus consists of facial, gestural, and postural shots. While the corpus material displays a wide range of emotions, the emotion is subtle and natural as it is embedded in the context and interactions of the subjects. The data consists of TV recordings, outdoor noisy data, as well as controlled laboratory data. The corpus is also annotated in a sophisticated manner. The database can be accessed via the HUMAINE Association portal (www.emotion-research.net).

5.2.4 FAU AIBO Emotion Corpus: Human–Machine Interaction

The AIBO corpus captures emotion in a human–machine interaction setup [14]. It consists of speech from 51 children aged 10–13 years interacting with Sony's pet robot Aibo. During recording, the robot was controlled remotely to introduce a mixture of obedient and disobedient behaviors. The data consists of 11 categories of emotions including joyful, surprised, emphatic, boredom, and others.

5.2.5 UT-Drive: Driving Induced Stress—Speech and Multi-Modal Data

The UT-Drive database is collected in real-world urban driving conditions [15–17]. The session routes chosen for data collection consist of a mixture of secondary, service, and main roads in residential and business districts of Richardson, Texas (USA). The data is being collected in a Toyota RAV4 which is equipped with microphones, CCD (charged coupled device) cameras, optical distance sensor, GPS (global positioning system), CAN-Bus (controller area network) OBD II (on board diagnostics) port for collecting vehicle speed, steering wheel angle, gas and brake inputs from driver, and gas/brake pedal pressure sensors. Each driving session includes a mixture of several secondary tasks that the driver is asked to perform while driving such as (i) sign reading, (ii) operating radio/AC, (iii) talking to a passenger, and (iv) calling automated dialog systems (American Airlines and Tell ME). The speech signal collected reflects driving induced stress, with artifacts such as distraction, frustration, *etc.*

5.2.6 SmartKom: Human–Computer Interaction

SmartKom [18] is a German and English language multi-modal corpus that focuses on emotion in the context of human–computer interaction. The corpus consists of Wizard-Of-Oz dialogs. The database consists of multiple audio channels and two video channels. The corpus captures seven broad emotional categories: neutral, joy, anger, helplessness, pondering, surprise along with some unidentifiable episodes. Each recording session is over 4 min in duration.

5.3 Detection of Speech Under Stress

A detection method of speech under stress can be effectively employed to improve speech processing systems. In particular, automatic speaker/speech recognition system drastically degrades in performance when the input speech is under stress, because the system is, in general, trained on neutral speech collected in clean conditions. The classification/detection algorithm for stressed/emotional speech due to cognitive/physical stress can be also utilized to improve the system performance for a wide range of applications. In this section, a number of research studies on detection of speech under stress conducted in past decades are considered.

In an effort to better understand the mechanism of human voice communication, researchers have attempted to determine reliable acoustic indicators of stress using such speech production features as (i) fundamental frequency (F0), (ii) intensity, (iii) duration, (iv) glottal spectral tilt, and (v) spectral structure including formant location, bandwidth, and the distribution of spectral energy, as well as others. A series of 200 features with over 10,000 statistical tests were performed across these domains [2–4]. While this previous work established knowledge of stressed speech production variations, further research was necessary for the detection of speech under stress.

In an earlier study [23], it was hypothesized that speech production involves a complex series of events which under neutral speech production conditions can sufficiently be characterized by a linear model/component, while under stressed

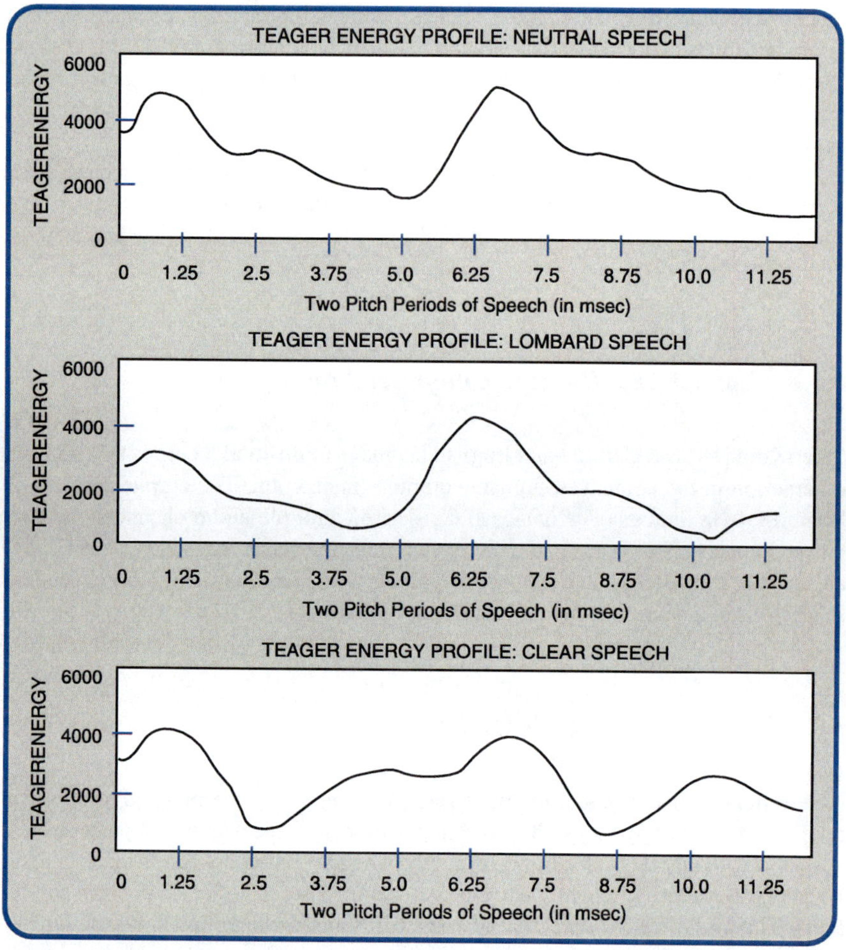

Fig. 5.3 Teager energy profiles [23] for neutral, Lombard, and clear speaking styles, using SUSAS [9] speech data

speech conditions the nonlinear component changes markedly between normal and stressed speech. To quantify the changes between normal and stressed speech, a classification procedure was developed based on the nonlinear Teager Energy Operator (TEO) [24]. The Teager Energy operator provides an indirect means of evaluating the nonlinear component of speech. Figure 5.3 shows an example of the Teager energy profile for neutral, Lombard, and clear utterances of the word "on". These plots illustrate the different forms of the Teager energy profile across speaking styles. In that study, the stress detection system was evaluated using VC and CVC utterances from native speakers of English across the following speaking styles; neutral, loud, angry, Lombard effect, and clear. Results of the system evaluation showed that loud and angry speech can be differentiated from neutral speech,

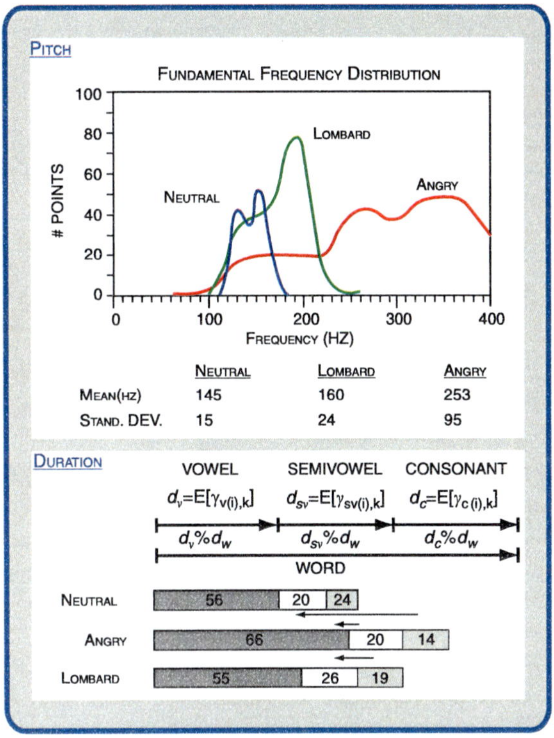

Fig. 5.4 Sample pitch (fundamental frequency) distributions and duration variation (mean phone class duration in % of overall word duration) for speech under neutral, angry, and Lombard effect stress conditions

while clear speech is more difficult to differentiate. Results also showed that reliable classification of Lombard effect speech is possible, but system performance varies across speakers.

Analysis and modeling of speech characteristics brought on by workload task stress, speaker emotion/stress or speech produced in noise (Lombard effect) were reported in [2, 25]. This study was conducted on over 200 parameters in the domains of pitch, duration, intensity, glottal source and vocal tract spectral variations. Figures 5.4 and 5.5 illustrate variations of pitch and vocal tract spectrum for a speech sample under neutral, Lombard, and angry stress conditions. This motivates the development of a speech modeling approach entitled Source Generator Framework [26] in which a formal structure was developed to characterize the deviation of speech dynamics of speech under stress. This framework provides an attractive means for performing feature equalization of speech under stress [27–31].

In [32], an array of speech features were considered as potential stress-sensitive relayers using the SUSAS database [4], including mel, delta-mel, delta-delta-mel, auto-correlation-mel, and cross-correlation-melcepstral parameters. In that study, an algorithm for speaker-dependent stress classification was formulated for the 11 stress conditions: *angry, clear, cond50, cond70, fast, Lombard, loud, normal, question, slow, and soft*. It was suggested that additional feature variations beyond neutral conditions reflect the perturbation of vocal tract articulator movement under

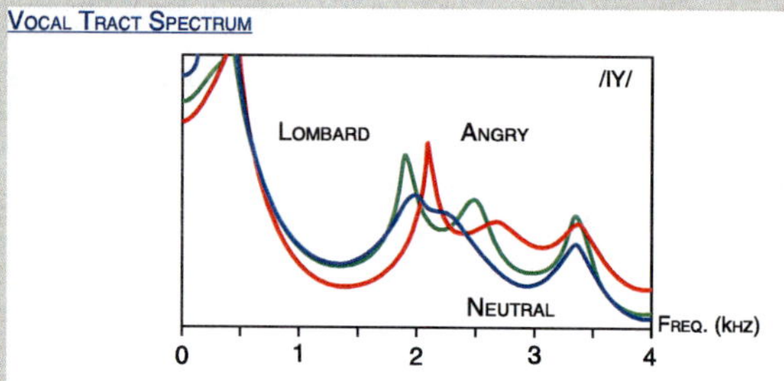

	F1 (Bw)	F2 (Bw)	F3 (Bw)	F4 (Bw)
Neutral	411 (52)	1970 (222)	2607 (496)	3368 (366)
Lombard	412 (73)	2006 (139)	2644 (250)	3376 (185)
Angry	586 (102)	2078 (166)	2661 (464)	3357 (392)

Fig. 5.5 Sample vocal tract spectra variation for speech under neutral, angry, and Lombard effect stress conditions. Mean formant location and bandwidths are also summarized

stressed conditions. Given a robust set of features, a neural network-based classifier was formulated based on an extended delta-bar-delta learning rule. Classification rates across 11 stress conditions were 79% for in-vocabulary and 46% for out-of-vocabulary tests, further confirming that the auto-correlation-mel (AC-MEL) parameters are the most separable feature set considered for detection of speech under stress.

Although some acoustic variables derived from linear speech production theory have been investigated as indicators of stress, they are not always consistent. In [33] three new features derived from signal processing advancements based on the nonlinear TEO were investigated for stress classification. It was suggested that these TEO based features are better able to reflect the nonlinear airflow structure when speech production occurs under adverse stressful conditions. The proposed features include (i) the TEO-decomposed FM variation (TEO-FM-Var), (ii) the normalized TEO autocorrelation envelope area (TEO-Auto-Env), and (iii) the critical band based TEO autocorrelation envelope area (TEO-CB-Auto-Env). Figure 5.6 shows a flow diagram of the feature extraction for the TEO-CB-Auto-Env. The proposed features were evaluated for the task of stress classification using simulated and actual stressed speech and it was shown that the TEO-CB-Auto-Env feature outperforms traditional pitch and mel-frequency cepstrum coefficients (MFCC) substantially. Performance for the TEO based features are maintained in both text-dependent and text-independent models, while performance of traditional features degrades in text-independent models. Overall neutral versus stress classification rates are also shown to be more consistent across different stress styles.

Such stress detection advancements were later extended, where a detection scheme based on weighted TEO features derived from critical bands frequencies was proposed in [34, 35]. This detection framework was evaluated on a military speech corpus collected in a Soldier of the Quarter (SOQ) paradigm. Heart rate, blood pressure, and blood chemical analysis/measurements of the SOQ corpus confirm subjects were under stress. Figure 5.7 shows classification sensitivity of each critical band for stress and neutral speech detection. It can be seen that bands 5,

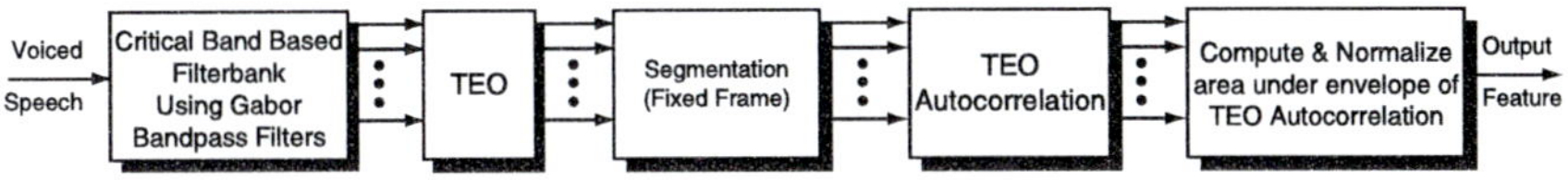

Fig. 5.6 TEO-CB-Auto-Env feature extraction for nonlinear speech analysis

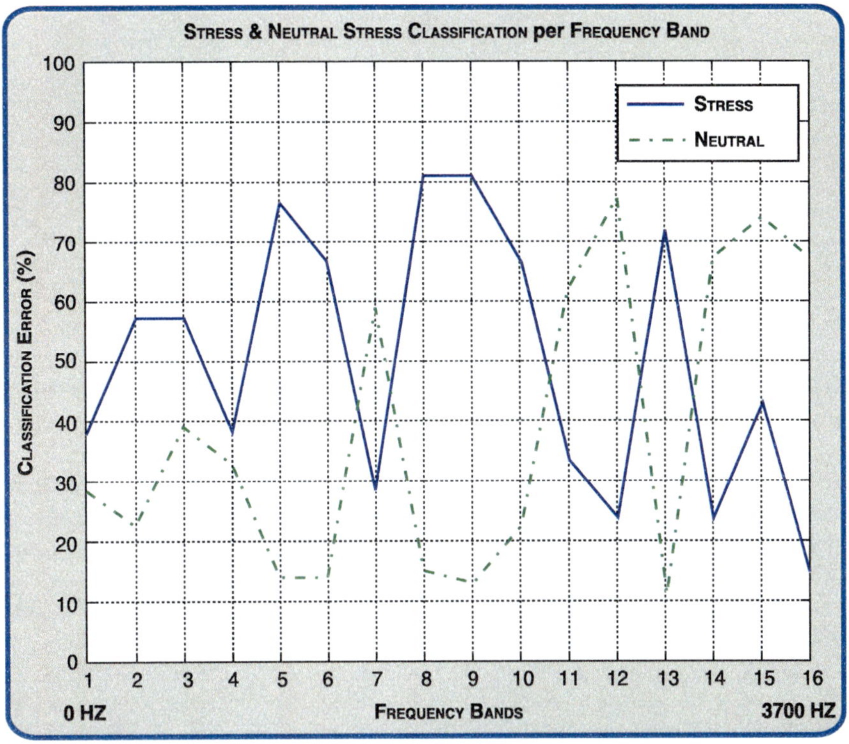

Fig. 5.7 Stress and neutral SOQ speech classification results for critical frequency bands [35]

8, 9, and 13 are sensitive to neutral speech (i.e., above 85% correct neutral classification), while bands 7, 12, 14 and 16 are sensitive to speech under stress (i.e., above 70% correct stress classification). Using the TEO-CB-AutoEnv feature with a Hidden Markov Model (HMM) trained stressed speech classifier, the error rates of 22.5% and 13% were obtained for stress and neutral speech detection respectively. With the formulated weighted sub-band detection scheme [35], detection error rates were reduced to 4.7% and 4.6% for stress and neutral detection, a relative error reduction of 79.1% and 64.6% respectively.

In [35, 36], the use of nonlinear TEO based features was further explored for classification of emotional/stressful speech, which is derived from multi-resolution sub-band analysis. A detection scheme was proposed for automatic sub-band

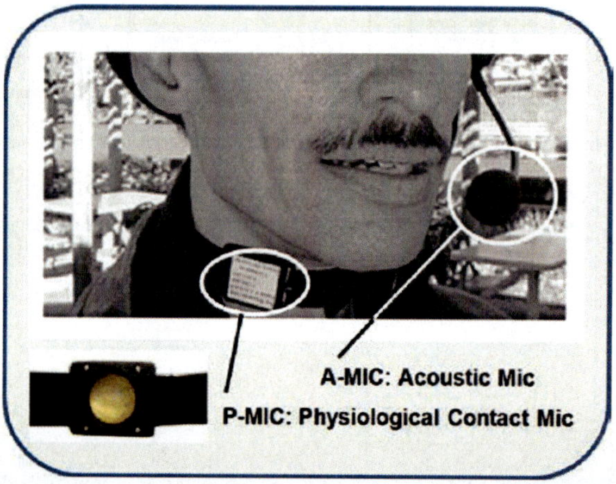

Fig. 5.8 Speaker set-up for physiological contact P-MIC, acoustic A-MIC

weighting in an effort towards developing a generic algorithm for understanding emotion or stress in speech. The proposed algorithm was evaluated also using the SOQ corpus. With the new frequency distribution based scheme, relative detection error reductions of 81.3% and 75.4% were obtained for stress speech and neutral speech respectively.

It is clear that speech produced under stress causes significant changes in both linear and nonlinear based features within the speech production process. However, in many of these situations high levels of background noise are present which can reduce the reliability of stress detection schemes. Alternative sensors which are less prone to acoustic background distortions are therefore potential options for both stress detection, as well as migration into effective speaker recognition systems.

The use of a contact physiological microphone (P-MIC) was explored as an alternative to traditional microphone sensors for stressed speech detection, since the acoustic microphone (A-MIC) suffers from limitations, such as sensitivity to background noise and relatively far proximity to speech production organs [37]. Figure 5.8 shows the experimental set-up for both A-MIC and P-MIC. In that study, an experimental evaluation of the TEO-CB-AutoEnv feature was first considered in an actual law enforcement training scenario. The feature relation to stress level assessment over time was considered. Next, the use of the physiological microphone was explored, which is a gel-based device placed next to the vocal folds on the outside of the throat used to measure vibrations of the vocal tract and minimize background noise. Both acoustic and physiological sensors were employed as stand-alone speech data collection devices as well as consider both sensors concurrently. A weighted composite decision scheme was devised using both the acoustic and physiological microphone data that yields relative average error rate reductions of 32% and 6% versus sole employment of acoustic and physiological microphone data, respectively, in a realistic stressful environment.

Table 5.1 Stress Detection Performance (% accuracy) of fused system for 12-speaker size model

Features	Close-mic (A-MIC)	P-MIC
MFCC	73.61	77.77
TEO	62.69	66.36
Fused system	*72.39*	*79.35*
Improvement (%)	*+9.43*	*+12.99*

A fusion method was also developed for more effective stress speech detection [38]. In this study, the focus was on physical stress, with specific emphasis on: (i) the number of necessary speakers used for speaker modeling, (ii) alternative audio/non-acoustic sensors, and (iii) fusion based stress detection using a new audio corpus (UT-Scope). A Gaussian Mixture Model (GMM) framework was used with the TEO-CB-AutoEnv features for neutral/physical stress detection. Stress detection performance was investigated for both acoustic (A-MIC) and non-acoustic (P-MIC) sensors. Evaluations showed that effective stress models can be obtained with 12 speakers out of a random size of 1–42 subjects, with stress detection performance of 62.96% (for close-talking A-MIC) and 66.36% (for P-MIC) respectively. Table 5.1 shows the detection performance of the proposed fused system.The TEO-CB-AutoEnv model scores were fused with traditional Mel-Frequency Cepstral Coefficients (MFCC) based stress model scores using the Adaboost algorithm, resulting in an improvement in overall system performance of 9.43% (absolute, for close-talking A-MIC) and 12.99% (absolute, for P-MIC) respectively. These three advances allow for effective stress detection algorithm development with fewer training speakers and/or alternative sensors in combined feature domains.

An extensive analysis of speech under physical task stress across several parameters was also performed to identify acoustic correlates [39]. Formal listener tests were also performed to determine the relationship between acoustic correlates and perception. To verify the statistical significance of all results, Student-t statistical tests were applied. It was determined that fundamental frequency decreases for many speakers, that utterance duration increases for some speakers and decreases for others, and that the glottal waveform is quantifiably different for many speakers. A plot of the vowels in the two dimensional space of the first two formants is shown in Fig. 5.9, where it can be seen that the formant space shifts inward. To test whether these shifts are statistically significant, a distribution was formed for each condition for each formant across all speakers, and the two distributions compared across conditions. Perturbation of two speech features, fundamental frequency and the glottal waveform, was applied in listener tests to quantify the degree to which these features convey physical stress content in speech. The listener test results showed that shifting the pitch of the physical stress utterances caused a statistically significant decrease in listener performance of more than 20%. Shifting the pitch of the neutral speech did not have an effect on performance. Such changes in speech under physical task stress would significantly impact speaker ID performance for any neutral trained system.

This section has thus far considered (i) analysis, (ii) feature development, and (iii) stress detection methods for speech under a range of stress/emotion/Lombard Effect conditions. The ability to detect such speech conditions is vital in the assess-

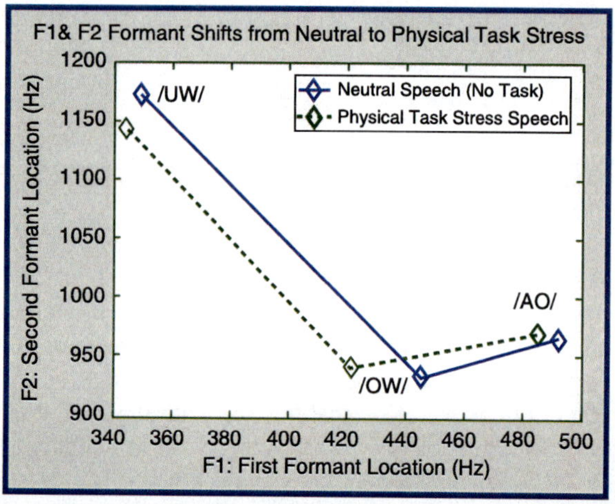

Fig. 5.9 Illustration of speech production changes for three vowels moving from neutral to physical task stress within the formant space

ment of speaker traits within audio forensics, and especially for audio lineups in voice authentication. A final domain is considered for stress due to driver distraction.

A recent study explored the impact of secondary task based stress related to changes in cognitive load and variations of emotional state on a subject's capability to control a vehicle while driving [40]. It is believed that availability of reliable cognitive load and emotion detection in drivers would benefit the design of active safety systems and other intelligent in-vehicle interfaces. In that study, speech produced by 68 subjects while driving in urban areas was analyzed. A particular focus was on speech production differences in two secondary cognitive tasks, interactions with a co-driver and calls to an automated spoken dialog systems (SDS), and two emotional states during the SDS interactions—neutral/negative. The task/emotion dependent mean values for fundamental frequency (F0), spectral center of gravity (SCG), and voiced segment durations are shown in Fig. 5.10, where the vertical bars represent 95% confidence intervals. Increases in F0, SCG, and voiced segment duration are observed when switching from co-driver interaction to the dialog system task. A parameter increase can be seen from neutral to negative emotions, and mostly also from no-repetition (Re0) to first repetition (Re1), and 2nd–6th repetition when engaging with the telephone based dialog system. It can be seen that there is distinct changes in selected cepstral and production based speech features for automatic cognitive task/emotion classification. A fusion of GMM and Support Vector Machine (SVM) classifiers showed an accuracy of 94.3% in cognitive task and 81.3% in emotion classification.

5.4 Speaker Recognition Under Stress

Automatic speaker recognition plays an important role in the area of forensics and security, but it is also important within speech communication such as recognizing a speaker for automatic speech recognition or dialogue systems. It is well known

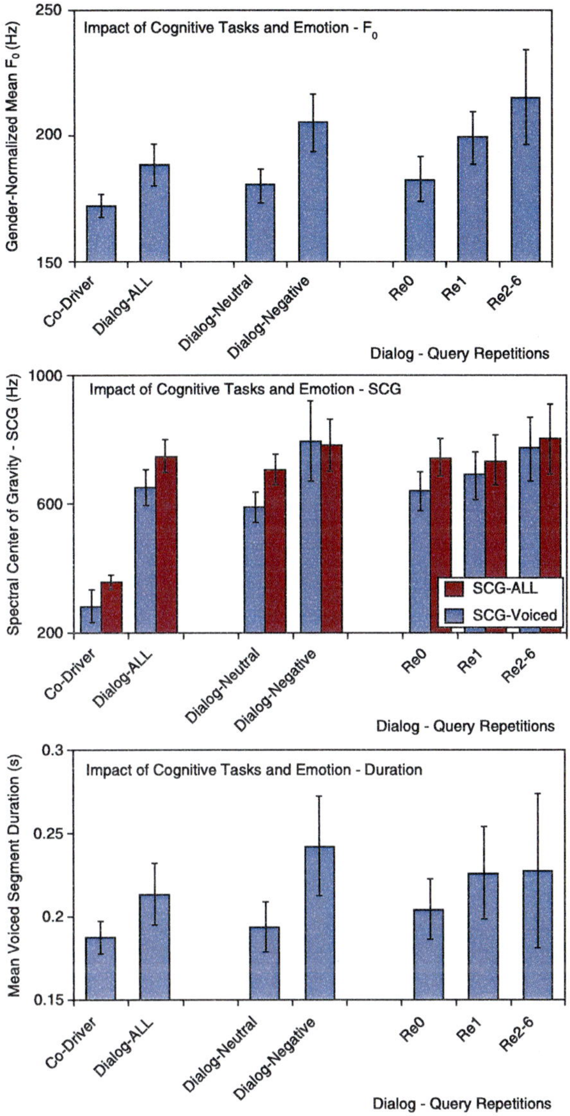

Fig. 5.10 Impact of secondary cognitive tasks and emotions during subject vehicle driving on speech production [40]

that stress and emotion effect impact speech production as shown in the previous section, and detecting such variations represents one important step in ensuring overall system robustness to speaker variability. Speech production in the presence of noise results in the Lombard Effect, which is also known to have a serious impact on speech recognition performance [2, 5, 25–28, 30, 31, 50]. In this section, the impact of speech under stress including Lombard effect on speaker recognition is considered. A number of previous research efforts specifically tailored to address the performance degradation are also introduced.

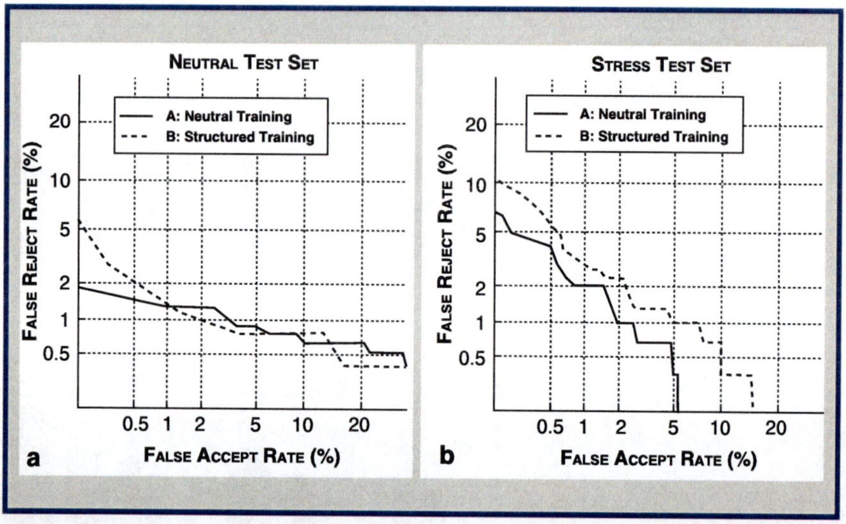

Fig. 5.11 DET curves for speaker ID system with conventional neutral training and with structured training, tested on two subsets of (a) neutral test set and (b) cognitive stressed test set [42]

An earlier research carried out some experiments to study within-speaker variations caused by Lombard effect and cognitive stress for speaker verification system as a project VeriVox [42]. In that study, a speaking style elicitation software was developed to induce speaker variation. The software elicits involuntary variation by means of an interactive module where subjects perform a sequence of tasks, which cause changes in speaking styles (e.g., normal, fast and loud). The tasks include (i) speaking in the presence of background white noise, (ii) speaking from memory with a time pressure and (iii) speaking while solving a logical reasoning and auditory recognition task, with background noise distraction (i.e., under high cognitive load), eliciting the recording of speech under cognitive stress. The database was used for acoustic analysis and also for speaker verification evaluation. The voluntary speech variations were used to build an enrolment set, which is called "structured training" and was compared to conventional "neutral training" including normal speech. As illustrated in Fig. 5.11, this research shows that the speaker verification performance increases when employing structured training without a performance decrease for normal speech test. It should be noted that the concept of "structured training" was previously developed by Lippmann, Martin, and Paul [43] under the framework of multi-style training [41], and also compared with compensation schemes [44, 50] for speech recognition. Limited research has been performed specifically for speaker recognition under stress.

A study was conducted which focused on the analysis of Lombard speech produced under different types and levels of noise [21, 45]. The speech used for the analysis forms a part of the UT-SCOPE [12, 46] database and consists of sentences from the well-known TIMIT corpus, spoken in the presence of highway, large crowd and pink noise. The analysis shows differences in the speech characteristics under these varying noise types. The in-set/out-of-set speaker identification trained on neutral speech degrades in Equal Error Rate (EER) system performance when tested with Lombard speech. A clear demarcation between the effect of noise and Lombard effect on noise is also given by testing with noisy Lombard speech. The experiments showed that test duration has no effect on the EER under Lombard effect (i.e., an increase from 3 to 12 s test sets). The average EER for 12 s test duration showed 7.2, 26.4, 45.8, 50.8% for neutral clean, clean Lombard, noisy neutral and noisy Lombard respectively.

Another study has investigated how Lombard effect impacts perceptual speaker recognition [47]. Experimental results were reported employing in-set/out-of-set speaker identification (ID) tasks performed by human subjects with a comparison to automatic algorithms. The results show that mismatch in reference and test data causes a significant decrease in human speaker ID accuracy. It is noted that Lombard speech contributes to higher accuracy for In-Set speaker ID, but interferes with correct detection of out-of-set speakers. In addition, it was observed that the mismatched conditions cause a higher false reject rate, and that the matched conditions result in higher false acceptance.

An experimental mechanism has also been explored, which enables the use of inherent stress-in-speech or speaking style information present in speech of a person as additional cues for speaker recognition [48]. In this research, the inherent stress included in speech was quantified using three speech features including pitch, amplitude and duration. It was observed that the employed feature vectors of similar phones in different words of a speaker are close to each other in the three-dimensional feature space. This study confirms that the impact of a speaker under stress on different syllables in their speech is unique to them.

In a recent research study, Lombard speech produced under different types and levels of noise was analyzed in terms of duration, energy histogram, and spectral tilt [21]. Acoustic-phonetic differences were shown to exist between different levels of Lombard speech based on analysis of trends from a Gaussian mixture model (GMM)-based Lombard speech type classifier. In this study, the impact of the different flavors of Lombard effect on speech system performance was shown with respect to an in-set/out-of-set speaker recognition task. System performance degraded from an EER of 7.0% under matched neutral training and testing conditions, to an average EER of 26.92% when trained with neutral and tested with Lombard effect speech. Furthermore, improvement in the performance of in-set/out-of-set speaker recognition was demonstrated by adapting neutral speaker models with Lombard speech data of limited duration. Figure 5.12 compares the DET curves for the in-set speaker ID system with Lombard speech with and without adaptation, along with the baseline DET curve. Here Lombard effect is for large crowd noise with different levels 70, 80, and 90 dB-SPL (i.e., LCR1, LCR2, and LCR3). The DET curves reveal the tremendous improvement in performance due to the compensation of the Lombard effect. Improved average EERs of 4.75% and 12.37% were achieved for matched and mismatched adaptation and testing conditions, respectively. At the highest noise levels, an EER as low as 1.78% was obtained by adapting neutral

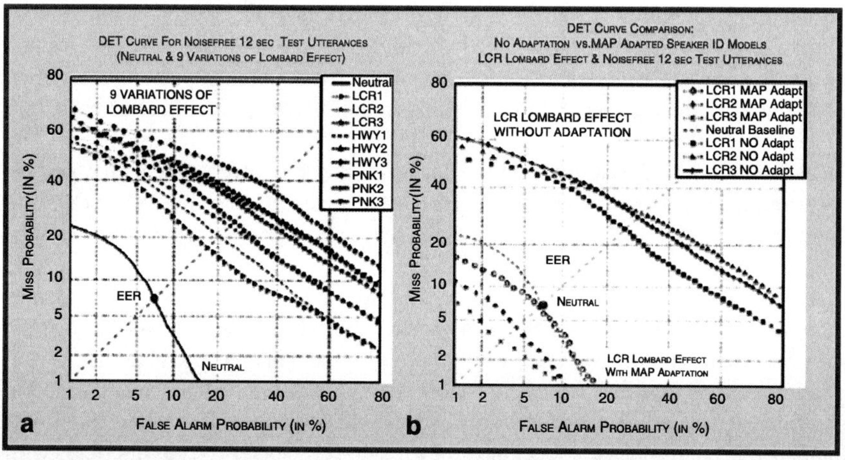

Fig. 5.12 Comparison of DET curves for in-set/out-of-set speaker ID (a) performance with neutral trained speaker ID system using nine different forms of Lombard effect speaker data; (b) performance improvement using MAP adaptation based schemes for LCR {large crowd noise} Lombard effect at three noise levels (Note: all Lombard effect conditions had subjects producing speech in clean conditions with background noise presented through open-air headphones; therefore all speaker ID data is noise free, but Lombard effect speaker test data contains various flavors of noise induced Lombard effect). EER for noise-free neutral speaker ID test condition is 7.0%, and EER for matched adaptation and test conditions vary from 1.29 to 9.35%; (LCR Lombard effect noise had EERs 3.12-5/62%))

speaker models with Lombard speech of limited duration. Therefore, speaker recognition under Lombard effect has been shown to be effectively addressed via dedicated adaptation schemes with limited data sets.

5.5 Future Directions

Effective automatic speaker recognition requires careful assessment of mismatch between training and test material when automatic algorithms are employed. The impact of speech under stress, emotion, and Lombard effect is extensive in the field of speech technology, and in fact much advancement has been made in the field over the past twenty-five years, especially for automated speech recognition. However, much less work has been dedicated to the field of speaker recognition under stressed/compromised conditions. Only within the past few years has there been any real effort to address mismatch for speaker ID/verification.The US lead NIST SRE continues to serve as the primary focus for speaker recognition for many groups, and therefore issues such as microphone and handset mismatch, and to a lesser extent language and session variability, are the prime research focus areas. The challenge to address speaker variability due to stress, emotion, and Lombard effect will continue to require creative advancements in feature development, modeling training, and classifier design. Such advancements are critically needed if speaker recognition technology is to find its place in everyday speech technology where users are engaged in a diverse range of contexts, most of which are never in the traditional "neutral" or "laboratory" conditions where training data are generally obtained.

References

1. NIST SRE USA National Institute of Standards and Technology (NIST) speaker recognition evaluation. http://www.itl.nist.gov/iad/mig/tests/sre/. Accessed 25 Jan 2011
2. Hansen JHL (1988) Analysis and compensation of stressed and noisy speech with application to robust automatic recognition. Ph.D. Thesis, School of Electrical and Computer Engineering, Georgia Institute of Technology, p 396
3. Hansen JHL, Clements M (1987) Evaluation of speech under stress and emotional conditions. J Acoust Soc Am 82(s1):S17–S18
4. Hansen JHL (1989) Evaluation of acoustic correlates of speech under stress for robust speech recognition. IEEE proceedings of the fifteenth annual northeast bioengineering conference, (invited paper), March 1989. Boston, Mass, pp 31–32
5. Hansen JHL, Clements M (1989) Stress compensation and noise reduction algorithms for robust speech recognition. IEEE proceedings international conference on acoustics, speech, and signal processing, May 1989. Glasgow, Scotland, pp 266–269
6. Hansen JHL SUSAS: speech under simulated and actual stress corpus. U.S. Linguistics Data Consortium(LDC).http://www.ldc.upenn.edu/Catalog/CatalogEntry.jsp?catalogId=LDC99S78
7. Hansen JHL SUSAS transcripts: speech under simulated and actual stress transcripts. U.S. Linguistics Data Consortium (LDC). http://www.ldc.upenn.edu/Catalog/CatalogEntry.jsp?catalogId=LDC99T33
8. Hansen JHL, Bou-Ghazale S (1997) Getting started with SUSAS: a speech under simulated and actual stress database, vol 4, Sept 1997. EUROSPEECH-97, Rhodes, pp 1743–1746
9. Hansen JHL, Swail C, South AJ, Moore RK, Steeneken H, Cupples EJ, Anderson T, Vloebergh CRA, Trancoso I, Verlinde P (2000) The impact of speech under 'stress' on military speech technology. NATO Research and Technology Organization RTO-TR-10, AC/323(IST) TP/5 IST/TG-01, March 2000 (ISBN: 92-837-1027-4)
10. Engbert IS, Hansen AV, (2007) Documentation of the Danish emotional speech database DES. Center for PersonKommunikation, Aalborg University, Denmark, Tech. Rep.
11. Burkhardt F, Paeschke A, Rolfes M, Sendlmeier W, Weiss B (2005) A Database of German Emotional Speech. ISCA Interspeech-05, Lisbon, pp 1517–1520
12. Ikeno A, Varadarajan V, Patil S, Hansen JHL (2007) UT-Scope: speech under lombard effect and cognitive stress. IEEE Aerospace Conference, March 2007, Big Sky, Montana, pp 1–7, 3–10
13. Douglas-Cowie E, Cowie R, Sneddon I, Cox C, Lowry O, McRorie M, Martin JC, Devillers L, Abrilian S, Batliner A (2007) The HUMAINE database: addressing the collection and annotation of naturalistic and induced emotional data. Affective Computing and Intelligent Interaction, pp 488-500
14. Steidl S (2009) Automatic classification of emotion-related user states in spontaneous children's speech. Logos, Berlin
15. Angkititrakul P, Hansen JHL (2008) UTDrive: the smart vehicle project. In-vehicle corpus and signal processing for driver behavior. Springer (Chapter 5)
16. Angkititrakul P, Petracca M, Sathyanarayana A, Hansen JHL (2007) UTDrive: driver behavior and speech interactive systems for in-vehicle environments. IEEE Intelligent Vehicle Symposium, 13–15 June 2007, Istanbul

17. Angkititrakul P, Hansen JHL (2007) Getting start with UTDrive: driver-behavior modeling and assessment of distraction for in-vehicle speech systems. ISCA INTERSPEECH-2007, Aug 2007, Antwerp, pp 1334–1337
18. Steininger S, Schiel F, Dioubina O, Raubold S (2002) Development of user-state conventions for the multimodal corpus in SmartKom. Workshop on Multimodal Resources and Multimodal Systems Evaluation, Las Palmas, pp 33–37
19. Douglas-Cowie E, Campbell N, Cowie R, Roach P (2003) Emotional speech: towards a new generation of databases. Speech Commun 40(1–2):33–60
20. Fernandez R, Picard RW (2002) Modeling drivers' speech under stress. ISCA Workshop (ITRW) on Speech and Emotion, Belfast
21. Hansen JHL, Varadarajan VS (2009) Analysis and normalization of Lombard speech under different types and levels of noise with application to in-set/out-of-set speaker recognition. IEEE Trans Audio Speech Lang Process 17(2):366–378
22. Patil S, Sangwan A, Hansen JHL (2010) Speech under physical stress: a production-based framework. IEEE ICASSP-2010: International Conference Acoustics, Speech, and Signal Processing, Dallas, pp 5146–5149
23. Cairns D, Hansen JHL (1994) Nonlinear analysis and detection of speech under stressed conditions. J Acoust Soc Am 96(6):3392–3400
24. Kaiser JF (1990) On a simple algorithm to calculate the 'energy' of a signal. IEEE ICASSP-1990, New Mexico, pp 381–384
25. Hansen JHL (1996) Analysis and compensation of speech under stress and noise for environmental robustness in speech recognition. Speech Commun 20(2):151–170
26. Hansen JHL (1993) Adaptive source generator compensation and enhancement for speech recognition in noisy stressful environments, vol II. IEEE ICASSP-93, April 1993, Minneapolis, pp 95–98
27. Hansen JHL, Womack B, Arslan L (1994) A source generator based production model for environmental robustness in speech recognition, vol 3. ICSLP-94: international conference spoken language processing, Sept 1994, Yokohama, pp 1003–1006
28. Bou-Ghazale S, Hansen JHL (1995) Improving recognition and synthesis of stressed speech via feature perturbation in a source generator framework. NATO-ESCA international tutorial and research workshop on speech under stress, Sept 1995, Lisbon, pp 45–48
29. Bou-Ghazale S, Hansen JHL (1995) A source generator based modeling framework for synthesis of speech under stress, vol 1. IEEE ICASSP-95: international conference on acoustics, speech, and signal processing, May 1995, Detroit, pp 664–667
30. Hansen JHL, Cairns D (1995) ICARUS: a source generator based real-time system for speech recognition in noise, stress, and Lombard effect. Speech Commun 16(4):391–422
31. Hansen JHL, Clements M (1995) Source generator equalization and enhancement of spectral properties for robust speech recognition in noise and stress. IEEE Trans Speech Audio Process 3(5):407–415
32. Hansen JHL, Womack B (1996) Feature analysis and neural network based classification of speech under stress. IEEE Trans Speech Audio Process 4(4):307-313
33. Zhou G, Hansen JHL, Kaiser JF (2001) Nonlinear feature based classification of speech under stress. IEEE Trans Speech Audio Process 9(2):201–216
34. Rahurkar M, Hansen JHL, Meyerhoff J, Saviolakis G, Koenig M (2002) Frequency band analysis for stress detection using a Teager energy operator based feature. ISCA INTERSPEECH-02/ICSLP-02, Denver, pp 2021–2024
35. Hansen JHL, Kim W, Rahurkar M, Ruzanski E, Meyerhoff J (2011) Robust emotional stressed speech detection using weighted frequency subbands. EURASIP J Advan Signal Processing. doi:10.1155/2011/906789
36. Rahurkar MA, Hansen JHL, Meyerhoff J, Saviolakis G, Koenig M (2003) Frequency distribution based weighted sub-band approach for classification of emotional/stressful content in speech. ISCA INTERSPEECH-03, Sept 2003, Geneva, pp 721–724
37. Ruzanski E, Hansen JHL, Finan D, Meyerhoff J (2005) Improved 'TEO' feature-based automatic stress detection using physiological and acoustic speech sensors. ISCA INTERSPEECH-05, Sept 2005, Lisbon, pp 2653–2656

38. Patil S, Hansen JHL (2008) Detection of speech under physical stress: model development, sensor selection, and feature fusion. ISCA INTERSPEECH-08, Sept 2008, Brisbane, pp 817–820
39. Godin KW, Hansen JHL (2008) Analysis and perception of speech under physical task stress. ISCA INTERSPEECH-08, Sept 2008, Brisbane, pp 1674–1677
40. Boril H, Sadjadi O, Kleinschmidt T, Hansen JHL (2010) Analysis and detection of cognitive load and frustration in drivers' speech. ISCA Interspeech-10, 26–30 Sept 2010, Makuhari, pp 502–505
41. Casale S, Russo A, Serrano S (2007) Multistyle classification of speech under stress using feature subset selection based on genetic algorithms. Speech Commun 49:801–810
42. Karlsson I, Banziger T, Dankovicov J, Johnstone T, Lindberg J, Melin H, Nolan F, Scherer K (2000) Verification with elicited speaking styles in the VeriVox project. Speech Commun 31:121–129
43. Lippmann R, Martin E, Paul D (1987) Multi-style training for robust isolated-word speech recognition. IEEE ICASSP-87, April 1987, pp 705–708
44. Chen Y (1988) Cepstral domain talker stress compensation for robust speech recognition. IEEE Trans Acoust Speech Signal Proc 36(4):433–439
45. Varadarajan VS, Hansen JHL (2006) Analysis of Lombard effect under different types and levels of background noise with application to in-set speaker ID systems. ISCA INTERSPEECH-06, Sept 2006, Pittsburgh, pp 937–940
46. Varadarajan VS, Hansen JHL,Ikeno A (2006) UT-Scope: a corpus for speech under cognitive/physical task stress and emotional. ELRA—LREC-2006: language resources and evaluation conference, May 22–26, 2006, Genoa
47. Ikeno A, Hansen JHL (2007) Lombard speech impact on perceptual speaker recognition. ISCA INTERSPEECH-07, Aug 2007, Antwerp, pp 414–417
48. Narayana ML, Kopparapu SK (2009) On the use of stress information in speech for speaker recognition. TENCON-2009, Jan 2009
49. Hansen JHL, Patil S (2007) Speech under stress: analysis, modeling and recognition. In: Müller C (ed) Speaker classification I: fundamentals, features, and methods. Springer, Berlin, pp 108–137
50. Boril H (2008) Robust speech recognition: analysis and equalization of Lombard effect in Czech corpora. PhD Thesis, Czech Technical University, Prague, p 149

Chapter 6
Speaker Identification over Narrowband VoIP Networks

Hemant A. Patil, Aaron E. Cohen and Keshab K. Parhi

Abstract Automatic Speaker Recognition (ASR) has been an active area of research for the past four decades with speech collected mostly in research laboratory environments. However, due to growing applications and possible misuses of Voice over Internet Protocol (VoIP) networks, there is a need to employ robust ASR systems over VoIP networks, especially within the context of internet security and law enforcement activities. There is, however, little systematic study on analyzing effects of several artifacts of VoIP (such as speech codec, packet loss, packet reordering, network jitter and foreign-cross talk or echo) on performance of an ASR system. This chapter investigates each of the issues of VoIP individually and trades it with the performance of the ASR system. In this chapter, a narrowband 2.4 kbps mixed-excitation linear prediction (MELP) codec is used over a VoIP network.

6.1 Introduction

Automatic Speaker Recognition (ASR) refers to the task of identifying a person from his or her voice with the help of machines. It is an economical method of voice biometrics because of availability of low cost computers and powerful processors. It finds applications in banking transactions, forensic science and homeland security [11, 15, 18, 72, 102]. Due to the rapid growth of Internet and e-commerce, there is a growing interest of using voice-based applications (e.g., speech or speaker recognition) over the World Wide Web. With the help of the Internet, a countless number of applications can be accessed from every PC, workstation or cellular phone. This is evidence of the power and flexibility of the Internet and motivates the integration of different types of applications and services (which is a key driving force for the telecom industry) [19].

Voice over Internet Protocol (VoIP) or Internet telephony is one of the most popular Internet services (originally developed by Cerf and Kahn [21]) which use Internet Protocol (IP). Web-based call centers, telephone banking, long-distance

H. A. Patil (✉)
Dhirubhai Ambani Institute of Information and Communication Technology, DA-IICT, Gandhinagar, India
e-mail: hemant_patil@daiict.ac.in

A. Neustein, H. A. Patil (eds.), *Forensic Speaker Recognition*,
DOI 10.1007/978-1-4614-0263-3_6,

communication between two people, etc. are some of the most promising VoIP applications. This may be due to their increased flexibility, lower costs and new capabilities. However, benefits of VoIP come at the cost of increased complexity, reliance on untested software and heightened risk of fraud. Specifically, the integration of VoIP and public switched telephone networks (PSTNs) presents new challenges for both law enforcement and consumer privacy that had not previously been an issue with corporate semi-private switch-isolated telephony networks. In addition, VoIP technology can be misused by criminals. Typical crimes requiring ASR technology are homicide, drug matters, kidnapping, bomb threat, rape, physical assault, obscene calls, extortion calls, false calls, white collar crimes, etc [62, 63]. For example, imagine a situation where an impostor may claim to be a telephone bank's account holder (i.e., genuine speaker) or there can be suspicious voice conversation (intercepted by Police) between two terrorists who are using VoIP [113]. In addition, there is a great need to protect homeland security. For example, forensic voice analysis investigative tools may be useful to determine the likelihood of a match between a suspect's voice and a criminal's voice (e.g., classic method of spectrogram matching [47, 48, 106]). In the context of speaker forensics, important ASR capabilities such as *text-independent* and *channel-independent* are desirable. In addition, ASR system needs to deal with the problem of speech source characterization, transmission channel variability, SNR-based decision, usable speech duration, etc. Furthermore, there is a need to intercept their voices from the VoIP network to identify them.

Hence combining legal interception of VoIP auditing and speaker recognition can assist the security agency of a country to investigate crime on VoIP network. This application can also be useful for home-parole monitoring, prison call monitoring and to collaborate with aural or spectral inspections of voice samples for forensic analysis [113]. These scenarios motivate the proposed research effort to develop ASR systems over VoIP networks. However, VoIP networks are not intended to transmit voice exclusively (they serve many other roles as well). They are inherently affected by challenges such as packet loss, packet reordering, delay, network jitter, foreign cross-talk, etc. [60]. In this context, an important issue which must be addressed for deployment of an ASR system in VoIP domain is to investigate effects of various issues (*viz.*, narrowband codec such as mixed-excitation linear prediction abbreviated as MELP, packet loss, packet reordering, network jitter or delay, foreign cross-talk, etc. in narrowband VoIP network) on the performance of ASR system. One of the motivations for employing narrowband codec (i.e., MELP) is its suitability and acceptability as a North Atlantic Treaty Organization (NATO) international standardization agreement. This codec is known to give best speech quality with least possible bit-rate (2.4 kbps) [58, 59] and hence used in the present chapter as well.

VoIP network and ASR have traditionally been researched *independently*; however, the above scenarios motivate to trade the effects of one with the other. This chapter is a step forward to achieve this goal and use the ASR technology for the benefit of society. Next, a brief review of related work is discussed and the objectives are summarized in Sect. 6.2 followed by a detailed discussion in Sects. 6.3 and 6.4, respectively. Finally, Sect. 6.5 presents experimental results on effects of various artifacts of VoIP on the performance of ASR system and Sect. 6.6 concludes the chapter with brief discussion on our future research directions.

6.2 Context and Related Work

Different issues in ASR such as mimic resistance (e.g., study on voices of twins [64, 78], professional mimics [9, 32, 33, 40, 75, 80, 84–86,121, 122, impostor transformation [13]), channel and session variability, mono-lingual and cross-lingual ASR tasks [91, 108], age variation [33], disguise [33, 85], text-dependence and text-independence [8, 38], phonological content [2, 79], dialects variation [72], fusion techniques [124], source, system and supra-segmental features [22, 87, 117, 124], population size [72, 99], speaker modeling [16, 38, 97, 107], training and testing speech durations, gender [81, 100], forensic conditions [2, 18], infant identification [73], etc. have been researched for decades [29, 30, 62, 69, 72, 74, 108]. In the last decade, there has been a growing interest to investigate ASR performance over different codecs. One of the early works by Reynolds and Rose found that the performance of ASR system is nearly 100% up to a population size of 630 speakers using the TIMIT speech database (clean speech) with about 24 s of training and 6 s of test utterances [97, 99]. The performance degraded significantly for telephone-quality speech and is nearly 60% for similar population size [14]. The investigation to alleviate the effect of the MELP codec on ASR performance is an even more challenging task.

There has been very little work published in analyzing effects of different issues in VoIP on ASR performance meanwhile there have been substantial efforts directed towards the area of speech recognition and speech compression [20, 37, 39, 44, 46, 115, 119, 120]. One of the early attempts in this area was by Quatieri et al. [93, 94]. They employed three codecs, *viz.*, GSM at 12.2 kbps, G.729 at 8 kbps and G.723.1 at 5.3 kbps and found that using speech synthesized from the three codecs, Gaussian Mixture Model (GMM)-based speaker verification and phone-based language recognition generally degrades with coder bit rate, i.e., from GSM to G.729 to G.723.1, relative to an uncoded baseline. Dunn et al. found that speaker verification for all codecs showed a performance decrease as the degree of mismatch between training and testing conditions increases [28]. Similar studies and observations have been reported by Gagan et al. for G.711, G.726 and GSM-FR [90].

Two approaches based on either extracting features from codec parameters (i.e., bit-stream) or G.729 residual were presented for improvements in performance degradation [94]. Similar approaches, of extracting feature parameters from encoder bit-stream, were proposed by Besacier et al. to improve the ASR performance with the GSM codec [7]. Gagan et al. suggested a modified RPE-LTP bit-allocation (keeping the same 13 kbps bit-rate) strategy by reducing bit-allocation for excitation pulses and using released bits to accommodate more log-area ratios (LARs) (thus keeping same bit-rate of GSM-FR 06.10) to improve the degradation in performance. It was also observed that a low order linear prediction (LP) coefficients in GSM codec is responsible for most performance degradations [89]. Other studies were reported independently by Besacier et al. on TIMIT database. They have also found that the adverse effects of packet loss (simulated using Gilbert model) alone are negligible, while the encoding of speech, particularly at a low bit-rate, coupled with packet loss, can reduce the verification accuracy considerably [5–7].

Borah and DeLeon observed significant performance degradation for test utterances acquired from VoIP which may have dropouts due to packet loss. They improved the performance by training the Gaussian mixture model (GMM) with lossy speech packets corresponding to the loss rate experienced by the speaker to be identified [14]. IBM T. J. Watson Research Center has presented the research on ASR from compressed VoIP packet stream [1].

Chen and Wang proposed prosodic and acoustic features to improve the performance of ASR with G.729A and GSM codecs [22]. Recently, Wang and Lin observed the degradation in ASR performance due to speech codec and packet loss on VoIP auditing for population of 15 speakers [113]. The earlier studies concentrated on the effect of speech codec or packet loss on the performance of ASR.In this chapter, effect of network jitter, foreign cross-talk or echo, packet reordering (out-of-order packets) in addition to speech codec and packet loss on ASR performance is analyzed systematically and **possible** techniques will be explored to improve ASR performance. To the best of the authors' knowledge this is first study of its kind proposed in the context of ASR.

6.3 Different Issues in VoIP Network

In this section, different issues in VoIP network that affect quality of service (QoS) and their relevance to ASR performance will be described. To transport speech over a network, speech samples must be coded, inserted into packets that have sequence numbers and creation of time-stamps, transported by the network, received in a playout buffer, decoded in a sequential order and played back as shown in Fig. 6.1. Next, some of the issues in the deployment of VoIP networks [24, 49, 60] and a brief summary of our contributions are presented in this chapter (with the details given in Sects. 6.4 and 6.5).

1. *Voice coders*—An efficient voice encoding and decoding (*codec*) is vital for using the packet-switched networks. A number of factors to take into account while employing a voice codec in a VoIP network are bandwidth usage, coding delay, silence compression, intellectual property, look-ahead and frame size, resilience to loss, resilience to bit errors, layered coding and fixed-point *vs.* floating point digital signal processor. In this chapter, the narrowband codec, *viz.*, MELP is employed which gives least bit-rate 2.4 kbps with best speech quality and as discussed in the introduction, MELP is an international North Atlantic Treaty Organization (NATO) standardization agreement [58, 59].
2. *Packet loss*—Network bandwidth is not the only criterion for speech quality. VoIP network cannot provide a guarantee that packets will be delivered at all,

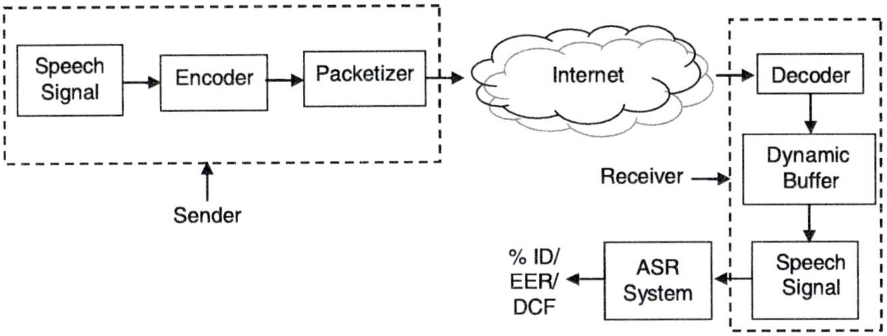

Fig. 6.1 Data flow in VoIP network followed by generic ASR system. (After Mehta and Udani [60])

much less in order. Packets will be dropped under peak load and during periods of network congestion. Thus packet loss is unavoidable. Packet losses greater than 10% are generally intolerable, unless encoding schemes provide extra-ordinary robustness. Earlier approaches used (referred as loss concealment techniques) to compensate for packet loss include interpolation of speech by replaying the last packet and sending redundant information, noise and silence substitution [114]. In Sect. 6.5.2, we propose a novel 1:3 *interleaving* scheme (based on philosophy of spreading the risks of loss, i.e., rearrangement of the original frames to ensure that previously consecutive frames are separated at transmission and rearranged back in the original sequence at the receiver part of VoIP network) to improve ASR performance with packet loss simulated (i.e., Gilbert model) by using different bandwidth in network simulator (NS-3) [111].

3. *Packet reordering (out-of-order packets)*—Due to network congestion, some packets may arrive in a different order than transmitted. In this chapter, a novel and interesting result on ASR performance for packet reordering with 3, 4, 5, 10 and 20 packets is presented. It is shown that ASR performance is almost unaffected by these packet reordering schemes and this experimental evidence suggests to extend the proposed work to build secure speaker identification in networked environments.
4. *Network jitter*—In VoIP networks, some delays are fixed such as coding and decoding while others depend upon network conditions. Delay due to the transport network is nondeterministic in nature and packet delay variance which is called *network jitter*. In Sect. 6.5.4, we present simulation results on ASR performance for different network jitter conditions (modeled through Gaussian normal distributions with difference variances). Our results indicate that jitter remains one of the serious problems in real-time implementation of VoIP networks in addition to deployment of ASR technology.
5. *Echo*—There are three types of echo that can impact VoIP network, *viz., Far End Crosstalk (FEXT), Near End Crosstalk (NEXT) and combined effect of FEXT and NEXT*. The FEXT is caused by the four-wire hybrid conversion. End users will hear their own voice signal bouncing off the remote central office's line-card hybrid. The second form of echo, i.e., NEXT occurs when a free-air

microphone and speakers are used (i.e., when two people are talking on different phones in the same room and hence some bleeding or pickup can occur on each other's phone.), as is the case of most PC endpoint VoIP architecture. The remote user's voice signal produced by the speakers is picked up by the microphone and echoed back to the remote user. We can simulate NEXT by having speech recordings and have canned speech played back at different distances away from the recording device. Afterwards, these recordings can be sent through a VoIP network. Our results on the effect of echo, on a VoIP network, on the performance of ASR indicate that the degradation in performance is a function of the amount of remote user's voice echoed back to the sender.

6.4 Technical Approach

ASR is a pattern recognition problem which consists of two main blocks, *viz.*, feature extractor and pattern classifier as shown in Fig. 6.2 [31]. The feature extractor performs the job of mapping the speech signal into the set of *speaker-specific* features called the *feature vector*, and in the pattern classifier module, speaker models are built for each speaker during the training phase of the machine. In order to recognize a speaker, testing speech is passed through the same feature extractor and the test features are compared with each of the stored models of speakers and the speech is assigned to the speaker whose model gives the highest score. Next, details of the ASR system used in this chapter are described [72].

6.4.1 DA-IICT Speech Corpus

To address forensic conditions (and to complement efforts of LDC [53]), Forensic Voice Database (FV1) of all male 50 speakers with variations in speaking mode, transmission, and session was developed at a law enforcement agency [63]. Recently a new corpus, *viz.*, Multilingual and Multichannel Speaker Recognition (MMSR) data is developed as a result of collaborative work between NIST and MIT Lincoln Laboratory and law enforcement agency [17]. Motivated by this study, a database of 100 speakers (46 male and 54 female, with age range 18–22 years) in three languages, *viz.*, English, Hindi (an Indian language) and mother tongue of subjects is developed for mono-lingual, cross-lingual and multi-lingual ASR at *DA-IICT Gandhinagar* (India) [82].

The recordings were done with the help of Creative Headset HS-300 noise canceling microphone. The speech data recorded with 22050 Hz sampling frequency and finally downsampled to 8,000 Hz. The subjects were not paid; their participation in the data collection was purely voluntary. The subjects were recorded in a single session mostly during the evening or night hours. A list was prepared in all the languages consisting of questions, isolated words, digits, combination-lock phrases and sentences. The speakers were asked to speak spontaneously on any

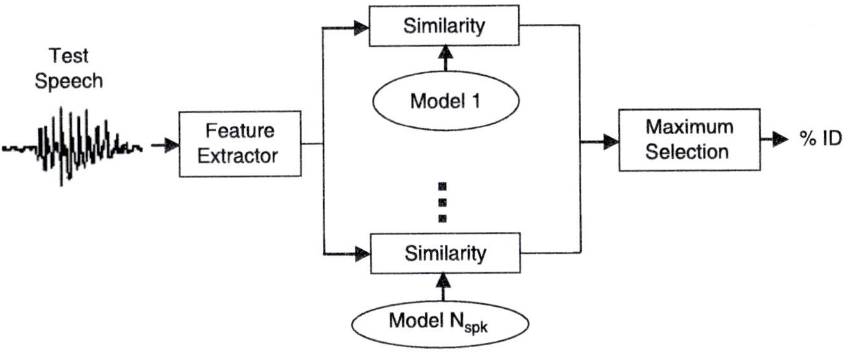

Fig. 6.2 Basic ASR system consists of test speech, feature extractor, similarity comparison models, and a maximum selection block. [72]

topic of their choice at the end of each recording for each language. The contextual speech consisted of a description of the speaker, his or her family and friends, native place or some memorable event. Due to the varied nature of the topics, the speech was mostly conversational. Speakers were posed with some questions about any of the above topics to motivate them to speak fluently. The interview started with a few questions to know about the speaker such as his or her name, age, education, profession, etc. After that the list was given to the speaker to read in his or her own way. During the contextual speech, speakers were also asked to produce unconventional sounds such as *whispers*, *whistle*, *cough*, *frication*, etc. (to mimic forensic scenarios) [65, 112]. Recently, there is a growing interest in whispered speech analysis [123]. The data was recorded with 10 repetitions except for the contextual speech and unconventional sounds, to track all the possible variations in speech. In this chapter, a subset of corpus in English language for 100 speakers is considered. Table 6.1 gives the details of corpus.

6.4.2 *Cepstral Features*

Cepstral analysis has been originally motivated by the problem of echo removal [12] and independent study in Oppenheim's doctoral work followed by work of Schafer [67, 68, 105]. In ASR, cepstral-based features are dominantly used due to their property of capturing vocal-tract based *spectral* information. Hence different spectral features such as Linear Prediction Cepstral Coefficients (LPCC) [4] and Mel frequency Cepstral Coefficients (MFCC) (originally motivated by the doctoral work of Pols [88] and experimental evaluation for ASR) are used [25, 98, 101, 104, 116]. Feature analysis was performed using *12th* order LPC on a 23.2 ms frame with an overlap of 50%. Each frame was pre-emphasized with the filter $1–0.97\, z^{-1}$, followed by Hamming window and then the mean value is subtracted from each speech frame. LPCC were extracted by computing roots of LP polynomial. The standard MFCC computations were performed as per method suggested in [25].

Table 6.1 Details of DA-IICT speech corpus

Item	Details
No. of speakers	100 (46 males and 54 females)
Speakers' experience	Most of the speakers studied in English medium
No. of sessions	1
Data type	Speech
Sampling rate	22,050 Hz down sampled to 8,000 Hz
Sampling format	1-channel, 16-bit resolution
Recording Software	Sony Sound Forge
Type of speech	Read sentences, isolated words and digits, combination-lock phrases, questions, contextual speech of considerable duration, unconventional sounds
Application	Text-independent speaker recognition system
Acoustic environment	Cafeteria, hostel, classroom, staff club, lab, garden
Training segments	30 s
Test segments	60 test segments varied from 1, 2, … 60 s)
Microphone	Creative Headset HS-300 noise canceling microphone
Genuine Trails	100 speakers × 60 trials per speaker = 6,000 genuine trials
Impostor Trials	60 × 100 × 100–6000 = 5,94,000 impostor trials

Recently, warped linear prediction coefficients (WLPC) (originally proposed by Strube [110]) are used in speech and audio coding applications because of their ability to warp LP spectrum and emphasize formant structure [26, 41]. The recursive form of WLPC is used in this project [51]. The features such as LPC, LPCC, MFCC, WLPC are called as spectral features or system features and they are known to represent vocal tract information and has relevance with the acoustics of speech production [34, 35, 70, 92, 95].

6.4.3 Polynomial Classifier

Initially, vector quantization (VQ)-based approaches were used for ASR followed by new proposal of MFCC-GMM system [97, 100]. Even though MFCC-GMM is state-of-the art ASR system, modern ASR task requires high accuracy at low complexity. These features are essential for embedded devices, where memory is at a premium. On server-based systems (e.g., VoIP), this property increases the capacity to handle multiple recognition tasks simultaneously [10, 16]. In this context, Campbell et al. proposed polynomial classifier for ASR which has several advantages such as best approximation to Bayes classifier, uses out-of-class data through

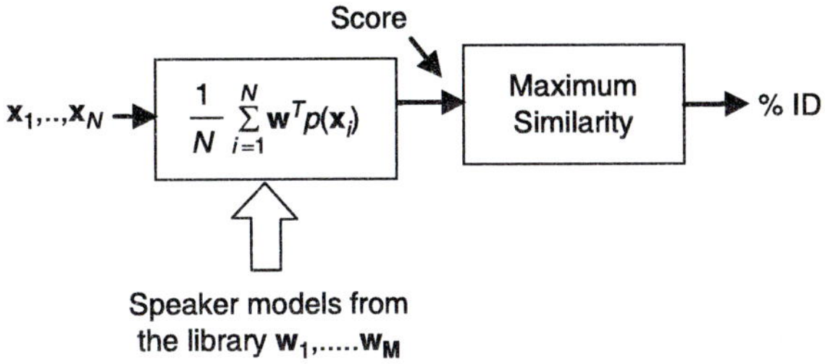

Fig. 6.3 The structure of polynomial classifier [16]

t*discriminative* training to optimize the performance as opposed to other statistical models such as Hidden Markov Model (HMM) or GMM, efficient multiply and add DSP structure for VLSI hardware implementation [71], generate small size speaker models, easy adaptation to new class addition. The basic structure of the classifier is shown in Fig. 6.3. The feature vectors are processed by the polynomial discriminant function. Every speaker i has a speaker-specific vector w_i, to be identified during training and the output of a discriminant function is averaged over time resulting in a score for every w_i given by

$$s_i = \frac{1}{N}\sum_{i=1}^{N} \mathbf{w}^T p(\mathbf{x}_i) = \frac{1}{N}\langle w, p(\mathbf{x})\rangle \tag{6.1}$$

where $x_i = i^{th}$ input test feature vector, w = speaker model and $p(x)$ = vector of polynomial basis terms of the input test feature vector. Thus during recognition, the score for each test segment is computed as *inner product* between the polynomial expansion of test segment feature vectors and speaker model for each hypothesized speaker. Polynomial structure of classifier helps to reduce *feature occupancy* in higher-dimensional feature space [75]. Other details of training algorithm are given in [16].

During recognition, the score for each test segment is computed as *inner product* between the polynomial expansion of test segment feature vectors and speaker model for each hypothesized speaker. For speaker identification, the test segment is assigned to the speaker whose score is maximum whereas for speaker verification, if a score for speaker is higher than a threshold, the claimant is accepted, otherwise it is rejected. In this chapter, polynomial classifiers of 2nd order approximation and training speech of 30 s duration is used here as the basis for all the experiments [16]. An interesting link of this classifier with artificial neural network (ANN) is reported for language and dialect recognition task in [76, 77].

6.4.4 *Performance Measures*

In this chapter, simulation results are shown for speaker verification and speaker detection tasks in closed set mode (i.e., machine has the knowledge of unknown speaker) and work in open set speaker classification is reported in [72]. Each task consisted of several thousand trials. A *trial* consisted of comparison between single hypothesized speaker and a specific test segment. The ASR system was required to make an actual (true or false) decision on whether the specified speaker was present in the test segment. Along with each actual decision, systems were also required to provide for each trial a likelihood score indicating the degree of confidence in the decision. Higher scores indicated greater confidence in the presence of the speaker. A trial where the hypothesized speaker was present in the test segment (correct answer "true") is referred to as a *target trial.* Other trials (correct answer "false") are referred to as *impostor trials* [55, 56]. In this chapter, for each speaker 60 test segments (varied from 1, 2, 3, and 60 s) are used. Each target speaker serves as a non-target example for each of the other target. So for a particular test segment we get 100×100 score matrix whose diagonal elements are scores obtained for *genuine* trials and off-diagonal elements are *impostor* trails. So for 60 test segments, we have $100 \times 60 = 6{,}000$ genuine trials and $60 \times 100 \times 100 – 6{,}000 = 5,\ 94{,}000$ impostor trials. The larger number of trials is considered to state the statistical significance of results.

For speaker verification, there are two types of errors, *viz.*, false alarm probability (FA) of an impostor and false rejection or missed probability (FR) of the genuine speaker. By varying the threshold, different FA and FR rates can be attained. This leads to a detection error trade-off (**DET**) curve (a variant of ROC curves using a normal deviate scale for each axis) for evaluating the ASR system performance by giving uniform treatment to both the errors. One of the performance measure used for the evaluation is the equal error rate (**EER**), which is the point at which the false acceptance and false rejection rates are equal. For speaker detection, used performance measure is the detection cost function (**DCF**) which is a weighted sum of FR and FA probabilities [57]:

$$\begin{aligned}\min DCF = {} & C_{Miss}.P(Miss|Target).P_{Target} \\ & + C_{FalseAlarm}.P(FalseAlarm|NonTarget).P_{NonTarget}\end{aligned}$$

where C_{Miss} is the cost of a missed detection, $C_{FalseAlarm}$ the cost of a false alarm, the a priori probability of a target speaker is P_{Target}, and the a priori probability of a non-target speaker is $P_{NonTarget}$. While the DET curve gives an overall intuitive system performance showing how the FR can be traded-off against FA in different decision thresholds, the DCF evaluates the system performance for particular decision conditions. In this chapter, results are reported for following choice of DCF parameters $C_{Miss} = 1$, $C_{FalseAlarm} = 1$, $P_{Target} = 0.5$, $P_{NonTarget} = 1 - P_{Target}$. In addition, *min. DCF* is scaled by factor 10^{-2} in the all ASR experiments reported in this chapter. NIST's DET curve plotting software, *viz.*, DETware is used to report experimental results in this chapter [27].

6.5 ASR Performance over Simulated VoIP Network

As mentioned in Sect. 6.1, VoIP network introduces several artifacts into speech which can affect ASR performance. If we record directly speech database from publicly available VoIP network such as Skype, Googletalk [103] or PJSIP [66] then the recorded data at the receiver will contain combined effects of speech codec, packet loss, packet reordering, network jitter and foreign cross-talk, etc. So it will be difficult to observe the relative effects on degradation of ASR performance. Hence in this work, the artifacts were considered (simulated) individually to determine which artifacts introduce the most errors into the ASR system and to investigate possible methods which can overcome the artifacts. Next, the effect of several artifacts of VoIP such as compressed speech, packet loss, packet reordering (out-of-order packets), network jitter and echo (foreign cross-talk) on the performance of ASR system are discussed.

6.5.1 MELP Codec

Speech database was transcoded using version 1.2 of the MELP codec [3]. This codec achieves 2.4 kbps narrowband voice transmission with best speech quality. Originally, MELP was selected by NIST and standardized in FIPS 137 and later NATO STANAG 4591. We used MELP codec PJSIP open source software to generate real world VoIP databases using SIP protocol. We modified PJSIP to include a MELP coder (analysis) and decoder (synthesis). Also we modified it to transmit only 1 frame per packet as the default implementation was unable to accommodate MELP's 22.5 ms frame time (or 180 samples or frame for 8 kHz sampling rate).

Some of the distinct features of MELP codec are as follows [58, 59]:

1. *Mixed pulse and noise excitation*—eliminates buzzy quality speech usually associated with LPC vocoders
2. *Periodic and aperiodic pulses*—to reproduce erratic glottal pulses without introducing tonal noises
3. *Adaptive spectral enhancement*—sharpens the formant resonances and improves the bandpass filtered waveform between synthetic and natural speech
4. *Pulse dispersion filter*—allows synthesizer to better match for synthetic and natural speech in regions away from the formant zones

Figure 6.4 shows DET curves for different features such as MFCC, LPCC and WLPC for ASR performance of MELP encoded speech *vs.* PCM (8 kHz) whereas Table 6.2 shows EER and min DCF. As the total number of genuine and impostor trials are 6 lacs, DET curves are close to ideal straight line [57] in all the ASR experiments reported in this chapter. From the DET curves, it is clear that performance of speaker verification and detection degrades significantly for MELP encoded speech (EER is around 26–30%). Degradation for MFCC is less (0.93%) as compared to WLP (2.18%) and LPCC (2.29%). The higher degradation in ASR performance is attributed to very low bit-rate of codec.

Fig. 6.4 DET curves for effect of MELP codec *vs.* PCM

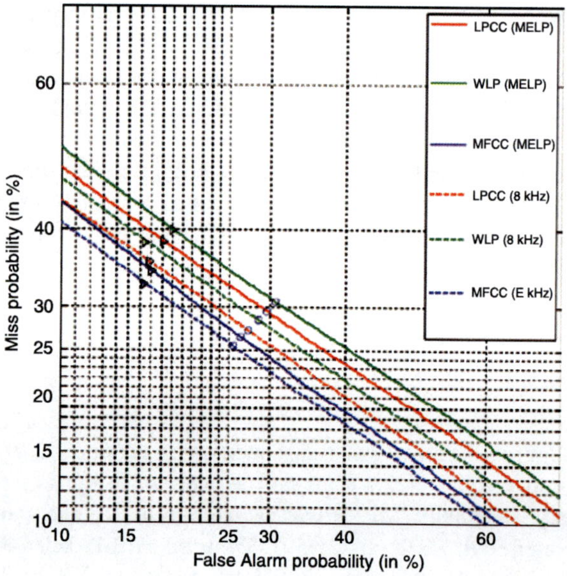

Table 6.2 Effect of MELP codec on ASR performance

FS	Codec			
	8 kHz (PCM)		MELP	
	EER	DCF	EER	DCF
LPCC	27.25	26.24	29.54	28.24
WLP	28.43	27.37	30.61	29.44
MFCC	25.50	24.59	26.43	25.71

6.5.2 Packet Loss

As discussed in Sect. 6.3, in VoIP networks, packet loss is caused by network congestion and/or router failure. This causes some packets to arrive too late or not at all at their destination. In this scenario, the signal information in that packet is lost and the synthesized speech must be recreated without that information by a packet loss concealment technique. An example of packet loss is shown in Fig. 6.5. Typical packet loss probabilities **(P_L)** are set as a percentage (e.g., $P_i = P_L = 0.01\%$). If more bursty errors are desired, then the packet loss probability is set as a function of the previous packet loss probability ($P_i = 0.25 * P_{i-1} + 0.75 * P_L$). Details of packet loss simulation are given in Appendix A. From the DET curves and EER and min DCF (shown in Fig. 6.6 and Table 6.3), it is clear that performance of speaker verification and detection degrades with packet loss. Degradation increases significantly with increase in % packet loss. Degradation for MFCC is less as compared to WLP and LPCC. Next, two databases are discussed that are generated from DA-IICT speech corpus (discussed in Sect. 6.4.1) with packet loss in *Network Simulator* NS-3 network simulator to investigate and ultimately improve ASR performance under packet loss [111].

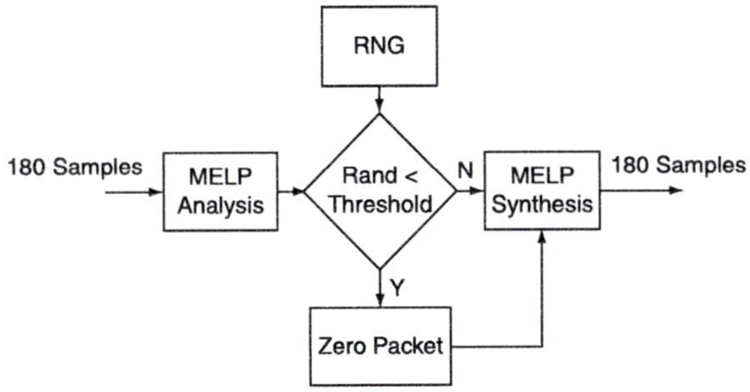

Fig. 6.5 Flowchart for simulation of packet loss. A pseudo random number generator is used along with a threshold probability to implement packet loss. If the random number generator generates a number that falls into the good region then the packet was successfully transmitted in the simulation. If the random number generator generates a number that falls into the bad region then the packet was lost in the simulation and silence is played back.

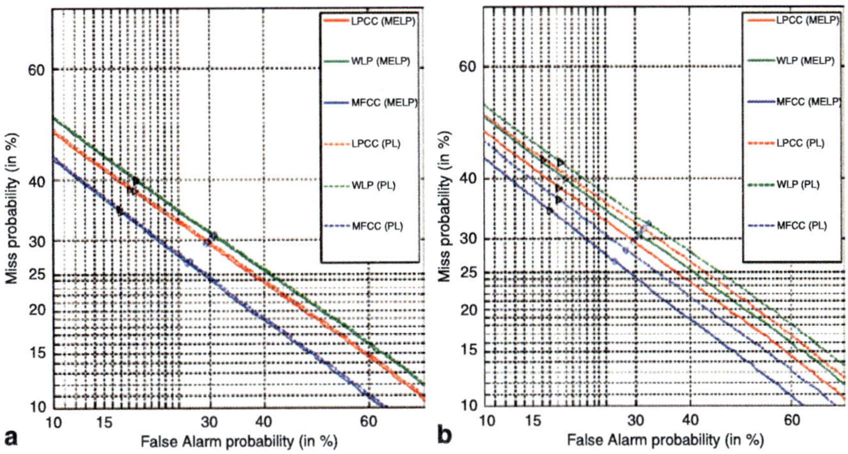

Fig. 6.6 DET curve for effect of Packet Loss **a** 1% and **b** 10%

Table 6.3 Effect of Packet Loss on ASR Performance

FS	Codec							
	MELP		PL (1%)		PL (5%)		PL (10%)	
	EER	DCF	EER	DCF	EER	DCF	EER	DCF
LPCC	29.54	28.24	29.78	28.46	30.58	29.09	31.43	29.79
WLP	30.61	29.44	30.83	29.56	31.48	30.04	32.26	30.65
MFCC	26.43	25.71	26.71	25.93	27.35	26.52	28.24	27.23

- ***NS-3 Database 1***

  This database contains training and testing files for 100 speakers with 25 k, 30 k, 35 k bandwidth (which is selected with *NS-3* simulator [111]). Since there is a 1 *frame per packet*, so there will be more packet transmission overhead caused

Table 6.4 Effect of Packet Loss on ASR Performance through NS3-database1 (1 frame per packet)

FS	Codec							
	MELP		PL (25 K BW)		PL (30 K BW)		PL (35 K BW)	
	EER	DCF	EER	DCF	EER	DCF	EER	DCF
LPCC	29.54	28.24	32.33	30.58	29.59	28.25	29.55	28.24
WLP	30.61	29.44	33.26	31.39	30.64	29.45	30.61	29.44
MFCC	26.43	25.71	29.45	28.17	26.45	25.72	26.43	25.72

Table 6.5 Effect of Packet Loss on ASR Performance through interleaving NS3-database2 (4 frames per packet)

FS	Codec							
	MELP		PL (5 K BW)		PL (8 K BW)		PL (11 K BW)	
	EER	DCF	EER	DCF	EER	DCF	EER	DCF
LPCC	29.54	28.24	29.93	28.57	29.71	28.38	29.54	28.38
WLP	30.61	29.44	31.01	29.78	30.78	29.56	30.61	29.56
MFCC	26.43	25.71	*26.75*	26.04	*26.55*	*25.82*	26.43	25.82

by the packet header. The selection of different bandwidths (i.e., 25 k, 30 k and 35 k) incorporates packet loss. There is *no interleaving* or replacement of previous packet. It has effect of packet loss through change in the bandwidth similar to the Gilbert model packet but not actually the Gilbert model. Table 6.4 shows EER and optimal DCF measure obtained using this database. It is evident from Table 6.4 that the ASR performance degrades significantly as the network bandwidth is reduced from 35 to 25 k, This is due to increase in packet loss for reduced network bandwidth.

- ***NS-3 Database 2***
  Recently, several interleaving schemes have been proposed to improve the PESQ (Perceptual Evaluation of Speech Quality [83]) score of speech degraded with packet loss [42]. We propose 1:3 interleaving scheme for improving ASR performance under packet loss scenarios. Packet interleaving for spreading the risk over larger time interval on MELP bit-stream is also prepared. In this scheme, 4 *frames per packet* and different bandwidths (5 k, 8 k and 11 k) are used to incorporate packet loss through network simulator [42, 111]. Tables 6.4 and 6.5 shows EER and DCF for ASR experiments on these two databases. It is clear that packet interleaving improves the ASR performance when packets have experienced loss using Gilbert Model (as shown in Tables 6.4 and 6.5) [118].

6.5.3 Packet Reordering (Out-of-Order Packets)

Due to network congestion, some packets may arrive in a different order than transmitted. In this scenario, the state information of the synthesizer would be incorrectly generated and the synthesized speech would be somewhat intelligible. In this

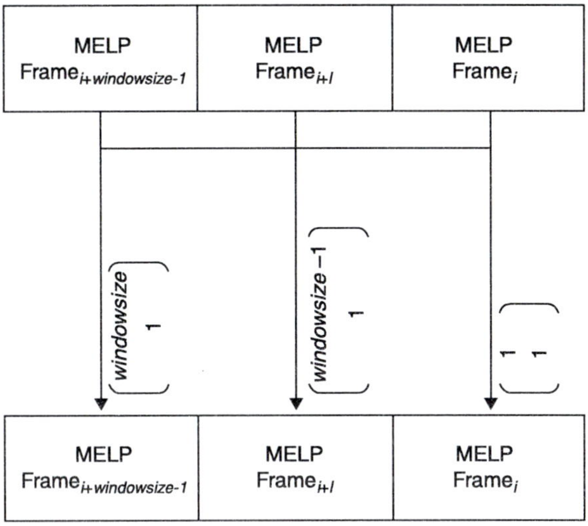

Fig. 6.7 Flowchart for simulation of packet reordering. The individual mappings are performed using (n choose 1) method from number counting where n decreases from left to right to account for selection. In this work n varies from windowsize to 1 and each item is randomly selected using a pseudo random number generator (RNG).

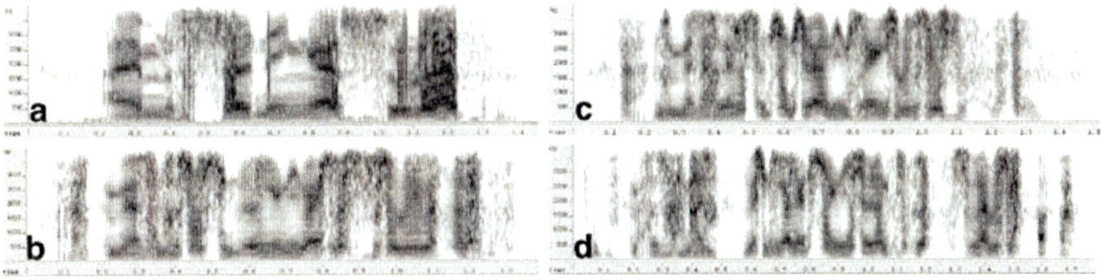

Fig. 6.8 Effect of packet reordering on spectrograms. **a** MELP coded speech. Packet ordered speech with **b** 5 packets, **c** 10 packets and **d** 20 packets

context, we are assuming that packets arrived are out of order and storage capacity of buffer at the receiver is not exceeded. Thus, we are trying to simulate only packet reordering here and not coupled packet loss. Figure 6.7 shows packet reordering scheme employed in this chapter whereas Fig. 6.8 demonstrates effects of this packet reordering scheme on spectrogram of speech signal. In Fig. 6.7, packet reordering assumes a 1 frame per packet example. In this scenario, frames in a specific window are reordered based on a randomly generated configuration. In mathematical terms, the first frame is chosen from all frames in the window. The second frame is chosen from all frames except the first frame that was chosen. This continues until all frames have been selected. It is evident from plots shown in Fig. 6.8 that packet reordering significantly affects formant contour and spectral energy distribution (over wide frequency range). This motivates us to investigate corresponding effect on performance of ASR system. Figure 6.9 shows DET curves corresponding to different spectral features for ASR system on speech database with packet reordering for 3 and 20 packets. In addition, EER and optimal DCF measures are reported in Table 6.6.

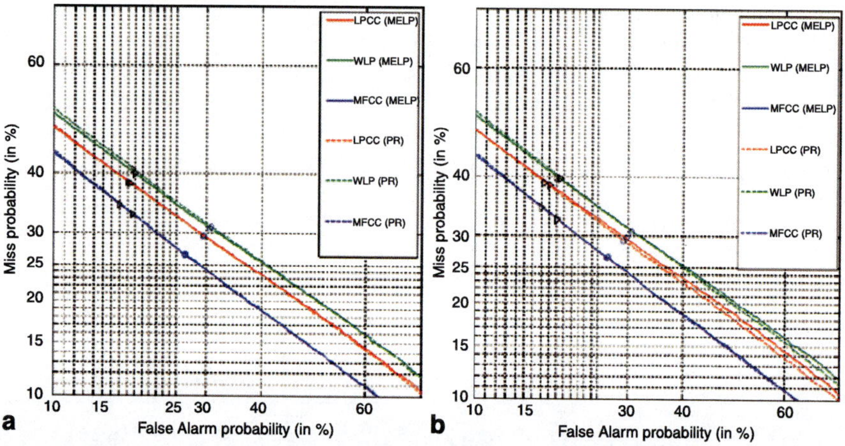

Fig.6.9 DET curve for effect of packet reordering (PR) **a** 3 packet and **b** 20 packets

Table 6.6 Effect of packet reordering (PR) on speaker recognition performance

Features	MELP		PR (3 packets)		PR (4 packets)		PR (5 packets)		PR (10 packets)		PR (20 packets)	
	EER	DCF	EER	DCF	EER	DCF	EER	DCF	EER	DCF	EER	DCF
LPCC	29.54	28.24	29.44	28.24	29.79	28.47	29.39	28.32	29.44	28.31	29.16	28.08
WLP	30.64	29.31	30.80	29.63	31.11	29.94	30.85	29.72	30.78	29.73	30.59	29.56
MFCC	26.70	25.80	26.45	25.72	26.73	26.01	26.56	25.79	26.61	25.85	26.51	25.79

Some of the observations from the results are as follows:

- From Table 6.6 and DET curves shown in Fig. 6.9, it is clear that performance of speaker verification and detection is not affected much by the packet reordering. On the other hand, it was interesting to observe that when the packet reordered signals were played back, then it was very difficult to identify original speaker by human listeners in addition to linguistic message. This shows the power of ASR system or machine recognition of speakers even though their speech packets are shuffled.
- Above mentioned results can be justified based on the fact that structure of polynomial classifier is developed based on concept of *discriminative* training which can be conceived as posterior probability estimation in Bayesian framework and hence equation for scoring, i.e., Eq. (6.1) can be approximated as log-likelihood estimate in *statistical* sense [16]. Thus, this means that the genuine and impostor probability distributions (i.e., histograms in reality) derived from spectral features (e.g., MFCC) will not change appreciably due to packet reordering and hence the distance between genuine and impostor distribution will be almost same (as for the case normal speech without packet reordering) to separate two classes in higher-dimensional feature space. In other words, speaker model does not depend on the *sequence information* of the speech feature vectors.

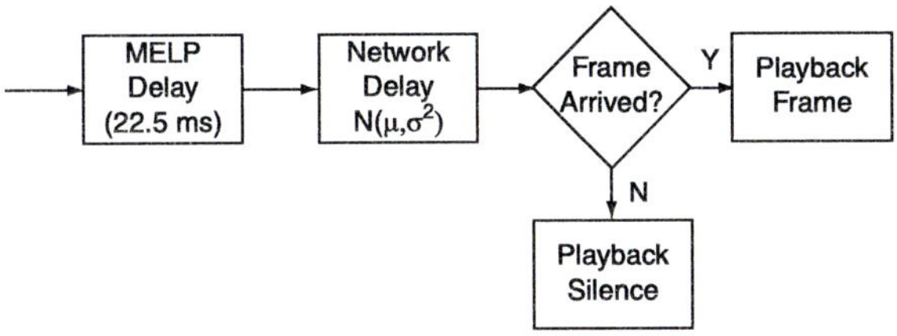

Fig. 6.10 Flowchart for simulation of network jitter. This simulation uses a network delay with a Gaussian normal distribution. If the delay is larger than a specified threshold then the frame arrived late and was not played back. If the delay was less than the threshold then the frame was buffered and played back.

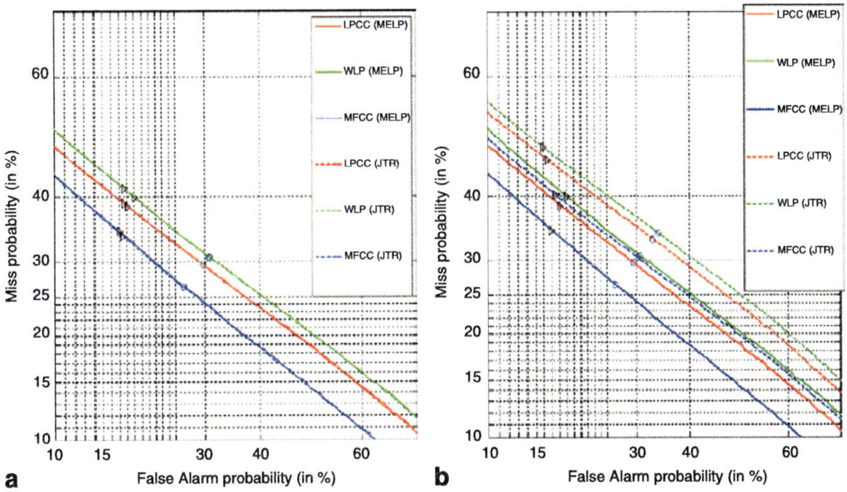

Fig. 6.11 DET curves for effect of network jitter (JTR) **a** variance=10 and **b** variance=100

- We can use this experimental evidence for developing interesting application where there is a requirement to transmit encrypted linguistics messages and hiding speaker's identity over the VoIP network.

6.5.4 *Network Jitter*

In VoIP, network jitter is caused by varying delays in the network. This causes some packets to arrive close to one another or farther away from one another. This is still one of the serious problems in successful deployment of real-time VoIP systems [43, 49] and several playout delay adaptation techniques are proposed to alleviate network jitter [96]. An example of jitter is shown in Fig. 6.10. Typical network jitter follows Gaussian normal distributions. Figure 6.11 shows DET curves corresponding to different spectral features for ASR system on speech database with network

Table 6.7 Effect of Network Jitter on ASR Performance

FS	Codec							
	MELP		Jitter (variance=10)		Jitter (variance=50)		Jitter (variance=100)	
	EER	DCF	EER	DCF	EER	DCF	EER	DCF
LPCC	29.54	28.24	29.54	28.25	31.39	29.83	33.01	31.18
WLP	30.64	29.31	30.69	29.47	32.39	30.90	34.01	32.00
MFCC	26.70	25.80	26.33	25.62	28.29	27.27	30.30	28.82

Fig. 6.12 Flowchart for simulation of foreign cross-talk (echo) (i.e., mic to speaker echo)

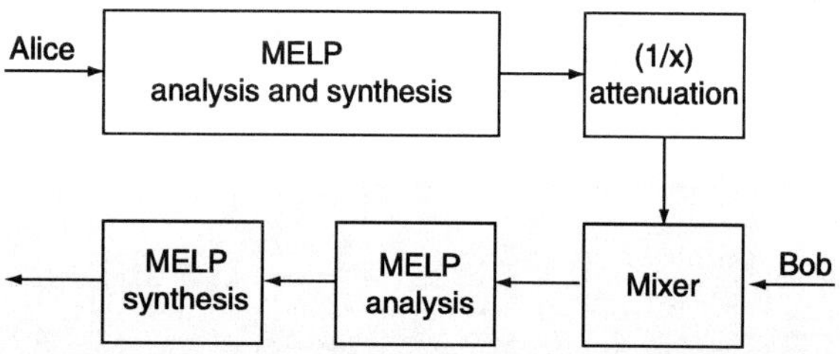

jitter for 10 and 100 as variance. In addition, EER and optimal DCF measures are reported in Table 6.7. From these DET plots and results reported in Table 6.7, it is clear that performance of speaker verification and detection degrades for network jitter. Degradation increases significantly with increase in variance in Gaussian distribution. Degradation for MFCC is less as compared to WLP and LPCC.

6.5.5 Echo—Far End Crosstalk (FEXT)

In VoIP, echo is caused by far end crosstalk (FEXT) from the receiver's speaker output to the receiver's microphone. In this scenario, there will be additional information due to the transmitter's voice on the received voice transmission on the transmitter's side. An example of echo is illustrated in Fig. 6.12. Related work on cross-talk estimation is reported in [50]. We have used open source FEXT program available at [36] to simulate FEXT. Figure 6.13 shows DET curves corresponding to different spectral features for ASR system on speech database with echo simulated for different attenuation factor. In addition, EER and optimal DCF measures are reported in Table 6.8.

Some of the observations from the results shown in Table 6.8 are as follows:

1. There is degradation in ASR performance due to echo or cross-talk.
2. The amount of degradation is a function of attenuation factor.
3. When the amplitude of voice for target speaker and background speaker are scaled with same attenuation factor, (either 1 or 0.5), then the degradations in ASR performance are almost similar.
4. When the amplitude of voice for target speaker and background speaker are scaled with different attenuation factor, then the degradation in ASR performance is maximum if attenuation factor for target speaker is less than that of background speaker and vice-versa.

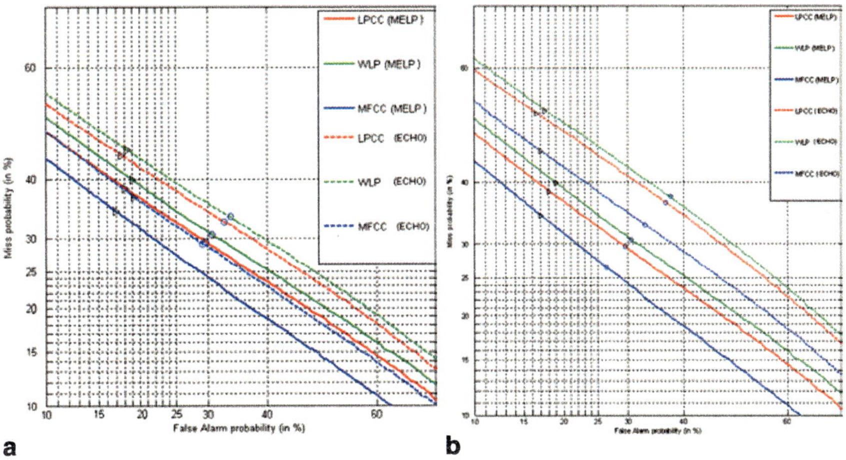

Fig. 6.13 DET curves for effect of echo with different attenuation factor **a** 1 × 1 and **b** 0.75 × 0.25

Table 6.8 Effect of Foreign cross-talk (Echo) on ASR Performance

FS	Codec											
	MELP		Echo (1 × 1)		Echo (0.5 × 0.5)		Echo (0.75 × 0.25)		Echo (0.25 × 0.75)		Echo (0.66 × 0.33)	
	EER	DCF	EER	DCF	EER	DCF	EER	DCF	EER	DCF	EER	DCF
LPCC	29.54	28.24	32.66	30.90	32.73	30.92	36.64	34.26	30.71	29.24	34.94	32.79
WLP	30.64	29.31	33.61	31.77	33.48	31.73	37.61	35.10	31.61	30.20	35.83	33.65
MFCC	26.70	25.80	29.15	28.09	29.23	28.09	32.95	31.20	27.41	26.62	31.23	29.76

5. We found ASR performance for 60 s training duration as well. It was observed that as the training speech duration increases from 30 s to 60 s, the results degrade for majority of the case of features. This is in contrast to traditional speaker recognition system. This may be due to the fact that as the training speech duration increases the amount of foreign cross-talk increases and hence confusion in discrimination for speakers increases which in turn results in performance degradation.

6.6 Summary and Conclusions

The intellectual merit of the work presented in this chapter lies in improving our understanding for several design issues in ASR systems for their deployment over narrowband (using MELP speech codec) VoIP network. Disjoint research efforts in networking and speaker identification are considered together to analyze the effects of several artifacts of VoIP on the performance of the ASR system. The broader impact of this chapter will be to enable the use of ASR technology in a networked environment for the benefit of society. Secure identification of the speaker will create new applications in voice enabled electronic transactions over networks. Our finding indicates that the performance of speaker recognition is unaffected by packet reordering (an artifact of VoIP network). This finding may help in design of robust and secure speaker identification systems over VoIP network where there is a requirement to transmit encrypted linguistic message and hiding speakers' identity over network.

One of the limitations of the work could be that the results are reported on simulated packet reordering techniques for testing and training done on *single* session. In addition, performance figures for ASR system show *large EER* primarily due to very low-bit rate speech codec (i.e., MELP) and due to use of speech corpus developed in realistic noisy conditions. Furthermore, the choice of number of packets to be reordered was arbitrary and hence needs to be selected based on standard or real-life VoIP scenarios. Future work will be directed towards investigation on using packet reordered data over real VoIP network such as Skype, GoogleTalk, or other SIP based VoIP networks with recordings for multiple sessions. In addition, in order to alleviate the degradation in the performance due to echo we would like to explore blind source separation (BSS) techniques. Furthermore, in order to alleviate the degradation in the performance due to network jitter, we would like to use different time-scale modification algorithms to alter the playout length of individual packets so as to adapt these speech segments to the varying delays within a talkspurt while still maintaining relatively smoother voice playout (i.e., *intratalkspurts* play out delay adaptation) [92, 96]. We plan to explore different error concealment techniques or some other sophisticated interleaving schemes to improve ASR performance in case of packet loss [23, 54, 61]. In addition, to alleviate the degradation due to echo, several blind source separation techniques will be explored [45, 52, 108, 109, 125]. Finally, the speaker recognition system will be analyzed on commercially available speech codecs for VoIP network such as G.711, G.729, G.723, CELP, AMBE, iLBC, etc.

Acknowledgements The first author would like to thank authorities of DA-IICT Gandhinagar for their kind support and grant from India Initiative at University of Minnesota, twin cities campus, Minneapolis, USA to carry out this research work. The authors would like to thank Dr. Jian-Hung Lin for his help to simulate packet loss via network simulator (NS-3) during this research work.

Appendix A: Simulation of Packet Loss

Let P_{LG} is the loss probability when the channel is in the good state and P_{LB} is the loss probability when the channel is in the bad state. Figure A.1 shows the diagram of finite state machine model of network channel states.

When simulating, the network channel state is used to determine if a packet is lost. This is done in the following manner:

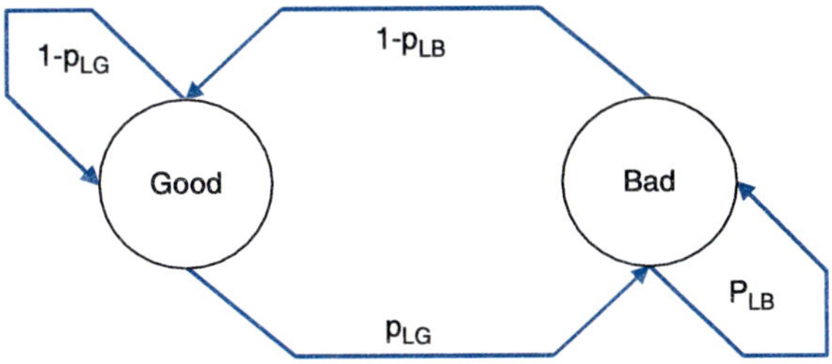

Fig. A.1 Finite state machine model of network channel states. *Good* represents successful transmission. *Bad* represents unsuccessful transmission

1. Generate new random number in range [0, MAXRANGE-1].
2. If previous state was good state
   a. Use threshold=P_{LG}*MAXRANGE
3. If previous state was bad state
   a. Use threshold=P_{LB}*MAXRANGE
4. Determine if random number is less than threshold=lost packet
   a. Set state to bad as the packet is lost
5. Determine if random number is greater than threshold=successfully transmitted packet
   a. Set state to good as the packet was successfully transmitted

where P_{LG} is the loss probability when the channel is in the good state and P_{LB} is the loss probability when the channel is in the bad state. Using the finite state machine from Fig. A.1, define P_i as the probability that the network channel is in the bad state where a packet has been lost. Then

$$P_i = P_{LG}(1-P_{i-1}) + P_{LB}P_{i-1}$$

$$P_i = (P_{LB} - P_{LG})P_{i-1} + P_{LG}.$$

The probability that the channel is in the good state where a packet has been successfully transmitted is $(1-P_i)$. To prevent the probability from growing to 1 or from shrinking to 0, care must be taken to ensure that for all P_i the probability is equal to some initial average loss probability (P_L). By substituting into the formula P_L for the state probabilities then

$$P_L = (P_{LB} - P_{LG})P_L + P_{LG}.$$

Next allowing $P_{LG}=xP_L$

$$P_L = (P_{LB} - P_{LG})P_L + xP_L.$$

Then

$$1 = (P_{LB} - xP_L) + \mathrm{x}$$

And finally solving for P_{LB} yields

$$P_{LB} = 1 - x + xP_L$$

The channel can be defined by selecting P_L and x. Then P_{LB} can be calculated to ensure the average probability does not change. Using the initial average loss probability, it is possible to remove the recursion which will lead to the following formula:

$$P_i = (P_{LB} - P_{LG})^i P_L + \left[\sum_{j=0}^{i-1} (P_{LB} - P_{LG})^j\right] P_{LG}$$

Although not readily apparent, if substitutions are made then

$$P_i = (1-x)^i P_L + \left[\sum_{j=0}^{i-1} (1-x)^j\right] x P_L$$

$$P_i = \left[(1-x)^i + \left[\sum_{j=0}^{i-1} (1-x)^j\right] x\right] P_L$$

$$P_i = \left[(1-x)^{i-1} - x(1-x)^{-1} + \left[\sum_{j=0}^{i-1} (1-x)^j\right] x\right] P_L$$

$$P_i = \left[1 - x\left[\sum_{j=0}^{i-1} (1-x)^j\right] + x\left[\sum_{j=0}^{i-1} (1-x)^j\right]\right] P_L$$

$$P_i = P_L$$

Therefore, the average loss probability is maintained.

References

1. Aggarwal C, Olshefski D, Saha D, Shae Z-Y, Yu P (2005) CSR: speaker recognition from compressed VoIP packet stream. IEEE Int. Conf. on Multimedia and Expo, ICME Amsterdam, The Netherlands, pp 970–973
2. Amino K, Arai T (2009) Speaker-dependent characteristics of the nasals. Forensic Sci Int 185(1–3):21–28
3. Analog-to-Digital Conversion of Voice by 2,400 BIT/Second Mixed Excitation Linear Prediction (MELP) MIL-STD-3005
4. Atal BS (1974) Effectiveness of linear prediction of the speech wave for automatic speaker identification and verification. J Acoust Soc Am 55(6):1304–1312
5. Besacier L (2008) Speech coding and packet loss effects on speech and speaker recognition. In: Tan ZH, Lindberg B (eds) Automatic speech recognition on mobile devices and over communication networks. Springer, London, pp 27–39
6. Besacier L, Grassi S, Dufaux A, Ansorge M, Pellandini F (2000) GSM speech coding and speaker recognition. ICASSP'00 2:1085–1088

7. Besacier L, Mayorga P, Bonastre J-F, Fredoulile C, Meignier S (2003) Overview of compression and packet loss effects in speech biometrics. IEE Proc Vision Image Signal Process 150(6):372–376
8. Bimbot F, Bonastre J-F, Fredouille C, Gravier G, Magrin-Chagnolleau I, Meignier S, Merlin T, Ortega-Garcia J, Petrovska-Delacretaz D, Reynolds DA (2004) A tutorial on text-independent speaker verification. EURASIP J Appl Signal Process JASP 4:430–451
9. Blomberg M, Elenius D, Zetterholm E (2004) Speaker verification scores and acoustic analysis of a professional impersonator. Proc. FONETIK, Stockholm University
10. Bocchieri E (2008) Fixed-point arithmetic. In: Tan ZH, Lindberg B (eds) Automatic speech recognition on mobile devices and over communication networks. Springer, London, pp 255–275
11. Boe LJ (2000) Forensic voice identification in France. Speech Commun 31(2–3):205–224
12. Bogert BP, Healy MJR, Tukey JW (1963) The quefrency alanysis of time series for echoes: cepstrum, pseudo-autocovariance, cross-cepstrum, and saphe tracking. In: Rosenblatt M (ed) Time series analysis. Wiley, New York, pp 209–243 (Ch 15)
13. Bonastre J-F, Matrouf D, Fredouille C (2007) Artificial impostor voice transformation effects on false acceptance rate. Proc Interspeech, pp 2053–2056
14. Borah DK, DeLeon P (2004) Speaker identification in the presence of packet loss. IEEE 11th Digital Signal Processing Workshop and IEEE Signal Processing Education Workshop, pp 302–306
15. Campbell JP Jr (1997) Speaker recognition: a tutorial. Proc IEEE 85(9):1437–1462
16. Campbell WM, Assaleh KT, Broun CC (2002) Speaker recognition with polynomial classifiers. IEEE Trans Speech Audio Process 10(4):pp 205–212
17. Campbell JP, Nakasone H, Cieri C, Miller D, Walker K, Martin AF, Przybocki MA (2004) The MMSR bilingual and cross channel corpora for speaker recognition research and evaluation. Proc. of the Speaker and Language Recognition Workshop, Odyssey'04, Toledo, Spain, pp 29–32
18. Campbell JP, Shen W, Campbell WM, Schwartz R, Bonastre J-F, Mastrouf D (2009) Forensic speaker recognition: a need for caution. IEEE Signal Process Mag 26(2):95–103
19. Carmen P-M, Ascension G-A, Fernando D-M (2001) Recognizing voice over IP: a robust front-end for speech recognition on the world wide web. IEEE Trans Multimedia 3(2):209–218
20. Carmona JL, Peinado AM, Pe'rez-Cordoba JL, Gomez AM, Sanchez V (2007) iLBC-based tansparametrization: a real alternative to DSR for speech recognition over packet networks. ICASSP, pp 961–964
21. Cerf VG, Kahn RE (1974) A protocol for packet network interconnection. IEEE Trans Commun 22(5):637–648
22. Chen SH, Wang HC (2004) Improvement of speaker recognition by combining residual and prosodic features with acoustic features. Proc IEEE Int Conf Acoustics, Speech and Signal Processing, ICASSP'04, Montreal, Canada
23. Chua TK, Pheanics DC (2006) QoS evaluation of sender-based loss-recovery techniques for VoIP, Nov--Dec 2006. IEEE Network, pp 14–21,
24. Davidson J, Peters J (2000) Voice over IP fundamentals. Cisco Press
25. Davis SB, Mermelstein P (1980) Comparison of parametric representations for monosyllabic word recognition in continuously spoken sentences. IEEE Trans Acoust Speech Signal Process ASSP 28(4):357–366
26. Deriche M, Ning D (2006) A novel audio coding scheme using warped linear prediction model and the discrete wavelet transform. IEEE Trans Audio Speech Lang Proc 14(6):2039–2048
27. DETware: DET curve-plotting software for use with MATLAB, http://www.itl.nist.gov/iad/mig//tools/
28. Dunn RB, Quatieri TF, Reynolds DA (2001) Speaker recognition from coded speech in matched and mismatched conditions. Proc. Speaker Recognition Workshop, 1, Grete, Greece, pp 115–120

29. Doddington GR (1985) Speaker recognition-identifying people by their voices. Proc IEEE 73:1651–1664
30. Doddington GR, Przybocki MA, Martin AF, Reynolds DA (2000) The NIST speaker recognition evaluation—overview, methodology systems, results, perspective. Speech Commun 31:225–254
31. Duda RO, Hart PE, Stork DG (2001) Pattern classification and scene analysis, 2nd edition. Wiley, New York
32. Eriksson A (2010) The disguised voice: imitating accents or speech styles and impersonating individuals. In: Llamas C, Watt D (eds) Language and identities. Edinburgh University Press, Edinburgh, pp 86–96
33. Endres W, Bambach W, Flösser G (1971) Voice spectrograms as a function of age, voice disguise, and voice imitation. J Acoust Soc Am 49:1842–1848
34. Fant G (1970) Acoustic theory of speech production. Mouton, The Hague
35. Flanagan JL (1972) Speech analysis, synthesis and perception. Springer, Berlin
36. FEXT: open source program sox, http://sox.sourceforge.net/
37. Gallardo-Antolin A, Pelaez-Moreno C, Diaz-De-Maria F (2005) Recognizing GSM digital speech. IEEE Trans Speech Audio Proc 13:1186–1205
38. Gish H, Schmidt M (1994) Text-independent speaker identification. IEEE Signal Process Mag 11:18–32
39. G′omez AM, Peinado AM, S′anchez V, Rubio AJ (2006) Recognition of coded speech transmitted over wireless channels. IEEE Trans Wireless Commun 5(9):2555–2562
40. Hair GD, Rekieta TW (1972) Mimic resistance of speaker verification using phoneme spectra. J Acoust Soc Am 51:131(A)
41. Harma A, Laine U (2001) A comparison of warped and conventional linear prediction coding. IEEE Trans Speech Audio Process 9(4):579–588
42. Harwell K, Scheets G, Weber J, Teague K (2009) A multilanguage study of the quality of interleaved MELP voice traffic over a lossy network. IEEE Signal Process Lett 16(7):565–568
43. Hassan M, Nayandoro A (2000) Internet telephony: services, technical challenges, and products. IEEE Commun Mag 38:96–103
44. Huerta JM, Stern RM (1998) Speech compression from GSM coder parameters. Proc Int Conf Spoken Lang Proc, ICSLP-98, vol 4, pp 1463–1466
45. Hyarinen A, Karhunen J, Oja E (2001) Independent component analysis. Wiley, New York
46. Ion V, Reinhold H-U (2008) A novel uncertainty decoding rule with applications to transmission error robust speech recognition. IEEE Trans Audio Speech Lang Proc 16(5):1047–1060
47. Kersta LG (1962) Voiceprint identification. Nature 196(4861):1253–1257
48. Koenig BE (1986) Spectrographic voice identification: a forensic survey. J Acoust Soc Am 79:2088–2090
49. Kostas TJ, Borella MS, Sidhu I, Schuster GM, Grabiec J, Mahler J (1998) Real-time voice over packet-switched networks. IEEE Network 12:18–27
50. Kuhlmann M, Sapatnekar S, Parhi KK (1999) Efficient crosstalk estimation. Proc of 1999 IEEE Int Conf on Computer Design, Austin
51. Lansky P, Steiglitz K (1981) Synthesis of timbral families by warped linear prediction. Comput Music J 5(3):45–49
52. Lee J, Lee YW, O'Clock G, Zhu X, Parhi K, Warwick W (2009) Induced respiratory system modeling by high frequency chest compression using lumped system identification method. Proc of 2009 IEEE Engineering in Medicine and Biology Society Conference, Minneapolis, MN, Sept 2009
53. Linguistic Data Consortium. http://www.ldc.upenn.edu/
54. Maheswari K, Punithavalli M (2010) Enhanced packet loss recovery in voice multiplex-multicast based VoIP networks. A2CWiC '10 Proceedings of the 1st Amrita ACM-W Celebration on Women in Computing in India
55. Martin A, Przybocki M (2000) The NIST 1999 speaker recognition evaluation—an overview. Digital Signal Process 10(1–3):1–18

56. Martin AF, Przybocki MA (2001) The NIST speaker recognition evaluations: 1996–2001. A Speaker Odyssey, A Speaker Recognition Workshop, Dec 2001
57. Martin AF, Doddington G, Kamm T, Ordowski M, Przybocki M (1997) The DET curve in assessment of detection task performance, vol 4. Proc Eurospeech'97, Rhodes, Greece, pp 1899–1903, Sept 1997
58. McCree AV, Barnwell TP III (1995) A mixed excitation LPC vocoder model for low bit rate speech coding. IEEE Trans Speech Audio Proc 3:242–250
59. McCree A, Truong K, George EB, Barnwell TP, Vzswanathanl V (1996) A 2.4 kbit/s MELP coder candidate for the new US. Federal standard, Proc Int Conf Acoust Speech Signal, ICASSP, pp 200–203
60. Mehta P, Udani S (2001) Voice over IP. IEEE Potentials 20:36–40
61. Merazka F (2008) Improved packet loss recovery using interleaving for CELP-type speech coders in packet networks. IAENG Int J Comput Sci 36:1, IJCS_36_1_08
62. Nakasone H (2003) Automated speaker recognition in real world conditions: controlling the uncontrollable. Proc Eurospeech
63. Nakasone H, Beck SD (2001) Forensic automatic speaker recognition. A speaker Odyssey-the speaker recognition workshop, Crete, Greece, 18–22 June, 2001
64. Nolan F, Oh T (1996) Identical twins, different voices. Forensic Linguist 3(1):39–49
65. Nolan JF (1983) The phonetic bases of speaker recognition. Cambridge University Press, Cambridge
66. Open source SIP stack and media stack for presence, instant messaging, and multimedia communication, http://www.pjsip.org
67. Oppenheim AV (1964) Superposition in a class of nonlinear systems. Ph.D. Dissertation, MIT, USA
68. Oppenheim AV, Schafer RW (1989) Discrete-time signal processing. Prentice-Hall, Englewood Cliffs
69. Ortega-Garcia J, Bigun J, Reynolds DA, Gonzalez-Rodriguez J (2004) Authentication gets personal with biometrics. IEEE Signal Process Mag 21(2):50–62
70. O'Shaughnessy D (2001) Speech communications: human and machine, 2nd edition. Universities Press
71. Parhi KK (2004) VLSI digital signal processing systems design and implementation. Wiley, New York
72. Patil HA (2005) Speaker recognition in Indian languages: a feature based approach. Ph.D. Thesis, Department of Electrical Engineering, IIT Kharagpur, India
73. Patil HA (2009) Infant identification from their cry. 7th Int Conf Advances in Pattern Recognition, ICAPR, ISI Kolkata, IEEE Comput Soc, 4–6 Feb 2009, pp 107–109
74. Patil HA, Basu TK (2007) Advances in speaker recognition: a feature based approach. Proc Int Conf Artificial Intelligence and Pattern Recognition, AIPR, Orlando, 9–12 July 2007, pp 528–537
75. Patil HA, Basu TK (2008) LP spectra vs. Mel spectra for identification of professional mimics in Indian languages. Int J Speech Technol 11(1):1–16
76. Patil HA, Basu TK (2008) A novel approach to language identification using modified polynomial networks. In: Prasad B, Prasanna SRM (Eds) Speech, audio, image and biomedical signal processing using neural networks, studies in computational intelligence, vol 83. Springer, pp 117–144
77. Patil HA, Basu TK (2009) A novel modified polynomial networks design for dialect recognition. 7th Int Conf Advances in Pattern Recognition, ICAPR, ISI Kolkata, IEEE Computer Society 4–6 Feb 2009, pp 175–178
78. Patil HA, Parhi KK (2009) Variable length Teager energy based Mel cepstral features for identification of twins. In: Chaudhury S et al. (eds) PReMI 2009, vol 5909, LNCS, Springer, pp 525–530
79. Patil HA, Parhi KK (2010) Novel variable length Teager energy based features for person recognition from their hum. In: Proc Int Conf Acoust, Speech and Signal Proc, ICASSP 2010, Dallas, 14–19 March 2010

80. Patil HA, Dutta PK, Basu TK (2006) Effectiveness of LP based features for identification of professional, mimics in Indian languages. Int Workshop on Multimodal User Authentication, MMUA06, Toulouse, France, 11–12 May 2006
81. Patil HA, Dutta PK, Basu TK (2006) On the investigation of spectral resolution problem for identification of female speakers in Bengali. Special session on person authentication: voice and other biometrics, IEEE Int Conf on Industrial Tech, IEEE ICIT'06, Mumbai, 15–17 Dec 2006
82. Patil HA, Sitaram S, Sharma E (2009) DA-IICT cross-lingual and multilingual corpora for speaker recognition. 7th Int Conf Advances in Pattern Recognition, ICAPR, ISI Kolkata, IEEE Computer Society, 4–6 Feb 2009, pp 187–190
83. Perceptual Evaluation of Speech Quality (PESQ) 2001, ITU-T, Recommendation.P.862.
84. Perrot P, Aversano G, Blouet R, Charbit M, Chollet G (2005) Voice forgery using ALISP: indexation in a client memory. In: Proc Int Conf Acoust Speech and Signal Process, ICASSP 2005
85. Perrot P, Aversano G, Chollet G (2007) Voice disguise and automatic detection: review and perspectives. In: Stylianou Y, Faundez-Zanuy M, Esposito A (eds) Progress in nonlinear speech processing. Springer, Berlin, pp 101–117
86. Perrot P, Razik J, Chollet G (2009) Vocal forgery in forensic sciences. Proc E Forensics—Adelaïde, Australia
87. Plumpe MD, Quatieri TF, Reynolds DA (1999) Modeling of the glottal flow derivative waveform with application to speaker identification. IEEE Trans Speech Audio Process 7(5):569–585
88. Pols LCW (1977) Spectral analysis and identification of Dutch vowels in monosyllabic words. Ph.D. thesis, Free University of Amsterdam
89. Porwal G, Patil HA, Basu TK (2004) Effect of GSM-FR coding standard on performance of text-independent speaker identification. Int Conf on Advanced Computing and Communications, ADCOM04, 13–15 Dec 2004
90. Porwal G, Patil HA, Basu TK (2005) Effect of speech coding on text-independent speaker identification. Int Conf on Intelligent Sensing and Information Processing, ICISIP0, 4–7 Jan 2005, pp 415–420
91. Prybocki MA, Martin AF, Le AN (2007) NIST speaker recognition evaluations utilizing the mixer corpora—2004, 2005, 2006. IEEE Trans Audio Speech Lang Process 15(7):1951–1959
92. Quatieri TF (2002) Discrete-time speech signal processing: principles and practices. Pearson Education
93. Quatieri TF, Singer E, Dunn RB, Rynolds DA, Campbell JP (1999) Speaker and language recognition using speech codec parameters, vol 2. Proc Eurospeech99, pp 787–790
94. Quatieri TF, Dunn RB, Reynolds DA, Campbell JP, Singer E (2000) Speaker recognition using G. 729 speech codec parameters. Proc Int Conf Acoust Speech and Signal Process, vol 2, ICASSP'00, pp 1089–1092
95. Rabiner LR, Schafer RW (1978) Digital processing of speech signals. Prentice-Hall, Englewood Cliffs
96. Ranganathan MK, Kilmartin L (2005) Neural and fuzzy computation techniques for playout delay adaptation in VoIP networks. IEEE Trans Neural Networks 16(5):1174–1194
97. Reynolds DA (1992) A Gaussian mixture modeling approach to text-independent speaker identification. Ph.D. Dissertation, Georgia Institute of Technology
98. Reynolds DA (1994) Experimental evaluation of features for robust speaker identification. IEEE Trans Speech Audio Process 2:639–643
99. Reynolds DA (1995) Large population speaker identification using clean and telephone speech. IEEE Signal Process Lett 2(3)46–48
100. Reynolds DA, Rose RC (1995) Robust text-independent speaker identification using Gaussian mixture models. IEEE Trans Speech Audio Process 3(1):72–83
101. Reynolds DA, Andrews W, Campbell J, Navratil J, Peskin B, Adami A, Jin Q, Klusacek D, Abramson J, Mihaescu R, Godfrey J, Jones D, Xiang B (2003) The SuperSID project:

exploiting high-level information for high-accuracy speaker recognition. Proc Int Conf Acoustics, Speech, and Signal Processing, 06–10 Apr 2003, ICASSP'03, Hong Kong, pp IV:784–787
102. Rose P (2002) Forensic speaker identification. Taylor and Francis, London
103. Sat B, Wah BW (2006) Analysis and evaluation of the Skype and Google-talk VoIP systems. ICME, pp 2153–2156
104. Sambur MR (1975) Selection of acoustic features for speaker identification. IEEE Trans Acoust Speech Signal Process ASSP-23;176–182
105. Schafer RW (1968) Echo removal by discrete generalized filtering. Ph.D. Dissertation, MIT, USA
106. Schwartz R (2006) Voiceprint in the United States—why they won't go away. Proc Int Association for Forensic Phonetics and Acoustics, Sweden
107. Soong FK, Rosenberg AE, Juang B-H (1987) A vector quantization approach to speaker recognition. AT&T Tech J 66(2):14–26
108. Special Section on Speaker and Language Recognition (2007) IEEE Trans Audio Speech Lang Proc 15(7):2104–2115
109. Stone JV (2004) Independent component analysis: a tutorial introduction. MIT Press, Boston
110. Strube HW (1980) Linear prediction on a warped frequency scale. J Acoust Soc Am 68(4):1071–1076
111. The NS-3 network simulator, http://www.nsnam.org
112. Tosi O (1979) Voice identification: theory and legal applications. University Park Press, Baltimore
113. Wang X, Lin J (2007) Applying speaker recognition over VoIP auditing. Proc of the 6th Int Conf on Machine Learning and Cybernetics, 19–22 Aug 2007, Hong Kong, pp 3577–3581
114. Wasem O, Goodman D, Dvorak C, Page H (1988) The effect of waveform substitution on the quality of PCM packet communications. IEEE Trans Acoust Speech Signal Process 36(3):342–348
115. Weerackody V, Reichl W, Potamianos A (2002) An error-protected speech recognition system for wireless communications. IEEE Trans Wireless Commun 1(2):282–291
116. Wolf JJ (1972) Efficient acoustic parameters for speaker recognition. J Acoust Soc Am 51:2030–2043
117. Yegnanarayana B, Prasanna SRM, Zachariah JM, Gupta ChS (2005) Combining evidence from source, suprasegmental and spectral features for a fixed-text speaker verification system. IEEE Trans Speech Audio Process 13(4):575–582
118. Yoma NB, Busso C, Soto I (2005) Packet-loss modeling in IP networks with state-duration constraints. IEE Proc Commun 152(1):1–5
119. Yoma NB, Molina C, Silva J, Busso C (2006) Modeling, estimating, and compensating low-bit rate coding distortion in speech recognition. IEEE Trans Audio Speech Lang Process 14(1):246–255
120. Yu AT, Wang H-C (1998) A study on the recognition of low-bit-rate encoded speech. Proc Int Conf Spoken Lang Proc ICSLP, pp 38–41
121. Zetterholm E (2005) Voice imitation: a phonetic study of perceptual illusions and acoustic success, PhD Abstract. Int J Speech Lang Law, vol 12, no 1
122. Zetterholm E (2007) Detection of speaker characteristics using voice imitation. In: Speaker classification II lecture notes in computer science, vol 4441, Springer, pp 192–205
123. Zhang C, Hansen JHL (2011) Whisper-island detection based on unsupervised segmentation with entropy based speech feature processing. IEEE Trans Audio Speech Lang Process 19:883–894
124. Zheng N, Lee T, Ching PC (2007) Integration of complementary acoustic features for speaker recognition. IEEE Signal Proc Lett 14(3):181–184
125. Zhu X, Parhi KK (2010) Underdetermined blind source separation based on continuous density Hidden Markov Model. Proc 2010 IEEE Int Conf Acoustics, Speech, and Signal Process, March 2010, Dallas, TX

Chapter 7
Noise Robust Speaker Identification: Using Nonlinear Modeling Techniques

Raghunath S. Holambe and Mangesh S. Deshpande

Abstract Session variability is one of the challenging tasks in forensic speaker identification. This variability in terms of mismatched environments seriously degrades the identification performance. In order to address the problem of environment mismatch due to noise, different types of robust features are discussed in this chapter. In state-of-the art features, the speech production system is modeled as a linear source-filter model. However, this modeling technique neglects some nonlinear aspects of speech production, which carry some speaker-specific information. Furthermore, the state-of-the art features are based on either speech production mechanism or speech perception mechanism. To overcome such limitations of existing features, features derived using non-linear modeling techniques are proposed in the chapter. The proposed features, Teager energy operator based cepstral coefficients (TEOCC) and amplitude-frequency modulation (AM-FM) based '*Q*' features show significant improvement in speaker identification rate in mismatched environments. The performance of these features is evaluated for different types of noise signals in the NOISEX-92 database with clean training and noisy testing environments. The speaker identification rate achieved is 57% using TEOCC features and 97% using AM-FM based '*Q*' features for 0 dB SNR compared to 25.5% using MFCC features, when the signal is corrupted by car engine noise. It is shown that, with the proposed features, speaker identification accuracy can be increased in presence of noise, without any additional pre-processing of the signal to remove noise.

7.1 Introduction

The field of speaker forensics is quite old (four to five decades) and embraces a number of areas. The area in which expert opinion is most frequently sought is that of speaker identification—the question of whether two or more recordings of speech (from a criminal suspect or a known perpetrator) are from the same speaker. Forensic speaker identification is aimed specifically at an application area in which

R. S. Holambe (✉)
Department of Instrumentation Engineering, SGGS Institute of Engineering and Technology, Nanded 431606, Maharashtra, India
e-mail: rsholambe@sggs.ac.in

A. Neustein, H. A. Patil (eds.), *Forensic Speaker Recognition*,
DOI 10.1007/978-1-4614-0263-3_7,

criminal intent occurs. This may involve espionage, blackmail, threats and warnings, suspected terrorist communications, etc. Civil matters, too, may hinge on identifying an unknown speaker, as in cases of harassing phone calls that are recorded.

State-of-the art automatic speaker recognition systems show very good performance in discriminating between voices of speakers under controlled recording conditions. However, the conditions in which recordings are made in investigative activities (e.g. anonymous calls and wire-tapping) cannot be controlled and pose a challenge to automatic speaker recognition. This necessitates a complex procedure, involving auditory and acoustic comparison of both linguistic and non-linguistic features of the speech samples in order to build up a profile of the speaker.

Recent research by Schmidt-Nielsen and Crystal [1] confirms that, while human listeners show tremendous individual variability in performance, on an average they tend to slightly outperform current state-of-the art speaker recognition systems. More importantly, they found that it is especially when conditions deteriorate as a result of differences in transmission channels, the presence of background noise and the like that human listeners are clearly superior to automatic speaker recognition algorithms. It is precisely these conditions that tend to prevail in the forensic context.

The interest in using automatic systems for forensic speaker recognition has increased in the last few years because of the improvement of accuracy in the speaker recognition technology [2] and a more comprehensive study about the role of automatic speaker recognition in forensic science [3]. Fully automatic systems are gradually being introduced in forensic casework on a relatively small scale [4–6]. However, speaker recognition technology has still important challenges to address. Session variability mismatch is one of the important challenges which introduce variation to different utterances of the same speaker, typically due to factors such as transmission channel, speaking style, speaker emotional state, environmental conditions, recording devices, etc. This variability seriously degrades the performance of a system, and its compensation has been the subject of recent research [7, 8]. In fact, this has been a major topic of research in recent speaker recognition evaluations (SRE) conducted by NIST [8].

In order to address the problem of mismatched environments, different types of features are discussed in this chapter. These features make use of nonlinear aspects of speech production model and outperform the most widely accepted mel frequency cepstral coefficient (MFCC) features. To evaluate the results, TIMIT and NOISEX-92 databases are used.

The organization of chapter is as follows. In Sect. 7.2, we first review the limitations of the state-of-the art features. In Sect. 7.3, the importance of different frequency bands for identifying speakers is investigated and a new filter structure is proposed. The Teager energy operator and it's use for noise suppression is discussed in Sect. 7.4. The performance of a speaker identification system under mismatched environments is evaluated using Teager energy operator based cepstral coefficient (TEOCC) features in Sect. 7.5. The AM-FM speaker modeling is introduced in Sect. 7.6 and AM-FM based features are developed in Sect. 7.7. The performance of the AM-FM features is evaluated in Sect. 7.8. In Sect. 7.9, the experimental result analysis is carried out.

7.2 Limitations of State-of-the Art Features for Speaker Recognition

The features like linear prediction cepstral coefficients (LPCC) and MFCC have proved highly successful in robust speech recognition where the primary aim is to extract the linguistic information which is mainly embedded in the first few formants. This information can be obtained from the magnitude spectrum, but the magnitude spectrum typically employed is highly sensitive to changes in speaking conditions such as changing channels and speaking style. These cepstral features were initially introduced for speech recognition and then adopted in speaker recognition [9–11]. However, the purpose of speech recognition is quite different from that of speaker recognition. The former task needs to emphasize linguistic information and suppress speaker individual information, whereas the later task needs more speaker individual information. It shows the need of a different type of features which will contain more speaker-specific information.

The LPCC features can well model the vocal tract by using an all pole model which reflects the vocal tract resonances in the acoustic spectra [12]. Basically it emphasizes the formant structure but ignores some significant details of individuals such as nasal, piriform fossa and other side branches [13]. On the other hand, the MFCC features take auditory nonlinear frequency resolution mechanism into consideration [10, 14]. In MFCC feature representation, the mel frequency scale is used to get high resolution in low frequency region, and low resolution in high frequency region. However, this kind of processing is good for obtaining stable phonetic information, but not suitable for speaker-specific features which are located in high frequency regions. The high frequency end of speech spectra (i.e., 3–8 kHz) has been suggested as robust for speaker recognition because this region of spectrum is less dependent on phonetic information than the lower F1-F3 range and more robust against echoes and hard to mimic [15, 16].

7.2.1 Linear and Nonlinear Aspects of Speech Production Model

Conventional theories of speech production are based on linearization of pressure and volume velocity relations. Furthermore, these variables are assumed constant within a given cross section of the vocal tract, i.e., a one-dimensional planar wave assumption. This linear assumption neglects the influence of any nonacoustic motion of the fluid medium. In the linear model, the output acoustic pressure wave at the lips is due solely to energy from injection of air mass at the glottis. It is known that, in this process, only a small fraction of the kinetic energy, in the flow at the glottis, is converted to acoustic energy propagated by compression and rarefaction waves [17]. The vocal tract acts as a passive acoustic filter, selectively amplifying some bands while attenuating others.

This linear, one-dimensional acoustic model is too tightly constrained to accurately model many characteristics of vocal tract. The LPCC and MFCC features are also based on this linear speech production model and assume that the airflow propagates in the vocal tract as a linear plane wave. However, there is an increasing collection of evidence suggesting that, non-acoustic fluid motion can significantly influence the sound field. For example, measurements by Teager [18] reveal the presence of separated flow within the vocal tract. The separated flow occurs when a region of fast moving fluid, -a jet- detaches from regions of relatively stagnant fluid. When this occurs, viscous forces (neglected by linear models) create a tendency for the fluid to 'roll up' into rotational fluid structures commonly referred to as *vortices*. Teager suggested that the presence of traveling vortices (i.e., smoke rings) could result in additional acoustic sources throughout the vocal tract. This contribution of nonlinear excitation sources is something neglected by source-filter theory [19]. Furthermore, the formation of the vortices depends on the physical structure of the vocal tract, which is speaker-specific.

Motivated by the measurements of Teager, Kaiser hypothesized that the interaction of the jet flow and the vortices with the vocal tract cavity is responsible for much of the speech fine structure, particularly at high formant frequencies. Then he proposed the need for time frequency analysis methods with greater resolution than short-time Fourier transform (STFT) for measuring fine structure within a glottal cycle. He further argues that the instantaneous formant frequencies may be more important than the absolute spectral shape.

Above discussion shows that, for simplicity of modeling purposes, speech production system is modeled as a linear system. However, one should not forget that the contribution of nonlinear sources also play an important role and it should be considered while modeling this system.

7.2.2 Importance of High Frequency Components in Speaker Identification

In the past, in many speaker recognition experiments, the frequency band of the speech signal was limited up to 4 kHz assuming that most of the speaker-specific information is contained in the low frequencies [62]. Limiting the frequencies up to 4 kHz is suitable only for speaker identification applications used over public switched telephone network (PSTN) which uses narrowband speech, nominally limited to about 200–3,400 Hz and sampled at a rate of 8 kHz. However recent research has shown that a rich amount of speaker individual information is contained in the high frequency band and is useful for speaker recognition [16, 20–22]. Today,

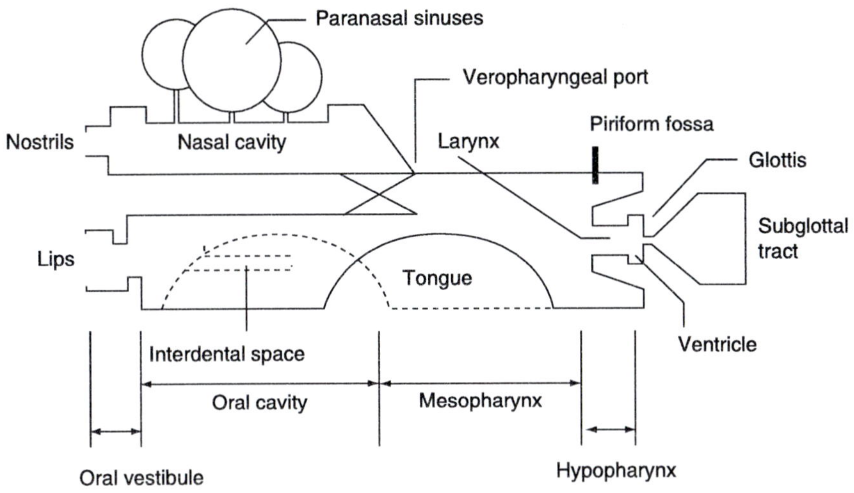

Fig. 7.1 Simplified model of vocal tract [23]

the increasing penetration of end-to-end digital networks such as the second- and third-generation wireless systems (2G and 3G) and voice over packet networks permit the use of wider speech bandwidth.

Traditional feature extraction methods focus on the large spectral peaks caused by the movements of the vocal tract and emphasize on the lower frequency bands. But several acoustic studies have revealed that individual differences are found both in lower and higher frequency regions. For example, the information of the glottis is mainly encoded in a low frequency band (between 100 and 400 Hz), the information of the piriform fossa in a high frequency band (between 4 and 5 kHz), the constriction of the consonants would be another factor in the higher frequency region around 7.5 kHz [13] etc. This kind of distribution of speaker-specific information in different frequency bands was also confirmed in [20].

Figure 7.1 shows the simplified model of the vocal tract with side branches [23]. The vocal tract anatomically divides into four segments: the hypopharyngeal cavities, the mesopharynx, the oral cavity, and the oral vestibule (lip tube). The hypopharyngeal part of the vocal tract consists of the supraglottic laryngeal cavity and the bilateral conical cavities of the piriform fossa. The mesopharynx extends from the aryepiglottic fold to the anterior palatal arch. The oral cavity is the segment from the anterior palatal arch to the incisors. The oral vestibule extends from the incisors to the lip opening [23]. In the nasal cavity, there are a number of paranasal cavities that contribute anti-resonances (zeros) to the transfer function of the vocal tract. Since the nasal cavity has a complicated structure and quite large individual differences, it also provides a lot of speaker-specific information. The piriform fossa is the entrance of the esophagus, and is shaped like twin cone-like cavities on the left and right sides of the larynx. Because of its obscure form and function, the piriform fossa has usually been neglected in many speech production models.

However, introducing the piriform fossa module into the production model causes spectral structure changes in frequency region between 4 and 5 kHz, which can fit the real acoustic speech spectrum well. In addition, the piriform fossa cavities are speaker dependent and less changed during speech production. In [24], Dang and Honda suggested that, piriform fossa should be regarded as one important 'cue' for finding speaker-specific features. Further they have tried to obtain such information using MRI measurements and noted that, the hypopharyngeal resonance, i.e., the resonance of the laryngeal cavity and the antiresonance of the piriform fossa, are more stable than other formants among vowels of each speaker, while they vary to a greater extent from speaker to speaker [25, 26]. Thus the hypopharyngeal cavity also plays an important role to determine individual characteristics.

All above facts motivate to investigate and incorporate the speaker-specific information contained in the high frequency components while extracting the features.

7.2.3 Importance of Phase

The cepstral features (MFCC and LPCC) in one way or another are based on the linear source-filter model of speech production [27, 28]. These cepstral features are computed from the magnitude spectrum of each frame of data and the phase spectra is neglected. The approach of considering only the magnitude and discarding the phase is accepted so far because the conventional analysis-synthesis methods believe that, the phase information is not important. However, the research [29–32] shows that phase information is also important for audio perception. Neglecting the phase spectra, results in a loss of all information in the spectral content changes that occurs within the duration of a single frame. The delta cepstral features, which are designed to capture spectral changes between frames, also fail to capture such changes.

To show the importance of phase, Lindemann et al. [33], compared two synthesized signals. One signal is the sum of harmonically related cosines, which is shown in Fig. 7.2a. It is equivalent to a bandlimited pulse train, which might be used to synthesize the voiced excitation of a linear predictive speech synthesizer. The other signal is a sum of harmonically related cosines with random initial phase. This second signal is shown in Fig. 7.2b. Both of these signals are perfectly periodic and have identically constant magnitude spectra, but it sound different. The random-phase signal sounds more 'active' in the high frequencies with a less pronounced fundamental. The authors shown that, the phase relationships between high frequency sinusoids in a critical band affect the signal envelope and, as a result, the firing rates of inner hair cells associated with the critical band. Therefore sounds with identical magnitude spectra can result in different firing patterns and causes the difference in perception. The approach to overcome the problem of neglecting the phase spectra is to use an AM-FM model [34]. In AM-FM model, the time varying speech signal is represented as a sum of AM and FM signal components. In this

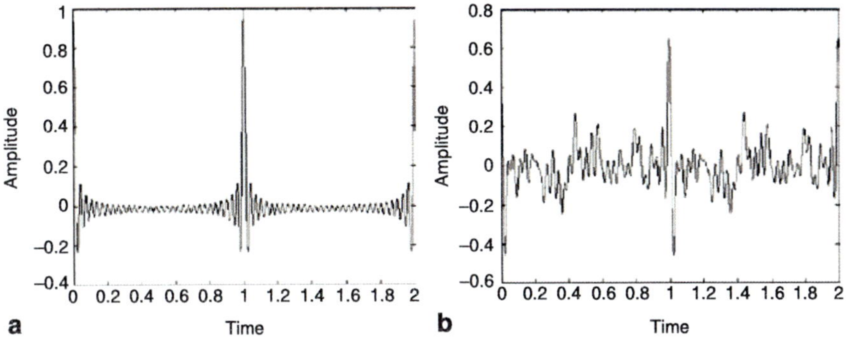

Fig. 7.2 **a** Sum of 32 cosines, **b** sum of 32 cosines with random phase

approach, a signal $x(t)$ is typically modeled as a real signal that has instantaneous amplitude, $A(t)$ and an instantaneous phase $\phi(t)$. It can be expressed as,

$$x(t) = A(t)\cos[\phi(t)] \tag{7.1}$$

where the AM is represented by the magnitude of $A(t)$ and is known as the envelope of the signal. The FM is represented by the derivative of the phase, $\phi(t)$, and is known as the instantaneous frequency of the signal. Numerous methods have been proposed in [35] to estimate the instantaneous amplitudes and frequencies.

In [36], Zeng et al. conducted listening tests using stimuli with different modulations in normal-hearing and cochlear-implant subjects. They found that although AM from a limited number of spectral bands may be sufficient for speech recognition in quiet; FM significantly enhances speech recognition in noise, as well as speaker recognition.

In this section, we have seen the limitations of the state-of-the art features and the necessity of a different type of features which will consider the contribution of nonlinear excitation sources, speaker-specific information in the high frequency bands and the information embedded in the phase. Actually, these are the aspects, neglected by the state-of-the art features.

7.3 Investigating Importance of Different Frequency Bands for Speaker Identification

To investigate the contribution of different frequency regions for speaker identification, we have conducted the following experiments. Qualitatively, these experiments show that the speaker-specific information lies not only in the range of frequencies commonly exploited for speaker and speech recognition (below 4 kHz), but extend to the higher part of the frequency scale, between 4 and 8 kHz. Similar conclusion was also drawn by Mishra et al. in [37] and Lu et al. in [21].

Table 7.1 Speaker identification performance evaluated on different frequency bands

Experiment	Frequency band (kHz)	Identification (%)
1	0–8	100
2	0–4	97.33
	4–8	94.66
3	0–2	54
	2–4	71.34
	4–6	69.34
	6–8	80.67
4	4–5	24
	5–6	36
	6–7	42.66
	7–8	46.67

Table 7.1 shows the speaker identification performance for four different experiments. To evaluate the results, 100 speakers (64 male and 36 female from eight different dialects) from TIMIT database are considered. TIMIT database consists of 630 speakers, 70% male and 30% female from eight different dialect regions in America. The speech was recorded using a high quality microphone at a sampling frequency of 16 kHz. The speech is designed to have rich phonetic contents. It consists of two dialect sentences (SA), 450 phonetically compact sentences (SX) and 1890 phonetically diverse sentences (SI). In the following experiments, training set consists of eight sentences, five SX and three SI (approximately 24 seconds) whereas two SA sentences (approximately 3 seconds each) are used for testing and average identification results are noted. In TIMIT database, the SA, SX and SI sentences are recorded in the same session. The classification engine used in all these experiments is based on a 32 mixtures Gaussian mixture model (GMM) classifier.

Experiment 1 In the feature extraction process, speech signal is first pre-emphasized using a pre-emphasize filter, $H(z)=1-0.97z^{-1}$. The pre-emphasized speech signal is then divided into 32 ms frames with 16 ms overlap. After multiplying with Hamming window, STFT of each frame is obtained.

As the speaker individual information is not distributed uniformly in each frequency band, the mel frequency analysis is not suitable for speaker individual information extraction [13, 21]. Therefore the mel scale warping is not used while extracting the features. To obtain cepstral features, eighty triangle-shaped band pass filters with linear frequency scale were used. Each filter band gives an output which integrates the frequency energy around the center frequency of the filter band. In this experiment, the speech signal used is of 8 kHz bandwidth and the interest lies in investigating speaker-specific information in different frequency regions. Therefore to capture such information in small frequency bands, more number of filters are used. After taking logarithm of energy of each filtered signal, the DCT is applied to get 36 order cepstral coefficient vectors. Figure 7.3a shows the block schematic of the feature extractor. This experiment shows 100% correct identification (as in Table 7.1).

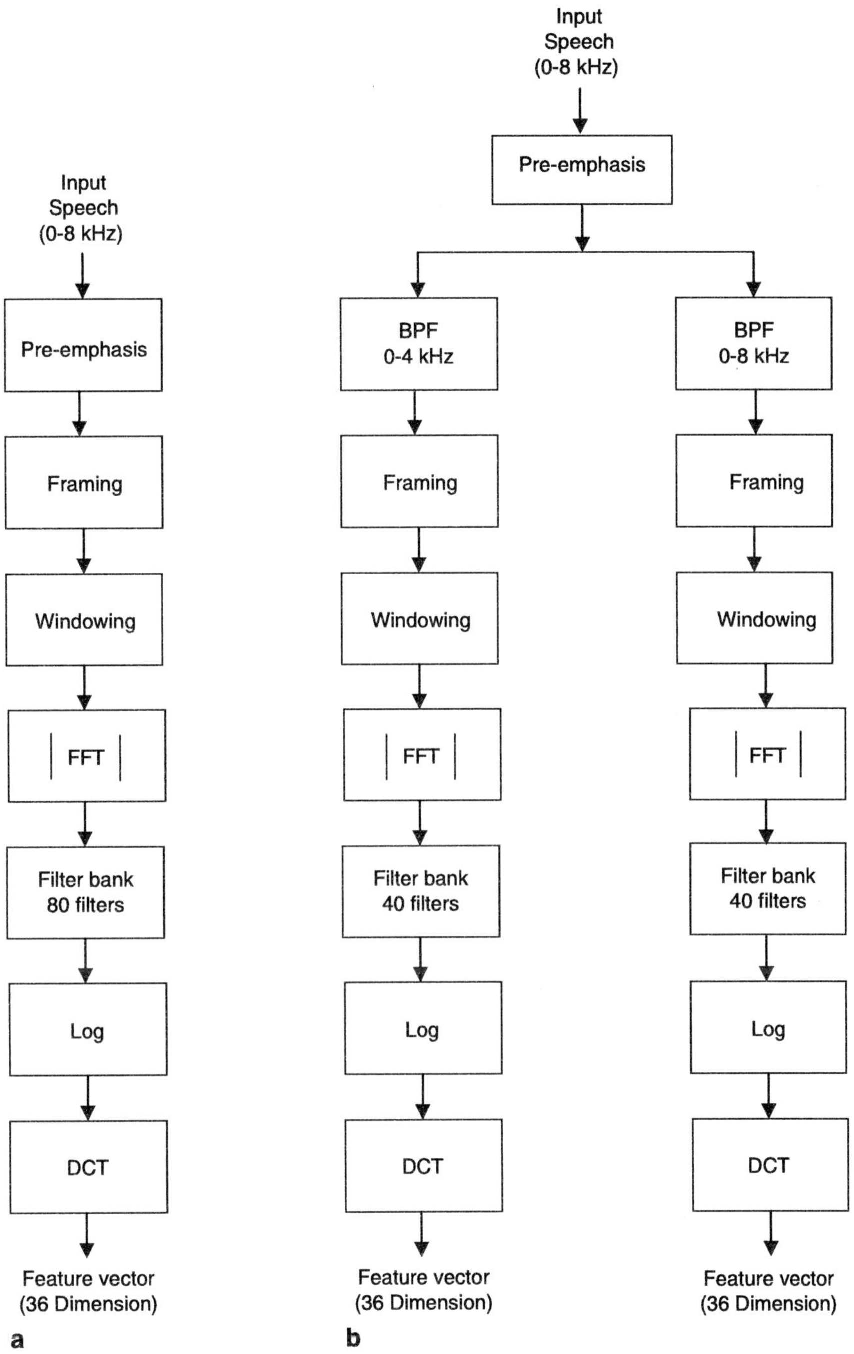

Fig. 7.3 Block schematic of feature extractor used in, **a** Experiment 1 and **b** Experiment 2

Experiment 2 In this experiment, the speaker identification rate is separately obtained for two frequency bands, 0–4 kHz and 4–8 kHz. This experiment is carried out to quantitatively evaluate the importance of the higher part of the frequency axis (frequency range 4–8 kHz). The speech signal is divided into two bands 0–4 kHz and 4–8 kHz by a pair of quadratic mirror filters implemented using 6th order Daubechies' orthogonal filters. Output of each filter is processed separately to obtain 36 dimensions feature vector from each band. The feature extraction procedure is same as in Experiment 1 except the number of filters used. The block schematic of the feature extractor is shown in Fig. 7.3b. In this experiment, a filter bank of 40 triangle shaped linearly spaced filters is used separately for 0–4 kHz and 4–8 kHz bands. Let us consider a band limited signal in the range 0–4 kHz. As maximum frequency component in this signal is 4 kHz, ideally it can be sampled with sampling frequency of 8 kHz. With 8 kHz sampling frequency, 32 ms frame contains only 256 samples (which are 512 in Experiment 1). Therefore it is required to reduce the number of filters used in the filterbank. The identification rate obtained is 97.33% and 94.66% for the frequency bands 0–4 kHz and 4–8 kHz respectively. It shows that both these frequency bands carry speaker-specific information.

Experiment 3 Experiment 2 shows that both, 0–4 kHz as well as 4–8 kHz frequency bands are equally important for speaker identification. Further to evaluate the importance of frequency bands with better resolution, in this experiment, the speech signal is passed through a bank of 4 bandpass filters each with bandwidth equal to 2 kHz with pass bands as 0–2 kHz, 2–4 kHz, 4–6 kHz and 6–8 kHz. The filter bank is implemented using 6th order Daubechies' orthogonal filters. The output of each filter is processed separately to obtain 18 dimensions feature vector from each frequency band. The feature extraction procedure is same as in Experiment 2 except the number of triangular filters used which are 20 in this case. From the identification rate given in Table 7.1, it is clear that speaker-specific information also lies in higher frequency bands, 4–6 kHz and 6–8 kHz.

Experiment 4 The main aim of all above experiments is to investigate the importance of high frequency components in identifying speakers. Experiment 3 shows that 4–6 kHz as well as 6–8 kHz bands are useful for speaker discrimination. Therefore in this experiment, only the higher frequency components, in the range 4–8 kHz are considered. The resolution is further increased by reducing the filter bandwidth to 1 kHz. The identification rate is obtained separately for four frequency bands, 4–5 kHz, 5–6 kHz, 6–7 kHz and 7–8 kHz. Each band pass filtered signal is further processed through a bank of 20 triangle shaped linearly spaced filters, to obtain 18 dimensions feature vector, similar to Experiment 3. The results show that the higher frequency bands 6–7 kHz as well as 7–8 kHz play an important role in speaker discrimination.

There are two different ways to emphasize the contribution of the different frequency bands with more speaker-specific information. The first solution is to assign weighting coefficients for different frequency bands, based on their impor-

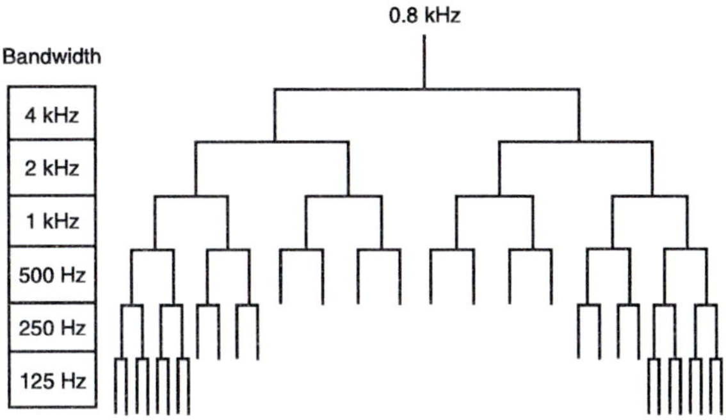

Fig. 7.4 A new fiilterbank structure, which takes into account the low as well as high frequency speaker-specific information

tance in identifying speakers. The second solution is to use non-uniform frequency warping, which applies different frequency resolutions to different frequency regions according to their importance in identifying speakers. The second approach, i.e., fine tuning the bandwidth of the filters is a standard practice found in many approaches. Based on the second approach, a non-uniform sub-band filterbank is designed to change frequency resolutions in different frequency regions. As MFCC features are the dominant features used in most of the state-of-the art speaker recognition systems, the mel filter-like structure (high resolution in low frequency region and low resolution in high frequency region) is used in the 0–4 kHz frequency band. It is implemented using wavelet transform [38]. A 6 level decomposition is applied to an interval of 0–1 kHz; a 5 level decomposition is applied to 1–2 kHz interval and a 4 level decomposition to 2–3 kHz and 3–4 kHz intervals. Experiment 4 shows that 7–8 kHz band carries more speaker related information compared to 4–5 kHz band, and the identification rate increases gradually from 4 to 8 kHz. Based on these results, decomposition of higher frequency bands is further carried out still we get noticeable energy. Typically, 7–8 kHz band is decomposed into eight frequency bands each of bandwidth equal to 125 Hz; 6–7 kHz band into four frequency bands with bandwidth of 250 Hz each; and 4–5 kHz as well as 5–6 kHz bands into two frequency bands with bandwidth of 500 Hz. Figure 7.4 shows this new filterbank structure with $L=32$ filters, which is used in Sect. 7.5.1 for feature extraction.

Many researchers have used discrete wavelet transform (DWT) and wavelet packet transform (WPT) for speech feature extraction [39–45]. They have used either wavelet coefficients with high energy as features or sub-band energies instead of mel filterbank sub-band energies. As MFCC features are the most widely used features, the filter structures proposed in [37, 42–45] follow the mel scale approximately, i.e., these filter structures follows the rule that, generally the frequency resolution is fine in the lower frequency bands while it gets considerably coarser in the higher frequency bands. Therefore, these filter structures are not suitable to capture high frequency speaker-specific information. Whereas, the proposed filter structure takes the advantage of mel scale (for low frequency region, up to 4 kHz) as well as capable to capture high frequency speaker-specific information.

7.4 Use of Teager Energy Operator for Noise Suppression

This section explains the TEO and it's use for noise suppression. The features such as LPCC and MFCC are based on the linear speech production models which assume that the airflow propagates in the vocal tract as a linear plane wave. This pulsatile flow is considered as the source of sound production in [46]. According to Teager [18], this assumption may not hold since the flow is actually separate and concomitant vortices are distributed throughout the vocal is actually the vortex flow interactions, which are nonlinear and a nonlinear model has been suggested based on the energy of airflow. Modeling the time varying vortex flow is a formidable task and Teager devised a simple algorithm which uses a nonlinear energy-tracking operator called as Teager energy operator (TEO). Jankowski used this operator for feature extraction in [47]. Further Jabloun et al. have used the TEO for noise suppression also [48].

7.4.1 *Teager Energy Operator*

In the conventional view of energy, we compute the sum of squared signal elements, i.e., the average of energy density. This means that, tones at 10 Hz and at 1,000 Hz with the same amplitude have the same energy. However, Teager observed that, the energy required to generate the signal at 1,000 Hz is much greater than that at 10 Hz. In an effort to reflect the instantaneous energy of nonlinear vortex-flow interactions, Teager developed an energy operator, with the supporting observation that, hearing is the process of detecting the energy. The simple and elegant form of the operator was introduced by Kaiser [49, 50] as

$$\begin{aligned}\Psi_c[x(t)] &= \left(\frac{d}{dt}x(t)\right)^2 - x(t)\frac{d^2}{dt^2}x(t)\\ &= [\dot{x}(t)]^2 - x(t)\,\ddot{x}(t)\end{aligned} \tag{7.2}$$

where $\Psi_c[x(t)]$ is the continuous time energy operator and $x(t)$ is a single component of the continuous speech signal. To discretize this continuous time operator, replace t with nT (T is the sampling period), $x(t)$ by $x(nT)$ or simply $x[n]$, $\dot{x}(t)$ by its first backward difference, $y[n] = \frac{x[n]-x[n-1]}{T}$ and $\ddot{x}(t)$ by $\frac{y[n]-y[n-1]}{T}$. Then, $\Psi_d(x[n])$, the discrete energy operator (the counterpart of the continuous-time energy operator $\Psi_c[x(t)]$) for discrete-time signal $x[n]$ is defined as,

$$\Psi_d(x[n]) = x^2[n] - x[n-1]x[n+1]. \tag{7.3}$$

An important property of the TEO in discrete time is that, it is nearly instantaneous and only three samples are required in the energy computation at each time instant: $x[n-1], x[n]$ and $x[n+1]$. This excellent time resolution is capable to capture energy fluctuations (in the sense of squared product of amplitude and frequency) within a glottal cycle. Henceforth the discrete time signal is considered and suffix 'd' is dropped in Eq. (7.3).

7.4.2 *Noise Suppression Using TEO*

It has been observed that the TEO could enhance the discrimination between speech and noise and further suppress noise components from noisy speech signals [48, 51]. Compared with traditional frequency domain noise suppression approaches, TEO is easier to implement.

Let $s[n]$ be a discrete-time wide-sense stationary random signal. The expected value of it's Teager energy is,

$$\begin{aligned} E\{\Psi(s[n])\} &= E\{s^2[n]\} - E\{s[n+1]s[n-1]\} \\ &= R_s(0) - R_s(2) \end{aligned} \tag{7.4}$$

where $R_s(.)$ is the autocorrelation function of $s[n]$. It shows that the TEO has filtering capability.

Let us consider an example of a car engine noise, $v[n]$, as shown in Fig. 7.5a. Figure 7.5b shows the Teager energy of $v[n]$ and its PSD is plotted in Fig. 7.5c. The car engine noise is mostly low pass in nature, as shown in Fig. 7.5c. The relation between the first three autocorrelation lags as obtained in [48] is $R_v(1) = 0.9999R_v(0)$, $R_v(2) = 0.9997R_v(0)$. Since $R_v(0) \approx R_v(1) \approx R_v(2)$, we have, $E\{\Psi(v[n])\} \approx 0$. Therefore the spectrum of $E\{\Psi\ (v[n])\}$ is almost negligible compared to the spectrum of $v[n]$, as shown in Fig. 7.5c. Figure 7.5d–f shows the similar plots for babble noise [52].

Now carry the similar analysis for a speech signal. It shows that, the first three autocorrelation lags are not as close to each other. For example, for /s/, the autocorrelation lags are, $R_s(1) = 0.1541R_s(0)$, $R_s(2) = 0.1805R_s(0)$. For the signal /f/, the autocorrelation lags are, $R_s(1) = 0.4936R_s(0)$, $R_s(1) = 0.2520R_s(0)$. For another signal /aa/, the values are, $R_s(1) = 0.9294R_s(0)$ and $R_s(1) = 0.7377R_s(0)$. Similar values for autocorrelation function for lag 0, 1 and 2 are obtained in [48]. Therefore, if we obtain a signal as a combination of speech and noise signals, then it is quite expected that, TEO should suppress noise.

Now, consider a speech signal $s[n]$ degraded by zero mean uncorrelated additive noise $v[n]$ as shown below

$$x[n] = s[n] + v[n] \tag{7.5}$$

The Teager energy of the noisy speech signal $x[n]$ is given by,

$$\Psi(x[n]) = \Psi(s[n]) + \Psi(v[n]) + 2\tilde{\Psi}(s[n], v[n]), \tag{7.6}$$

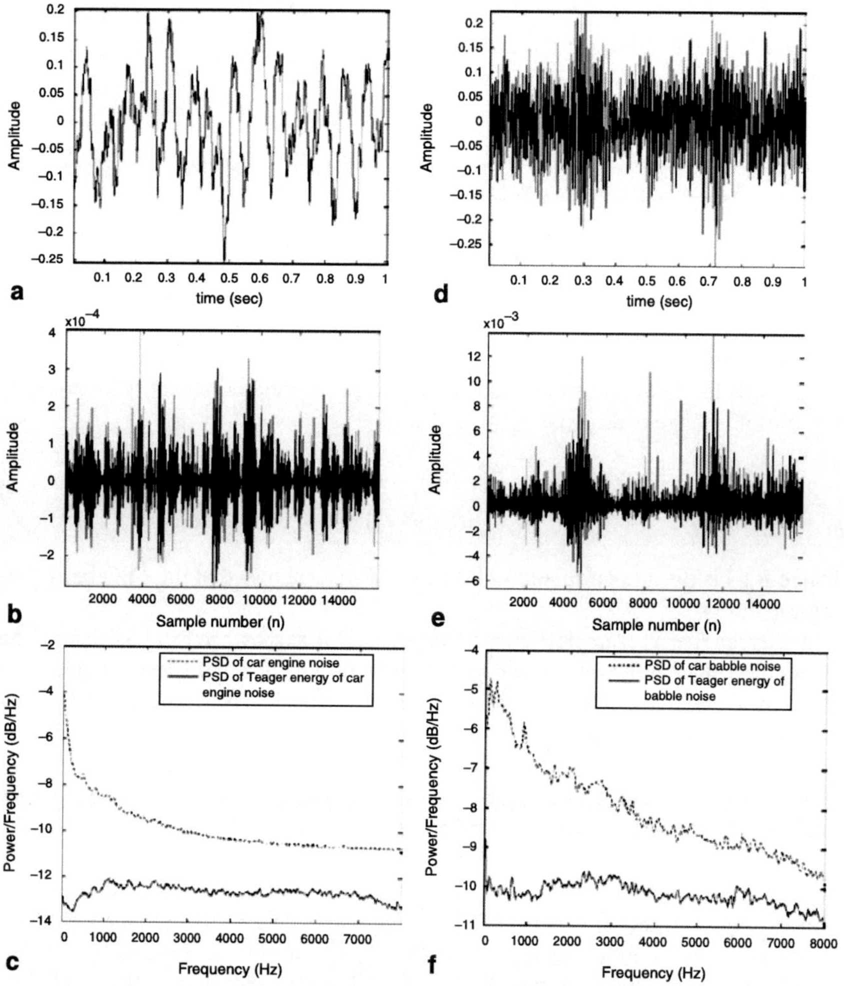

Fig. 7.5 **a** The plot of car engine noise signal, **b** Teager energy profile of car engine noise, **c** the PSD of car engine noise and it's Teager energy, **d** the plot of babble noise signal, **e** Teager energy profile of babble noise, **f** the PSD of babble noise and it's Teager energy

where, $\Psi\,(s[n])$ and $\Psi\,(v[n])$ are the Teager energies of the speech signal and the additive noise, respectively. $\tilde{\Psi}\,(s[n], v[n])$ is the cross- Ψ energy of $s[n]$ and $v[n]$ such that

$$\begin{aligned}\tilde{\Psi}(s[n], v[n]) = s[n]v[n] \\ - (1/2)s[n-1]v[n+1] \\ - (1/2)s[n+1]v[n-1].\end{aligned} \tag{7.7}$$

Since $s[n]$ and $v[n]$ are independent, the expected value of their cross- Ψ energy is zero. Therefore we can write,

$$E\{\Psi(x[n])\} = E\{\Psi(s[n])\} + E\{\Psi(v[n])\} \tag{7.8}$$

Moreover, $E\{\Psi(v[n])\}$ is negligible compared to $E\{\Psi(s[n])\}$. Hence,

$$E\{\Psi(x[n])\} \approx E\{\Psi(s[n])\}. \tag{7.9}$$

Therefore, it is expected that the TEO based features can improve the identification rate than the regular energy based features in the presence of noise.

7.5 Performance Evaluation Using TEO Based Features

This section explains how we can use TEO to extract features. The idea behind using TEO instead of the commonly used instantaneous energy is to take advantage of the modulation energy tracking capability of the TEO. This leads to a better representation of formant information in the feature vector and further as shown in Sect. 7.4.2, it can suppress the noise.

7.5.1 TEOCC Features

The input signal $x[n]$ is divided into 32 subband signals, $x_l[n]$, $l = 1, \ldots, L = 32$ using the filter structure discussed in Sect. 7.3, implemented using 6th order Daubechies' orthogonal filters. For every sub-signal, the average Teager energy, e_l is estimated as,

$$e_l = \frac{1}{N_l} \sum_{n=1}^{N_l} |\Psi[x_l(n)]|; \quad l = 1, \ldots, L \tag{7.10}$$

where N_l is the number of samples in the l th band.

Finally, log compression and IDCT computation is applied to obtain the coefficients as,

$$TC(k) = \sum_{l=1}^{L} \log(e_l) \cos\left[\frac{k(l-0.5)\pi}{L}\right]; \quad k = 1, \ldots, N \tag{7.11}$$

The idea of computing cepstrum of TEO profile of speech signal is also explored by Jankowski in [47]. These coefficients are called as Teager energy operator based cepstral coefficients (TEOCC) [48]. The feature vector consists of first 24 *TC* coefficients.

To evaluate the performance, noise from the NOISEX-92 database [52] has been added to clean speech database (TIMIT) with different SNRs. NOISEX-92 is a

Table 7.2 Speaker identification performance in percent with different SNRs evaluated for MFCC and TEOCC features; test speech is corrupted by car engine noise

Features	Speaker identification rate (%)					
	Clean	SNR=30 dB	SNR=20 dB	SNR=10 dB	SNR=5 dB	SNR=0 dB
MFCC	99	98.75	95	63.25	42.5	25.5
TEOCC	99	98.25	96	86.5	78.5	57

noise database which provides various noise signals recorded in real environments. In many real life applications, the test speech may be corrupted with noise. Car engine noise and bubble noise signals are frequently encountered in the real life. Therefore, in the experiments performed, 'car' and 'babble' noise have been used. The car noise is recorded inside a Volvo-340 on an asphalt road, in rainy conditions. In case of babble noise, the source of the babble is 100 people speaking in a canteen. The room radius is over two meters; therefore, individual voices are slightly audible.

In order to compare the performance of the TEOCC features, MFCC features are used as a baseline. A filter bank with 32 triangular filters is designed according to the mel scale warping to account the 8 kHz speech bandwidth. Then 24 dimension feature vectors are obtained excluding the zeroth coefficient.

7.5.2 *Performance Evaluation for Noisy Speech: Speech Corrupted by Car Engine Noise*

Car engine noise is added to the testing speech utterances to obtain the SNR of 20, 10, 5 and 0 dB. Table 7.2 shows the performance evaluated using MFCC and TE-OCC features for car engine noise. It shows that at high SNR values or with clean speech only, MFCC features work better. When the speech signal is corrupted by car engine noise, which is mostly low pass in nature as shown in Fig. 7.5c; it is the low frequency components of the speech which are affected the most. The MFCC features follow the mel scale, which gives better frequency resolution at low frequencies; hence the speaker identification accuracy is poor using MFCC features in noisy environment. It shows that the TEOCC features perform better than the MFCC features for noisy speech.

7.5.3 *Performance Evaluation for Noisy Speech: Speech Corrupted by Babble Noise*

Similar experiment is performed using babble noise. Table 7.3 shows the results obtained. Compared with the results in Table 7.2, speaker identification is particularly poor for the non-stationary noise like babble noise at 0 dB SNR. At higher SNR values, the TEOCC features work equally well as that of the MFCC features and

Table 7.3 Speaker identification performance in percent with different SNRs evaluated for MFCC and TEOCC features; test speech is corrupted by babble noise

Features	Speaker identification rate (%)				
	SNR=30 dB	SNR=20 dB	SNR=10 dB	SNR=5 dB	SNR=0 dB
MFCC	99	96.25	77	49.25	18.5
TEOCC	98	96.75	83	57.5	22.75

Table 7.4 Speaker identification rate (ID) as a function of TEOCC feature vector dimensions (Dim) at different SNRs

SNR (dB)	Speaker ID (%)	
	24 Dim	32 Dim
30	98.25	98.25
20	96	96.25
10	86.5	91.5
5	78.5	86.5
0	57	76.25

at low SNR values, the identification accuracy is better than the MFCC features. Above experiments show that the TEOCC features are robust and we can achieve better speaker identification rate in noisy environment without additional processing of the signal to remove the noise.

7.5.4 *Effect of Feature Vector Dimensions*

Let us see the effect of feature vector dimensions on the identification accuracy. To observe this effect, 24 as well as 32 dimensions TEOCC feature vectors are obtained. The performance evaluated using car engine noise at different SNRs is shown in Table 7.4. It shows that increasing the feature vector dimensions from 24 to 32, speaker identification performance is improved especially at low SNRs. For 0 dB SNR, the identification rate increases from 57 to 76.25%.

This section shows that, TEO's filtering capability can be effectively used for noise suppression to improve the speaker identification accuracy under mismatched training and testing conditions.

In the next section, we will be discussing about a different approach of speaker modeling, which is based on amplitude and frequency modulation techniques.

7.6 Speaker Modeling Using Amplitude and Frequency Modulation (AM-FM) Based Techniques

AM-FM model is a technique used especially by electrical engineers in the context of frequency modulated signals, such as FM radio signals. It can be effectively used for modeling the speech production system and identifying the speakers [53–56].

7.6.1 AM-FM Model

In speech production system, vocal tract resonances can change rapidly both in frequency and amplitude, even within a single pitch period. This may be due to rapidly varying and separated speech airflow in the vocal tract [35]. The effective air masses in vocal tract cavities and effective cross sectional areas of the airflow vary rapidly, causing modulations of air pressure and volume velocity. This leads to the actual speech signal, $s(t)$ composed of a sum of N resonances as,

$$s(t) = \sum_{i=1}^{N} R_i(t) \tag{7.12}$$

where $R(t)$ is a single speech resonance, which can be represented as an AM-FM signal,

$$R(t) = a(t)\cos\left[2\pi\left(f_c t + \int_0^t q(\tau)d\tau\right) + \theta\right] \tag{7.13}$$

where f_c is the center value of the resonance (formant) frequency, $q(t)$ is the frequency modulating signal and $a(t)$ is the time varying amplitude. The individual resonances can be isolated by band-pass filtering the speech signal. The instantaneous resonance frequency signal is defined as,

$$f_i(t) = f_c + q(t). \tag{7.14}$$

The estimation of the amplitude envelope and instantaneous frequency components, i.e., the demodulation of each resonant signal, can be done with the energy separation algorithm (ESA), or utilizing the Hilbert transform demodulation (HTD) algorithm as described in the next section.

Above discussion shows that, the speech signal (obtained through a speech production system) is of the type of AM-FM signal. Further, from speech perception viewpoint, the hypothesis given by Saberi and Hafter [57] for the measurement of frequency modulation by the auditory system is that the cochlear filters, and perhaps higher level neurophysiological tuning curves, use transduction of frequency modulation (FM) to amplitude modulation (AM); the instantaneous frequency of the FM sweeps through the nonflat passband of the filter, thus inducing a change in the amplitude envelope of the filter output. Psychoacoustic experiments by Saberi and Hafter indicate that FM and AM may be transformed into a common neural code in the brain stem [57]. Therefore, it is quite expected that, if we obtain a feature set based on speech production as well as speech perception mechanism, it will be more robust. This approach of feature extraction is discussed in Sect. 7.7.

7.6.2 Multiband Filtering and Demodulation

Numerous techniques have been proposed in the literature to perform the demodulation [35, 58, 59]. Although the digital energy separation algorithm (DESA) is computationally less expensive, the HTD can give smaller error and smoother frequency estimates [35, 60].

In order to characterize a (single) instantaneous frequency for a real-valued signal, an analytic signal is first constructed. It is a transformation of the real signal into the complex domain via Hilbert transform. More formally, given a speech signal $s(t)$, its analytic signal $s_a(t)$ can be computed as,

$$s_a(t) = s(t) + j\hat{s}(t), \tag{7.15}$$

where $\hat{s}(t)$ is the Hilbert transform of $s(t)$. We can decompose the analytical signal $s_a(t)$ as follows:

$$s_a(t) = a(t)e^{j\phi(t)}, \tag{7.16}$$

where

$$a(t) = |s_a(t)| \tag{7.17}$$

is called the *instantaneous amplitude* (or Hilbert envelope) of the signal, and

$$\phi(t) = \angle s_a(t) \tag{7.18}$$

is the *instantaneous phase*. This phase is in time domain and quite different from Fourier transform phase because, it is derived from analytic signal concept. The *instantaneous frequency* $f(t)$ is computed from the phase $\phi(t)$ as follows:

$$f(t) = \frac{1}{2\pi}\frac{d\phi(t)}{dt}. \tag{7.19}$$

Equations (7.17) and (7.19) show how to obtain instantaneous amplitude and frequency using HTD. The instantaneous frequency estimation is one of the effective methods to detect and track frequency changes of a mono-component signal. But, in the case of multi-component signals, the result becomes meaningless without breaking the signal down into its components [61]. As discussed in [61], the decomposition of a signal is not unique if its frequency components coincide at some points in the time–frequency plane. This is the case for speech, e.g., formants are well known to have points in the time–frequency plane where they appear to join or split. In this case, the decomposition is heuristic in nature and its optimal form will depend on the specific application. In order to estimate the center formant frequency we need to separate that formant from the speech signal using a proper filter. In the optimal case, the bandwidth of this filter does not overlap on the neighboring formants and the filter center frequency is set at the center frequency of the formant.

For speech signal, to obtain a single resonance signal $R(t)$, a filtering scheme can be used before demodulation, which is referred as multiband demodulation analysis

(MDA). The MDA yields rich time-frequency information. It consists of a multiband filtering scheme and a demodulation algorithm. First, the speech signal is band-pass filtered using a filterbank, then each band-pass waveform is demodulated and it's instantaneous amplitude and frequency is computed. In MDA, the filterbank plays an important role. While extracting the features, as discussed in Sect. 7.7, we have used a filterbank consisting of a set of Gabor band-pass filters. Gabor filters are chosen because they are optimally compact and smooth in both the time and frequency domains. This characteristic guarantees accurate amplitude and frequency estimates in the demodulation stage [58] and reduces the incidence of ringing artifacts in the time domain [60].

The following steps are adopted to demodulate the speech signal and to extract the features.

- The speech signal $s(t)$ is band-pass filtered and a set of waveforms, $w_k(t)$ is obtained (k denotes the k th filter in the filterbank).
- For each band-pass waveform $w_k(t)$, its Hilbert transform $\hat{w}_k(t)$ is computed.
- The instantaneous amplitude, $a_k(t)$ for each band-pass waveform is computed as,

$$a_k(t) = \sqrt{w_k^2(t) + \hat{w}_k^2(t)}. \tag{7.20}$$

- The instantaneous frequency, $f_{ik}(t)$ for each band-pass waveform is computed as the first time derivative of the phase $\phi_k(t)$ as,

$$\begin{aligned} f_{ik}(t) &= \frac{1}{2\pi}.\frac{d\phi_k(t)}{dt} \\ &= \frac{1}{2\pi}.\frac{d}{dt}[\arctan(\hat{w}_k(t)/w_k(t))]. \end{aligned} \tag{7.21}$$

After obtaining the instantaneous amplitude and frequency by demodulating each resonant signal, a short-time analysis is performed. The instantaneous amplitude and frequency can also be obtained using the DESA algorithm which is based on TEO, however, here we have used HTD algorithm.

7.6.3 Short-time Estimates

Simple short-time estimates for the frequency F and bandwidth B are the unweighted mean, F_{iu} and standard deviation, B_{iu} of the instantaneous frequency signal $f_i(t)$, i.e.,

$$F_{iu} = \frac{1}{\tau} \int_{t_0}^{t_0+\tau} f_i(t)dt, \tag{7.22}$$

$$[B_{iu}]^2 = \frac{1}{\tau} \int_{t_0}^{t_0+\tau} [f_i(t) - F_{iu}]^2 dt, \tag{7.23}$$

where t_0 and τ are the start (point of reference) and duration of the analysis frame, respectively. Alternative estimates are the first and second weighted moments of $f_i(t)$ [60, 63]. Using the squared amplitude, $a_i^2(t)$ as the weight, the first and second moments are,

$$F_{iw} = \frac{\int_{t_0}^{t_0+\tau} \left[f_i(t).a_i^2(t) \right] dt}{\int_{t_0}^{t_0+\tau} \left[a_i^2(t) \right] dt} \tag{7.24}$$

$$[B_{iw}]^2 = \frac{\int_{t_0}^{t_0+\tau} \left[\dot{a}_i^2(t) + (f_i(t) - F_{iw})^2.a_i^2(t) \right] dt}{\int_{t_0}^{t_0+\tau} \left[a_i^2(t) \right] dt}. \tag{7.25}$$

The adoption of a mean amplitude weighted instantaneous frequency and bandwidth as in Eqs. (7.24) and (7.25) is motivated by the fact that it provides more accurate frequency estimates and is more robust for low energy and noisy frequency bands when compared with an unweighted frequency mean [60, 64].

7.7 AM-FM Technique Based Features

The speech signal is first pre-emphasized using a pre-emphasis filter with transfer function, $H(z) = 1 - 0.97z^{-1}$. The pre-emphasized speech signal is divided into 32 ms frames with 16 ms overlap and multiplied by Hamming window. Then three different experimental set-ups are used to extract the features.

7.7.1 Set-up1: Combining Instantaneous Frequency and Amplitude

As discussed in Sect. 7.6.2, MDA was performed to compute the instantaneous frequency and amplitude of speech resonating signals. To perform MDA, two different filterbanks are used. The first filterbank (uniform) consists of 40 Gabor filters with uniformly spaced center frequencies and constant bandwidth of 200 Hz. The second filterbank (non-uniform) consists of 40 Gabor filters which are non-uniformly spaced and the bandwidth vary according to mel-scale. This filterbank is very similar to the filterbank used in conventional MFCC feature extraction technique. The only difference is, instead of using triangular filters, Gabor filters are used. After obtaining the instantaneous amplitude and frequency using HTD, the short-time mean amplitude weighted instantaneous frequency estimate is obtained using Eq. (7.24).

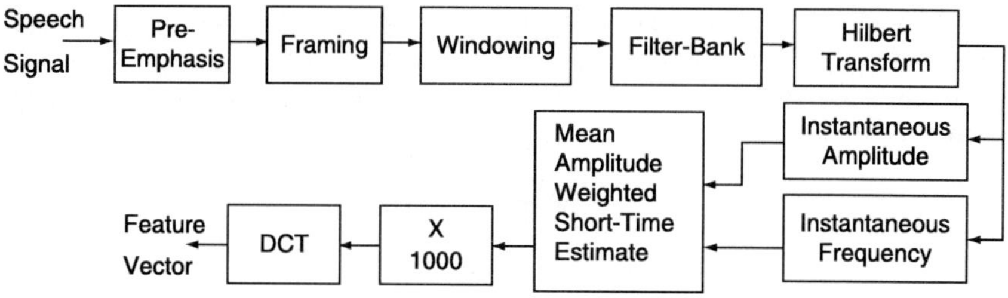

Fig. 7.6 Set-up 1, feature extraction scheme

The estimate of short-time instantaneous frequency is expressed in kilohertz in order to overcome the problem associated with the nodal variances of the GMM. Finally DCT is applied and only first 24 coefficients excluding zeroth coefficient are used to construct a feature vector. The feature vectors obtained using uniform filterbank are referred as F-1 and that of non-uniform filterbank as F-2. This feature extraction scheme is shown in Fig. 7.6.

7.7.2 Set-up 2: Combining Instantaneous Frequency and Bandwidth

In this set-up, only the non-uniform filterbank is used, because it was observed that the non-uniform filterbank shows improved results compared to uniform filterbank. After performing the demodulation using Hilbert transform, the instantaneous amplitude and frequency are combined together to obtain a mean-amplitude weighted short-time estimates, F_{iw} and B_{iw}, of the instantaneous frequency and bandwidth respectively. The estimate of short-time instantaneous frequency and bandwidth are expressed in kilohertz. Considering the basilar membrane as a bank of resonators; the quality factor 'Q' of each of these resonating filters is obtained as,

$$Q_i = \frac{F_{iw}}{B_{iw}}. \tag{7.26}$$

Finally DCT is applied and only first 24 coefficients excluding zeroth coefficient are used to construct a feature vector. These features are referred as F-3.

7.7.3 Combining Instantaneous Frequency, Bandwidth and Post Smoothing

This set-up is similar to set-up 2 except the introduction of post smoothing part. Sometimes, it is observed that, the estimates of instantaneous amplitude and frequency have singularities and spikes. To obtain robust Q features, the demodulation algorithm should provide smooth and accurate estimates. Therefore, a post-processing scheme is applied which employs a median filter with a short window (5-point). These features are referred as F-4. Here the feature extraction scheme is same as in set-up 2 with the addition of a median filter.

7.8 Performance Evaluation Using AM-FM Based Features

In this section we have compared the performance of the features obtained using three different set-ups as described in Sect. 7.7 with that of the most widely used MFCC features. In order to evaluate the performance of the features under mismatched training and testing conditions, the GMM speaker models (with 32 mixtures) are trained using clean speech and noise is added to the test data. Car engine noise and babble noise with SNR of 20, 10, 5 and 0 dB levels are added to the testing speech utterances.

7.8.1 Set-up 1 Performance Evaluation

Table 7.5 shows speaker identification rate in percent for the features obtained using set-up 1 (F-1 and F-2 features) and the MFCC features under mismatched conditions. The test speech signal is corrupted by car engine noise with different SNR values. The results show that speaker identification rate decreases with decreasing SNR for MFCC as well as F-1 features. Whereas the F-2 features show 96.62% average speaker identification rate irrespective of the SNR value. It shows that the F-2 features are the robust features when the noise is a car engine noise. Table 7.6 shows similar results for babble noise. It also shows that, speaker identification rate decreases with decreasing SNR. Further, speaker identification rate is particularly poor for the non-stationary noise like babble noise as compared to the car engine noise at SNR of 0 dB. At higher SNR values, the F-2 features work equally well compared to the MFCC features and at low SNR values, the identification accuracy is better than the MFCC features. It confirms that, the MFCC features are well suited only when the training and testing speech is clean (noise free) and recorded in the same environment. Furthermore, MFCC takes into account only the speech perception mechanism and not the speech production mechanism. Whereas, the features F-2, considerers both speech production (using AM-FM approach) as well as perception (non-uniform filter bank) mechanism, hence more robust compared to MFCC features.

7.8.2 Set-up 2 and 3 Performance Evaluation

Table 7.7 shows speaker identification rate in percent for the features obtained using set-up 2 and set-up 3 (F-3 and F-4 features) and the MFCC features under mis-

Table 7.5 Speaker identification rate obtained with the addition of car engine noise in the test speech utterances at different SNRs using MFCC, F-1 and F-2 features

Features	Speaker identification rate (%)			
	SNR=20 dB	SNR=10 dB	SNR=5 dB	SNR=0 dB
MFCC	95	63.25	42.5	25.5
F-1	86.5	77.75	70.5	57.5
F-2	97	96.75	96.5	96.25

Table 7.6 Speaker identification rate obtained with the addition of babble noise in the test speech utterances at different SNRs using MFCC, F-1 and F-2 features

Features	Speaker identification rate (%)			
	SNR=20 dB	SNR=10 dB	SNR=5 dB	SNR=0 dB
MFCC	96.25	77	49.25	18.5
F-1	89.75	79	59.25	28.5
F-2	96.75	88.25	66.5	37.75

Table 7.7 Speaker identification rate obtained with the addition of car engine noise in the test speech utterances at different SNRs using MFCC, F-3 and F-4 features

Features	Speaker identification rate (%)			
	SNR=20 dB	SNR=10 dB	SNR=5 dB	SNR=0 dB
MFCC	95	63.25	42.5	25.5
F-3	95.5	95	94	93
F-4	98.5	98	98	97

Table 7.8 Speaker identification rate obtained with the addition of babble noise in the test speech utterances at different SNRs using MFCC, F-3 and F-4 features

Features	Speaker identification rate (%)			
	SNR=20 dB	SNR=10 dB	SNR=5 dB	SNR=0 dB
MFCC	96.25	77	49.25	18.5
F-3	93.5	82	63.5	35
F-4	98	92	72.5	39

matched training and testing conditions. The test speech signal is corrupted by car noise with different SNR values. It shows that F-3 as well as F-4 features are more robust in the presence of car engine noise compared to MFCC features. Table 7.8 shows the similar results for babble noise. It also shows that, F-3 and F-4 features outperform the MFCC features. While obtaining F-3 features, the quality factor of each of the filters of the non-uniform filter bank is considered. This filter bank approximately represents the basilar membrane and it is known that, the basilar membrane acts as a tuned filter bank. Further improvement can be seen in the speaker

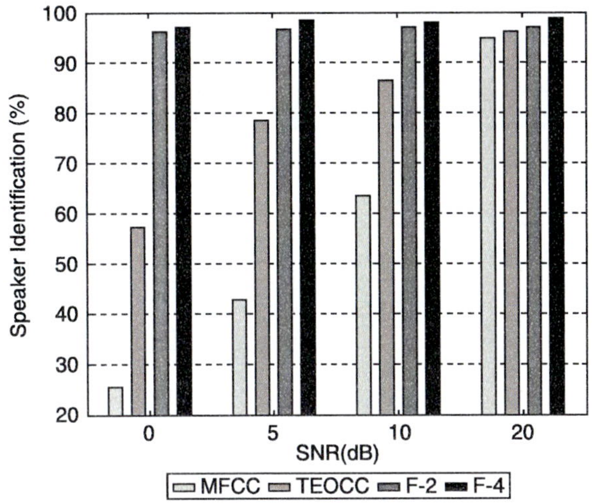

Fig. 7.7 The speaker identification performance obtained using TEOCC features, AM-FM features obtained using non-uniform filterbank (F-2), '*Q*' features obtained using post smoothing technique (F-4) and the MFCC features for test speech signal corrupted by car engine noise

identification rate using F-4 features compared to F-3 features, for both types of noise signals. It shows that the quality factor of each of these filters can be speaker-specific and plays an important role in identifying speakers in noisy environment.

7.9 Experimental Results

The features presented in this chapter can solve some of the problems of speaker identification under mismatched conditions. Figure 7.7 shows the speaker identification performance obtained using TEOCC features, AM-FM features obtained using non-uniform filterbank (F-2), '*Q*' features obtained using post smoothing technique (F-4) and the MFCC features for test speech signal corrupted by car engine noise. Figure 7.8 shows the speaker identification performance obtained using TEOCC, F-2, F-4 and the MFCC features for test speech signal corrupted by babble noise. Both these graphs show that, the TEOCC and AM-FM model based features outperform the MFCC features.

MFCC features are based on the speech perception mechanism, which follow the logarithmic scale. Therefore, the filterbank used in the MFCC resolves low frequency components in better manner compared to high frequency components. However, as discussed in Sect. 7.3, the speaker-specific information lies in high frequency components also and MFCC fails to capture such information. This high frequency speaker-specific information combined with the information obtained from low frequencies using mel scale to derive a new filter structure as discussed in Sect. 7.3. Instead of using the conventional energy definition while extracting the features, the Teager energy shows better speaker identification rate in noisy environment because TEO provides filtering capability to suppress noise. Therefore

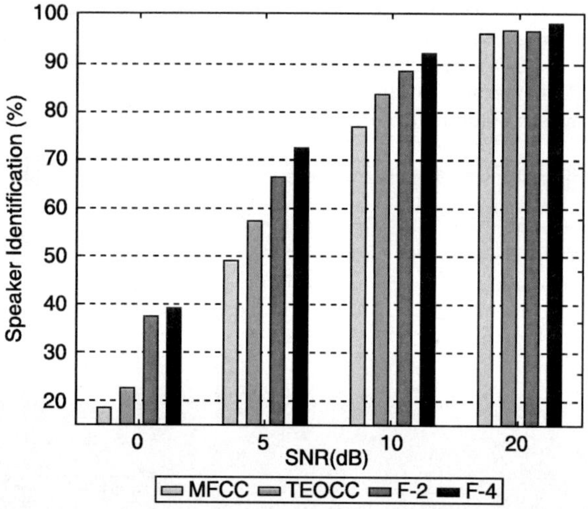

Fig. 7.8 The speaker identification performance obtained using TEOCC features, AM-FM features obtained using non-uniform filterbank (F-2), '*Q*' features obtained using post smoothing technique (F-4) and the MFCC features for test speech signal corrupted by babble noise

speaker identification performance can be improved without additional processing of the noisy speech signal to suppress noise.

Further improvement in speaker identification rate can be seen by using AM-FM modeling technique. If features are derived by considering both, the speech production and speech perception mechanism, better identification rate can be achieved. Therefore the features derived from the quality factor of the resonating filters (F-2 and F-4) show significant improvement in the identification rate at low SNR. Further, it can be seen from these plots that, the proposed AM-FM based features are more robust for stationary noise like car engine noise compared to non-stationary noise (babble noise).

7.10 Summary

In this chapter, we have discussed the nonlinear modeling techniques for noise robust speaker identification. To overcome the limitations of the state-of-the art features like LPCC and MFCC, which are based on the linear source-filter model of speech production system, new feature extraction techniques using TEO and AM-FM model are proposed.

Earlier research was more focused on the low frequency band (0–4 kHz), because of the impact of speech recognition algorithms. However it is investigated that the high frequency region also contains speaker-specific information which is useful for speaker identification. This high frequency component speaker-specific information combined with low frequency component information obtained using mel scale approach improves the speaker identification rate. However, this range is not preserved over the telephone and is not robust to noise, due to weaker speech energy at high frequencies [62]. It is shown that TEO possesses noise suppression capability and its use in feature extraction improves the speaker identification rate in noisy environment without any additional processing of noisy speech signal to suppress noise.

Finally, the AM-FM speaker modeling is discussed. It is shown that the '*Q*' factor of the resonating filters across the basilar membrane is also speaker specific. After knowing the speech production and perception mechanisms, one can derive a robust feature set based on the combination of speech production and perception systems. This feature set will consider the speaker-specific cues from both the production and perception viewpoint and hence can be more robust. It is required to investigate such parameters and some way of fusing these parameters to increase speaker identification accuracy in noisy environments.

References

1. Schmidt-Nielsen A, Crystal TH (1998) Human vs machine speaker identification with telephone speech. Proceedings ICSLP '98, pp 1–4
2. Przybocki MA, Martin AF, Le AN (2007) NIST speaker recognition evaluations utilizing the mixer corpora-2004, 2005, 2006. IEEE Trans Audio Speech Lang Process 15(7):1951–1959
3. Gonzalez-Rodriguez J, Rose P, Ramos D, Doroteo TT, Ortega-Garcia J (2007) Emulating DNA: rigorous quantification of evidential weight in transparent and testable forensic speaker recognition. IEEE Trans Audio Speech Lang Process 15(7):2072–2084
4. Marescal F (1999) The forensic speaker recognition method used by the French Gendrmerie. Internal publication, IRCGN, Paris
5. González-Rodriguez J, Ortega-García J, Lucena-Molina J (2001) On the application of the Bayesian framework to real forensic conditions with GMM-based systems. A Speaker Odyssey, Crete, Greece, pp 135–138
6. Nakasone H, Beck SD (2001) Forensic automatic speaker identification. A Speaker Odyssey, Crete, Greece, pp 139–144
7. Kenny P, Boulianne G, Ouellet P, Dumouchel P (2007) Speaker and session variability in GMM-based speaker verification. IEEE Trans Audio Speech Signal Process 15(4):1448–1460
8. Vogt R, Sridharan S (2007) Explicit modelling of session variability for speaker verification. Comput Speech Lang 22(1):17–38
9. Atal BS (1974) Effectiveness of linear prediction of the speech wave for automatic speaker identification and verification. J Acoust Soc Am 55(6):1304–1312
10. Davis SB, Mermelstein P (1980) Comparison of parametric representations for monosyllabic word recognition in continuously spoken sentences. IEEE Trans Acoust Speech Signal Process ASSP-28(4):357–366
11. Raynolds DA (1994) Experimental evaluation of features for robust speaker identification. IEEE Trans Speech Audio Process 2:639–643
12. Rabiner LR, Juang BH (1993) Fundamentals of speech recognition. Prentice-Hall, New Delhi
13. Lu X, Dang J (2008) An investigation of dependencies between frequency components and speaker characteristics for text-independent speaker identification. Speech Commun 50:312–322
14. Gold B, Morgan N (2002) Speech and audio signal processing. Wiley, New York
15. Shauhnessy DO (2001) Speech communications: human and machine, 2nd edn. University Press, Hyderabad

16. Lin Q, Jan E-E, Che D-S, Flanagan J (1997) Selective use of speech spectrum and a VQGMM method for speaker identification. Proc. European conf. speech communication and technology, pp 2415–2418
17. Quatieri TF (2004) Discrete-time speech signal processing, principles and practice. Pearson Education, Delhi
18. Teager HM (1980) Some observations on oral air flow during phonation. IEEE Trans Speech Audio Process 28(5):599–601
19. Hansen JHL, Liliana G-C, Kaiser JF (1998) Analysis method with application to vocal fold pathology assessment. IEEE Trans Biomed Eng 45(3):300–313
20. Hayakawa S, Itakura F (1994) Text-dependent speaker recognition using the information in the higher frequency band. Proc. IEEE international conference on acoustic speech and signal Processing (ICASSP '94), Adelaide, Australia, pp 137–140
21. Lu X, Dang J (2007) Physiological feature extraction for text independent speaker identification using non-uniform subband processing. Proc. IEEE international conference on acoustic speech and signal processing (ICASSP '07), Adelaide, Australia, IV-461-IV-464
22. Wu J-D, Lin B-F (2009) Speaker identification using discrete wavelet packet transform technique with irregular decomposition. Expert Syst Appl 36:3136–3143
23. Honda K (2008) Physiological processes of speech production. In: Benesty J et al (eds) Springer handbook of speech processing. Springer, Berlin
24. Dang J, Honda K (1997) Acoustic characteristics of the piriform fossa in models and humans. J Acoust Soc Am 101(1):456–465
25. Kitamura T, Takemoto H, Adachi S, Mokhtari P, Honda K (2006) Cyclicity of laryngeal cavity resonance due to vocal fold vibration. J Acoust Soc Am 120(6):2239–2249
26. Dang J, Honda K (1996) An improved vocal tract model of vowel production implementing piriform fossa resonance and transvelar nasal coupling. Proc. ICSLP1996, pp 965–968
27. Rabiner LR, Shafer RW (1989) Digital signal processing of speech signals. Prentice-Hall, Englewood Cliffs
28. Rao A, Kumaresan R (2000) On decomposing speech into modulated components. IEEE Trans Speech Audio Process 8(3):240–254
29. Patterson RD (1987) A pulse ribbon model of monoaural phase perception. J Acoust Soc Am 82(5):1560–1586
30. Paliwal K, Arslan L (2003) Usefulness of phase spectrum in human speech perception. EUROSPEECH '03, Geneva, pp 2117–2120
31. Paliwal K, Alsteris LD (2005) On the usefulness of STFT phase spectrum in human listening tests. Speech Commun 45(2):153–170
32. Alsteris LD, Paliwal K (2006) Further intelligibility results from human listening tests using the short-time phase spectrum. Speech Commun 48(6):727–736
33. Lindemann E, Kates JM (1999) Phase relationships and amplitude envelopes in auditory perception. Proc. IEEE workshop on applications of signal processing to audio and acoustics, New Paltz, New York, pp 17–20
34. Loughlin PJ, Tacer B (1996) On the amplitude and frequency modulation decomposition of signals. J Acoust Soc Am 100(3):1594–1601
35. Maragos P, Kaiser JF, Quatieri TF (1993) Energy separation in signal modulations with application to speech analysis. IEEE Trans Signal Process 41(10):3024–3051
36. Zeng F-G, Nie K, Stickney GS, Kong Y-Y, Vongphoe M, Bhargave A, Wei C, Cao K (2005) Speech recognition with amplitude and frequency modulations. Proc Natl Acad Sci U S A 102(7):2293–2298
37. Mishra H, Ikbal S, Yegnanarayana B (2003) Speaker specific mapping for text-independent speaker recognition. Speech Commun 39:301–310
38. Deshpande MS, Holambe RS (2009) Improving speaker identification in noisy environment. Proceedings of the 4th Indian International Conference on Artificial Intelligence (IICAI-09), Tumkur, Bangalore, pp 1687–1700
39. Farooq O, Datta S (2001) Mel filter like admissible wavelet packet structure for speech recognition. IEEE Signal Process Lett 8(7):196–199

40. Hsieh CT, Lai E, Wang YC (2002) Robust speech features based on wavelet transform with application to speaker identification. IEE Proc Image Signal Process 149(2):108–114
41. Torres MH, Rufiner HL (2002) Automatic speaker identification by means of mel cepstrum, wavelets and wavelets packets. In Proc. IEEE international conference, EMBS, Chicago, IL, pp 978–981
42. Sarikaya R, Pellon BL, Hansen JHL (1998) Wavelet packet transforms features with application to speaker identification. IEEE nordic signal processing symp., pp 81–84
43. Sarikaya R, Hansen JHL (2000) High resolution speech feature parameterization for monophone based stressed speech recognition. IEEE Signal Process Lett 7(7):182–185
44. Patil HA, Dutta PK, Basu TK (2006) The wavelet packet based cepstral features for open set speaker classification in Marathi. In: Spiliopoulou M et al (eds) Studies in classification, data analysis, and knowledge organization. Springer, Berlin, pp 134–141
45. Patil HA, Basu TK (2004) Comparison of subband cepstrum and Mel cepstrum for open set speaker classification. In IEEE INDICON, IIT Kharagpur, pp 35–40
46. Zhau G, Hanscn JHL, Kaiser JF (2001) Non-linear feature based classification of speech under stress. IEEE Trans Speech Audio Process 9:201–216
47. Jankowski CR (1996) Fine structure features for speaker identification. PhD thesis, MIT, USA
48. Jabloun F, Cetin AE, Erzin E (1999) Teager energy based feature parameters for speech recognition in car noise. IEEE Signal Process Lett 6(10):159–261
49. Kaiser JF (1993) Some useful properties of Teagers energy operator. Proc. IEEE int. conf. acoustics, speech, and signal processing, vol 3, pp 149–152
50. Kaiser JF (1990) On a simple algorithm to calculate the energy of a signal. Proc. IEEE Int. Conf. acoustics, speech, and signal processing, Albuquerque, NM, pp 381–384
51. Deshpande MS, Holambe RS (2009) Teager energy operator based robust speaker identification in noisy environment. International conference on VLSI and communication (ICVcom-2009), Kottayam, pp 541-545
52. Noisex-92. (Online) http://www.speech.cs.cmu.edu/comp.speech/Section1/Data/noisex.html
53. Deshpande MS, Holambe RS (2009) Speaker identification based on robust AM-FM features. Proceedings of second IEEE international conference on emerging trends in engineering and technology (ICETET-2009), Nagpur, pp 880–884
54. Deshpande MS, Holambe RS (2009) Robust Q features for speaker identification. Proceedings of IEEE international conference on Advances in Recent Technologies in Communication and computing (ARTCom-2009), Kottayam, Kerala, pp 209–213
55. Grimaldi M, Cummins F (2008) Speaker identification using instantaneous frequencies. IEEE Trans Audio Speech Lang Process 16(6):1097–1111
56. Quatieri TF, Hanna TE, O'Leary GC (1997) AM-FM separation using auditory-motivated filters. IEEE Trans Speech Audio Process 5(5):465–480
57. Saberi K, Hafter ER (1995) A common neural code for frequency and amplitude-modulated sounds. Nature 374:537–539
58. Potamianos A, Maragos P (1994) A comparison of the energy operator and the Hilbert transform approach to signal and speech demodulation. Signal Process 37:95–120
59. Francesco G, Giorgio B, Paolo C, Claudio T (2007) Multicomponent AM-FM representations: an asymptotically exact approach. IEEE Trans Audio Speech Lang Process 15(3):823–837
60. Potamianos A, Maragos P (1996) Speech formant frequency and bandwidth tracking using multiband energy demodulation. J Acoust Soc Am 99(6):3795–3806
61. Boashash B (1992) Estimating and interpreting the instanteneous frequency of a signal—Part 1: fundamentals. Proc IEEE 80(4):519–538
62. Jankowski CR, Quatieri TF, Reynolds DA (1995) Measuring fine structure in speech: application to speaker identification. Proc. IEEE int. conf. acoustics, speech, and signal processing, pp 325–328

63. Potamianos A, Maragos P (1995) Speech formant frequency and bandwidth tracking using multiband energy demodulation. Proc. IEEE int. conf. acoustics, speech, signal processing, pp 784–787
64. Dimitriadis DV, Maragos P, Potamianos A (2005) Robust AM-FM features for speech recognition. IEEE Signal Process Lett 12(9):621–624

Chapter 8
Robust Speaker Recognition in Noisy Environments: Using Dynamics of Speaker-Specific Prosody

Shashidhar G. Koolagudi, K. Sreenivasa Rao, Ramu Reddy, Vuppala Anil Kumar and Saswat Chakrabarti

Abstract In this chapter, we propose speaker-specific prosodic features for improving the performance of speaker recognition in noisy environments. This approach can be especially useful in the forensic analysis of speech. Degradation in speaker recognition is a common phenomenon observed due to transmission and channel impairments, microphone variability and background noise. In this work spectral features are used to perform speaker recognition in the first stage and dynamic aspects of speaker-specific prosody are used to improve the performance in the second stage. For this task, speech corpus is collected at Indian Institute of Technology, Kharagpur, using 50 speakers recorded over the mobile phone. Background noise is simulated using additive white random noise from Noisex database. Speech enhancement techniques are used to improve the speaker recognition performance in the case of noisy speech. Gaussian mixture models (GMMs) and support vector machines (SVMs) are used for developing speaker models. Performance of the speaker recognition system is observed to be 55 and 66% using prosodic and spectral features respectively, for TIMIT speech at 15 dB SNR. The speaker recognition performance of around 73% is achieved using the combination of spectral and prosodic features for noisy speech after speech enhancement.

8.1 Introduction

Introduction section is dealt with two subsections. The first one deals with the introduction to automatic speaker recognition systems and the second one deals with the robustness of the prosodic features toward automatic speaker recognition.

S. G. Koolagudi (✉)
School of Information Technology, Indian Institute of Technology Kharagpur, Kharagpur 721302, West Bengal, India
e-mail: koolagudi@yahoo.com

A. Neustein, H. A. Patil (eds.), *Forensic Speaker Recognition*,
DOI 10.1007/978-1-4614-0263-3_8,

8.1.1 Introduction to Automatic Speaker Recognition

There has been a long desire to be able to identify a person on the basis of his or her voice. For many years, judges, lawyers, detectives, and law enforcement agencies have wanted to use forensic voice authentication to investigate a suspect or to confirm a judgment of guilt or innocence [1, 2]. However, in 2003, several scientific institutions reported on the status of the use of automatic speaker recognition technologies in the forensic field. They concluded by sending a clear need-for-caution message, including statements such as, *currently, it is not possible to completely determine whether the similarity between two recordings is due to the speaker or to other factors* ... [3]. The reason for this is, at present, there is no scientific process that enables one to uniquely characterize a person's voice or to identify with absolute certainty an individual from his or her voice [3]. Identifying a voice using forensic-quality samples is generally a challenging task for automatic, semiautomatic, and human based speaker recognition methods. The speech samples being compared may be recorded in different situations; for instance, one sample could be a yelling over the telephone, whereas the other might be a whisper in an interview room. A speaker could be disguising his or her voice, ill, or under the influence of drugs, alcohol, or stress in one or more of the samples. The speech samples will most likely contain noise, may be very short, and may not contain enough relevant speech material for comparative purposes [4].

Speaker or voice recognition is a biometric modality that uses an individual's voice for recognition purpose. Speaker-specific attributes vary due to the difference in (a) physiological characteristics of speech production organs and (b) learned habits of speaking, acquired through sociocultural aspects and (c) influence of the first native language. Physiological difference in the shape and size of oral cavity, nasal tract, vocal folds and trachea can lead to the differences in vocal tract dynamics and excitation source characteristics [5]. The performance of speaker recognition systems depends upon individual's speech production mechanism, the behavioral characteristics of the speaker, the influence due to recording devices, channel properties and background noise. Present speaker identification systems are primarily based upon modeling short term spectral features. Essentially they capture the speaker-specific information about individual's vocal tract system and produce very good results for clean speech [6], however, they fail to capture idiosyncratic long term pattern in speaker's habitual speaking style, duration and pause patterns, intonation and use of particular phrases [7]. Prosodic features are known to capture the speaker-specific behavioral characteristics, such as speaking rate, speaking style, speech quality, duration and pausing patterns, intonation style, rhythm, melody, loudness and so on. It is also known that prosodic features are less affected by the impairments caused due to channel, microphone and noise properties [8].

The task of automatic speaker recognition is a classical example of pattern recognition problem. It basically contains 2 issues: (a) speaker identification where, system takes speech signal of the speaker as an input and gives the name of the speaker from the list of enrolled speakers. (b) Speaker verification where, system

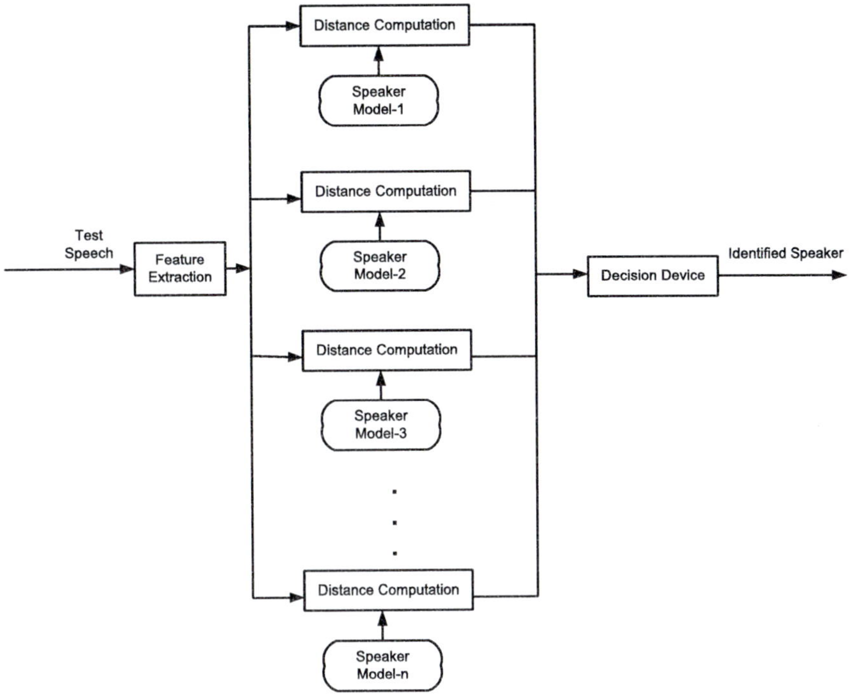

Fig. 8.1 Typical speaker recognition system

approves or disproves the identity claim of the speaker of an input voice. Speaker recognition can be performed using known fixed text, which is called as text dependent or constrained model. On the other hand text independent or unconstrained model performs speaker recognition using any unknown text. Collection of speech samples from different speakers for developing speaker recognition system is known as an enrollment phase. Individual speaker models are developed by using suitable features from the collected speech utterances. After enrollment, during recognition phase, the same features are extracted from the test speech sample and are compared with all the models. The feature vector set from the input voice sample and enrolled models are compared to calculate likelihood ratio, indicating the likelihood that the input sample has come from the claimed or the hypothesized speaker. If the voice input belongs to the identity claimed or hypothesized, then the score reflects the sample to be more similar to the claimed or hypothesized identity's model than to the anti-speaker models. The block diagram of typical speaker recognition system is shown in Fig. 8.1.

The most frequently used speech features for speaker recognition are short time spectral (LPCCs, MFCCs, PLPs) [9–11], prosodic (voice quality, duration of sound units, intensity, pitch, dynamics of these contours) [8, 12, 13] features and their combinations. Hidden Markov models (HMM), Gaussian mixture models (GMM), neural networks (NN) are some of the commonly used classifiers for developing speaker models. Some of the applications of automatic speaker recognition are person authentication through voice (telephone banking, online credit card purchase), voice based access control (accessing a machine or websites), personalization (intelligent answering machine), speech data management (voice mail browsing, speech skimming), general surveillance and forensic applications.

During the early phase of speaker recognition (up to 1960), aural and spectrogram matching techniques were used. Researchers have used template matching till early 1970s. Dynamic time warping and vector quantization were the predominant technologies during mid 1970s to mid 1990s. GMM and HMM based speaker recognition systems were popular till mid of the first decade of this century. Now-a-days systems are being developed using the long term features in combination with the spectral features, using non-linear classifiers such as: support vector machines and artificial neural networks [6, 14]. The evolution in the use of databases is observed from small, clean, controlled speech database to their large, realistic and unconstrained versions. In 1960 Gunnar Fant, a Swedish professor, published a model describing the physiological components of acoustic speech production, based on X-rays captured during production of different sound units [15]. After 10 years, in 1970, Dr Joseph Perkell used motion X-rays, included tongue and jaws, to expand the Fant's model [16]. In the beginning, speaker recognition was done using average outputs of several analog filters [17–19]. Prototype of speaker recognition system was developed in 1976 by Texas Instruments [16]. In 1980s NIST (National Institute of Standard and Technology) started the research in speech tasks, since then NIST speech group has hosted several projects related to speaker recognition. A wide range of voice characteristics have been used in the literature for speaker recognition and verification tasks. These are pitch, intensity, linear prediction coefficients, vocal tract resonances and area functions, spectral parameters, cepstral coefficients, etc. [20–22]. Recently some study has been done to know the contribution of excitation source information toward speaker recognition [23]. Some interesting studies on speaker recognition, chosen from the literature, are briefed out in Table 8.1.

8.1.2 *Robustness of Prosodic Features*

In the literature, duration, pitch and energy parameters are widely used as prosodic features. Figure 8.2 shows these prosodic features for an utterance. The speech signal is taken from a broadcast news data in Telugu language read by a male speaker. The waveform shown in Fig. 8.2a represents the time domain representation of a speech signal, X-axis indicates the timing information and Y-axis indicates the magnitude information. The transcription (Fig. 8.2b) consists of sequence of sound units and their boundaries. This gives the information about the sound units present in a speech signal and their durations. The labels of a sequence of sound units are obtained by listening to a speech signal segment by segment manually. Energy contour (Fig. 8.2c) indicates how energy is distributed in a speech signal and roughly

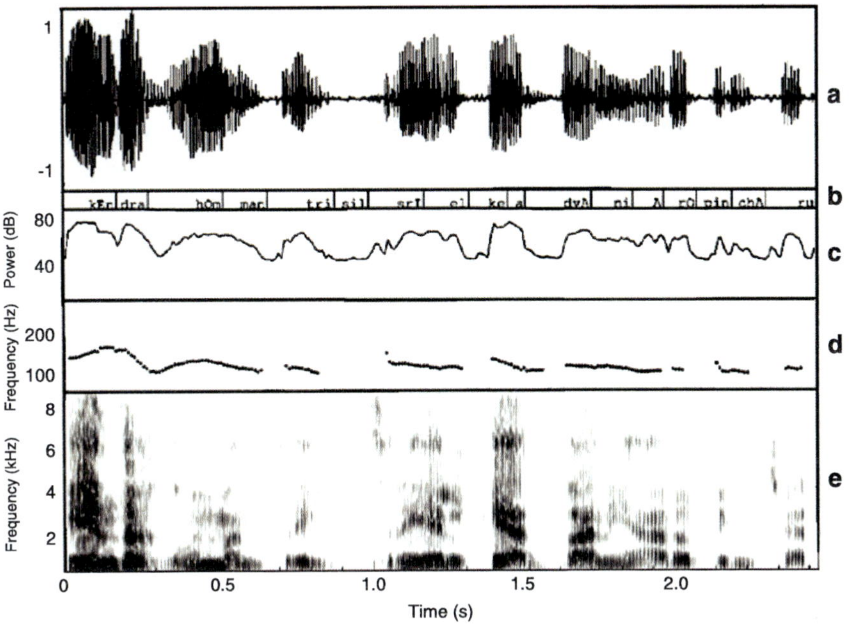

Fig. 8.2 **a** Speech signal, **b** transcription of the utterance ('kEndra hOm mantri srI el ke advAni arOpinchAru'), **c** energy contour, **d** pitch contour and **e** wideband spectrogram

indicates the voiced and unvoiced, silence and non-silence regions. Pitch contour plot (Fig. 8.2d) indicates the global and local intonation patterns associated to a speech signal. Global intonation patterns refers to the characteristics of a whole sentence or phrase. That is a rising intonation pattern at global level indicates that the sentence (phrase) is interrogative in nature and a declining intonation pattern indicates a declarative sentence. Local fall-rise patterns indicate the nature of words and basic units. The spectrogram (Fig. 8.2e) is used to represent the frequency components present in the speech signal. It is a three dimensional representation. X-axis represents the timing information, Y-axis shows the frequency components present in the speech signal and the darkness indicates the energy present in speech signal at that frequency. The dark bands in the spectrogram represents the resonances of a vocal tract system for the given sound unit. These resonances are also called as formant frequencies which represents the high energy portions in the frequency spectrum of a speech signal. The shape of the dark bands indicates, how the vocal tract shape changes from one sound unit to the other. Along with this, speech signal also contains the information about semantics, speaker identity, emotional state of the speaker and language [24].

The effect of channel distortions and background noise on the performance of speaker identification systems is a very serious concern. However it is clear from the published results, studies [12, 25] and the citations given in those published studies, that the prosodic information can be used effectively to improve the per-

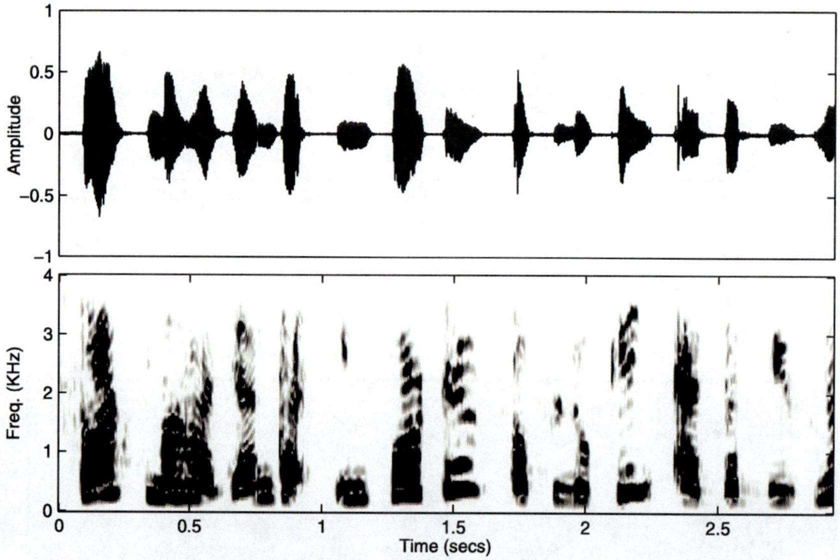

Fig. 8.3 Speech signal and its spectrogram in clean environment

formance level and add robustness to the speaker recognition systems [26, 27]. The vocal tract characteristics of different sound units can be appropriately modeled using spectrum analysis. Hence, the associated parameters such as linear prediction coefficients (LPC) and cepstral coefficients (CC) are used to optimally represent the vocal tract response of the sound units. The prosodic characteristics such as the duration and intonation patterns for the sequence of sound units are difficult to handle automatically by a machine. However, human beings mostly exploit the prosodic characteristics of speech for performing different speech tasks. It is known that prosodic characteristics are robust to different types of degradations, whereas spectral characteristics are sensitive to degradations. The proposed method of speaker recognition using dynamics of speaker prosody may be useful in forensic applications. Presently forensic linguistics or forensic speaker recognition focuses on high-level features that distinguish speakers and relies primarily on a human analyst for processing and interpretation. Most of the speech, available for forensic analysis is degraded due to back ground noise, channel losses and so on. Therefore, the proposed robust speaker recognition system developed using speaker-specific prosody is suitable for forensic applications.

Figures 8.3 and 8.4 show the speech signal and its spectrogram in clean and noisy environments. The speech signal in Fig. 8.3 is recorded in lab environment using a close talking microphone. The speech signal in Fig. 8.4 is obtained by adding additive white Gaussian noise (AWGN) to the speech signal shown in Fig. 8.3. The spectrogram in Fig. 8.4 shows the effect of AWGN on speech signal as a uniform dark cloud. The clear formant structure visible in Fig. 8.3 is completely immersed in the noise (as shown evident from the spectrogram of Fig. 8.4). However, the prosodic features like duration, pitch remain the same as they are not affected by the ex-

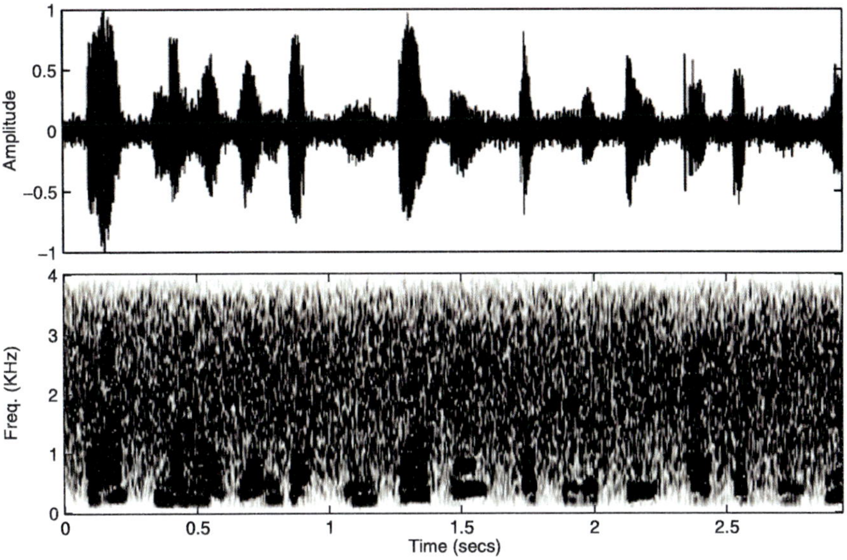

Fig. 8.4 Speech signal and its spectrogram in noisy environment (15 dB SNR)

ternal noise. Hence, the conventional spectral features extracted from the degraded speech will severely affect the performance of speaker recognition systems. Since, the prosodic features are robust against degradation, performance of the systems, developed using these features does not deteriorate much as compared to that of the systems developed using spectral features alone.

Most of the studies, mentioned in the Table 8.1 have used global prosodic features extracted from the entire speech of the particular speaker. Global prosodic features, usually represents the statistical parameters of the frame level prosodic features. For example global prosodic parameters consists of average duration of utterance, speaking rate (# of words spoken per minute), mean, maximum, minimum, median of pitch and energy values. However, these global prosodic features are unable to capture the prosodic variations with respect to time. The ambiguity in discrimination of speaker-specific information due to global prosodic features can be resolved by using dynamics of prosodic features (time varying prosodic features). So far no systematic speaker recognition study is carried out using the dynamics of prosodic features. Therefore, in this work, we have proposed speaker-specific dynamics of prosody for speaker recognition. In this work, speaker-specific dynamic prosody is represented using (1) duration contour (sequence of syllable durations of an utterance), (2) pitch contour (sequence of pitch values) and (3) energy contour (sequence of energy values).

Section 8.2 briefly discusses the design and collection of speech database used in this study. Sect. 8.3 explains the motivation behind this study. Basic principles of GMM and SVM classifiers are discussed in Sect. 8.4. Discussion of experimental setup and analysis of results is done in Sect. 8.5. The chapter concludes with summary and scope for future work.

Table 8.1 Literature review of Speaker recognition

Sl. no.	Features	Databases	Models/classifiers	Contributions and performance (in %)	Reference
Speaker recognition using spectral features					
01	Spectral features	NIST-SRE	GMM-UBM	Score normalization and handset detection techniques have greatly improved the speaker recognition performance	[9]
02	Mel frequency cepstral coefficients (MFCCs)	Authors collected their own database	Dynamic warping	Word recognition studies using two speakers is 96.5% and 95%	[28]
03	MFCCs, LPCCs,LFCCs (Linear frequency spaced filter bank cepstral coefficients) PLPCs (Perceptual linear prediction cepstral coefficients)	Clean, King speech database (conversational telephone speech database)	GMM - maximum likelihood classifier	Performance difference between basic features is small however the major gains are due to channel compensation	[10]
04	33 Inverse filter spectral coefficients (using LPC model) Combination of inverse and normal spectral coefficients	Authors collected their own database (12 Canadian male speakers)		Speaker recognition accuracy is 97.4 Speaker recognition accuracy is 98.32	[11]
Speaker recognition performance using spectral and/or prosodic features (only prosodic features are given below as spectral features are basic in every study)					
05	Pitch and energy contours	NIST-1995 evaluation test	HMM	Mean and variance of pitch periods in voiced region is useful in speaker identification. 30% improvement over 1995 NIST speaker evaluation tests	[12]
06	20 dimensional F_0 contour 20 dimensional F_0 contour and duration	10 Female speakers		97% recognition is achieved 98% recognition is achieved	[29]
07	F_0 and energy trajectories	Switch board-I, NIST extended data evaluation	Dynamic time warping	3.7% EER is achieved	[8]

Table 8.1 (continued)

Sl. no.	Features	Databases	Models/classifiers	Contributions and performance (in %)	Reference
08	19 Prosodic features are extracted from duration-9 (word, phone and segment) pause duration and frequency-5 pitch related features-8	NIST-2001 extended data task	k-nearest neighbor	Pure prosodic features yielded an EER of less than 10%	[30]
09	Syllable level pitch, energy and duration values SNERF-gram features	2-Party, spontaneous English telephone conversation data from the Fisher corpus	SVM	Speaker recognition performance decreases monotonically for pitch duration and energy when used independently	[6]
10	F_0, duration and energy values are extracted from syllable like units, identified using VOPs. Features used: F_0 mean, Change in F_0 distance of F_0 peak WRT VOP, amplitude tilt duration tilt, change in log energy	2003-NIST speaker recognition extended data	AANN (7L-14N-3N-14N-7L)	After combining spectral and prosodic features EER of 9.3 has been achieved	[5]
11	Prosodic and lexical patterns	2003-NIST speaker recognition extended data	GMM	EER of 7.07% (without T-norm) EER of 4.62% (with T-norm)	[7]

8.2 Speech Corpora

In this work, three speech databases are used for analyzing the performance of speaker recognition (SR) system for mobile devices. Database-1 consists of speech utterances of 50 speakers chosen randomly from TIMIT database. Database-2 contains the speech corpus recorded by 50 speakers of Indian Institute of Technology, Kharagpur, using microphone. All the speakers are graduate students of the institution, with the age group of 23–25 years. The duration of speech given by each speaker is about 10 min. Database-3 contains the speech corpus recorded by a mobile phone. For realizing the effects of varying background environment, speech coding and wireless channels, the database-3 is recorded at the receiving end of the mobile phone. Databases 2 and 3 have been recorded simultaneously using microphone and mobile phone receiver, respectively. Database-3 is recorded by a mobile phone receiver, which is at the remote place. The sequence of steps for recording the database-3 is as follows: (1) The connection between speaker's mobile phone and destination mobile phone has to be established, and (2) Speaker's voice will be recorded at the destination mobile phone. For simulating the varying background environment, noisy samples of NOISEX [31] data are played using loud speaker close to microphone/mobile phone. Different SNRs are achieved by shifting the position of loud speaker and varying the volume control. In this work, SR performance is analyzed for speech with 15, 10, 5, 0, −5 dB SNRs. The background noise considered in this work is white random noise [32].

8.3 Motivation

Humans use several prosodic features like pitch, duration, energy, speaking rate and speaking style obtained at suprasegmental level for recognizing speakers. Duration, intonation and intensity patterns of speech are unique to speaker. Therefore, researchers have used these features for speaker recognition along with spectral features. Atal [29] proposed a speaker recognition method using pitch contours. Significance of long-term features like pitch and energy for speaker recognition is discussed in [12]. Statistical features of pitch, pitch tracks and local dynamics in pitch are also used in speaker verification [33]. However, dynamics of prosodic features also plays a crucial role in deciding the speaker's identity. Perceptually it is difficult to identify the speaker, if dynamics of prosody is removed from the utterances. The following study has been conducted to analyze the speaker-specific information present in the prosody dynamics. The subjective listening tests are conducted on the speech utterances, where the dynamics of the prosody, specific to the speaker are removed.

In this study the speech utterances of 5 famous Hindi movie actors (all males) are considered. Five utterance of duration 3 s each, are chosen from every actor, for subjective evaluation. For each of these 25 (5 $uttrances \times 5\ actors$) utterances (s), corresponding 4 speech utterances are generated by (1) replacing all natural

Table 8.2 Speaker recognition performance using subjective listening tests for (i) original sentences (S), (ii) sentences after replacing the dynamics in syllable durations by their average duration (S_{dur}), (iii) sentences after replacing the dynamics in the sequence of pitch values by their average (S_{pit}), (iv) sentences after replacing the dynamics in the sequence of frame energies with their average frame energy (S_{erg}), and (v) sentences after replacing the dynamics in durations of syllables, sequence of pitch values and sequence of frame energies together with their respective average values (S_{dpe})

Speaker class	Speaker recognition performance (%)				
	S	S_{dur}	S_{pit}	S_{erg}	S_{dpe}
Speaker-1	78	43	22	62	20
Speaker-2	80	45	31	65	23
Speaker-3	89	48	35	61	32
Speaker-4	85	51	36	58	41
Speaker-5	62	40	28	55	21

syllable durations with their average value (S_{dur}). (2) Replacing all natural frame wise pitch values with their average value (S_{pit}). (3) Replacing frame wise natural energy values with their average (S_{erg}). (4) Replacing all natural syllable durations, frame wise pitch and energy values with corresponding averages (S_{dpe}). These 5 sets (25 utterances in each set) are separately played to 20 research students with basic knowledge of Hindi movies. They are asked to recognize the speakers. For each speaker there are five sets of speech files. each set contains five sentences. Speaker recognition performance is evaluated with respect to each set. Each set of speech files specific to a speaker are evaluated by all 20 listeners (5 *speech files* × 20 *listeners*), which provides 100 test cases. Correctness of listeners choice is indicated by the recognition performance in Table 8.2.

From the results, it is observed that subjects could recognize speakers well, when original utterances are played. Out of 3 prosodic features, *energy dynamics* contains least speaker-specific information, and hence there is no considerable reduction in the speaker recognition results, even after energy dynamics is removed from the utterances. However, *pitch dynamics* plays important role in recognizing the speaker. In general, listeners are not able to clearly recognize the speaker's voice, when the dynamics of prosody, specific to the speaker are removed. The recognition performance is observed to be very poor, if the dynamics of all prosodic parameters are removed together. Note here that, there is no change in the spectral characteristics of the speech, as only prosodic features are varied in this study.

8.4 Classification Models

In this work, GMMs and SVMs are used for developing the speaker models. GMMs are used for developing the speaker models using spectral features. SVMs are used for developing the speaker models using prosodic features. The following subsections provide the basic philosophy of the classification models used in this study.

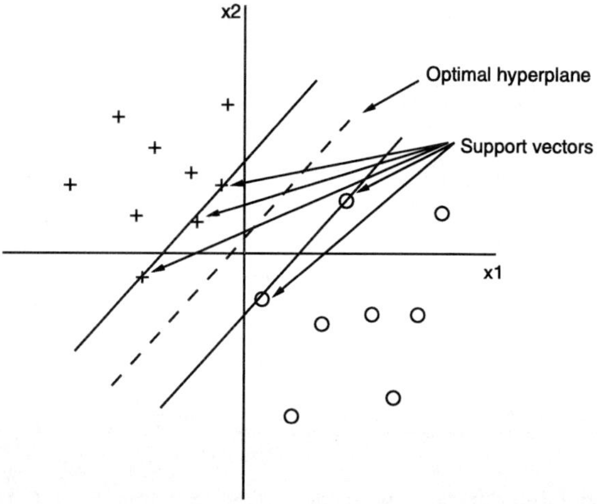

Fig. 8.5 Classification mechanism of support vector machine

8.4.1 Support Vector Machines (SVM)

SVMs are designed for two-class pattern classification. Multi-class (n-class) pattern classification problems can be solved using a combination of binary (2-class) support vector machines. One-against-the-rest approach is used for decomposition of n-class pattern classification problem into n two-class classification problems. The classification system consists of *n* SVMs. The set of training examples $\left\{\{(x_i, k)\}_{i=1}^{N_k}\right\}_{k=1}^{n}$ consists of *Ni*~ number of examples belonging to the *k*th class, where the class label $k \in \{1, 2, 3, \ldots, n.\}$. All the training examples are used to construct the SVM for a class. The SVM for the class *k* is constructed using the set of training examples and their desired outputs, $\left\{\{(x_i, y_i)\}_{i=1}^{N_k}\right\}_{k=1}^{n}$. The desired output y_i for the training example x_i is defined as follows:

$$y_i = \begin{cases} +1 & \text{if } x_i \in k^{th} \text{ class} \\ -1 & \text{otherwise} \end{cases}$$

The examples with $y_i=+1$ are called positive examples, and those with $y_i=-1$ are negative ones. An optimal hyperplane is constructed to separate positive examples from negative ones. The separating hyperplane (margin) is chosen in such a way as to maximize its distance from the closest training examples of different classes. Figure 8.5 illustrates the geometric construction of hyperplane for two dimensional input space. The support vectors are those data points that lie closest to the decision surface, and therefore the most difficult to classify. They have a direct bearing on the optimum location of the decision surface [24]. For a given test pattern x, the evidence $D_k(x)$ is obtained from each of the SVMs [34]. In the decision logic, the class label *k* associated with SVM, which gives maximum evidence is hypothesized as the class (*C*) of the test pattern, that is

$$C(x) = argmax_k(D_k(x))$$

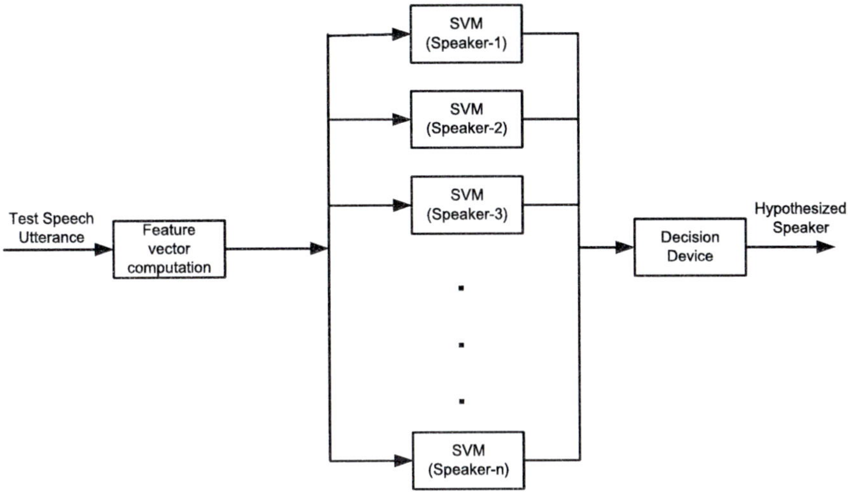

Fig. 8.6 Speaker recognition system using support vector machines

For developing the SVM model for the specific speaker, feature vectors derived from the speech of desired speaker are used as positive examples, and the feature vectors derived from the speech of all other speakers (other than the desired speaker) are used as negative examples. The block diagram of the Speaker Recognition (SR) system using SVM models is shown in Fig. 8.6. For evaluating the performance of the SR system, the feature vectors derived from the test utterances are given as input to all SVM models. The output of the each model is given to decision logic. Decision logic determines the speaker based on the highest score among the evidence provided by the speaker models. Gaussian kernels are used to develop SVM based speaker recognition systems. The parameters like *standard deviation* are determined empirically. Support vector machines (SVMs) are known to capture the discriminative information present among the feature vectors. Performance of SVMs is critically dependent on the number of discriminative feature vectors (known as support vectors) rather than the total number of feature vectors. Trained SVM achieves structural risk minimization by reducing generalized error along with training error. Therefore, SVMs are employed to develop speaker recognition models using prosodic features. In forensic applications, we generally have less number of feature vectors as shorter speech samples are available for analysis. Open source tool *SVM Torch* is used for implementing the SVMs [34–36].

8.4.2 Gaussian mixture models (GMM)

Gaussian Mixture Models (GMMs) are among the most statistically mature methods for clustering (they are also used intensively for density estimation). GMMs are used as a classification models in this task. They belong to the class of pattern recognition systems. They model the probability density function of observed data points using a multivariate Gaussian mixture density. Given a set of inputs, GMM refines the weights of each distribution through expectation- maximization algorithm. Mixture models are the type of density models which comprise number of component functions, usually Gausses. These component functions are combined to provide a multi-modal density. They are able to smooth over gaps resulting from sparse sample data and provide tighter constraints in assigning object membership to cluster regions. Once a model is generated, conditional probabilities can be computed for test patterns (unknown data points). An expectation-maximization (EM) algorithm is used in statistics for finding maximum likelihood estimates of parameters in probabilistic models. Expectation Maximization is an iterative method which alternates between performing an expectation (E) step, which computes an expectation of the log likelihood with respect to the current estimate of the distribution for the latent variables, and a maximization (M) step, which computes the parameters that maximize the expected log likelihood found on the E step. These parameters are then used to determine the distribution of the latent variables in the next E step.

Number of Gausses in the mixture model is known as number of components. They indicate the number of clusters in which data points are to be classified. In this work one GMM is developed to capture the information about one speaker. The components within each GMM capture finer level details among the feature vectors of each speaker. Depending on the number of data points, number of components may be varied in each GMM. Presence of few components in GMM, used for classifying large number of data points may lead to more generalized clusters, failing to capture specific details related to each class. On the other hand over fitting of the data points may happen, if too many components represent few data points. Obviously the complexity of the models increases, if they contain higher number of components. Therefore a trade-off has to be reached between the complexity and the accuracy of the classification results required. In this work, GMMs are designed with 64 components and iterated for 30 times to attain convergence. Gaussian mixture models (GMMs) are known to capture distribution of input data points in a feature space. GMMs especially perform well, when number of input data point is high. In the case of frame wise feature vector extraction, there are sufficient number of feature vectors. Therefore, GMMs are employed to develop speaker recognition models using spectral features. Open source netlab functions are used to implement GMMs [37].

8.5 Performance Evaluation

In this work, 10 minutes of speech data per speaker is considered for analyzing the time varying variations of prosodic parameters for discriminating the speakers. Speaker recognition performance using prosodic and spectral features is studied in isolation as well as in combination. Three types of speech data are used in this study for speaker recognition: (a) TIMIT, (b) 50 speaker microphone data collected

Table 8.3 Performance of the speaker recognition systems developed using spectral features. The entries in the table indicate the percentage of recognition. Noisy: Enrollment-clean speech, Testing-noisy data; Enhanced speech: Enrollment-clean speech, Testing-enhanced speech

Speech SNR (dB)	Speaker recognition performance using spectral features (in %)					
	TIMIT		50 speaker-microphone		50 Speaker-cellphone	
	Noisy	Enhanced speech	Noisy	Enhanced speech	Noisy	Enhanced speech
15	43.97	55.28	32.12	54.26	25.92	41.65
10	28.97	41.03	22.77	34.23	15.60	31.33
5	21.00	29.50	11.29	24.95	9.92	20.26
0	21.23	27.87	20.12	22.32	7.80	19.13
−5	20.23	26.90	12.32	19.35	7.30	14.22

at IIT-KGP by the graduate students, (c) 50 speaker cell phone data collected simultaneously at IIT-KGP by the same graduate students. For simulating noisy data, an additive white random noise, with different intensities has been added to the clean, microphone and cell phone data. Spectral and temporal speech enhancement techniques are used on noisy data, to get enhanced speech.

Eight minutes of speech per speaker is used for developing the models, and 2 minutes of speech per speaker is used for validation. As a baseline system, GMM based speaker recognition systems are developed using spectral features. Twenty one (21) mel frequency cepstral coefficients (MFCC) are extracted from a frame size of 20 ms, using 10 ms of frame shift. This set of feature vectors represents speaker-specific spectral properties. Spectral features are extracted from clean speech and enhanced noisy speech. Noisy speech is simulated by adding white random noise of different intensities (measured in dB). Samples of white random noise are taken from noisex database. Table 8.3 shows the performance of speaker recognition using spectral features. Two types of studies are undertaken on each of the three databases. *'Clean'* indicates the speaker recognition performance obtained by testing noisy speech data on the models developed using clean speech. *'Speech enhanced'* indicates the speaker recognition results of enhanced speech data tested on the models developed using clean speech. The rows in the table indicate the speaker recognition performance obtained for different noise intensities. Noise intensities are shown in the form of SNR (Signal to noise to ratio). There is a clear indication of improvement in the speaker recognition, if the noisy speech is enhanced before speaker recognition is performed. In this work speech enhancement is performed using temporal and spectral level enhancement techniques [32, 38, 39]

Temporal enhancement is done using the following steps: (1) For every voiced region the coherently-added covariance signal is smoothed using a window of first 40 samples of 80 sample hamming window. (2) The smoothed signal is normalized with its running mean. (3) The normalized signal is mapped using a sigmoidal function to obtain the fine weight function. (4) The fine weight function computed for every voiced region is used to modify the LP residual. (5) The modified LP residual is used to re-synthesize the speech enhanced at temporal level.

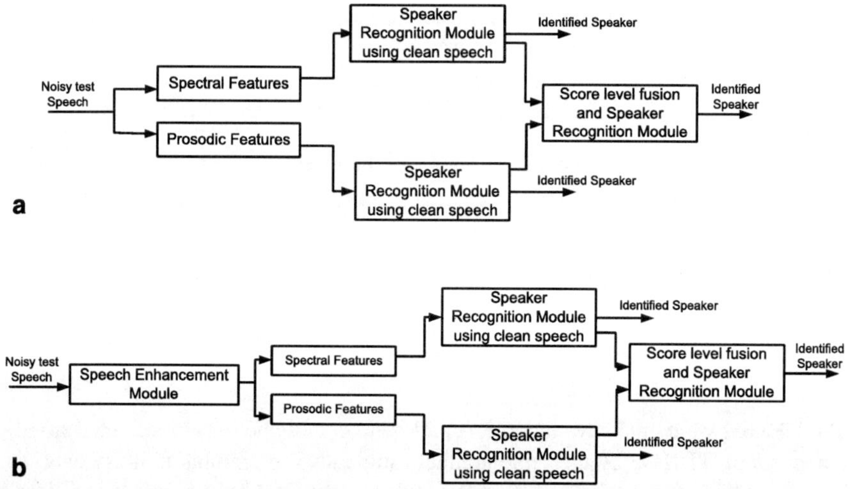

Fig. 8.7 Speaker recognition systems **a** without speech enhancement and **b** with speech enhancement

Spectral level enhancement is done using the following steps: (1) The Autoregressive (AR) modeling using autocorrelation analysis is performed on speech signal enhanced at temporal level to obtain IIR filter coefficients. (2) The poles corresponding to the first three formants are identified. (3) The bandwidths of the poles are modified to give equal emphasis to all the three formants. (4) The set of modified filter coefficients are obtained. (5) The gain of the filter is set to the value which is the ratio of energy of speech frame before filtering operation to energy after filtering operation. (6) The speech enhanced at temporal level is passed through this time varying filter to obtain speech enhanced at spectral level. Speaker recognition systems with and without speech enhancement at the front end are shown in Fig. 8.7.

Usually, long term prosodic features such as, duration, intensity and intonation patterns do not get distorted by normal speech degradation. However, in presence of severe degradations, extraction of these prosodic features may be difficult. In this work, dynamics of individual prosodic parameters are separately explored for discriminating the speakers. Later, we studied the feature and score level fusion techniques for improving the performance. Three speaker recognition systems (SRS) developed using dynamics of individual prosodic features are:

1. SRS-1: SR system using the sequence of durations of syllables within the utterance.
2. SRS-2: SR system using the sequence of fundamental frequency (F_0) values (pitch values) of the utterance.
3. SRS-3: SR system using the sequence of frame level energies of the utterance.

Syllable level durations are used as the parameters representing the duration. Durations of syllables are derived using ergodic hidden Markov models (EHMM). Using

EHMM, in the first step, phone boundaries are obtained by using *forced alignment.* From the phoneme boundaries syllable boundaries are determined using syllabification rules. In this work, we use 3 states to represent phoneme in developing EHMM [40, 41]. Sequence of fifteen consecutive duration values of syllables forms a feature vector representing the duration pattern of the utterance, and it is also viewed as duration contour. For representing the intonation pattern of an utterance, the sequence of pitch values of all voiced frames of the speech is used. The sequence of 100 consecutive pitch values is considered as a feature vector. In this work, pitch contours (sequence of pitch values) are derived by performing the autocorrelation on Hilbert envelope of LP residual signal [42–44]. Similarly for representing energy contour, sequence of voiced frame energies is used. Pitch and energy values are derived from speech signal, using a frame size of 20 ms and a frame shift of 10 ms. For representing the durations of the sequence of syllables, a 15-dimensional feature vector is used. For representing the sequence of pitch values (pitch contour) and energy values (energy contour), feature vectors of different sizes are explored. The optimal size of the feature vector to capture the variations in pitch and energy contours for recognizing the speakers is observed to be 100.

The performance of the SR systems developed using individual prosodic features is given in Table 8.4. The numbers shown in the table indicate the percentage of speaker recognition performance. Speaker recognition performance using duration, pitch and energy parameters are observed to be around 41, 61 and 39% respectively for TIMIT database in the case of enhanced speech using SVM models. It is observed from the Table 8.4 that, pitch values contain better speaker discriminative information compared to other two prosodic features. To improve the speaker recognition performance using prosodic features alone, feature and score level fusion techniques are explored. All the prosodic features are simply concatenated to perform feature level fusion. Duration (15), Pitch (100) and energy (100) features are concatenated, to form a feature vector of dimension 215. Then SVMs are used for the classification. Mere concatenation of features does not show much improvement in speaker recognition performance. Therefore score level fusion is performed by summing the weighted confidence scores (evidences) derived from the SR systems developed using individual prosodic features. The weighting rule for combining scores of individual modalities is as follows: $c^m = \frac{1}{m}\sum_{i=1}^{m} w_i c_i$, where c^m is the multi-modal confidence score, w_i and c_i are weighting factor and confidence score of the ith modality, and m indicates number of modalities used for combining the scores.

In this work, we have combined three modalities: (1) Model developed using durational features (SRS-1), (2) Model developed using sequence of pitch values (SRS-2) and (3) Model developed using sequence of frame energies (SRS-3). In our study, one of the weights (w_i) is varied in steps of 0.1 from 0 to 1, and the other weights are determined using the formula: $w_j = \frac{1-w_i}{m-1}$, where $j=1$ to m and $j \neq i$, $i=1$ to m. In this study, weighting factors associated to each system is varied from 0 to 1, with the steps of 0.1. With this we get a total of 33 combinations (11 combinations with respect to each weighting factor) of weighting factors. The block diagram of speaker recognition system using score level fusion of prosodic features is shown

Table 8.4 Performance of the speaker recognition systems developed using the features representing (1) Duration contour (SRS-1), (2) Pitch contour (SRS-2) and (3) Energy contour (SRS-3). The entries in the table indicate the percentage of recognition. Noisy : Enrollment-clean speech, Testing-noisy speech; Enhanced speech : Enrollment-clean speech, Testing-enhanced speech

Speech Corpus	Speaker recognition performance using prosodic features (in %)									
	Duration (SRS-1)		Pitch (SRS-2)		Energy (SRS-3)		Feature level fusion		Score level fusion	
	Noisy (15 dB)	Enhanced speech	Noisy (15 dB)	Enhanced speech	Noisy (15 dB)	Enhanced speech	Noisy (15 dB)	Enhanced speech	Noisy (15 dB)	Enhanced speech
TIMIT	32.14	41.45	43.15	61.42	26.32	38.61	46.13	63.12	45.29	65.67
50 speaker microphone	29.67	39.23	40.67	59.00	23.45	31.72	40.32	60.04	35.62	62.63
50 speaker cellphone	25.30	38.46	40.23	51.67	19.67	26.23	43.12	55.16	28.67	58.18

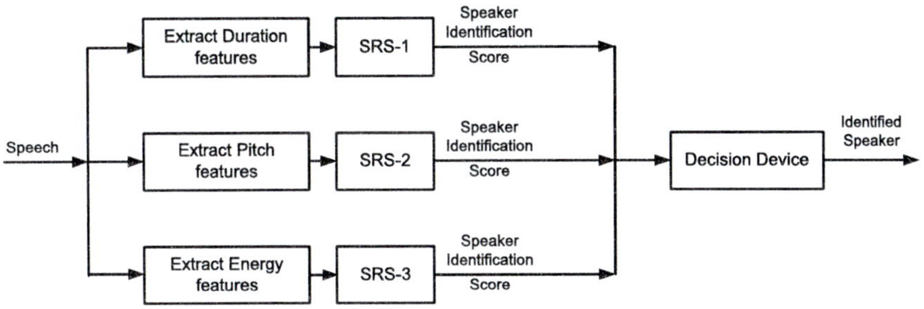

Fig. 8.8 Score level fusion of prosodic parameters

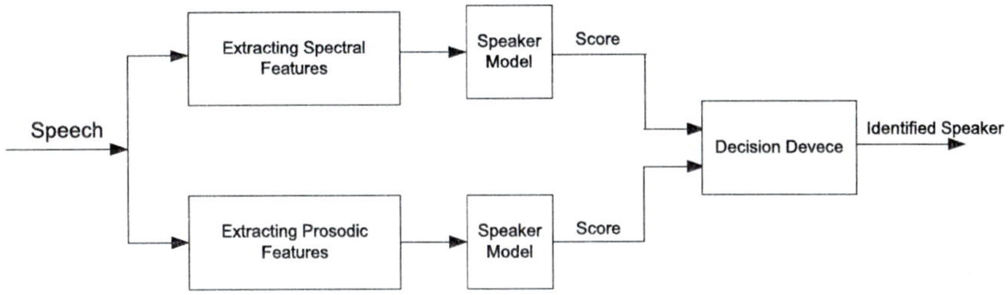

Fig. 8.9 Score level fusion of spectral and prosodic parameters

in Fig. 8.8. It is observed that the best recognition performance is about 66% for the weighting factors 0.2, 0.6 and 0.2 for the confidence scores of duration, pitch and energy contour based SR systems respectively. The details of the recognition performance are shown in Table 8.4. Score level fusion of all prosodic features improved the speaker recognition performance by around 4%, in the case of TIMIT database.

The combination of spectral and prosodic features may improve the speaker recognition performance due to their complimentary nature. Combining the evidences obtained from SR systems developed using spectral and prosodic features has shown an improvement in the recognition performance by more than 15%. The block diagram of speaker recognition system using score level fusion of spectral and prosodic features is shown in Fig. 8.9. The weighting factors of 0.6 and 0.4 for spectral and prosodic evidence have proved to perform better compared to the other combinations of weights. Table 8.5 shows the results obtained using score level fusion of spectral and prosodic features.

8.6 Summary and Conclusions

Due to high variability and diversified source of speech available for forensic analysis, it is difficult to perform speech analysis for forensic applications [4]. High speaker recognition rate is a normal phenomenon, in the case of clean data, using

Table 8.5 Performance of the speaker recognition systems developed using spectral and prosodic features. The entries in the table indicate the percentage of recognition. Noisy: Enrollment-clean speech, Testing-noisy data; Enhanced speech: Enrollment-clean speech, Testing-enhanced speech

Speech SNR (dB)	Speaker recognition performance using score level fusion of spectral and prosodic features (in %)					
	TIMIT		50 Speaker-microphone		50 Speaker-cellphone	
	Noisy	Enhanced speech	Noisy	Enhanced speech	Noisy	Enhanced speech
15	61.83	72.67	64.12	73.82	60.42	67.83
10	60.06	71.32	62.89	69.68	58.21	64.62
5	58.43	68.46	61.31	66.72	55.89	64.08
0	57.21	68.02	59.88	65.81	52.12	61.02
−5	56.54	66.38	57.23	64.32	51.46	60.82

spectral features. But most of the time speech data is noisy, short and degraded when it is collected through different channels for forensic analysis. Achieving high accuracy of speaker recognition in the case of noisy data using conventional methods is a real challenge, as spectral features are highly sensitive to the noise. It is observed from the results shown in Table 8.3 that speaker recognition performance decreases drastically in the case of the impairments caused due to background noise and channel (transmission) constraints. In this work we have proposed dynamics of speaker-specific prosodic features to improve speaker recognition performance in the case of noisy data. In this study, different intensities of noise from noisex database are added to the speech signal for simulating noisy speech with different SNRs. The simulated noisy speech resembles the speech samples used in forensic applications. Spectral and prosodic features are separately explored for recognizing the speaker in the first stage. Later, improvement in speaker recognition performance is achieved by combining spectral and prosodic features at score level. This approach of speaker recognition may be used in forensic applications, as it is performing better in the case of simulated noisy speech.

We have used TIMIT, 50 speaker microphone and cell phone speech databases for performing speaker recognition task. Speech enhancement is performed for noisy speech, before speaker recognition is performed, to reduce the effect of noise. Around 12% improvement is observed in speaker recognition using enhanced speech. Further the evidence from dynamics of prosodic features are fused with that of spectral features to improve the speaker recognition performance. Overall speaker recognition of around 73% is achieved for the speech under noisy conditions. With this, one may conclude that prosodic features add robustness to the speaker recognition systems, in particular for noisy speech. This is because of less susceptibility of prosodic features toward noisy conditions. These studies may be further extended to address different speech degradations such as: distance speech, whisper, telephone tapped speech, party conversation, reverberant speech and so on, in isolation and in combination. This study restricts itself to simulated noisy speech of different intensities. Studies may be further extended to the speech database with different actual noisy backgrounds than the simulated ones presented here.

References

1. Bolt RH, Cooper FS, Green DM, Hamlet SL, McKnight JG, Pickett JM, Tosi O, Underwood BD, Hogan DL (1979) On the theory and practice of voice identification, tech. rep. National Research Council, National Academy of Sciences, Washington
2. Rose P (2002) Forensic speaker identification. Taylor & Francis, London
3. Bonastre JF, Bimbot F, Boe LJ, Campbell JP, Reynolds DA, Magrin-Chagnolleau I (2003) Person authentication by voice: a need for caution. In: Proceedings of Eurospeech, ISCA, Geneva, pp 33–36, Sept 2003
4. Campbell JP, Shen W, Campbell WM, Schwartz R, Bonastre J-F, Matrouf D (2009) Forensic speaker recognition: a need for caution. IEEE Signal Process Mag 26:95–103
5. Mary L, Yegnanarayana B (2008) Extraction and representation of prosodic features for language and speaker recognition. Speech Commun 50:782–796
6. Shriberg E, Ferrer L, Kajarekar S, Venkataraman A, Stolcke A (2005) Modeling prosodic feature sequences for speaker recognition. Speech Commun 46(3–4):455–472
7. Kajarekar S, Ferrer L, Venkataraman A, Sonmez K, Shriberg E, Stolcke A, Bratt H, Gadde RR (2003) Speaker recognition using prosodic and lexical features. IEEE workshop on automatic speech recognition and understanding—ASRU 03, pp 19–24. doi:10.1109/ASRU.2003.1318397
8. Adami AG, Mihaescu R, Reynolds DA, Godfrey JJ (2003) Modeling prosodic dynamics for speaker recognition. IEEE international conference on acoustics, speech, and signal processing (ICASSP'03), pp 788–791. doi:10.1109/ICASSP.2003.1202761
9. Reynolds DA, Quatieri TF, Dunn RB (2000) Speaker verification using adapted Gaussian mixture models. Digital Signal Process 10:19–41. doi:10.1006/dspr.1999.0361
10. Reynolds DA (1994) Experimental evaluation of features for robust speaker identification. IEEE Trans Speech Audio Process 2:639–643
11. Shridhar M, Mohankrishnan N (1982) Text-independent speaker recognition: a review and some new results. Speech Commun 1:257–267 (North-Holland Publishing Company)
12. Carey MJ, Parris ES, Lloyd-Thomas H, Bennett S (1996) Robust prosodic features for speaker identification. Fourth international conference on spoken language-ICSLP, IEEE, October 1996, Philadelphia, pp 1800–1803. doi:10.1109/ICSLP. 1996.607979
13. Mary L, Yegnanarayana B (2006) Prosodic features for speaker verification. Proceedings of international conference on spoken language processing, Pittsburgh, pp 917–920, Sep 2006
14. Yegnanarayana B, Prasanna S, Zachariah J, Gupta C (2005) Combining evidence from source, suprasegmental and spectral features for a fixed-text speaker verification system. IEEE Trans Speech Audio Process 13:575–582
15. Fant G (1960) Acoustic theory of speech production with calculations based on X-ray studies of Russia articulation. Mouton and Co. N.V. Publishers, the Hague
16. Woodward JD, Orlans JNM, Higgins PT (2003) Biometrics. McGrawHill, Osborne
17. Pruzansky S (1963) Pattern-matching procedure for automatic talker recognition. J Acoust Soc Am 26:403–406
18. Bricker P, Pruzansky S (1966) Effects of stimulus content and duration on talker identification. J Acoust Soc Am 40:1441–1449
19. Stevens K et al (1968) Speaker authentication and identification: a comparison of spectrographic and auditory presentations of speech material. J Acoust Soc Am 44:1596–1607
20. Atal B (1976) Automatic recognition of speakers from their voices. Proc IEEE 64:460–475
21. Rosenberg AE (1976) Automatic speaker verification: a review. Proc IEEE 64:475–487
22. Mary L, Murty KSR, Prasanna SM, Yegnanarayana B (2004) Features for speaker and language identification. In: ODYSSEY04—the speaker and language recognition workshop, Toledo, 31 May–3 June 2004
23. Prasanna SM, Gupta CS, Yegnanarayana B (2006) Extraction of speaker- specific excitation information from linear prediction residual of speech. Speech Commun 48:1243–1261

24. Rao KS (2005) Acquisition and incorporation prosody knowledge for speech systems in Indian languages. PhD thesis, Dept. of Computer Science and Engineering, Indian Institute of Technology Madras, Chennai, May 2005
25. Bartkova K, Le-Gac D, Charlet D, Jouvet D (2002) Prosodic parameter for speaker identification. In: Proc Int Conf Spoken Language Processing, pp 1197–1200
26. Mary L, Rao KS, Yegnanarayana B (2005) Neural network classifiers for language identification using syntactic and prosodic features. In: IEEE Int Conf Intelligent Sensing and Information Processing (ICISIP-05), Chennai, Jan 2005
27. Mary L, Yegnanarayana B (2005) Consonant-vowel based features for language identification systems. In: Int Conf Natural Language Processing, IIT-Kanpur, India, pp 103–106, Dec 2005
28. Devis SB, Mermelstein P (1980) Comparison of parametric representations for monosyllabic word recognition in continuously spoken sentences. IEEE Trans Acoust Speech Signal Process 28:357–366
29. Atal B (1969) Automatic speaker recognition based on pitch contours (a). J Acoust Soc Am 45:309–309
30. Peskin B, Navratil J, Abramson J, Jones D, Klusacek D, Reynolds DA, Xiang B (2003) Using prosodic and conversational features for high performance speaker recognition: report from jhu ws'02. In: IEEE international conference on acoustics, speech, and signal processing (ICASSP '03), pp 792–795, June 2003
31. Varga A, Steeneken HJM (1993) Assessment for automatic speech recognition: II. NOISEX-92: a database and an experiment to study the effect of additive noise on speech recognition systems. Speech Commun 12:247–251
32. Rao KS, Vuppala AK, Chakrabarti S, Dutta L (2010) Robust speaker recognition on mobile devices. In: IEEE international conference on signal processing and communication (SPCOM), IISC Bangalore, July 2010
33. Sonmez MK, Shriberg E, Heck L, Weintraub M (1998) Modeling dynamic prosodic variation for speaker verification. Int Conf Spoken Language Processing, Sydney, Nov–Dec 1998
34. Burges CJ (1998) A tutorial on support vector machines for pattern recognition. Data Mining Knowl Discov 2:121–167
35. Collobert R, Bengio S (2001) A support vector machine for large-scale regression and classification problems. IDIAP
36. Collobert R, Bengio S, Williamson C (2001) SVMTorch: support vector machines for large-scale regression problems. J Machine Learn Res 1:143–160
37. Website: http://code.google.com/p/bnt/downloads/detail?name=FullBNT-1.0.7.zip
38. Krishnamoorthy P, Prasanna SRM (2009) Reverberant speech enhancement by temporal and spectral processing. IEEE Trans Speech Audio Lang Process 17:253–266
39. Chaitanya N (2005) Single channel speech enhancement. Master's thesis, Dept of Computer science and Engineering, Indian Institute of Technology Madras, Chennai
40. Yuvaraja S, Keri V, Pammi SC, Prahallad K, Black AW Building a Tamil voice using HMM segmented labels. http://researchweb.iiit.ac.in/~santhoshy/tamtts.pdf
41. Mporas I, Lazaridis A, Ganchev T, Fakotakis N (2009) Using hybrid hmm-based speech segmentation to improve synthetic speech quality. In: 13th panhellenic conference on informatics
42. Ananthapadmanabha TV (1978) Epoch extraction of voice speech. PhD thesis, Indian Institute of Science, Bangalore
43. Ananthapadmanabha TV, Yegnanarayana B (1979) Epoch extraction from linear prediction residual. IEEE Trans Acoust Speech Signal Process ASSP-27:309–319
44. Prasanna SRM, Yegnanarayana B (2004) Extraction of pitch in adverse conditions. In: Proc IEEE Int Conf Acoust Speech Signal Processing, Montreal, May 2004

Chapter 9
Characterization of Noise Associated with Forensic Speech Samples

Jiju P. V., C. P. Singh and R. M. Sharma

Abstract For speech enhancement, different methods have been developed in the past decades. This study has been carried out for characterization of various noises associated with forensic speech samples and their classification to find specific set of filtering technique for speech recognition and speaker identification. Noisy speech samples are collected from the exhibits received in case examination in the laboratory for this study. The experiment is performed in a two-fold way: enhancing the speech for (i) speech recognition and (ii) speaker identification. The original and simulated samples are subjected to various filtering techniques, namely, FFT Filter, noise reduction, noise gate, notch filter, bandpass, butterworth filter, digital equalizer and parametric equalizer for speech recognition. For speaker identification, noise reduction, noise gate, notch filter, bandpass and butterworth filter are applied to the noisy speech samples. Characterization of noise embedded with the noisy speech samples were attained based on the application of these filtering techniques and subsequent analysis performed on them using Computerized Speech Laboratory (CSL). For speech recognition, maximum SNR improvement was achieved by FFT filter on samples Noisy Speech-I (Direct Recording), Noisy Speech-II (Telephonic Landline Recording) and Noisy Speech-III (Mobile Phone Recording). The corresponding improvements in SNR for original and simulated samples were 3.81, 7.57, 5.62 dB and 4.39, 6.26, 5.57 dB respectively. FFT filter, when applied to the Noisy Speech-I, Noisy Speech-II and Noisy Speech-III of original noisy speech samples, have given an improvement of 75, 71 and 48%, whereas simulated noisy speech samples gave an improvement of 82, 78 and 52%. For speaker identification, maximum improvement was achieved by noise reduction filter when applied to the Noisy Speech-I, Noisy Speech-II and Noisy Speech-III of original noisy speech samples, have given an improvement of 60, 64 and 52% whereas simulated noisy speech samples gave an improvement of 64, 70 and 54%. Statistical study of improvised original noisy speech and simulated noisy speech samples after filtering have

---

R. M. Sharma (✉)
Department of Forensic Science, Punjabi University, Patiala
147002, Punjab, India
e-mail: rmsforensics@gmail.com

A. Neustein, H. A. Patil (eds.), *Forensic Speaker Recognition*,
DOI 10.1007/978-1-4614-0263-3_9, © Springer Science+Business Media, LLC 2012

revealed the degree of efficiency of different filters for Speaker Identification and how far they are dependable in forensic adverse contexts. For Speech Recognition, the degree of efficiency of filters in enhancing the speech signal is found to be in a descending order; viz. FFT Filter, Noise reduction, Noise gate, Notch filter, Bandpass, Butterworth filter, Digital equalizer and Parametric equalizer. The degree of efficiency of filters in enhancing the speech signal for Speaker Identification is found to be in a descending order; viz. Noise Reduction, Noise Gate, Notch filter, Bandpass, and Butterworth filter.

9.1 Introduction

9.1.1 Overview

The laws of nature that define each and every fact involved in any process of the universe, even its existence itself, are called Science. When they are applied to find out the truth behind any offence against the laws and conventions of society it is called Forensic Science. That is, Forensic Science—an ever-expanding branch of applied science and technology, made its evolution to fulfill the need of Law enforcement agencies to provide useful information to the investigators in solving crime cases and to the Courts of Law to help them in the justice delivery process. The term 'Forensic Audio' may be defined as the application of audio technologies and methodologies to Law.

Present day criminals and antisocial elements are using communication systems like telephone, radio, mobile phone and tape recorder while committing crime and thus the voice of an individual is often the only available clue for identification. The recorded conversations among criminals with the complainant(s) or victim(s) can be utilized by the law enforcement agents in relevant investigations to find the truth behind any offence towards the law. Recorded audio may also be a part of video recording of dispute that the law enforcement agency or criminal justice system requires as evidence against the culprit. Detailed plan of the crime to be committed (or which has been committed) i.e., the specific information about the participants, the place of occurrence, its nature and future plan of action to be undertaken by them, all such information are important to the law enforcement agents to solve the dispute. The recorded data very often contain such valuable and decisive information. Such information can be utilized to prevent crime and to assist investigators in identifying and capturing the criminals. Also it can be used in the court-of-law to prove the involvement of the accused. Moreover, many tape recordings contain not only speech but also other materials that can be utilized effectively for crime countermeasure purposes.

Speech is a complex combination of information from different levels (discourse, semantics and syntax, phonological, phonetic and acoustics) that is used to convey a message. The speech signal can be considered as time series resulting from complex nonlinear process in the larynx.

The conventional methods in forensic science used for individual identification have been proved to be successful in helping the Law Enforcement agencies in many challenging situations. In audio-forensics, speech recognition, speaker identification and tape authentication are potentially high enough to give efficient and sound interpretation of the audio evidence in justice delivery system. It not only helps in nailing down the individual by biometric identification but also help in recognizing the speech of the context and also to check the authenticity of the tape.

Adverse situation is often faced by an audio expert while examining the recorded speech whose recording is carried out under noisy situation or some other signal distortions. Hence, speech recognition and/or speaker identification and/or tape authentication from such recordings become erroneous. The credibility as well as accuracy of an expert's testimony are tested in such situations which can create lose of faith in scientific interpretation of evidences in justice delivery system. Hence, the need of the hour is an improvised and efficient method and technique to reduce the effect of noise embedded in the recorded speech for efficient speech recognition and/or speaker identification and/or tape authentication.

The present study is focused on the characterization of the noise associated with forensic speech samples for speech recognition and speaker identification.

9.1.2 Speaker Identification and Noisy Speech Problem

There are two methods adopted in speaker identification: auditory analysis and spectrographic method. In auditory analysis, speech sample is subjected for critical listening based on which the perceptual features of a person's voice are studied. Finally, the linguistic characteristics of a person in specimen speech sample are compared with that of the person in questioned speech sample. In the case of spectrographic method, voice of a person is converted into electric signal and the spectrogram of both the questioned as well as specimen speech samples is produced. An expert compares and evaluates these spectrograms and then incorporating the results from the auditory analysis give a final opinion whether the voice in the disputed speech sample is matching with that in the specimen sample or not. In speaker identification process, there are parameters with which the uniqueness of a person's voice is established and they are called speaker dependent parameters.

The presence of noise in speech signals could contribute to a higher degree of mismatch in performance of speech processing systems for speaker identification as well as speech recognition. Identifying a speaker's voice from a noisy speech is a very challenging task. When a speech is embedded with noise from various sources, the speaker dependent parameters that act as the key for speaker identification changes drastically. A change in any of the speaker dependent features can affect the examination of such noisy speech, resulting into a false identification or false elimination.

The performance of systems for speaker identification is quite satisfactory with clear speech than with noisy speech or distorted speech exemplars. Considering human/machine interfaces as a major area of application, it is obvious that the signal becomes more challenging as the acoustic environment becomes more complex and hostile.

Before any kind of examination, as a primitive measure it is essential to evaluate the recording of the unknown voice to make sure that the recording has a sufficient amount of speech with which to work and also the quality of the recording is of adequate clarity within the frequency range required for analysis [1].

Over the past decade, the developments in the digital signal processing have produced wide variety of techniques for removing the noise on degraded speech, depending on the type of the application and the character of noise. The broadband noise and a very low SNR deteriorate the intelligibility of most of the recorded speech samples. Noise reduction is the process of removing noise from a signal. Noise reduction techniques are conceptually very similar regardless of the signal being processed.

9.1.3 Speech Recognition

The process through which the speech contents of a disputed speech sample are recognized is called speech recognition. In speech recognition, a forensic expert is interested in recognizing the words that have been uttered by the speaker.

As speech recognition is mainly concerned about the intelligibility of speech, automatic systems are preferred to give perform this task. The performance of speech recognition systems is usually specified in terms of accuracy and speed. Accuracy may be measured in terms of performance accuracy which is usually rated with Word Error Rate (WER), whereas speed is measured with the real time factor. Other measures of accuracy include Single Word Error Rate (SWER) and Command Success Rate (CSR).

Most of the speech recognition software can achieve accuracy between 98 and 99% if operated under optimal conditions. "Optimal conditions" usually assume that users: have speech characteristics which match the training data, can achieve proper speaker adaptation, and work in a clean noise environment (e.g. quiet office or laboratory space).

9.1.4 Speech Enhancement

As recording is performed usually in uncontrolled environments the recorded speech evidences may contain noise. To perform efficient speaker identification and speech recognition, forensic audio specialist requires speech sample with high signal to noise ratio (SNR). A recorded signal with low SNR is capable of producing erroneous result.

Processing of unintelligible or noisy recordings for the purpose of increased signal intelligibility, attenuation of noise, improvement of understanding the recorded material and improvement of the quality or the ease of hearing the signal of interest is termed as speech enhancement.

The goal of speech enhancement is to facilitate the understanding, communication and processing of speech signals by suppressing the noise distortion from noisy signals. As such, noise reduction can improve the perception of speech signals and increase the comfort of the listener in a variety of applications like e.g. teleconferencing, mobile phones and hearing aids.

9.1.4.1 Present Scenario

For speech enhancement, different methods have been developed in the past decades. Scientists were successful in enhancing the noisy speech to a certain limit. Some of the important methods developed for speech enhancement are namely, Cepstrum Mean Normalization (CMN), Relative Spectral processing of speech (RASTA), Spectral Subtraction technique, Short-Time Spectral Attenuation (STSA), Artificial Neural Networks (ANN), Deconvolution method and Normalized Least Mean Square (NLMS) etc.

Attempts have been made by scientists to solve the problem of retrieving the signal from the noise-masked signal by filtering out the noise. Atal in 1974 has developed effective technique to filter the slowly varying background noise by using Cepstrum Mean Normalization (CMN) technique [2].

Another approach is based on the technique involving suppression of constant additive offsets in every log spectral component of the short-term spectrum and the technique is called Relative Spectral processing of speech (RASTA) [3].

Boll [4] in 1979 developed a novel method for suppression of acoustic noise in speech using Spectral Subtraction. In case of Spectral Subtraction technique, it requires only an estimate of the noise power spectrum and this can be viewed as a fundamental limitation in that, it does not utilise the statistics and the distributions of the signal processing [5]. Due to random variations in noise spectrum, Spectral Subtraction technique also results in distortion of the signal. Also typical Short-Time Spectral Attenuation (STSA) speech enhancement algorithms are ineffective in the presence of highly non-stationary noise due to difficulties in the accurate estimation of the local noise spectrum [6].

Use of Artificial Neural Networks (ANN) and Fuzzy Inference Systems (FIS) along with other techniques for Noise Cancellation has given encouraging results in the early and mid 1990s. Multilayer Perceptrons (MLPs) using Back Propagation (BP) algorithm in ANN with time delay and Adaptive Networks has been used for Noise Reduction, echo cancellation. But the main disadvantage with this method is that the network requires noisy speech and the corresponding clear speech for its training which is not simultaneously available in many practical situations [7].

Cole et al. in 1997 shown that the waveform deconvolution (inversion) approach is impractical in any situation where, either source or receiver is mobile, due to variation of room impulse response with position. Although the modulation envelope deconvolution method has been claimed to be position independent, it reduces the intelligibility of the speech signal [8].

Another method called Normalized Least Mean Square (NLMS) also gets it performance degraded significantly in the presence of high levels of background noise [9].

Manohar and Rao [10] introduced a post processing scheme that can be applied to the output of an STSA speech enhancement algorithm to improve the speech quality in the presence of random noise bursts, characteristic of many environmental sounds. Typical short time spectral attenuation (STSA) speech enhancement algorithms are ineffective in the presence of highly nonstationary noise due to difficulties in the accurate estimation of the local noise spectrum. The post processing algorithm based on spectral properties of the noise in order to detect noisy-time-frequency regions which are then attenuated using a suitable SNR-based rule. The objective measures indicate that post-processing has a greater influence at low SNRs relative to that at high SNRs. They developed a frequency domain post-processing algorithm in order to improve STSA enhanced speech quality in the presence of random noise bursts.

Healy et al. [11] conducted two experiments for studying the effect of smoothing filter slope and spectral frequency on temporal speech information. In the first experiment low pass filter was applied and temporal smoothing filter slope at two cut off frequencies 16–100 Hz were examined. Smoothing filter slope angle was found to have large improvement in intelligibility with the shallowest slope employed and dropped precipitously as slope angle increased. In the second experiment temporal extraction and modulation were performed at 400 Hz for one of the spectral channels and at 16 Hz for the remaining two channels. They also prepared four band conditions 100–750, 750–1500, 1500–3000 and 3000–8000 Hz to confirm the results at higher overall levels of performance. Intelligibility was higher when all three bands possessed higher temporal rate.

The different temporal smoothing cut-offs employed for the different component bands caused noticeable phase shifts (time delays) across the constituent bands. To eliminate this asynchrony across channels, temporal extraction was performed using a pair of cascaded fourth order Butterworth LP filters. The first pass was performed on each rectified speech partition in usual fashion, and then a subsequent pass was made on the time reversed version of this filtered signal.

In their work, Bai et al. [12] a comprehensive study was undertaken to compare various audio spatializers for use with dual loudspeaker handsets in the contexts of inverse filtering strategies. The required inverse filters were designed with two deconvolution method, frequency domain method and the time domain method. Different approaches to design audio spatializers with the HRTF, CCS: Head Related Transfer Function and Crosstalk Cancellation System and their combination are compared. In particulars, two modified CCS approaches are suggested. Comprehensive objective and subjective tests were conducted to investigate regularization, complex smoothing and structures of inverse filters of audio spatializers.

Buera et al. [13] presented a novel noise and speech predictive models. The FPM-NE model is based on a time varying comb filter and uses a non-stationary noise estimator jointly with VTS speech enhancement. The FPM-SE model uses the fine pitch model, together with conditional priors for the enhancement parameters to directly enhance the speech signal.

Interesting speech recognition results have been obtained in both cases against the Aurora 2 database. The best results are from the FPM-SE model, which achieves a 50.91% WER reduction on average, and more than 70% WER reduction in SNR conditions above 10 dB. These high SNR conditions are the most interesting for building real applications.

Xi et al. [14] tackled the problem of noise reduction for chaotic signal. They defined a new parameter which can effectively measure the determinacy of chaotic signals. The parameter can be used to construct stop condition for existing noise reduction techniques. As an example, an improved local projection algorithm is proposed and tested with real world data. Their results show that the parameter gives a good measurement of the signal determinacy and thus also the noise reduction performance.

Based on the concepts of reconstructed phase-space and dimension embedding, Pinter [15] developed an algorithm in 2008. In this, proposed algorithm separates the speech from noise using a non-linear transformation in a transformed domain. The algorithm is based on dimension embedding, and assumes the separability the speech sub-space and the noise sub-space in the Euclidean space, determined by the covariance matrix of the data set after embedding. The Euclidean space in question is spanned by the eigenvectors of the covariance matrix above and the eigenvectors have been computed using Jacobi's algorithm. The data set has been determined after periodic extension of the speech segment, which differs from the published methods.

Shannon and Paliwal [16] proposed and investigated estimating the STFT phase spectrum independently from the STFT magnitude spectrum for speech enhancement applications. In their experiments they estimated STFT magnitude spectrum from degraded speech and a STFT phase estimate from clean speech, little improvement in speech quality could be observed. This set-up was similar to Wang and Lim's [17] and confirmed their findings. When they replaced the window function used to estimate the phase spectrum to one with a lower dynamic range, a substantial increase in noise reduction and speech quality could be observed.

Lyons and Paliwal [18] studied the modulation-based speech enhancement method and investigated the role of the compression function applied to the short-time power spectrum before modulation filtering. They used objective as well as subjective measurements to measure the quality and intelligibility of the enhanced speech. For objective measurements, PESQ is used for speech quality and STI for speech intelligibility. For subjective measurements, human listening tests are used.

Falk et al. [19] have applied bandpass filtering to the temporal trajectories of short-time magnitude spectrum to achieve enhancement. Drullman et al. [20] have reported a study where the speech signal is split into a number of frequency subbands, the temporal envelope of each subband is lowpass filtered, and the original carriers and filtered envelopes of all the subbands are combined to reconstruct the output speech signal. Using these reconstructed signals, they have shown that modulation frequencies below 16 Hz are important for intelligibility. In a similar study carried out using highpass modulation filters, Drullman et al. [21] have shown the modulation frequencies above 4 Hz are important for intelligibility. Arai et al. [22] have applied filters to the time trajectories of LPC cepstrum, showing that applying a bandpass filter with passband between 1 and 16 Hz does not impair speech intelligibility. They have also shown that some modulation frequencies are more important than others, with the region around 4 Hz being the most important for intelligibility.

Hermansky et al. [23, 24] in their study propose a speech enhancement procedure where FIR filters are applied to the time trajectories of the cubic root compressed short-time power spectrum to achieve better speech quality in the presence of additive noise.

Analysis of information on tape recordings constitutes one of the single most powerful tools that criminal investigators have at their command. With improved knowledge and techniques it could be explored to a far greater extent than it is currently. In turn, the amount of information that could be captured and ultimately utilized could be expanded by a substantial factor. That is the latest technology and procedures when properly utilized would permit considerably more effective use of tape recordings in finding the truth behind it.

Unfortunately, the equipment available/used, the techniques employed and/or the field situation itself often lead to conditions where the messages (speech) on the tape recordings are obscured by masking signals or are distorted in such a manner that the speech is unintelligible—or at least, difficult to decode [25].

On the other hand, there are techniques that often permit the intelligibility of the speech to be improved—sometimes to an extent that the relevant information can be reclaimed. Masking signals recorded onto the tape along with the speech causes the most common form of speech signal degradation. The accuracy of parametric values determined depends upon the level of noise in speech signal [26]. It is to be conceded that only primitive techniques are presently available that permit enhancement of the intelligibility of speech recorded on tape in the presence of noise and/or other distorting conditions.

Researchers often resort to less formal listening tests to assess the quality of an enhanced signal, and they use automatic speech recognition tests to assess the intelligibility of that signal. Quality and intelligibility are also hard to quantify and express in a closed form that is amenable to mathematical optimization. The design of speech enhancement systems is often based on mathematical measures that are somehow believed to be correlated with the quality and/or intelligibility of the speech signal. A popular example involves estimation of the clean signal by minimizing the mean square error (MSE) between the logarithms of the spectra of the original and estimated signals [27]. This criterion is believed to be more perceptually meaningful than the minimization of the MSE between the original and estimated signal waveforms [28].

Another difficulty in designing efficient speech enhancement systems is the lack of explicit statistical models for the speech signal and noise process. In addition, the speech signal, and possibly also the noise process, are not strictly stationary processes. Common parametric models for speech signals, such as an autoregressive process for short-term modelling of the signal, and a hidden Markov process (HMP) for long-term modelling of the signal, have not provided adequate models for speech enhancement applications. A variant of the expectation-maximization (EM) algorithm, for maximum likelihood (ML) estimation of the autoregressive parameter from a noisy signal, was developed by Lim and Oppenheim [29] and tested in speech enhancement. Several estimation schemes, which are based on hidden Markov modelling of the clean speech signal and of the noise process, were developed over the years, see, e.g., Ephraim [30]. In each case, the HMP's for the speech signal and noise process were designed from training sequences from the two processes, respectively. While autoregressive and hidden Markov models have proved extremely useful in coding and recognition of clean speech signals, respectively, they were not found to be sufficiently refined models for speech enhancement applications.

Classified techniques are required to make it effectively less time consuming as well as to preserve the characteristics of the signal while Noise Attenuation is carried out and the proposed study finds ground here. Choosing of appropriate filter configuration for any noisy speech signal is a highly time consuming process. Speaker's idiosyncratic speech should not be affected when the noise reduction is carried out; otherwise speaker identification becomes highly erroneous. The proposed study is being carried out for characterization of various noises associated with forensic speech samples and their classification to find specific set of filtering technique with special consideration to the prime objectives of 'forensic audio'—speech recognition and speaker identification. The study refers to understand the noise characteristics for finding appropriate filtering technique/s so as to obtain sufficiently clear speech samples for speaker identification.

9.2 Materials and Methods

9.2.1 Methods

9.2.1.1 Collection of Speech Samples and Sampling

In the study, noisy speech samples, clear speech samples and simulated noisy speech samples are collected from the exhibits received in case examination in the laboratory. Clear speech or speech signal under studio conditions of high quality has been recorded in the laboratory. Simulated samples have been prepared from the noise speech samples and recorded clear speech samples. All the samples are sampled for duration of 20s as standard for this study. This specific length of sample is so chosen because it is found sufficient enough to contain the characteristic features of all the three types of noisy speech signals for the study. This was determined after preliminary examination of samples.

Noisy Speech Samples

One hundred and fifty noisy speech samples received in actual crime cases at Central Forensic Science Laboratory, Chandigarh, were collected for the study. The samples were classified into direct (tape recorded), telephonic (land line) and mobile phone conversation based upon their mode of recording (hereafter referred as Noisy Speech-I, Noisy Speech-II and Noisy Speech-III respectively). Such categorization is being done for the purpose of identifying the sources of the noise as well. The samples collected are used for identifying the speaker by comparing with corresponding specimen samples as well as for recognizing the speech in the context.

Speech Samples Recorded in Ideal Condition

Specimen voices recorded were also collected from the actual crime cases received in the laboratory. This has been carried out for comparison purpose for speaker identification from the corresponding collected questioned noisy speech samples. In order to keep a Reference Speech Signal (hereafter referred as RSS) for the second phase of experiment, speech samples have been recorded in studio condition in the laboratory. RSS is recorded directly in digital form using hardware and software of Computerized Speech Laboratory (CSL). A unidirectional microphone of Shurie make is used in this process.

Simulated Speech Samples

Noise portion from the actual case exemplars for each sample is collected and kept as Reference Noise Signal (hereafter referred as RNS). This contains both additive as well as convolutive noise. With the help of Adobe Audition software (version 1.0.) the corresponding RNS for each noisy speech sample is mixed with RSS for preparation of simulated noisy samples in the laboratory. This is carried out by superimposing the RNS of each noisy speech sample and RSS to produce corresponding simulated noisy speech samples. The simulated speech samples are used for conducting the experiment for speech recognition as well as for speaker identification purposes. Figure 9.1 plots the waveform for the word 'Particular' in noisy condition in three different mode of recordings.

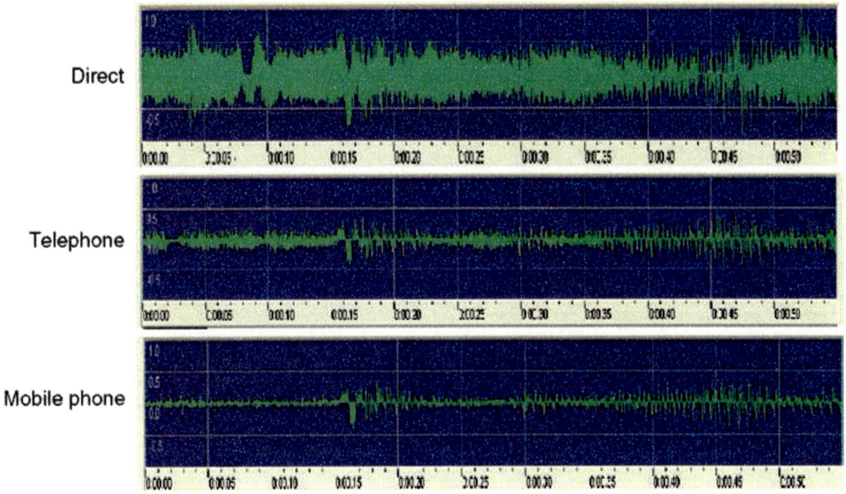

Fig. 9.1 Noisy speech signal of the utterance 'Particular'

9.2.1.2 Methodology

Different kinds of noise associated with each mode of recording are identified and grouped according to their basic features. For this, non-speech regions containing noise have been collected from the sample and are mixed with Reference Speech Signal (RSS) recorded in ideal condition for preparation of simulated noisy samples to enable a comparative study of systematic noise filtering. Both the original noisy speech and simulated noisy speech are subjected to various filtering techniques. The improvements produced by various filtering techniques were studied in a statistical way. A comparative study of systematic noise filtering by comparing the distortion level of the filtered simulated signal with RSS is carried out. This helps in deciding degree of quality of the speech signal retrieved for speech recognition and speaker identification in noisy conditions. This is carried out by comparing the distortion level and SNR of the filtered speech signal with RSS. From the resultant signal, speech is reclaimed to a maximum extent by nullifying the effect of embedded noise with the help of specific filter(s). This in turn helps to decide the degree of quality of speech signal retrieved for speaker identification under noisy conditions in terms of SNR and perceptual features.

Thus, various groups of noise embedded with the speech samples are subjected for suitable filtering technique(s) for efficient noise reduction. The SNR of simulated speech samples before and after applying the filters are also compared and studied. Then the effect of filter application on SNR and speaker dependent features of speech samples before and after the application of filters are also studied.

Table 9.1 Mean Opinion Score (MOS) definition

Score	Opinion	Impairment scale
1	Unsatisfactory	Annoying and unacceptable
2	Poor	Annoying
3	Fair	Slightly annoying
4	Good	Perceptible but acceptable
5	Excellent	Imperceptible

$$\mathrm{SNR(dB)} = 10\ \mathrm{Log}_{10}(P_{\mathrm{signal}}/P_{\mathrm{noise}}) = 20\ \mathrm{Log}_{10}(A_{\mathrm{signal}}/A_{\mathrm{noise}}) \tag{9.1}$$

SNR Signal to noise ratio
P_{signal} Power of the speech signal
P_{noise} Power of the noise
A_{signal} Amplitude of the speech signal
A_{noise} Amplitude of the noise

As perceptual features are very much important in the case of Forensic Speaker Identification, listening tests are conducted to ensure that the perceptual features of the original noisy speech are preserved while applying filters. For this purpose, 13 listeners of age group of 25–30 were trained with the original noisy speech until they are familiar enough to follow the perceptual features of each speaker. The filtered speech samples are then subjected to critical listening by the trained listeners. Finally, an opinion is made based on the method of Mean Opinion Score (MOS); the most widely accepted method for speech quality evaluation and a simplest subjective measure for the assessment that gives an overall opinion of the performance. The standard and possible set of score for MOS is presented in the Table 9.1.

A decision on the processed samples is made by estimating SNR with the help of various analyses conducted with Computerized Speech Laboratory (CSL) and Mean Opinion Score (MOS) test.

9.2.1.3 Instrumentation

Digitization of Noisy Speech Samples

Sony model number HCD-V515, HI FI audio system MHC-V777 is used to play the recorded tape and the output is fed into the computer through a connector. Digitization of noisy speech samples are conducted with the help of sound blaster card of creative audigy make installed on the computer along with the Adobe audition software (version 1.0.). The noisy speech samples in analogue mode are converted in PCM format with 16bit quantization (unsigned) and 22,050 Hz sampling rate. Further a down sampling of the data at a sampling rate of 11,025 Hz has been carried out during the experiment.

Computerized Speech Laboratory (CSL)

Computerized Speech Laboratory (CSL) 4300B is used for the recording and analysis of the speech samples. It is Windows based software that is used for the acquisition, acoustic analysis, display and playback of speech signals.

Signal acquisition, storing speech to disk memory, graphical and numerical display of speech parameters, audio output, signal editing are the main operations of CSL with options like spectrographic analysis, pitch contour analysis, Linear Predictive Coding (LPC) Analysis, Cepstrum analysis, Long Term Average (LTA), FFT and Energy contour analysis etc. can be performed with CSL.

9.2.1.4 Experiment

As a preliminary test, the classified digital noisy speech samples are subjected for critical listening. This is done in order to understand the perceptual characteristics of noise embedded with the speech. A specific duration of noise only portion is selected and is kept as Reference Noise Signal (RNS) for the particular sample throughout the experiment. The sample containing more than one type of noise, are segregated. From these segregated samples with each having unique kind of noise structure, a specific duration of noise only portion are selected. The selected noise only part which is prominent in the sample is kept as Reference Noise Signal (RNS) for the particular sample throughout the experiment. This is used to carry out the study exclusively on noise, its structure and its effect when embedded on speech. Thereafter the degree of distortion is compared as the factor of applying different filter(s).

Filters Used

The classified noisy speech exemplars are subjected to undergo application of various filtering techniques. Filters are applied upon original noisy speech samples and simulated samples in a twofold manner. The filters available in the softwares such as Adobe Audition and Goldwave are used in this study.

The filtering is carried out for speech recognition and speaker identification both separately. The filters applied for Speech recognition are FFT filter, noise reduction, noise gate, notch filter, band-pass, butterworth filter, digital equalizer and parametric equalizer. For speaker identification filters applied are noise reduction, noise gate, notch filter, band-pass and butterworth filter. Such unique identification of software/filters is necessary to preserve the speaker-specific information in the case of speaker identification. For such reason corresponding filters is chosen keeping the speaker dependent characteristics unaltered at any level of the signal. The outcome of each filter upon noisy speech samples is studied and analysed. Preliminary experiments have proved the suitability of these filters [31, 32].

The FFT-filters are basically FIR filters, but the filtering is not done in the time domain. The input signal is transformed from the time- into the frequency domain (using the FFT), the spectrum multiplied with the filter's frequency response, and the result is transformed back into the time domain (using the inverse FFT). Noise reduction helps to eliminate unwanted noise within a sound, such as a background hiss, a power hum, or random interference. A reduction envelope is used to remove noise. The envelope can be created in four different ways, depending on the Reduction envelope setting.

Noise gate removes background hiss from quiet parts of the selection. For example, in a recorded noisy speech, the noise gate can remove the noise any place where the speaker paused or stopped talking. Noise gate do not remove background hiss from louder parts of the selection. The Digital Equalizer boosts or reduces certain ranges of frequencies and thus controls the noise. The Parametric Equalizer is a flexible tool for reducing or enhancing ranges of frequencies thus reducing the effect of noise. Bandpass filters block all frequencies outside the specified range, keeping only frequencies within the range. Notch filters block all frequencies within the specified range, keeping all other frequencies outside the range. The notch filter is similar to bandstop filter provided it has multi narrow bandwidth selection of frequencies. Those selected bandwidths are not allowed and this is used in removing various noises of different frequency ranges.

Speech is reclaimed to a maximum extent by nullifying the effect of embedded noisy signal from the resultant signal by the application of filters. The degree of distortion upon speech signal due to embedded noise is also studied by comparing the results obtained by the application of filters.

Analysis of Speech Exemplars

Various analyses are performed using Computerized Speech Laboratory. In the experiment, blackman windowing technique is used with a frame length of 20 ms.

Both time domain and frequency domain analysis, namely, energy contour analysis, FFT analysis, Long Term Average (LTA) analysis and Linear Predictive Coding (LPC) analysis are carried out in order to study the effects of embedded noisy signal upon speech signal for original noisy speech and simulated noisy speech before and after applying various filters.

The energy contour analysis is carried out with a sampling rate of 11,025 Hz, frame length of 20 (ms), total 1013 samples with no smoothing Level. The blackman window weighing technique is used for the analysis.

While taking the FFT power spectrum, the sampling rate is kept at 11,025 Hz with 512 FFT size and zero pre-emphasis level. Blackman window weighting technique with no smoothing level and 257samples is used for the study.

During the Long Term Average (LTA) analysis, sampling rate is kept at 11,025 Hz with 512 FFT size, zero pre-emphasis level. Using Blackman window weighting of no smoothing level and 257 samples, Long Term Average (LTA) for original noisy speech samples and simulated noisy speech samples were calculated.

Formant values were extracted from Linear Predictive Coding (LPC) spectra. The sampling rate is kept at 11,025 Hz with frame length of 20 ms with 221 points. Autocorrelation analysis method is used with filter order of 12. Blackman window weighting technique with pre-emphasis of 0.90 for 201 samples is used for the analysis.

Based on the inference from the subsequent study conducted upon the results of various analyses, appropriate filters for each class of noise associated with forensic speech samples is identified and discussed for their efficiency in enhancing the speech for speech recognition and speaker identification.

Statistical Study

Statistical study is conducted over the results produced in terms of Signal to Noise Ratio (SNR) by the various filtering techniques and the subsequent analysis by CSL. Thus the improvement produced by various filters upon original noisy speech and simulated noisy speech samples is studied. The percentage of improvement by each of the filters is calculated and compared for both original noisy speech and simulated noisy speech samples. Based on this study, characterization of noise and their classification are performed. Thus, a systematic characterization of noise has been achieved by which it is possible to classify the same.

9.3 Results and Discussion

This section of the chapter deals with the findings of the research work pertaining to the application of filters on noisy speech samples for the purpose of speech recognition and speaker identification.

The first part discusses the effect of filter application on the noisy samples, viz, Noisy Speech-I, Noisy Speech-II and Noisy Speech-III for Speech recognition and the second part for the speaker identification respectively.

9.3.1 Discussion of the Effect of Application of Filter for Speech Recognition

Time domain analyses followed by Mean Opinion Score Test of samples after the application of various filters are carried out for Speech Recognition purpose.

9.3.1.1 Time Domain Analysis—Energy Contour Analysis

Energy contour analysis is carried out in order to find out the energy distribution of the original noisy speech samples (Noisy Speech-I, Noisy Speech-II and Noisy Speech-III) and simulated noisy speech samples (Noisy Speech-I, Noisy Speech-II and Noisy Speech-III) in the time domain. This in turn helped in estimat-

Table 9.2 SNR improvement after applying FFT filter

Samples		Average improvement in dB
Noisy Speech-I	Original	3.81
	Simulated	4.96
Noisy Speech-II	Original	7.57
	Simulated	4.95
Noisy Speech-III	Original	5.62
	Simulated	5.53

ing the Signal to Noise Ratio (SNR) of both the original noisy speech samples and simulated noisy speech samples before and after the filter application. Subsequently the difference in Signal to Noise Ratio (SNR) is found out for both the original noisy speech samples and simulated noisy speech samples. Thus, the improvement in Signal to Noise Ratio (SNR) as a result of the application of various filters is determined.

Energy contour analysis calculates the sum of the absolute amplitude values in a frame of data divided by the number of points in the frame. The energy is first computed and then converted to decibels of sound pressure levels (dB SPL).

Further, in the case of simulated samples, the analysis is carried out in order to find out the degree of distortion occurred on the speech samples by embedding the noise. Analysis carried out after applying filters helped to estimate the quality of the speech reclaimed from the noise.

FFT Filter Application

Improvement in Signal to Noise Ratio (SNR)

When FFT filter is applied, of the total number of original noisy speech samples 68% of Noisy Speech-I, 64% of Noisy Speech-II and 44% of Noisy Speech-III could produce improvement in Signal to Noise Ratio (SNR). In the case of simulated noisy speech samples, 82% of Noisy Speech-I, 78% of Noisy Speech-II and 52% of Noisy Speech-III could produce improvement in Signal to Noise Ratio (SNR). The corresponding average values of Signal to Noise Ratio (SNR) for Noisy Speech-I, Noisy Speech-II and Noisy Speech-III of both original noisy speech and simulated noisy speech samples after FFT filter application are given in the Table 9.2.

MOS Test

In case of Noisy Speech-I, Noisy Speech-II and Noisy Speech-III of original noisy speech samples and simulated noisy speech samples, the results based on the Mean Opinion Score Test (MOS Test) after applying FFT filter is illustrated in the Tables 9.3 and 9.4 respectively.

Original Samples

Table 9.3 Average MOS test score for noisy speech-I, noisy speech-II and noisy speech-III

Score	Opinion	Impairment scale	Percentage of improvement
1	Unsatisfactory	Annoying and unacceptable	–
2	Poor	Annoying	–
3	Fair	Slightly annoying	10.67%
4	Good	Perceptible but acceptable	30.67%
5	Excellent	Imperceptible	58.67%

Simulated Samples

Table 9.4 Average MOS test score for noisy speech-I, noisy speech-II and noisy speech-III

Score	Opinion	Impairment scale	Percentage of improvement
1	Unsatisfactory	Annoying and unacceptable	–
2	Poor	Annoying	–
3	Fair	Slightly annoying	7.33%
4	Good	Perceptible but acceptable	22.00%
5	Excellent	Imperceptible	70.67%

Noise Reduction Filter Application

Improvement in Signal to Noise Ratio (SNR)

When noise reduction filter is applied, of the total number of original noisy speech samples 66% of Noisy Speech-I, 74% of Noisy Speech-II and 56% of Noisy Speech-III could produce improvement in Signal to Noise Ratio (SNR). In the case of simulated noisy speech samples, 70% of Noisy Speech-I, 78% of Noisy Speech II and 58% of Noisy Speech-III could produce improvement in Signal to Noise Ratio (SNR). The corresponding average values of Signal to Noise Ratio (SNR) for Noisy Speech-I, Noisy Speech-II and Noisy Speech-III of both original noisy speech and simulated noisy speech samples after noise reduction filter application are given in the Table 9.5.

MOS Test

In case of Noisy Speech-I, Noisy Speech-II and Noisy Speech-III of original noisy speech samples and simulated noisy speech samples, the results based on the Mean

Table 9.5 SNR improvement after applying Noise Reduction filter

Samples		Average improvement in dB
Noisy Speech-I	Original	3.71
	Simulated	4.85
Noisy Speech-II	Original	7.52
	Simulated	4.92
Noisy Speech-III	Original	5.54
	Simulated	5.41

Opinion Score Test (MOS Test) after applying noise reduction filter is illustrated in the Tables 9.6 and 9.7 respectively.

Original Samples

Table 9.6 Average MOS test score for noisy speech-I, noisy speech-II and noisy speech-III

Score	Opinion	Impairment scale	Percentage of improvement
1	Unsatisfactory	Annoying and unacceptable	–
2	Poor	Annoying	–
3	Fair	Slightly annoying	8%
4	Good	Perceptible but acceptable	26.67%
5	Excellent	Imperceptible	65.33%

Simulated Samples

Table 9.7 Average MOS test score for noisy speech-I, noisy speech-II and noisy speech-III

Score	Opinion	Impairment scale	Percentage of improvement
1	Unsatisfactory	Annoying and unacceptable	–
2	Poor	Annoying	–
3	Fair	Slightly annoying	8%
4	Good	Perceptible but acceptable	23.33%
5	Excellent	Imperceptible	68.67%

Noise Gate Application

Improvement in Signal to Noise Ratio (SNR)

When noise gate filter is applied, of the total number of original noisy speech samples 48% of Noisy Speech-I, 60% of Noisy Speech-II and 52% of Noisy Speech-III

Table 9.8 SNR improvement after applying Noise Gate filter

Samples		Average improvement in dB
Noisy Speech-I	Original	3.50
	Simulated	4.31
Noisy Speech-II	Original	7.13
	Simulated	4.85
Noisy Speech-III	Original	5.38
	Simulated	5.56

could produce improvement in Signal to Noise Ratio (SNR). In the case of simulated noisy speech samples, 52% of Noisy Speech-I, 70% of Noisy Speech-II and 56% of Noisy Speech-III could produce improvement in Signal to Noise Ratio (SNR). The corresponding average values of Signal to Noise Ratio (SNR) for Noisy Speech-I, Noisy Speech-II and Noisy Speech-III of both original noisy speech and simulated noisy speech samples after noise gate filter application are given in the Table 9.8.

MOS Test

In case of Noisy Speech-I, Noisy Speech-II and Noisy Speech-III of original noisy speech samples and simulated noisy speech samples, the results based on the Mean Opinion Score Test (MOS Test) after applying noise gate filter is illustrated in the Tables 9.9 and 9.10 respectively.

Original Samples

Table 9.9 Average MOS test score for noisy speech-I, noisy speech-II and noisy speech-III

Score	Opinion	Impairment scale	Percentage of improvement
1	Unsatisfactory	Annoying and unacceptable	–
2	Poor	Annoying	–
3	Fair	Slightly annoying	17.33%
4	Good	Perceptible but acceptable	29.33%
5	Excellent	Imperceptible	53.33%

Simulated Samples

Table 9.10 Average MOS test score for noisy speech-I, noisy speech-II and noisy speech-III

Score	Opinion	Impairment scale	Percentage of improvement
1	Unsatisfactory	Annoying and unacceptable	–
2	Poor	Annoying	–
3	Fair	Slightly annoying	15.33%
4	Good	Perceptible but acceptable	25.33%
5	Excellent	Imperceptible	59.33%

Table 9.11 SNR improvement after applying Notch filter

Samples		Average improvement in dB
Noisy Speech-I	Original	3.19
	Simulated	3.77
Noisy Speech-II	Original	6.85
	Simulated	4.70
Noisy Speech-III	Original	5.10
	Simulated	5.44

Notch Filter Application

Improvement in Signal to Noise Ratio (SNR)

When notch filter is applied, of the total number of original noisy speech samples 45% of Noisy Speech-I, 12.50% of Noisy Speech-II and 62.50% of Noisy Speech-III could produce improvement in Signal to Noise Ratio (SNR). In the case of simulated noisy speech samples, 45% of Noisy Speech-I, 37.50% of Noisy Speech-II and 62.50% of Noisy Speech-III could produce improvement in Signal to Noise Ratio (SNR). The corresponding average values of Signal to Noise Ratio (SNR) for Noisy Speech-I, Noisy Speech-II and Noisy Speech-III of both original and simulated samples after notch filter application are given in the Table 9.11.

MOS Test

In case of Noisy Speech-I, Noisy Speech-II and Noisy Speech-III of original noisy speech samples and simulated noisy speech samples, the results based on the Mean Opinion Score Test (MOS Test) after applying notch filter is illustrated in the Tables 9.12 and 9.13 respectively.

Original Samples

Table 9.12 Average MOS test score for noisy speech-I, noisy speech-II and noisy speech-III

Score	Opinion	Impairment scale	Percentage of improvement
1	Unsatisfactory	Annoying and unacceptable	–
2	Poor	Annoying	–
3	Fair	Slightly annoying	16.67%
4	Good	Perceptible but acceptable	36.11%
5	Excellent	Imperceptible	47.22%

Simulated Samples

Table 9.13 Average MOS test score for noisy speech-I, noisy speech-II and noisy speech-III

Score	Opinion	Impairment scale	Percentage of improvement
1	Unsatisfactory	Annoying and unacceptable	–
2	Poor	Annoying	–
3	Fair	Slightly annoying	16.67%
4	Good	Perceptible but acceptable	33.33%
5	Excellent	Imperceptible	50.00%

Table 9.14 SNR improvement after applying Band-pass filter

Samples		Average improvement in dB
Noisy Speech-I	Original	2.98
	Simulated	3.58
Noisy Speech-II	Original	5.07
	Simulated	3.18
Noisy Speech-III	Original	3.66
	Simulated	3.89

Band-Pass Filter

Improvement in Signal to Noise Ratio (SNR)

When band-pass filter is applied, of the total number of original noisy speech samples 48% of Noisy Speech-I, 44% of Noisy Speech-II and 30% of Noisy Speech-III could produce improvement in Signal to Noise Ratio (SNR). In the case of simulated noisy speech samples, 52% of Noisy Speech-I, 48% of Noisy Speech-II and 32%of Noisy Speech-III could produce improvement in Signal to Noise Ratio (SNR). The corresponding average values of Signal to Noise Ratio (SNR) for Noisy Speech-I, Noisy Speech-II and Noisy Speech-III of both original noisy speech and simulated noisy speech samples after band-pass filter application are given in the Table 9.14

MOS Test

In case of Noisy Speech-I, Noisy Speech-II and Noisy Speech-III of original noisy speech samples and simulated noisy speech samples, the results based on the Mean Opinion Score Test (MOS Test) after applying band-pass filter is illustrated in the Tables 9.15 and 9.16 respectively.

Original Samples

Table 9.15 Average MOS test score for noisy speech-I, noisy speech-II and noisy speech-III

Score	Opinion	Impairment scale	Percentage of improvement
1	Unsatisfactory	Annoying and unacceptable	–
2	Poor	Annoying	–
3	Fair	Slightly annoying	24.67%
4	Good	Perceptible but acceptable	34.67%
5	Excellent	Imperceptible	40.67%

Simulated Samples

Table 9.16 Average MOS test score for noisy speech-I, noisy speech-II and noisy speech-III

Score	Opinion	Impairment scale	Percentage of improvement
1	Unsatisfactory	Annoying and unacceptable	–
2	Poor	Annoying	–
3	Fair	Slightly annoying	22.67%
4	Good	Perceptible but acceptable	33.33%
5	Excellent	Imperceptible	44%

Table 9.17 SNR improvement after applying Butterworth filter

Samples		Average improvement in dB
Noisy Speech-I	Original	2.22
	Simulated	2.99
Noisy Speech-II	Original	5.40
	Simulated	3.25
Noisy Speech-III	Original	3.27
	Simulated	3.19

Butterworth Filter

Improvement in Signal to Noise Ratio (SNR)

When butterworth filter is applied, of the total number of original noisy speech samples 40% of Noisy Speech-I, 24% of Noisy Speech-II and 36% of Noisy Speech-III could produce improvement in Signal to Noise Ratio (SNR). In the case of simulated noisy speech samples, 48% of Noisy Speech-I, 28% of Noisy Speech-II and 40% of Noisy Speech-III could produce improvement in Signal to Noise Ratio (SNR). The corresponding average values of Signal to Noise Ratio (SNR) for Noisy Speech-I, Noisy Speech-II and Noisy Speech-III of both original noisy speech and simulated noisy speech samples after butterworth filter application are given in the Table 9.17.

MOS Test

In case of Noisy Speech-I, Noisy Speech-II and Noisy Speech-III of original noisy speech samples and simulated noisy speech samples, the results based on the Mean Opinion Score Test (MOS Test) after applying butterworth filter is illustrated in the Tables 9.18 and 9.19 respectively.

Original Samples

Table 9.18 Average MOS test score for noisy speech-I, noisy speech-II and noisy speech-III

Score	Opinion	Impairment scale	Percentage of improvement
1	Unsatisfactory	Annoying and unacceptable	–
2	Poor	Annoying	–
3	Fair	Slightly annoying	16%
4	Good	Perceptible but acceptable	50.67%
5	Excellent	Imperceptible	33.33%

Simulated Samples

Table 9.19 Average MOS test score for noisy speech-I, noisy speech-II and noisy speech-III

Score	Opinion	Impairment scale	Percentage of improvement
1	Unsatisfactory	Annoying and unacceptable	–
2	Poor	Annoying	–
3	Fair	Slightly annoying	16%
4	Good	Perceptible but acceptable	45.33%
5	Excellent	Imperceptible	38.67%

Digital Equalizer

Improvement in Signal to Noise Ratio (SNR)

When digital equalizer is applied, of the total number of original noisy speech samples 34% of Noisy Speech-I, 24% of Noisy Speech-II and 28% of Noisy Speech-III could produce improvement in Signal to Noise Ratio (SNR). In the case of simulated noisy speech samples, 46% of Noisy Speech-I, 28% of Noisy Speech-II and 36% of Noisy Speech-III could produce improvement in Signal to Noise Ratio (SNR). The corresponding average values of Signal to Noise Ratio (SNR) for Noisy Speech-I, Noisy Speech-II and Noisy Speech-III of both original noisy speech and simulated noisy speech samples after digital equalizer application are given in the Table 9.20.

Table 9.20 SNR improvement after applying Digital Equalizer

Samples		Average improvement in dB
Noisy Speech-I	Original	1.66
	Simulated	2.46
Noisy Speech-II	Original	3.84
	Simulated	2.05
Noisy Speech-III	Original	1.83
	Simulated	2.21

MOS Test

In case of Noisy Speech-I, Noisy Speech-II and Noisy Speech-III of original noisy speech samples and simulated noisy speech samples, the results based on the Mean Opinion Score Test (MOS Test) after applying digital equalizer is illustrated in the Tables 9.21 and 9.22 respectively.

Original Samples

Table 9.21 Average MOS test score for noisy speech-I, noisy speech-II and noisy speech-III

Score	Opinion	Impairment scale	Percentage of improvement
1	Unsatisfactory	Annoying and unacceptable	–
2	Poor	Annoying	–
3	Fair	Slightly annoying	37.33%
4	Good	Perceptible but acceptable	34%
5	Excellent	Imperceptible	30.67%

Simulated Samples

Table 9.22 Average MOS test score for noisy speech-I, noisy speech-II and noisy speech-III

Score	Opinion	Impairment scale	Percentage of improvement
1	Unsatisfactory	Annoying and unacceptable	–
2	Poor	Annoying	–
3	Fair	Slightly annoying	32.67%
4	Good	Perceptible but acceptable	30.67%
5	Excellent	Imperceptible	36.67%

Parametric Equalizer

Improvement in Signal to Noise Ratio (SNR)

When parametric equalizer is applied, of the total number of original noisy speech samples 20% of Noisy Speech-I, 36% of Noisy Speech-II and 26% of Noisy Speech-III could produce improvement in Signal to Noise Ratio (SNR). In the case

Table 9.23 SNR improvement after applying Parametric Equalizer

Samples		Average improvement in dB
Noisy Speech-I	Original	1.11
	Simulated	1.58
Noisy Speech-II	Original	2.45
	Simulated	1.87
Noisy Speech-III	Original	1.95
	Simulated	2.05

of simulated noisy speech samples, 24% of Noisy Speech-I, 50% of Noisy Speech-II and 34% of Noisy Speech-III could produce improvement in Signal to Noise Ratio (SNR). The corresponding average values of Signal to Noise Ratio (SNR) for Noisy Speech-I, Noisy Speech-II and Noisy Speech-III of both original noisy speech and simulated noisy speech samples after application of parametric equalizer are given in the Table 9.23.

MOS Test

In case of Noisy Speech-I, Noisy Speech-II and Noisy Speech-III of original noisy speech samples and simulated noisy speech samples, the results based on the Mean Opinion Score Test (MOS Test) after applying filter is illustrated in the Tables 9.24 and 9.25 respectively.

Original Samples

Table 9.24 Average MOS test score for noisy speech-I, noisy speech-II and noisy speech-III

Score	Opinion	Impairment scale	Percentage of improvement
1	Unsatisfactory	Annoying and unacceptable	–
2	Poor	Annoying	–
3	Fair	Slightly annoying	24.67%
4	Good	Perceptible but acceptable	48%
5	Excellent	Imperceptible	28.67%

Simulated Samples

Table 9.25 Average MOS test score for noisy speech-I, noisy speech-II and noisy speech-III

Score	Opinion	Impairment scale	Percentage of improvement
1	Unsatisfactory	Annoying and unacceptable	–
2	Poor	Annoying	–
3	Fair	Slightly annoying	21.33%
4	Good	Perceptible but acceptable	42.67%
5	Excellent	Imperceptible	36%

Table 9.26 Average values of SNR improvement by various filters in dB

Filters	Noisy Speech-I (in dB)			Noisy Speech-II (in dB)			Noisy Speech-III (in dB)		
	Original	Simulated	Average	Original	Simulated	Average	Original	Simulated	Average
FFT Filter	3.81	4.96	4.39	7.57	4.95	6.26	5.62	5.53	5.57
Noise reduction	3.71	4.85	4.28	7.52	4.92	6.22	5.54	5.41	5.47
Noise gate	3.50	4.31	3.90	7.13	4.85	5.99	5.38	5.56	5.47
Notch filter	3.19	3.77	3.48	6.85	4.70	5.78	5.10	5.44	5.27
Bandpass	2.98	3.58	3.28	5.07	3.18	4.13	3.66	3.89	3.78
Butterworth filter	2.22	2.99	2.61	5.40	3.25	4.33	3.27	3.19	3.23
Digital equalizer	1.66	2.46	2.06	3.84	2.05	2.95	1.83	2.21	2.02
Parametric equalizer	1.11	1.58	1.34	2.45	1.87	2.16	1.95	2.05	2.00

9.3.1.2 Results Based on Improvement in Signal to Noise Ratio (SNR)

The number of samples qualified for speech recognition is found out based on improvement in Signal to Noise Ratio (SNR) and MOS test conducted on the filtered original and simulated samples.

The Table 9.26 shows the SNR improvement by each filter for Noisy Speech-I Noisy Speech-II Noisy Speech-III of both original noisy speech and simulated samples noisy speech in terms of intensity respectively. Their average value of improvement in Signal to Noise Ratio (SNR) infers about the filters suitable for particular noisy sample recorded in particular mode of recording. Figure 9.2 implies the suitable Filters for Noisy Speech-I Noisy Speech-II and Noisy Speech-III respectively.

9.3.1.3 Results Based on MOS Test

The subjective quality of the speech is also taken care in the study. The perceptual features of speech are important in speech recognition. The results from the Mean Opinion Score Test are used to characterize the noise in terms of the subjective quality of the speech. Thus, it is also possible to find out appropriate filters for each group of noise recorded in three modes of recordings. Such characterization and the specific filters suitable for improving the intelligibility for efficient speech recognition is summed up, as shown in Table 9.27. Figure 9.3 gives the filter implications for noisy speech samples based on MOS Test.

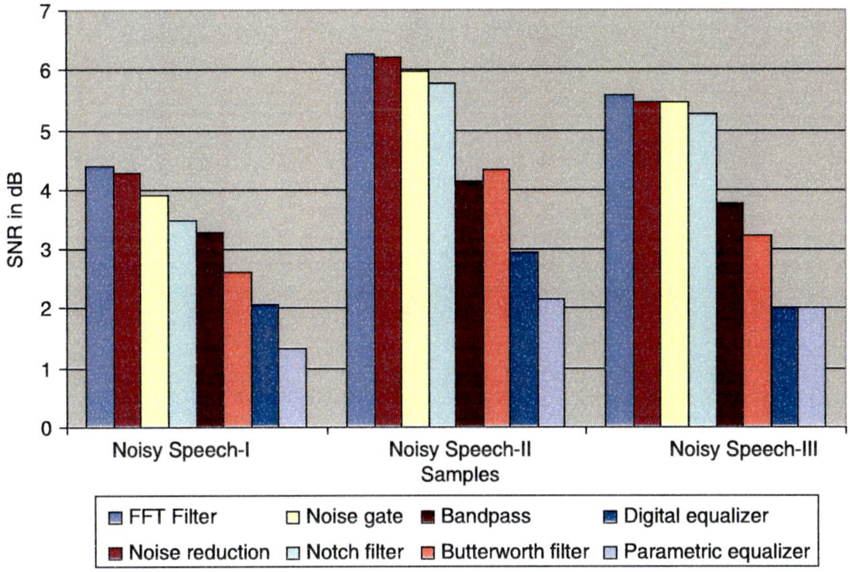

Fig. 9.2 Filter implication for noisy samples based on SNR improvement

Table 9.27 Average values of MOS test score for each sample for each filter

Filter	Noisy Speech-I			Noisy Speech-II			Noisy Speech-III		
	Original	Simulated	Average	Original	Simulated	Average	Original	Simulated	Average
FFT Filter	68%	82%	75%	64%	78%	71%	44%	52%	48%
Noise reduction	66%	70%	68%	74%	78%	76%	56%	58%	57%
Noise gate	48%	52%	50%	60%	70%	65%	52%	56%	54%
Notch filter	45%	45%	45%	12.50%	37.50%	25%	62.50%	62.50%	62.50%
Bandpass	48%	52%	50%	44%	48%	46%	30%	32%	31%
Butterworth filter	40%	48%	44%	24%	28%	26%	36%	40%	38%
Digital equalizer	34%	46%	40%	24%	28%	26%	28%	36%	32%
Parametric equalizer	20%	24%	22%	36%	50%	43%	26%	34%	30%

9.3.1.4 Statistical Study

The improvements produced by various filtering techniques were studied in a statistical way. The average number of noisy speech samples improved in intelligibility and thus qualified for Speech recognition is given in percentage as follows.

FFT Filter, when applied to the original noisy speech samples, 66.67% of the samples have shown improvement whereas simulated noisy speech samples gave an improvement of 70.67%.

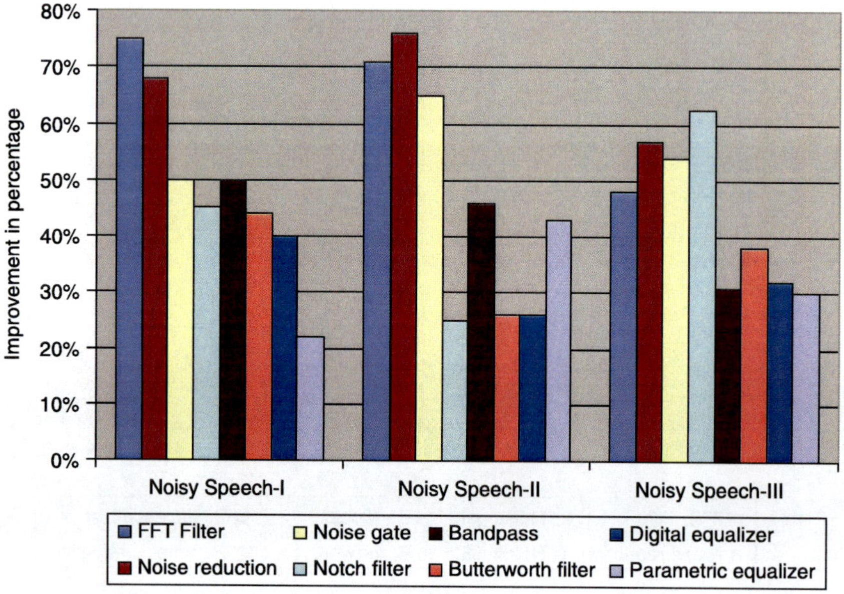

Fig. 9.3 Filter implications for noisy speech samples based on MOS Test

An improvement of 65.33% in original noisy speech and 68.67% in simulated noisy speech samples were observed when noise reduction filter is applied.

Noise gate filter improved the original noisy speech samples by 53.33% and simulated noisy speech samples by 56.67% to qualify for speech recognition.

Selective application of notch filter gave an improvement of 47.22% in the original noisy speech samples and 50% improvement in simulated noisy speech samples.

The application of bandpass filter produced an improvement of 40.67% in original noisy speech and 44% in simulated noisy speech samples.

Butterworth filter application on original noisy speech samples produced 33.33% of improvement and when applied on simulated noisy speech samples it gave 38.67% of improvement.

Digital equalizer could able to produce better results for 30.67% of the original noisy speech samples and 36.67% of the simulated noisy speech samples.

Lastly, parametric equalizer, on their application on original noisy speech samples and simulated noisy speech samples results in an improvement of 28.67 and 35.33% respectively thus qualifying for speech recognition.

In terms of Signal to Noise Ratio (SNR) FFT filter is able to produce maximum SNR improvement in both the original noisy speech sample as well as simulated noisy speech samples. It is noticed that an estimation of noise only portion can enhance the performance of FFT filter in improving the intelligibility of the speech signal. The filters are ranked and tabulated based on their performance on improving the Signal to Noise Ratio (SNR) and thus intelligibility of the original noisy speech and simulated noisy speech samples in all the three modes.

The results obtained by MOS Test conducted are verified by energy level of the speech signal in the energy contour of the samples before and after applying the filters.

9.3.2 Discussion of the Effect of Application of Filter for Speaker Identification

Selective and controlled filtering is carried out on original noisy speech and simulated noisy speech samples for the purpose of Speaker Identification. Preliminary experiments have shown that the application of FFT filter, digital equalizer and parametric equalizer can produce variations in speaker dependent characteristics. In order to preserve the speaker dependent characteristics, the applications of these filters are restricted.

9.3.2.1 Frequency Domain Analysis

Frequency domain analysis is carried out on both the original noisy speech and simulated noisy speech noisy speech samples before and after applying the filters. Frequency domain analysis is helpful in order to estimate the variation occurring in the frequency component due to noise. For this FFT analysis, Long Term Averaging (LTA) and LPC analysis were used in the study.

FFT Analysis

More accuracy in characterization and classification of noise is attained by comparing the data obtained by FFT analysis before and after the application of filters on both the original noisy speech and simulated noisy speech samples. Thus, FFT is successfully used as a tool in determining the specific filter with optimum parametric values for efficient speech enhancement for speaker identification.

There have been very high variations in speaker dependent characteristics by the application of FFT filter, digital equalizer and parametric equalizer. This is established by comparing the values obtained by the FFT analysis followed by analyzing the subjective quality of the speech by critical listening.

Long Term Average (LTA) Analysis

Long-term average spectrum (LTAS) is one of the basic characteristics of a digital signal and it shows the energy distribution over the frequency band. It is one of the basic statistical features for the arbitrary digital signal. To obtain the energy distribution of the whole sample in frequency domain, the best way is to calculate the long-term average spectrum (LTAS).

The LTA analysis for all the three modes of recordings namely, Noisy Speech-I, Noisy Speech-II, and Noisy Speech-III were carried out.

Original Samples

The energy distribution over the frequency domain of each samples were obtained. A comparative study is conducted over the values obtained for original noisy speech and simulated noisy speech samples.

This analysis helped in determining the optimum parametric values to be assigned for each filtering techniques by comparing the standard deviation of the randomness of the signal before and after filtering. LTA analysis and the comparative study of the values obtained for original noisy speech and simulated noisy speech samples results in determining the specific filter suitable for particular group of noise.

Thus the analysis helped in characterizing the noise in terms of the filter suitable for reducing the particular noise. A final decision in this is taken by studying the formant frequency values obtained for the original noisy speech and simulated noisy speech samples before and after applying the filters.

Linear Predictive Coding (LPC) Analysis

The analysis is carried out for original noisy speech, filtered original noisy speech, reference speech signal (RSS), simulated noisy speech and filtered simulated noisy speech for all the three modes of recordings, namely, Noisy Speech-I, Noisy Speech-II, Noisy Speech-III.

The formant values in each case are compared in order to find out the best suitable filter that produced minimum shift. In the case of noise reduction filter the analysis indicates that the controlled application of this filter can produce enhanced samples qualified for speaker identification with negligible shift in the formant frequencies, namely, F1, F2 and F3.

Noisy Speech-I

Formant values namely, F1, F2 and F3 of original noisy speech samples, reference speech signal (RSS) and simulated samples before and after the application of filters are found out by the LPC analysis. The values are tabulated and represented graphically. Based on the shift occurred in formant values of the speaker, efficiency of the filters are decided.

The controlled application of five filters namely, noise reduction, noise gate, notch filter, butterworth and band-pass produced improvement in the noisy speech. The purpose of applying these filters is to identify the speaker. It is checked and established that the formant values F1, F2 and F3 are affected to a minimum. For this, results from the experiment on original noisy speech samples are compared with the results obtained on Reference Speech Signal (RSS) and simulated speech sample. The results are illustrated in the Tables 9.28, 9.29, 9.30, 9.31, 9.32, 9.33 and Figs. 9.4, 9.5, 9.6, 9.7, 9.8, 9.9, 9.10, 9.11, 9.12, 9.13, 9.14, 9.15.

Original Samples

Table 9.28 Formant analysis of Noisy speech-I (representative sample) before and after applying filters

Samples		Original	Noise Reduction	Noise Gate	Notch Filter	Butterworth	Band Pass
Formant (Hz)	F1	1168.78	1168.77	1201.94	1283.85	879.88	882.86
	F2	2152.39	2152.38	2287.65	2881.75	1456.33	1450.01
	F3	3739.20	3739.08	3705.54	5490.61	2144.92	2075.14
Bandwidth (Hz)	B1	756.70	756.66	563.31	337.69	200.39	175.81
	B2	72.10	72.10	628.11	355.91	128.57	103.08
	B3	1215.24	216.06	457.08	373.57	45.26	386.89

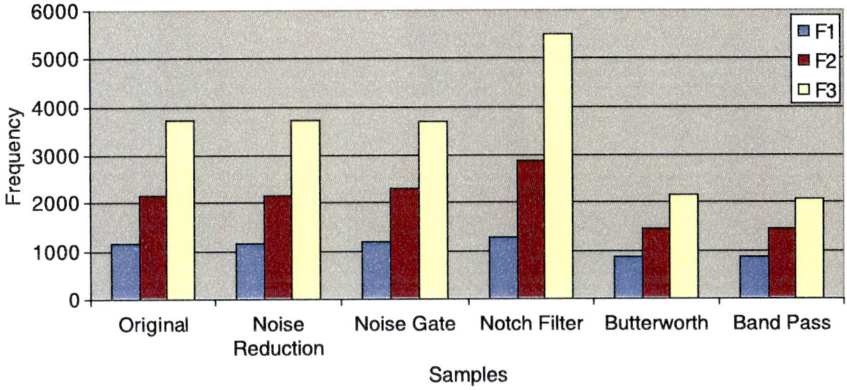

Fig. 9.4 Formant frequencies of Noisy Speech-I and filtered signal

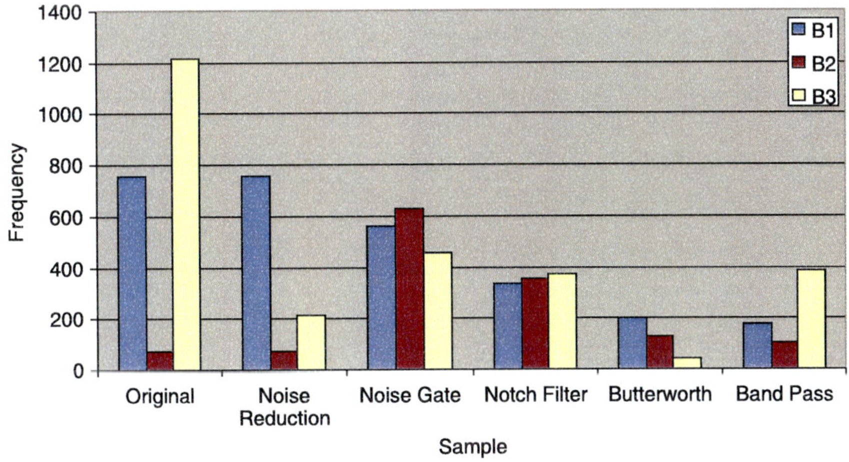

Fig 9.5 Bandwidth of Noisy Speech-I and filtered signal

Table 9.29 Formant analysis of Noisy speech-II (representative sample) before and after applying filters

Samples		Original	Noise Reduction	Noise Gate	Notch Filter	Butterworth	Band Pass
Formant (Hz)	F1	1166.28	1165.97	1198.94	1280.85	875.88	877.86
	F2	2149.89	2149.58	2284.65	2878.75	1452.33	1445.01
	F3	3736.7	3736.28	3702.54	5487.61	2140.92	2070.14
Bandwidth (Hz)	B1	744.70	743.66	547.31	320.69	191.39	169.81
	B2	60.10	59.10	612.11	338.91	119.57	97.08
	B3	1203.24	203.06	441.08	356.57	36.26	380.89

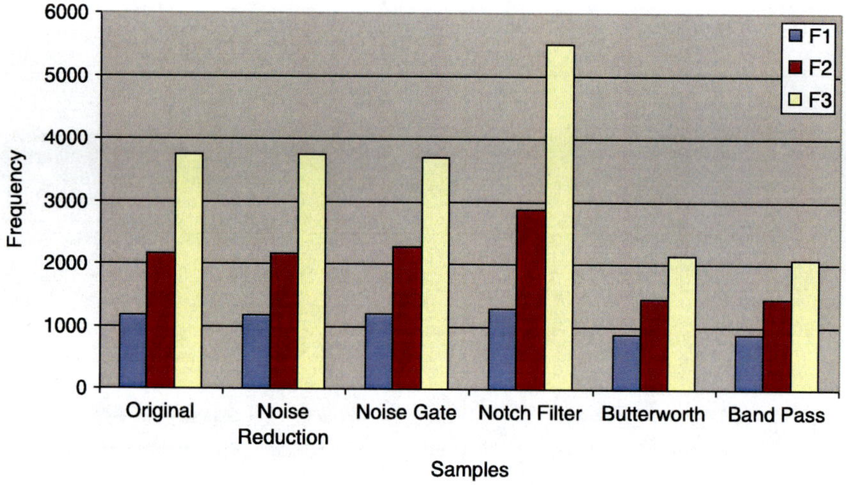

Fig. 9.6 Formant frequencies of Noisy Speech-II and filtered signal

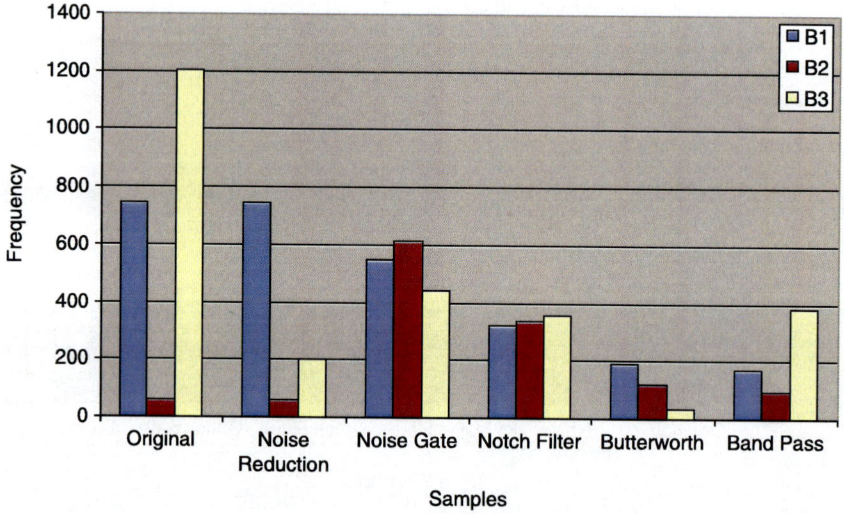

Fig. 9.7 Bandwidth of Noisy Speech-II and filtered signal

Table 9.30 Formant analysis of Noisy speech-III (representative sample) before and after applying filters

Samples		Original	Noise Reduction	Noise Gate	Notch Filter	Butterworth	Band Pass
Formant (Hz)	F1	1143.78	1143.67	1171.94	1251.85	843.88	824.86
	F2	2127.39	2127.28	2257.65	2849.75	1420.33	1392.01
	F3	3714.20	3713.98	3675.54	5458.61	2108.92	2017.14
Band-width (Hz)	B1	717.70	716.86	518.31	277.69	180.39	255.81
	B2	33.10	32.30	583.11	295.91	108.57	183.08
	B3	1176.24	176.26	412.08	313.57	25.26	466.89

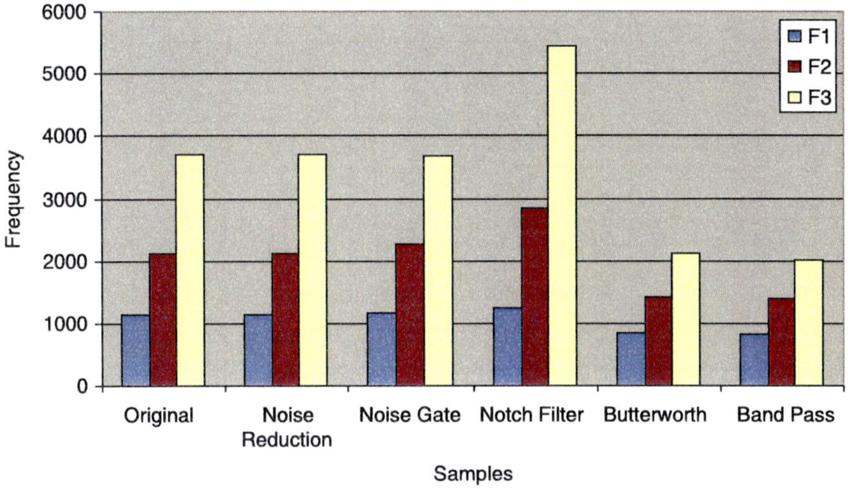

Fig. 9.8 Formant frequencies of Noisy Speech-III and filtered signal

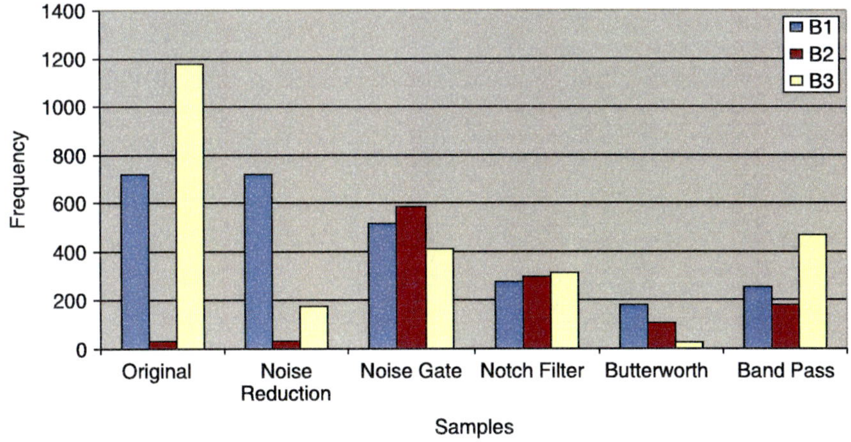

Fig. 9.9 Bandwidth of Noisy Speech-III and filtered signal

Simulated Samples

Table 9.31 Formant analysis of RSS and simulated Noisy Speech-I (representative sample) before and after applying filters

Samples		Reference Speech Signal (RSS)	Simulated	Noise Reduction	Noise Gate	Notch Filter	Butterworth	Band Pass
Formant (Hz)	F1	1169.16	1170.52	1170.48	1204.59	1290.43	885.48	890.61
	F2	2152.53	2150.24	2150.41	2291.38	2892.25	1496.52	1451.40
	F3	3740.57	3741.45	3741.26	3714.72	4021.37	2127.21	2093.01
Bandwidth (Hz)	B1	735.09	760.10	760.23	612.14	380.94	232.43	201.26
	B2	110.15	70.36	70.46	663.57	374.11	143.76	122.10
	B3	358.55	1220.48	456.78	478.03	381.2	83.20	3986.29

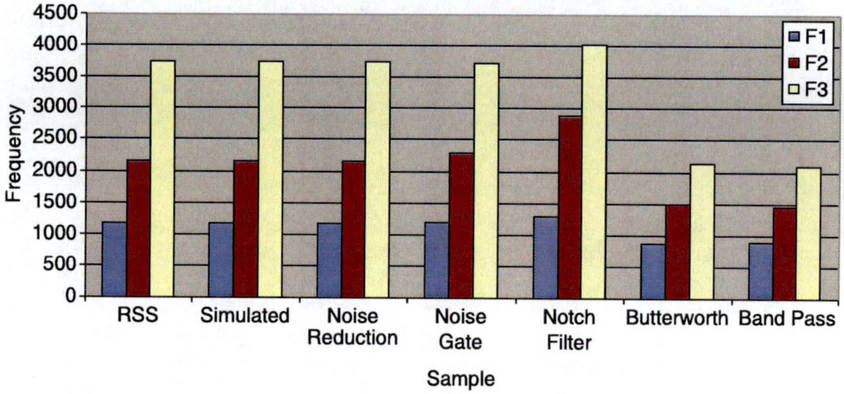

Fig. 9.10 Formant frequencies of RSS and Noisy Speech-I and filtered signal

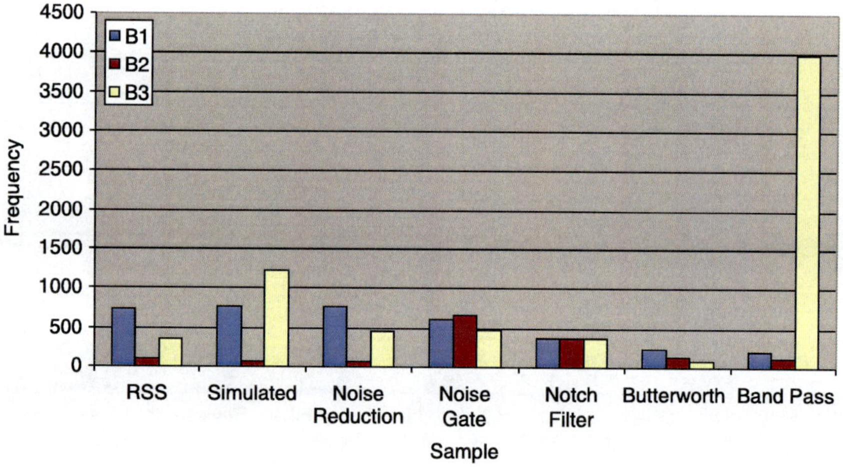

Fig. 9.11 Bandwidth of RSS and Noisy Speech-I and filtered signal

Table 9.32 Formant analysis of RSS and simulated Noisy Speech-II (representative sample) before and after applying filters

Samples		Reference Speech Signal (RSS)	Simulated	Noise Reduction	Noise Gate	Notch Filter	Butter-worth	Band Pass
Formant (Hz)	F1	1166.19	1166.52	1166.28	1199.59	1284.43	875.48	876.61
	F2	2146.6	2146.24	2146.21	2286.38	2886.25	1486.52	1437.40
	F3	3738.69	3737.45	3737.06	3709.72	4015.37	2117.21	2079.01
Bandwidth (Hz)	B1	794.46	745.10	743.23	593.14	358.94	209.43	196.26
	B2	74.46	55.36	53.46	644.57	352.11	120.76	117.10
	B3	465.78	1205.48	439.78	459.03	359.20	60.20	3981.29

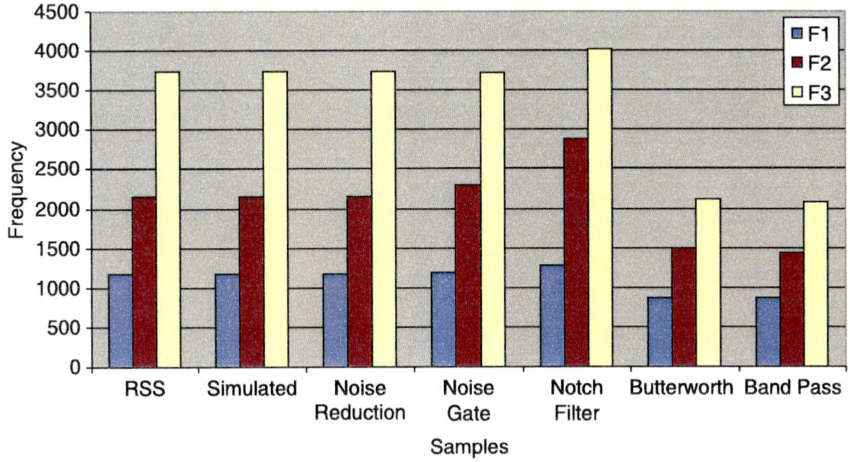

Fig. 9.12 Formant frequencies of RSS and Noisy Speech-II and filtered signal

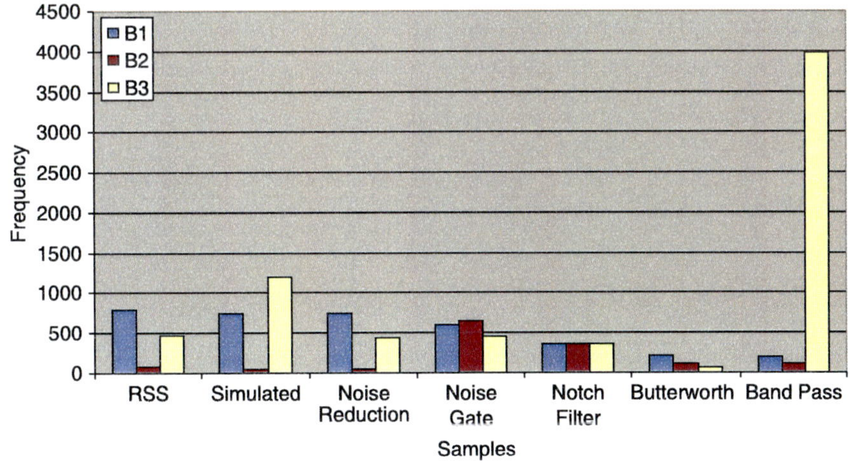

Fig. 9.13 Bandwidth of RSS and Noisy Speech-II and filtered signal

Table 9.33 Formant analysis of simulated RSS and Noisy Speech-III (representative sample) before and after applying filters

Samples		Reference Speech Signal (RSS)	Simu-lated	Noise Reduc-tion	Noise Gate	Notch Filter	Butter-worth	Band Pass
Formant (Hz)	F1	1206.14	1205.52	1205.84	1222.59	1374.43	794.48	877.61
	F2	2185.45	2185.24	2185.77	2309.38	2976.25	1405.52	1438.4
	F3	3776.74	3776.45	3776.62	3732.72	4105.37	2036.21	2080.01
Bandwidth (Hz)	B1	839.23	734.1	810.23	548.14	337.94	160.43	140.26
	B2	60.46	44.36	120.46	599.57	331.11	71.76	61.1
	B3	291.78	1194.48	506.78	414.03	338.2	11.20	3925.29

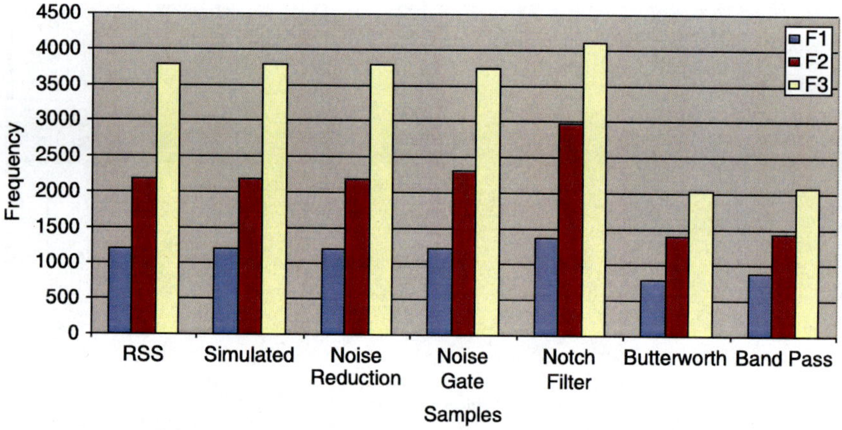

Fig. 9.14 Formant frequencies of RSS and Noisy Speech-III and filtered signal

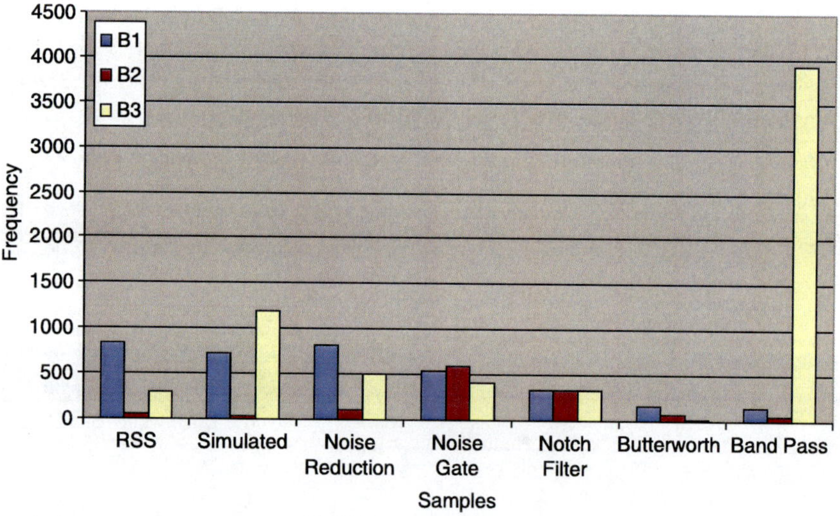

Fig. 9.15 Bandwidth of RSS and Noisy Speech-III and filtered signal

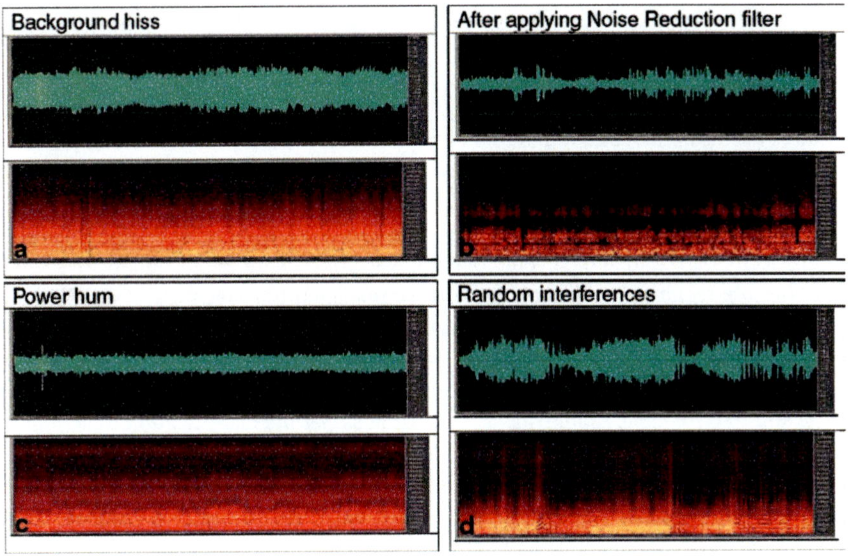

Fig. 9.16 Waveform and their corresponding spectrograms for **a** background hiss, **b** after applying noise reduction filter, **c** power hum and **d** random interferences

Plots of several practical noise signals such as background hiss, power hum, random interferences, etc. came across during the experiment are given below. In addition, plot of hiss noise when passed through the noise reduction filter and their corresponding spectrograms are also shown in Fig. 9.16a–d.

As perceptual features are very much important in the case of speaker identification, listening tests are conducted to ensure that the perceptual features of the original noisy speech are preserved while applying filters. The filtered speech samples are then subjected to critical listening by the trained listeners. Finally, an opinion is made based on the method of Mean Opinion Score (MOS); the most widely accepted method for speech quality evaluation and a simplest subjective measure for the assessment that gives an overall opinion of the performance.

Noise reduction filter has been widely accepted for removing the background hiss, power hum or random interference. 60% of Noisy Speech-I, 64% of Noisy Speech-II and 52% of Noisy Speech-III has become useful for the analysis by the application of this technique. Out of one hundred and fifty exemplars of simulated samples after applying this filtering technique an improvement of 64, 70 and 54% for Noisy Speech-I, Noisy Speech-II and Noisy Speech-III respectively were achieved.

Application of noise gate is effective only in removing the background hiss from quiet parts (i.e., any place where the speaker paused or stopped talking). Results from our experiment have also emphasized this fact. 44% of the Noisy Speech-I, 52% of the Noisy.

Speech-II and 46% of the Noisy Speech-III noisy speech samples became useful for the analysis by the application of this technique among the original noisy samples. In the case of simulated noisy speech samples, 48% of the Noisy Speech-I, 58% of the Noisy Speech-II and 50% of the Noisy Speech-III noisy speech samples became useful for the analysis. Noise gate is not efficiently applicable if the speech region is not embedded by the characteristics of the noise in the gap and it does not remove background hiss from louder parts of the selection.

Notch filter has been widely accepted for removing a particular frequency component from the signal. 40% of Noisy Speech-I, 12.50% of Noisy Speech-II and 58.33% of Noisy Speech-III has become useful for the analysis by the application of this technique. A total of seventy two noisy samples were found suitable to apply this filtering technique. For the simulated samples, an improvement of 45, 25 and 58.33% for Noisy Speech-I, Noisy Speech-II and Noisy Speech-III respectively were achieved in the total 72 samples.

In order to preserve maximum speaker specific information, when band-pass filter is applied, frequency range has been chosen as per the classification of samples (Noisy Speech-I, Noisy Speech-II and Noisy Speech-III). In the original noisy samples, 42% of noisy samples of Noisy Speech-I, made significant improvement. In case of Noisy Speech-II and Noisy Speech-III, there were 40 and 28% improvement of samples respectively. In the simulated noisy speech samples, for Noisy Speech-I, 46% of noisy samples made significant improvement. In case of Noisy Speech-II, 44% of improved version is obtained. The improvement in Noisy Speech-III speech samples is 30% when subjected to band-pass filtering.

As this filtering technique is flexible it does not affect the desired frequency band and hence it is of great value for speaker identification purposes. The promiscuous speech exemplars in which the noise falls in the broadband, this technique is found to be unsuitable. The experiment with exclusively noisy exemplars of corresponding speech exemplars stands as the ground for this.

Butterworth filter has been applied to both the original noisy speech and simulated noisy speech samples. 36% of Noisy Speech-I, 22% of Noisy Speech-II and 32% of Noisy Speech-III has qualified for speaker identification by the application of this technique. For the simulated samples, an improvement of 42, 24 and 36% for Noisy Speech-I, Noisy Speech-II and Noisy Speech-III respectively were achieved in the total one hundred and fifty samples.

Corresponding filters have been identified based on the contributed enhancement upon noisy speech signal, which amount to the fruitful speaker identification test. Among the filters applied, noise reduction filter is most successful in improving the Signal to Noise Ratio (SNR) as well as preserving speaker dependent features of the speech signal, than any other filters when applied to noisy speech samples and corresponding simulated samples.

It is ensured that the speaker-specific information is preserved during the process of noise reduction by various filters. A comparative study upon the performance of each filter is conducted and specific filters are identified for speech enhancement for the noisy speech samples.

Based on the inference from the subsequent study conducted upon the results of various analyses, appropriate filters for each class of noise associated with forensic speech samples is identified and recommended for speaker identification. The char-

Table 9.34 Average values of improvement by various filters in percentage

Filter	Noisy Speech-I			Noisy Speech-II			Noisy Speech-III		
	Original	Simulated	Average	Original	Simulated	Average	Original	Simulated	Average
Noise reduction	60.00%	64.00%	62.00%	64.00%	70.00%	67.00%	52.00%	54.00%	53.00%
Noise gate	44.00%	48.00%	46.00%	52.00%	58.00%	55.00%	46.00%	50.00%	48.00%
Notch filter	40.00%	45.00%	42.50%	12.50%	25.00%	18.75%	58.33%	58.33%	58.33%
Bandpass	42.00%	46.00%	44.00%	40.00%	44.00%	42.00%	28.00%	30.00%	29.00%
Butterworth filter	36.00%	42.00%	39.00%	22.00%	24.00%	23.00%	32.00%	36.00%	34.00%

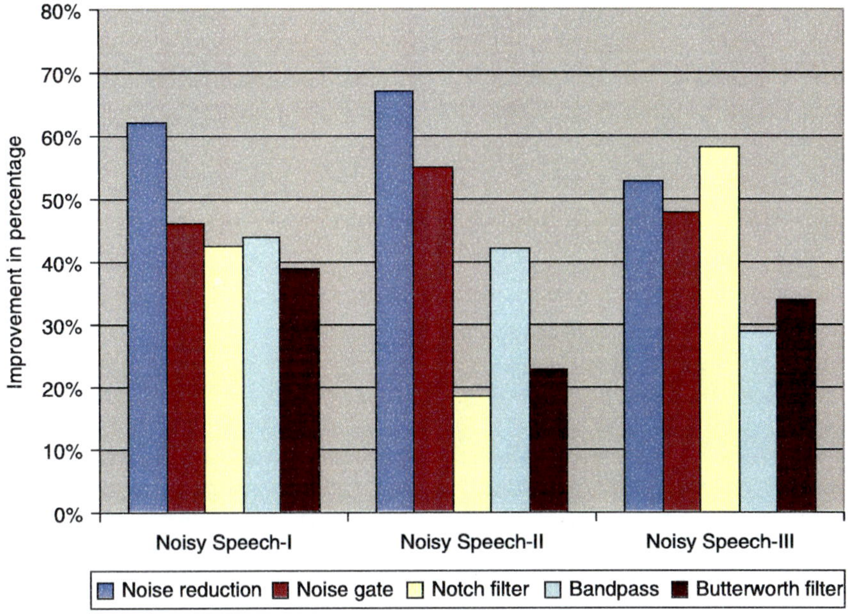

Fig. 9.17 Average Improvements by various filters in percentage

acterization of noise associated with the forensic speech samples and their classification are performed. When Noise reduction, Noise gate, Notch filter, Bandpass and Butterworth filters are applied to original and simulated samples of Noisy Speech-I, Noisy Speech-II Noisy Speech-III, percentage of improvement made in number of samples are tabulated in Table 9.34. A graphical representation of percentage of improvement is given in Fig. 9.17.

Plots of noise having room transfer function, music, speech of no interest, traffic noise, factory noise and noise due to mechanical fault in the recording device came across during the experiment are given below. In addition, plot of a typical real life noise before and after filtering through the notch filter and their corresponding spectrograms are also shown below in Fig. 9.18a–g.

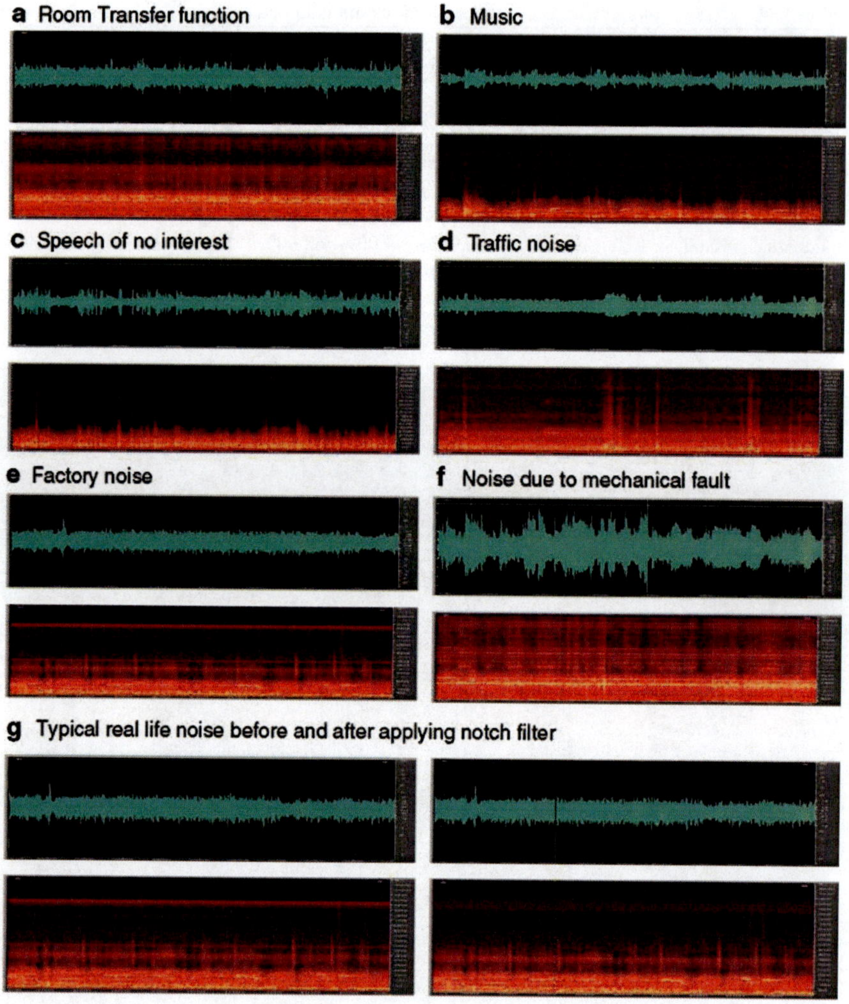

Fig. 9.18 Waveform and their corresponding spectrograms of **a** room transfer function, **b** music, **c** speech of no interest, **d** traffic noise, **e** factory noise, **f** noise due to mechanical fault, **g** typical real life noise before and after applying notch filter

While applying the filters to the Noisy Speech-I, i.e., the noisy speech recorded in direct recording mode, it is observed that the major part of the convolutional noise which is the product of the transfer function embedded with the recorded speech signal falls in the white Gaussian noise region. Such noise falls in the entire frequency range of the spectrum. In addition to this it is observed that random noise also occur frequently in such recordings. The critical listening revealed the fact that in forensic context, the noise embedded with the Noisy Speech-I is the sum of the environmental noise, a product of the room transfer function, music, speech of no interest, random noise, traffic noise, factory noise and noise due to the mechanical fault in the recording device etc. Hence the filtering of the Noisy Speech-I produced the least improvement in Signal to Noise Ratio (SNR) as compared to Noisy Speech-II and Noisy Speech-III. This is verified by applying the filters to the Noisy Speech-I of the simulated noisy speech samples and comparing the values of the analysis.

The experiment on Noisy Speech-II show that the noise embedded with the speech recorded through telephone channel is mainly due to the noise produced by the communication channel itself. The analysed noisy speech samples contain additive noise which is the product of channel distortion. The major part of the noise in this type of noisy speech signal is humming and hissing in nature. It is also observed that at the speaker's end, background noises such as music, conversation between other people in the background, cross talk, instrumental fault (due to mechanical fault the instrument produces impulse response in a random manner) etc. contribute to the noise in the channel. In addition to this the near far problem that reduces the Signal to Noise Ratio of the recording is observed in such recordings.

The main contribution of noise in Noisy Speech-III is from the interference during the transmission. Adjacent Channel Interference, Co-channel Interference, Intermodulation Interference and Intersymbol Interference are observed in this type of recording. It is observed that the recordings carried out in heavily crowded area results in poor signal quality in terms of clipping of particular frequency, cross talk, reduced amplitude and thus low Signal to Noise Ratio (SNR). The results from the application of various filters indicate that FFT filter, noise reduction filter and notch filter are efficiently able to reduce the noise for speech recognition. In the case of speaker identification noise reduction filter and noise gate filter are able to enhance the speech most efficiently.

Thus the characterization of noise associated with forensic speech samples is possible for all the three modes of recordings, namely Noisy Speech-I, Noisy Speech-II and Noisy Speech-III for speech recognition and speaker identification purposes. Based on this study appropriate filters are identified for different types of embedded noise in the Noisy Speech-I, Noisy Speech-II and Noisy Speech-III for speech enhancement. The Tables 9.26 and 9.27 is self explanatory in deciding specific filter for a particular group of noise for efficient speech enhancement for speech recognition.

Thus a twofold approach of noise characterization is successfully carried out for efficient speech enhancement for speech recognition in noisy recordings produced from three modes of recordings, namely, direct, telephonic and mobile.

9.3.2.2 Statistical Study

Statistical study is conducted over the results produced by the various filtering techniques and the subsequent analysis by CSL. The percentage of improvement by each filters are calculated and compared for both original noisy speech and simulated noisy speech samples.

A comparative study of systematic noise filtering by comparing the distortion level of the filtered simulated signal with Reference Speech Signal (RSS) is carried out. This helps in deciding degree of quality of the speech signal retrieved for speaker identification in noisy conditions. From the resultant signal, speech is reclaimed to a maximum extent by nullifying the effect of embedded noise. This in turn helps to decide the degree of quality of speech signal retrieved for speaker identification under noisy conditions in terms of speaker dependent characteristics.

The subsequent improvements in the noisy speech are as under. Noise reduction filter produced 58.67% improvement in original noisy speech samples and 62% of improvement in simulated noisy speech samples.

The application of Noise gate is able to produce 47.33% of improvement in original noisy speech samples and 50.67% in simulated noisy speech samples.

Notch filter, upon their application on selective samples both in original noisy speech and simulated noisy speech samples improved 43.06 and 47.22% respectively.

Bandpass filters when applied to both original noisy speech and simulated noisy speech samples it gave 36.67 and 40.67% improvement respectively.

An improvement of 30% in the original noisy speech samples and 34% in the simulated noisy speech samples were observed when butterworth filter is applied.

Such comparative and statistical study of improvised original noisy speech and simulated noisy speech samples after filtering have revealed the degree of efficiency of different filters for speaker identification and how far they are dependable in forensic adverse contexts.

9.4 Conclusion

Application of different filtering techniques (e.g., FFT Filter, noise reduction, noise gate, notch filter, bandpass, butterworth filter, digital equalizer and parametric equalizer) for noise reduction (or suppression) is found to be of great help for improving SNR and MOS. Furthermore, this finding will help to improve the performance of speech and speaker recognition systems under degraded signal conditions.

The outcome of the study on the basis of improvement in Signal to Noise ratio (SNR) and Mean Opinion Score Test (MOS Test) is significant. For successful speech recognition, it is very important that in order for speech to be recognized it must be intelligible.

In the case of speaker identification, limitation of filtering technique is also considered; depending on various parameters of the noise, which in turn can help in retaining the speaker dependent features. The specific filters for enhancement of recorded speech from different characterized noise for speaker identification is recommended based on their effect on perceptual and acoustic features. For speech recognition, the degree of efficiency of filters in enhancing the speech signal is found to be in a descending order; viz. FFT filter, noise reduction, noise gate, notch filter, bandpass, butterworth filter, digital equalizer and parametric equalizer. The degree of efficiency of filters in enhancing the speech signal for speaker identification is found to be in a descending order; viz. noise reduction, noise gate, notch filter, bandpass, and butterworth filter. Though there is improvement in the auditory perceptual features, distortion in speaker dependent characteristics is observed when digital equalizer, parametric equalizer and FFT filter are applied. Thus, the applications of digital equalizer, parametric equalizer and FFT filter are not recommended for speaker identification task.

FFT filter is found to be the most promising technique when only speech has to be recognized from a noisy recorded speech in any of the three modes of recordings (direct, telephonic and mobile) by improvising the SNR and acoustic features of the recorded speech. The noise reduction filtering technique for enhancing the noisy speech recorded in any of the three modes of recordings (direct, telephonic and mobile) proved to be efficient in terms of improving the SNR and preserving the speaker dependent features of the speech signal and hence is the most appropriate filter capable of qualifying the noisy speech for speaker identification purpose. Intelligibility and speaker dependent characteristics are found to be inversely proportional to each other in a non-linear manner when filters are applied. For better performance it is recommended to apply filter(s) for speech recognition and speaker identification both separately.

The credibility of increasing speech intelligibility on applying filters to the noisy speech samples is observed to be dependent on the sample, each in a unique manner. The near perfection in this effort is achieved by treating their noise only counterparts with the same filters. When filter(s) is/are applied on noisy speech samples, the elimination of noisy part containing information is unpredictable. In addition, speaker dependent characteristics and intelligibility are found to be inversely proportional to each other in a non-linear way when filter(s) is/are applied. Noise reduction filter is found to be the most promising technique for enhancing the noisy speech recorded in any of the three modes of recordings (direct, telephonic and mobile) by preserving the perceptual and acoustic features. By this study appropriate filter(s) can be decided to detect and reduce/eliminate the different classes of noises embedded in the speech signal for speaker identification.

The number of samples that qualified for speech recognition is found out based on SNR improvement and Mean Opinion Score Test (MOS Test) conducted on the filtered original noisy speech and simulated noisy speech samples. Based on this study appropriate filters are identified for different types of embedded noise.

Noisy Speech-I which is predominantly of the convolutional noise as a product of transfer function embedded with the recorded speech signal falls in the white Gaussian Noise region. Random noise does occur in some of the noisy speech samples and application of filter on the Noisy Speech-I results comparatively least improvement in the Signal to Noise Ratio (SNR).

Noisy Speech-II characterized as a noise produced by the communication channel as well the additive random noise similar to the Noisy Speech-I. Application of Noise removal filters namely, FFT filter for speech recognition and noise reduction filter for speaker identification respectively.

The results from the application of various filters indicate that FFT filter, noise reduction filter and notch filter are efficiently able to reduce the noise for speech recognition. In the case of speaker identification noise reduction filter and noise gate filter are able to enhance the speech most efficiently.

Thus the characterization of noise associated with forensic speech samples is possible for speech recognition purpose. The results from the Mean Opinion Score Test (MOS Test) are used to characterize the noise in terms of subjective quality of the speech. Thus it is also possible to find out appropriate filters for each group of noise recorded in three modes of recordings. Such characterization and the specific filters suitable for improving the intelligibility for efficient speech recognition are attained.

Thus a twofold approach of noise characterization is successfully carried out for efficient speech enhancement for speech recognition in noisy recordings produced from three modes of recordings, namely, direct, telephonic and mobile.

Based on the inference from the subsequent study conducted upon the results of various analyses, namely, appropriate Filters for each class of noise associated with forensic speech samples is identified and recommended for speaker identification. The characterization of noise associated with the forensic speech samples and their classification are performed. The result of the study is also relevant to other forensic audio problems such as audibility analysis, Authenticity analysis, event sequence analysis and other signal analysis.

Preference is given to the intelligibility of the speech signal while carrying out the steps for speech recognition. In the case of speaker identification, limitation of filtering technique is also considered; depending on various parameters of the noise, which in turn can help in retaining the speaker dependent features. The specific filters for enhancement of recorded speech from different characterized noise for speaker identification is recommended based on their effect on perceptual and acoustic features. For speech recognition, the degree of efficiency of filters in enhancing the speech signal is found to be in a descending order; viz. FFT filter, noise reduction, noise gate, notch filter, bandpass, butterworth filter, digital equalizer and parametric equalizer. The degree of efficiency of filters in enhancing the speech signal for speaker identification is found to be in a descending order; viz. noise reduction, noise gate, notch filter, bandpass, and butterworth filter. Though there is improvement in the auditory perceptual features, distortion in speaker dependent characteristics is observed when digital equalizer, parametric equalizer and FFT filter are applied. Thus, the applications of digital equalizer, parametric equalizer and FFT filter are not recommended for speaker identification task.

FFT filter is found to be the most promising technique when only speech has to be recognized from a noisy recorded speech in any of the three modes of recordings (direct, telephonic and mobile) by improvising the SNR and acoustic features of the recorded speech. Noise reduction filtering technique for enhancing the noisy speech recorded in any of the three modes of recordings (direct, telephonic and mobile); proved to be efficient in terms of improving the SNR and preserving the speaker dependent features of the speech signal and hence is the most appropriate filter capable of qualifying the noisy speech for speaker identification purpose. In addition, Intelligibility and speaker dependent characteristics are found to be inversely proportional to each other in a non-linear manner when filters are applied. For better performance it is recommended to apply filter(s) for speech recognition and speaker identification both separately. The result of the study is also relevant to other forensic audio problems such as audibility analysis, authenticity analysis, event sequence analysis and other signal analysis.

9.4.1 Future Research Directions

Keeping this study as a ground, specific customized Software tools also can be developed for the automatic detection of class of noise in the Forensic speech samples and to reduce/eliminate the same. The study presented in this chapter may be of help for the automatic detection of class of noise in the forensic speech samples and to reduce/eliminate the same.

Also the formant frequencies and –3 dB bandwidths reduce drastically (as opposed to clean speech) for the case Butterworth and Bandpass filters. A future study has to be carried out in order to ascertain what exactly happens in such case. The above mentioned studies will help in producing a user friendly and lesser time consuming facility in Speech Forensics.

9.4.2 Limitations of the Study

The method is ideal for noisy speech with uniform noise and for samples with varying noise structure; it will have to be treated by separating different noises and then applying corresponding techniques.

References

1. Koenig BE (1986) Spectrographic voice identification: a forensic survey. J Acoust Soc Am, 79(6):2
2. Atal B (1974) Effectiveness of linear prediction characteristics of the speech wave for automatic speaker identification and verification. J Acoust Soc Am 55:1304–1312
3. Hermansky H, Morgan N, Bayya A, Kohn P (1991) Compensation for the effect of communication channel in auditory-like analysis of speech (RASTA-PLP). Proceedings of European conference on speech technology, Genova, Italy, pp 1367–1371
4. Boll SF (1979) Suppression of acoustic noise in speech using spectral subtraction, institute of electrical and electronics engineers. Trans ASSP 27(2)113–120
5. Vaseghi SV (1996) Advanced signal processing and digital noise reduction publisher. Wiley & Teubner, West Sussex

6. Manohar K, Rao P (2004) Reduction of burst noises in STSA speech enhancement. Proceedings of international symposium on speech technology and processing systems and oriental COCOSDA-2004, New Delhi, vol 2, pp 254–257
7. Pal SK, Saxena PK (2000) Enhancement of highly noisy speech signals. J Discrete Math Sci Cryptogr, 3(1–3), 157–172
8. Cole D, Moody M, Sreedharan S (1997) Robust enhancement of reverberant speech using iterative noise removal, proceeding of ESCA, Eurospeech, Rhodes, Greece, ISSN 1018-4074, 2603–2606
9. Filiz B, Kumar S, Srinivas N (2000) Noise reduction and echo cancellation front-end for speech codecs, institute of electrical and electronics engineers. Trans Speech Audio Process 11(1):1–13
10. Manohar K, Rao P (2005) Reduction of burst noises in STSA speech enhancement proceedings of COCOSDA, 254257
11. Healy EW et al (2007) The effect of smoothing filter slope and spectral frequency on temporal speech information. JASA 121(2):1177–1181
12. Bai MR et al (2007) Comparative study of audio spatializers for dual loudspeaker mobile phones. JASA 121(1):298–309
13. Luis B, Jasha D, Alex A (2008) Speech enhancement using a pitch predictive model 1-4244-1484-9/08/ ©2008 IEEE, ICASSP, pp 4885–4888
14. Xi J, Lin Z, Yang Z, Chicharo C (2004) Noise reduction for chaotic signals based on new approach of measuring the signal determinacy. Proceedings of EUSIPCO, pp 293–296
15. Istvan P (2008) Speech enhancement in the reconstructed phase-space Info Commun J LXIII(7):41–45
16. Shannon BJ, Paliwal KK (2006) Role of phase estimation in speech enhancement. INTERSPEECH–ICSLP, pp 1423–1426
17. Wang DL, Lim JS (1982) The unimportance of phase in speech enhancement. IEEE Trans Acoust Speech Signal Process 30:679–681
18. Lyons G, Paliwal KK (2008) Effect of compressing the dynamic range of the power spectrum in modulation filtering based speech enhancement. INTERSPEECH, pp 387–390
19. Falk T, Stadler S, Kleijn WB, Chan G (2007) Noise suppression based on extending a speech-dominated modulation band. Proceedings of ICSLP, pp 970–973
20. Drullman R, Festen JM, Plomp R (1994) Effect of temporal envelope smearing on speech reception. JASA 95:2670–2680
21. Drullman R, Festen JM, Plomp R (1994) Effect of reducing slow temporal modulations on speech reception. JASA 95:2670–2680
22. Arai T, Pavel M, Hermansky H, Avendano C (1996) Intelligibility of speech with filtered time trajectories of spectral envelopes. Proceedings of ICSLP, pp 2490–2493
23. Hermansky H, Wan EA, Avendano C (1995) Speech enhancement based on temporal processing. Proceedings of ICASSP, pp 405–408
24. Hermansky H, Wan E, Avendao C (1994) Noise suppression in cellular communications. 2nd IEEE Workshop IVTTA, pp 85–85
25. Hollien H, Fitzgerald JT (1977) Speech enhancement techniques for crime lab use. Proceedings, international conference on crime countermeasures, science and engineering, Oxford
26. Maithani S (2004) Noisy speech analysis for speech recognition and enhancement. Proceedings of international symposium on speech technology and processing systems and oriental COCOSDA-2004, New Delhi, vol 2, pp 192–197
27. Ephraim Y, Malah D (1985) Speech enhancement using a minimum mean square error log-spectral amplitude estimator. IEEE Trans Acoust Speech Signal Process ASSP-33:443–445
28. Lim JS, Oppenheim AV (1979) Enhancement and bandwidth compression of noisy speech. Proc IEEE 67:1586–1604
29. Lim JS, Oppenheim AV (1978) All-pole modelling of degraded speech. IEEE Trans Acoust Speech Signal Process ASSP–26:197–210
30. Ephraim Y (1992) Statistical model based speech enhancement systems. Proc IEEE 80:1526–1555

31. Singh CP, Jiju PV (2007) Noise handling in forensic acoustics-encountering with real noise. CBI Bull XV(1–3):25–33
32. Jiju PV, Singh CP, Sharma RM (2009) Study on the selection of specific filters for enhancement of recorded speech for speaker Identification. Open Forensic Sci J 2:29–33, 1874-4028/09 Bentham Open

Chapter 10
Speech Processing for Robust Speaker Recognition: Analysis and Advancements for Whispered Speech

John H. L. Hansen, Chi Zhang and Xing Fan

Abstract In the field of voice forensics, the ability to perform effective speaker recognition from input audio streams is an important task. However, in many situations, individuals may prefer to lower their risk of being heard in public settings via whisper mode during communications. It is in precisely these conditions that speaker recognition should remain effective. Limited formal research has been performed in this domain to date. Whisper is an alternative speech production mode used by subjects in public conversation to protect content privacy or identity. Due to the profound differences between whisper and neutral speech in terms of spectral structure, the performance of speaker identification systems trained with neutral speech degrade significantly. In this chapter, studies that address acoustic analysis of whisper will be reviewed. Next, an effective data collection procedure for both spontaneous and read whisper speech will be introduced. An algorithm for whisper speech detection, which is a crucial front-end for whisper speech processing algorithms, will be presented. Finally, a seamless neutral/whisper mismatched closed-set speaker recognition system will be introduced. In the evaluation, a traditional MFCC-GMM system is employed as the baseline speaker ID system. An analysis of both speaker and phoneme variability in speaker ID performance using neutral trained GMMs is provided, which forms the basis for a final combined whisper based speaker ID system is presented. Experimental results are also provided followed by directions for future work.

10.1 Introduction

The field of voice forensics, the ability to identify an individual using automatic speaker recognition techniques can be an effective tool in removing or including an individual within a subject pool. For automatic speaker recognition algorithms

J. H. L. Hansen (✉)
Department of Electrical Engineering, Center for
Robust Speech Systems (CRSS), Erik Jonsson School of Engineering and Computer
Science, The University of Texas at Dallas, Richardson, TX 75080-3021, USA
e-mail: john.hansen@utdallas.edu

A. Neustein, H. A. Patil (eds.), *Forensic Speaker Recognition*,
DOI 10.1007/978-1-4614-0263-3_10,

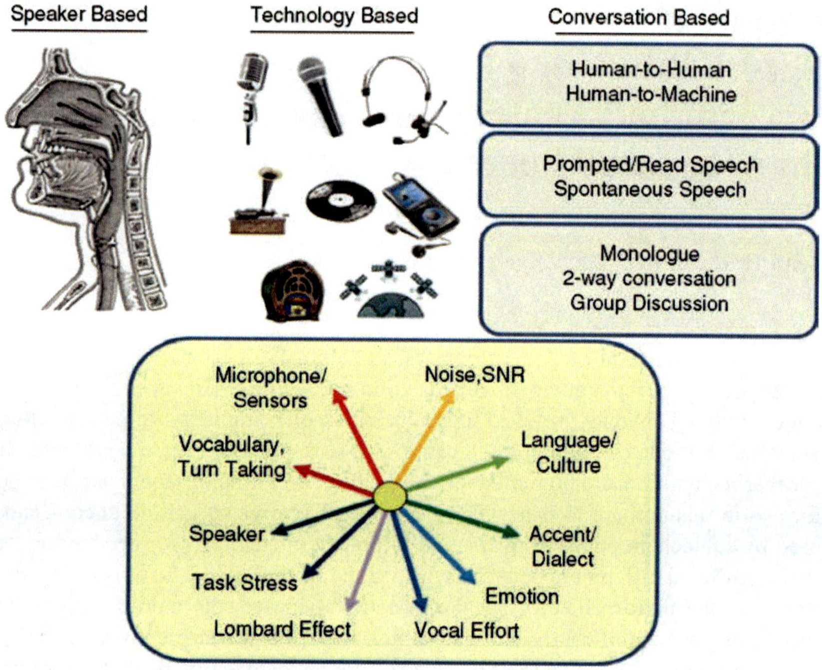

Fig. 10.1 Range of speech, speaker, technology and environmental factors that can influence speech for speaker recognition within audio forensics

however, great care is necessary to ensure there is limited mismatch between training and testing conditions in the audio stream. In the main speaker recognition evaluation (SRE) organized by the U.S. National Institute of Standards (NIST-SRE), which are held biannually, a number of mismatch conditions have been considered (NIST SRE [25]). From 2006 to 2010, these have focused primarily on noise-free audio data for which the primary mismatch is (1) microphone (one of approximately 14 mics), (2) handset (standard landline telephone, cordless telephone, cellphone), and (3) language (for NIST SRE-2008). Little if any work has been focused on speakers under non-neutral conditions (i.e., speech under stress, Lombard effect, or emotional). The range of mismatch due to speaker based variability, speech from human–human/human–machine variations, and technology or environmental factors is quite extensive. Figure 10.1 illustrates a broad perspective across these domains. The focus of this chapter is to consider mismatch due to vocal effort, and in particular whisper, for speaker recognition. Another chapter will consider speech production differences due to stress and Lombard effect.

The range of speaker variability, technology introduced mismatch, and variations in conversation based engagement causes changes to the audio stream that impacts speaker recognition performance. For Speaker Based mismatch, these include (1) speech under stress (physical, psychological/cognitive, speaking style including emotion, etc.), (2) vocal effort (whisper to shouted), (3) emotion, (4) Lombard effect (speech produced in the presence of noise), (5) accent/dialect (differences within a given language). For Technology Based mismatch, these include (1) microphone, (2) handset/phone, (3) communications/channel effects, (4) audio recording/storage/compression effects, (5) environmental including room noise and echo. Finally, Conversation Based mismatch includes (1) various forms of human-to-human communications, (2) human–machine interactions, (3) read versus spontaneous speech, (4) single person spontaneous monologue, (5) 2-way person-to-person communication, either face to face or via telephone system, (6) multiple speaker scenarios including group discussions, debates, etc. Speech captured in a controlled noise-free setting and used for training a speaker recognition model will not perform as well when test materials are drawn from these mismatched conditions. In addition, a recent study by Clark and Foulkes [4] showed that speakers undergoing some form of "disguise", either intentionally or through the use of computer/communications technology, will cause human perception of speaker identify be lowered. In this chapter, the focus will be on Whisper as a form of altered vocal effort.

Vocal Effort: Previous work [29] has shown that speech produced on different vocal effort levels has a profound impact on speech technology, and in particular speaker recognition systems. Vocal effort here refers to whisper, soft, loud, and shouted speaking conditions based on speaker ID systems trained on speech produced under neutral conditions. While the NATO RSG.10 study (Hansen et al. [14]) showed that closed set speaker ID performance under stress, including soft and loud spoken speech, was impacted, the study by Zhang and Hansen [29] represents one of the first which performed analysis of the production differences and quantified the loss in performance using a well organized speech corpus. While soft, loud, and shouted speech all result in significant mismatch between neutral speech conditions, whisper is more pronounced due to the fundamental differences in physical speech production.

Whisper Speech Mode: Whisper as one mode of vocal effort which represents a major challenge for effective speaker recognition in audio forensics. As a natural mode of speech production, whispered speech can be employed in public situations in order to protect the content of the speech information, or in some settings reduce the probability of being able to identify the speaker. For example, when asked to provide their credit number, bank account or other personal information in a public space, subjects using a cell-phone may prefer to whisper. When making hotel, flight, or car reservations by telephone in a public setting, customers may also employ whisper to provide information regarding their date of birth, full name, or billing address. Doctors may whisper if it is necessary to discuss patient medical records in public settings to maintain patient confidentiality. In the other hand, under the circumstances, in which the neutral speech is prohibited, such as library, formal conference, and theatre, whisper will be employed to convey the information between the speaker and listener avoiding heard by other people nearby. Furthermore, in some voice pathology cases, a change in the vocal fold structure or physiology or muscle control due to disease of the vocal system, such as functional aphonia, laryngeal cancer, functional voice disorder [13] or alteration of the vocal folds as a result of medical operations, may cause whisper as well [32].

In this chapter, the term "neutral speech" refers to modal speech produced at rest in a quiet sound booth. In neutral speech, voiced phonemes are produced by a

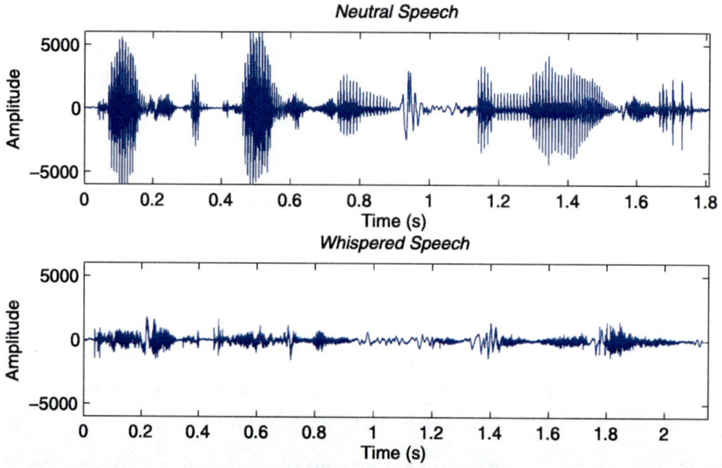

Fig. 10.2 Waveforms of neutral and whispered speech. [9]

periodic vibration of the vocal folds, which regulates air flow into the pharynx and oral cavities. However, for whispered speech, the shape of the pharynx is adjusted such that the vocal folds do not vibrate, resulting in a continuous air stream without periodicity [11, 12, 23, 28]. Figure 10.2 [9] shows the dramatic difference between neutral and whispered speech waveforms of the sentence "Guess the question from the answer" from the same speaker. It can be seen that the whispered speech waveform is much lower in the overall amplitude contour, lacks periodic segments, and is generally more aperiodic in nature.

Due to the different production mechanism, the resulting speech spectra between whisper and neutral speech are significantly different. The differences between whispered and neutral speech include [16, 19, 22, 24, 31]:

- A complete absence of periodic excitation or harmonic structure in whispered speech;
- A shift of lower formant locations;
- A change in the overall spectral slope;
- A shifting of the boundaries of vowel regions in the F1–F2 frequency space;
- A change in both energy and duration characteristics.

Given that speaker dependent whisper adaptation data is generally not available in real scenarios; these differences present a major challenge in maintaining effective speaker ID system performance.

10.2 Background

In recent years, whisper speech processing has attracted several research studies. The investigation of whispered speech focuses on from a theoretical point of view in speech production and perception [5, 10, 16, 18, 20, 21, 27], and for practical reasons in whispered speech recognition as well as whispered speaker recognition [6, 7, 9, 16, 31]. Recently, several studies have considered compensation of whisper for neutral/whisper mismatch in speaker ID systems. Those studies can be classified into two categories, where the first assumes that whisper adaptation data is available and a model adaptation method, such as Maximum a Posterior (MAP) or Maximum Likelihood Linear Regression (MLLR) can be applied; and the second assumes no adaptation data with a focus on robust features and modeling. In the first category, for example, an 8–33% relative improvement in speaker ID was achieved using 5–15 s of whispered speech adaptation data per speaker [17]. The second category assumes that no whisper adaptation data is available, and therefore these methods usually focus on alternative feature extraction approaches or compensation in the front-end process. For example, in [6], compensation strategies based on frequency warping, score competition were formulated for whisper based speaker ID. The study in [7] suggested that features based on a short-time analysis fail to capture the most salient speaker ID information for whisper frames with low signal-to-noise ratios (SNR). Speaker ID performance using model adaptation (i.e., the first category) generally performs better as more speaker dependent data is required for model construction. However, advancements in the second category can be more important since they deal with practical issues where whisper speech is generally not available *a priori*. As such, the remainder of this chapter will focus on the second category, where no adaptation data is available. Due to the importance of establishing effective scientific research on whisper speech data, the next section will concentrate on best practices for formulating a whisper speech data corpus, which includes a discussion of two sample collected corpora.

10.3 Data Collection

In order to analyze differences between whisper and neutral speech, and thereby formulate effective algorithms that can address the performance degradation for speech/language systems, whisper speech corpora with session variability are needed. In [32], two whisper based corpora were formulated, each with different foci: UT-VocalEffort (UT-VE) I & II. Due to the low energy property of whisper, as well as avoiding any data collected with too low a signal-to-noise ratio, whisper data collection should be carried out in a sound resistant booth. Also, careful monitoring of the subject is important since speakers often switch between true whisper (e.g., no vocal fold excitation) and soft spoken speech (e.g., contains vocal fold excitation, but glottal pulse shape is low and smooth resulting in a steep glottal spectral slope with limited high frequency content). Both UT-VocalEffort corpora were collected in an ASHA certified, single walled sound booth using a multi-track FOSTEX 8-channel synchronized digital recorder with gain adjustments for individual channels. The critical component here for vocal effort data collection is that a calibration test tone (1 kHz at 75 dB-SPL), should always (and was employed here), be employed for all recordings to ensure ground-truth in absolute dB sound levels for all speech. Therefore, as part of the recording phase for each subject, a 1 kHz test

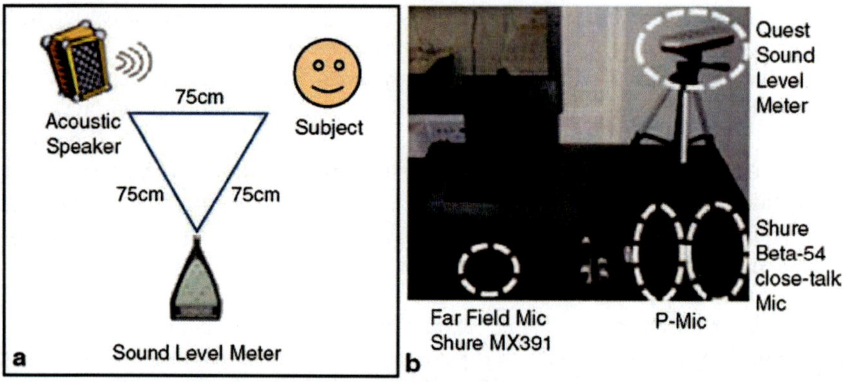

Fig. 10.3 **a** Table setting of data collection for UT-VE-I (measurement of dBA sound level for whisper production); **b** Setting of data collection for UT-VE-I in ASHA Certified Sound Room (microphone placement included). [32]

tone at 75 dB-SPL is produced with a sound level meter for all recordings. In this manner, any signal processing analysis will have a ground-truth, this tone (i.e., many A/D converters employ either manual or automatic gain control (AGC) can alter recording levels between neutral and whisper speech mode, making direct comparison impossible; having this test-tone with a fixed input recording sensitivity (no AGC) is the proper method for speech capture of vocal effort speech data).

UT-VocalEffort I: For UT-VE-I, a total of 12 male, native English-speaking subjects participated in data collection. All subjects are native American English speakers with no history of speech or hearing limitations/disorders. For each subject, speech was recorded for a series of tokens using three positioned microphones: a P-Microphone (physiological microphone) (Patil and Hansen [26]), a SHURE Beta-54 close-talking microphone and a SHURE MX391/S far field microphone. A 1 kHz sinusoid signal generated by an NTI analog audio generator was played through an ALTEC speaker at the same physical location as the calibration test tone and included in all recordings. At the beginning of each token, the volume of the test tone was carefully adjusted so the dBA sound pressure level (SPL) of the test tone measures 75 dB using a QUEST sound level meter (SLM). The test tone was recorded for all three microphones. The position of the subject, the location of the calibration test tone speaker, and the location of the sound level meter were all positioned in an equi-distant triangle separated by 75 cm. The table recording setting is illustrated in Fig. 10.3a, along with an image of the ASHA certified 13′×13′ sound room in Fig. 10.3b.

The data collection procedure was divided into 3 phases for each subject. Phase I consisted of 2 sessions with 5 tokens corresponding to the five speech modes (e.g., whisper, soft, neutral, loud, and shouted). In each token, 5 sentences from the TI-MIT database were spoken in one of five speech modes and recorded. Each subject was prompted to read the particular sentences from a laptop display positioned in

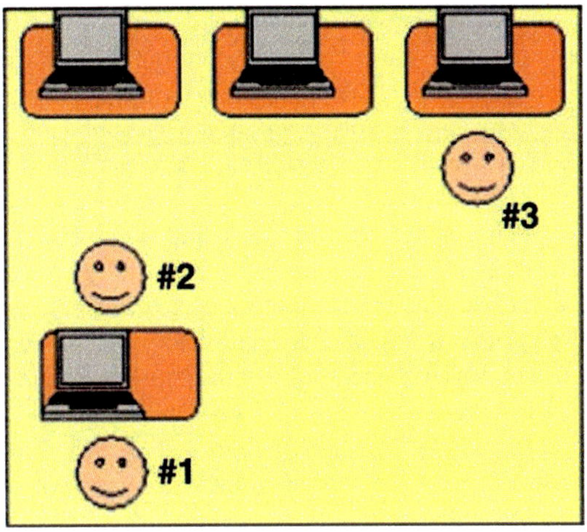

Fig. 10.4 Setting of data collection for UT-VE-II. [32]

front of the subject. Phase II consisted of 20 sentences which were all read sequentially in the neutral speech mode. Phase III includes spontaneous speech of one-minute duration in each of five vocal modes (e.g., whisper, soft, neutral, loud, and shouted). Human transcribers were used to verify speech and vocal effort content for all recordings for UT-VE-I.

UT-VocalEffort II: While it is possible to have cases where subjects produce sustained whisper throughout an entire conversation, it is quite difficult for a subject to sustain pure whisper for great lengths of time. The reason is that subjects will migrate to soft speech (i.e., some form of vocal fold excitation, since producing vowels, liquids, glides, in a non-voiced speech production mode requires additional cognitive processing). It is rare for someone to produce a sustained whisper conversation for 10–15 min, and therefore whisper islands will usually be embedded within a neutral speech stream. Those whisper "islands" cause a loss in performance for speech processing systems which are generally designed for neutral speech. In order to explore the development of algorithms for whisper island detection, a much larger corpus termed UT-VE-II was constructed in the same acoustic environment as UT-VE-I.

For UT-VE-II, 112 subjects (37 male/75 female) participated to produce whispered and neutral speech in spontaneous and read sessions. The UT-VE-II Corpus is focused on neutral speech embedded with whispered speech islands.

For the spontaneous session, the collection scenario was explained to the subject to emulate a cyber café scenario. Three subjects were positioned in the ASHA $13' \times 13'$ sound room as shown in Fig. 10.4. Subject 1 and 2 engage in a conversation (seated across from each other, where a laptop is placed in front of Subject 1). Here, Subject 1 is the volunteer who is producing neutral and whispered speech;

Subject 2 is the data collector and second party listener for Subject 1; and Subject 3 is a cyber cafe participant attempting to listen-in on the conversation between Subject 1 and 2 while using their computer. In order to achieve completely natural human-to-human conversation, the data collector (Subject 2) was instructed to keep their conversation engaged (e.g., between Subject 1 and 2). In order to satisfy IRB requirements, we did not want to record personal information regarding Subject 1 (e.g., their name, credit card or phone numbers, names of family/friends, etc.). To solve this challenge, random names were generated by selecting different first and last names from the Dallas telephone directory, along with company names and addresses pieced together from directory listings (e.g., "Acme Trucking" and "Dallas Furniture Mart" becomes "Acme Furniture" and "Dallas Trucking"). Each person or business name was printed on sheets of paper, with key parts for whisper production highlighted as sensitive information. The Subject 1 was instructed to be certain that when using the names/information in conversation, any highlighted parts needed to be kept confidential between Subject 1 and 2, so Subject 3 should not hear this information. The list of information, including names, addresses, phone numbers or credit card numbers, was given to Subject 1. Key information was randomly chosen to be spoken in whisper mode from the list by Subject 1. Furthermore, Subject 1 was told that Subject 3 is trying to pick up as much key information as possible, and thus Subject 1 was persuaded to produce the speech as low a volume as he/she can but to convey the key information to Subject 2 in conversation. By doing this, when Subject 1 introduces the information from the list to Subject 2, Subject 1 would be able to work into the audio rather than be required to produce whispered speech for key information in the neutral phonated conversation.

In the read part of UT-VE II, only Subject 1 was enrolled and **required** to read material in either neutral or whispered modes. Three types of read materials were used in the read part. The first type consists of sentences selected from the TIMIT database. Here, 41 TIMIT sentences were produced alternatively in neutral and whispered mode. The second material type consists of two paragraphs selected from a local newspaper. For each paragraph, four whisper-islands were produced, with each island consisting of 1–2 sentences. The third type of material consists of the same paragraphs as those of the second type. However, for each paragraph, five phrases were read in whispered mode, with each phrase 2–3 words in duration.

For different research foci, corpora of whisper speech with different content may be needed. For constructing any whisper speech corpus, the following points are suggested as guidelines:

- Data collection should be carried out in a sound resistant environment to avoid outside noise interference;
- A carefully measured test tone should be recorded at the beginning of each collection session (75 dBA SPL based on ASHA/AAA[1] commonly used procedures);

[1] ASHA—represents the American-Speech-Hearing Association; AAA—represents the American Academy of Audiology. In the field of speech and hearing research, a 1 kHz test tone at 75 dBA SPL is commonly used to ensure ground-truth of the audio recording for later presentation or analysis.

Table 10.1 Closed-set speaker recognition performance for GMMs systems

Speech mode		Accuracy (%)		
Training	Testing	Static MFCCs	PLPs	Static+delta MFCCs
Neutral	Neutral	99.10	96.97	99.27
Neutral	Whisper	79.29	43.91	60.04

- P-microphone [26] is suggested to use for research to explore the ground truth of vocal folds vibration for whisper;
- It is important to monitor true vocal effort throughout the sessions—since migration from Whisper to Soft spoken speech is possible;
- To ensure IRB (Internal Review Board) compliance, selecting non-personal information as target material is recommended;
- A reasonable subject engagement scenario will help subjects produce whisper speech more effectively in a spontaneous manner;
- A phoneme balanced set of materials should also be used in a read context as well.

10.4 Speaker ID Baseline System

Zhang and Hansen [29] showed that changes in vocal effort result in a significant impact on speaker identification (Speaker-ID) performance, and of the four non-neutral vocal effort conditions (whisper, soft, loud, shouted), whisper results in the most serious loss of system performance. In [9], difference sets of features were considered for use with a standard Gaussian Mixture Model (GMMs) closed-set speaker recognition scenario. The results are cited as in Table 10.1. It was found that the best performance is achieved using static Mel-Frequency Cepstral Coefficients MFCCs for mismatched neutral/whisper training/testing condition. The Perceptual Linear Predictive (PLP) feature, which is another popular front-end processing method used for speaker ID, shows a significant degradation in speaker ID recognition results for this condition. Due to the duration difference between whispered and neutral speech, the static+delta MFCCs feature set also shows degradation compared with only static MFCCs.

10.5 Advancement in Whispered Island Detection

From the previous section, current speaker ID systems have dramatic loss in performance when employing whisper speech. Therefore, detection of whisper in neutral speech streams becomes necessary and crucial. As a front-end step, whisper detection may help indicate when a whisper-dedicated algorithm or whisper adaption of a neutral trained system should be used. Furthermore, since whisper has a high probability of conveying sensitive information, with the help of a whisper detection algorithm, a speech document retrieval system or call center monitor system

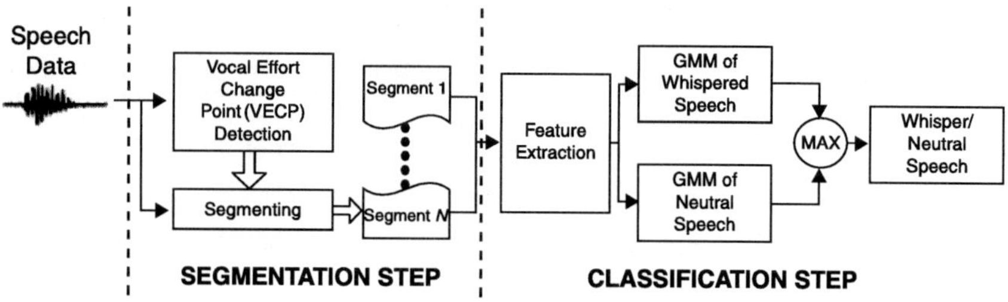

Fig. 10.5 Flow diagram of whisper-island detection. [32]

can identify where potential confidential or sensitive information occurs within a neutral speech audio stream.

In this section, an algorithm for whisper-island detection is introduced to identify where whisper is located within a given neutral speech audio stream. The algorithm is developed using an entropy-based feature set: Whisper Island Detection (WhID), which is sensitive to vocal effort changes between whisper and neutral speech. The WhID feature set is integrated within T^2-BIC segmentation [33] for vocal effort change point (VECP) detection and utilized for whisper island detection. Figure 10.5 shows the high level flow diagram of the whisper island detection algorithm.

10.5.1 Algorithm Development for Whisper Island Detection

The potential vocal effort change points (VECPs) of the input speech data embedded with whisper-islands are first detected in the segmentation step (left part of Fig. 10.5). Based on the sequence of potential detected VECPs, the speech stream is divided into segments. An improved T^2-BIC algorithm is incorporated to detect the potential VECPs between whisper and neutral speech. The T^2-BIC algorithm, developed by Zhou and Hansen [33] and also described in [15, 30, 31], is an unsupervised model-free scheme that detects acoustic change points based on the input feature data. A range of potential input features for the T^2-BIC algorithm can be used to detect input acoustic changes within the audio stream. Here, the T^2-BIC algorithm is considered as a potential method to detect the VECPs between whisper and neutral speech if an effective feature for vocal effort change is employed.

In the classification step (right part of Fig. 10.5), a GMM based vocal effort classifier is developed to label the vocal effort of each speech segment obtained from the previous step. GMMs of whisper and neutral speech are respectively trained with whisper and neutral speech data. The scores obtained by comparing the detected segment with two vocal effort models are sorted, and the model with the highest score is identified as the model which best fits the vocal effort of the current segment.

10.5.2 WhID Feature Set

The entropy-based 4-dimension feature set WhID is calculated for each 20 ms speech frame. It can be formulated as:

- 1-D spectral information entropy ratio (ER);
- 2-D spectral information entropy (SIE);
- 1-D spectral tilt (ST).

The entropy ratio (ER) is calculated from the ratio between a high frequency band (2800–3000 Hz) entropy and a low frequency band (450–650 Hz) entropy. The 2-D SIE feature were obtained from the SIE of two even frequency band (300–4150 and 4150–8000 Hz). The final dimension feature spectral tilt was calculated from the power spectrum of each frame. The details of spectral information entropy are detailed in [32].

10.5.3 The T^2-BIC Segmentation Algorithm

Segmentation can be reformulated as a model selection task between two nested competing models. Bayesian Information Criterion (BIC), a penalized maximum likelihood model selection criterion, is employed for model selection [3]. The segmentation decision is derived by comparing BIC values. As a statistical data processing method, BIC requires no prior knowledge concerning acoustic conditions and no prior model training is needed. Instead of choosing a hard threshold for the segmentation decision, BIC statistically finds the difference in the acoustic features of the input frames to determine a point which can separate the data within the processing window into two models.

However, the BIC-based segmentation algorithm has quadratic computational complexity, and in the study by Zhou and Hansen [33], the T^2 value (where the Hotelling's T^2-statistic is a multivariate analog of the well-known *t*-distribution was employed [1]) was calculated for frame $b \in (1; N)$ in order to find candidate boundary frames in the region near $\hat{b}$, thus significantly improving computational performance as well as increased accuracy for short duration turns. Next, BIC value calculations are performed only on the frames in the neighborhood of $\hat{b}$ to find the best frame breakpoint and verify the decision of the boundary as original BIC algorithm [33]. A later study [15] also explored trade-offs in performance for various speech features for T^2-BIC segmentation. Since that study focused on neutral speech, it is not entirely clear if the best features for neutral would provide consistent performance for whisper speech. Therefore, other features are considered in this investigation. For increased accuracy and reliable detection, the BIC calculation was performed within the range $[(\tilde{b} - 50); (\tilde{b} + 50)]$ after the T^2 statistic algorithm was used to detect the possible VECP $\tilde{b}$. Here, the T^2-Statistic was integrated within the BIC algorithm in this manner for processing shorter audio streams, while the tradi-

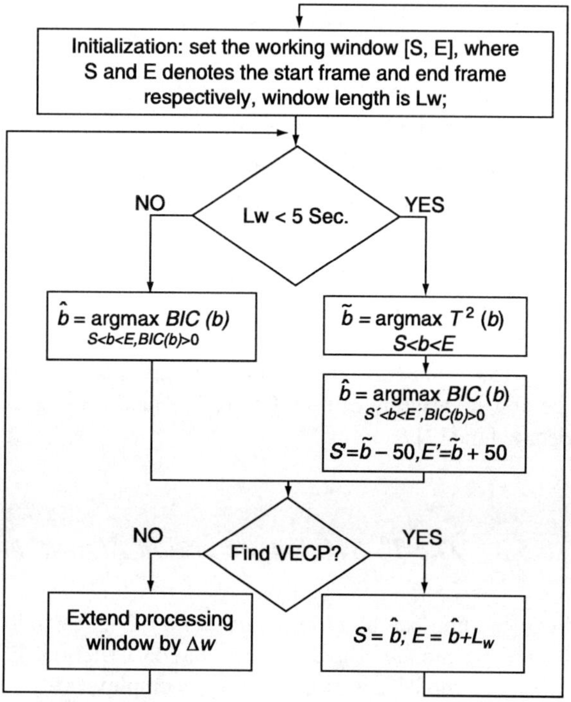

Fig. 10.6 BIC/T^2-BIC segmentation for VECP detection. [32]

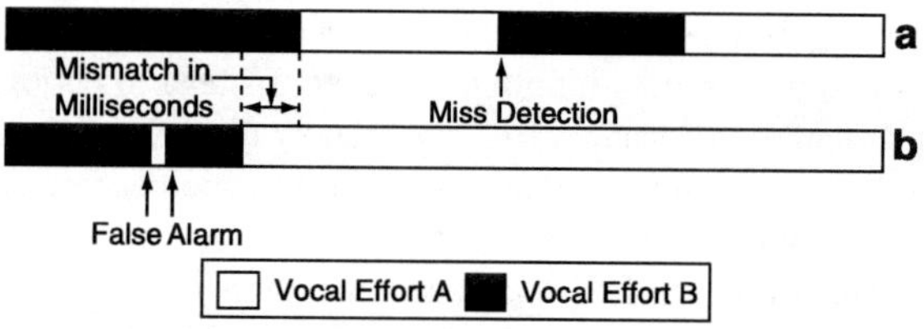

Fig. 10.7 Three types of segmentation error

tional BIC algorithm was used to process long duration blocks. The BIC algorithm was used for a process window L_w larger than 5 sec, and T^2-BIC was used when L_w was less than 5 sec. The implementation of the overall proposed segmentation algorithm for vocal effort change point (VECP) detection is described in Fig. 10.6.

10.5.4 Whisper-Island Detection Performance

In [32], the Multi-Error Score (MES) was developed and introduced to evaluate performance of acoustic features for detection of VECPs. The MES consists of 3 error types for segmentation mismatch: miss detection rate, false alarm rate and average mismatch in milliseconds normalized by dual-segment duration. Figure 10.7 illustrates these three types of error.

Table 10.2 Evaluation for Vocal Effort Change Points Detection

Feature type	MDR (%)	FAR (%)	MMR (%)	MES
13-D MFCC	1.13	27.44	2.63	36.09
4-D WhID	0.00	8.13	1.69	11.51

The calculation of MES can be illustrated by the following equation:

$$\begin{aligned}\text{MES} &= 1 \times \text{False Alarm Rate} + 2 \times \text{Mismatch Rate} \\ &\quad + 3 \times \text{Miss Detection Rate}\end{aligned} \tag{10.1}$$

The mismatch rate is obtained by calculating the percentage of the mismatch in milliseconds versus the total duration of the two segments corresponding to the actual breakpoints. More details concerning the MES can be found in [32]. Miss detection rate and mismatch rate are more costly errors for whisper island detection, so these errors are scaled by 3 and 2 respectively. MES is bounded by 0, for all 3 error rates at 0% and 600 for all 3 error rates at 100%. A score of 90 occurs when all 3 error rates are 15%.

To illustrate the effectiveness of WhID in VECP detection using the BIC/T^2-BIC algorithm, an experiment was carried out, with experimental results evaluated using the Multi-Error Score. For each subject, the speech audio from UT-VE II, which consists of 41 TIMIT sentences read by the subject alternatively in whisper and neutral mode, was used in the experiments. The vocal effort and onset and offset time of whisper/neutral segment within each speech audio were manually labeled and saved as a text file corresponding to the speech audio. The audio files from 59 subjects were employed in the present experiment to detect the VECPs. The transcript files of these audio streams were used to compare with VECP detection results obtained from the BIC/T^2-BIC algorithm using the proposed feature, so that the MES can be calculated. In addition to these features, the classic 13-D MFCC feature (without energy feature) was also used within our algorithm for experiments as a reference. The MES score and detailed error scores using MFCC and the proposed WhID feature are summarized in Table 10.2.

The reduction of MES from 36.09 to 11.51, as well as reduction in MDR, FAR, and MMR, is quite remarkable. From the experimental result, the zero value of MDR denotes that all the VECPs within the audio stream were detected, with 1.69% of mismatch rate compared to the real VECPs. Among the total VECPs in the detection result, 8.13% of change points are false alarms which will be compensated in the subsequent classification step.

Based on the VECPs detected in the segmentation step, the audio stream is partitioned into several segments which are then to be classified by a GMM based vocal effort classifier. There are 4 training scenarios for 64 GMMs of whisper and neutral vocal efforts for classifier. Table 10.3 shows the details of all four experimental scenarios. The round-robin technique was deployed in experimental Scenarios A&B to obtain the average performance of the vocal effort classification. The same audio streams used in the previous subsection are employed here. The subjects,

Table 10.3 Training scenarios for GMM based classifier

Scenario	Testing subjects	Training subjects	Feature
A	Each of 59	Rest 58 of 59	13-D MFCC
B	Each of 59	Rest 58 of 59	4-D WhID
C	20 Male	39 Female	4-D WhID
D	39 Female	20 Male	4-D WhID

Table 10.4 Evaluation for overall whisper island detection

Scenario	Detected number	Detection rate (%)
A	572	48.39
B	1182	100
C	400	100
D	782	100

used include a combination of male and female (scenarios A&B), as well as gender dependent (Scenarios C&D). The detection results for each subject's speech was compared with the human transcript file having manually labeled vocal effort to calculate the detection rate of whisper-island. Table 10.3 shows the performance for the overall system in terms of detection rate of whisper-islands.

From Tables 10.3 and 10.4, it can be observed that the algorithm presented in this section has 0.00% miss detection rate with low multi-error score and 100% detection rate, respectively indicating all VECPs between whisper and neutral speech can be detected and all whisper islands can be identified.

10.6 Advancement in Speaker ID for Whispered Speech

10.6.1 Speaker Variability

It would be useful to consider an analysis of the distribution of recognition performance among all speakers, since some speakers may convey more speaker dependent structure in whisper than others. A confusion matrix for the static MFCC-GMM speaker ID (SID) system is shown for the neutral/whisper train/test scenario in Fig. 10.8 [9]. The grey scale key to the right indicates that dark shades denote higher SID probability for the input speaker recognized as the corresponding correct speaker. Thus, darker diagonal entries indicate stronger system performance. The number along the x and y axis is the closed-set speaker index. It can be observed that the accuracy varies significantly across speakers. For example, for Speakers 4 and 12, an accuracy of 100.0% is achieved. However, for Speaker 18, only 46.0% accuracy is obtained.

Figure 10.8 illustrates an interesting property of whispered speech: some speakers maintain sufficient speaker dependent structure under whisper condition; so that no additional processing or system changes are needed when neutral speech models

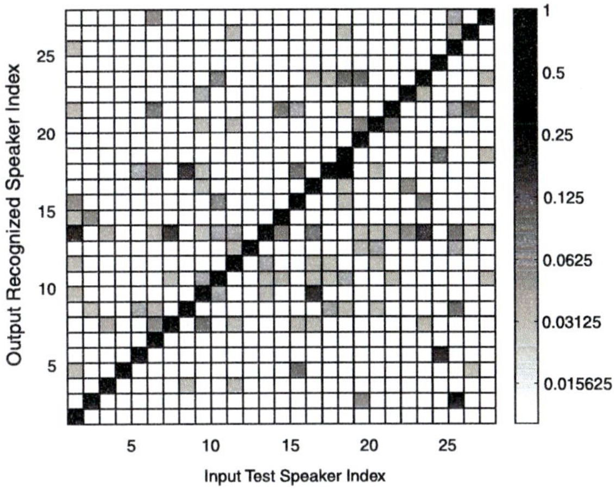

Fig. 10.8 Confusion matrix of neutral trained MFCC-GMM system when tested on whispered utterances (28 speakers in total). [9]

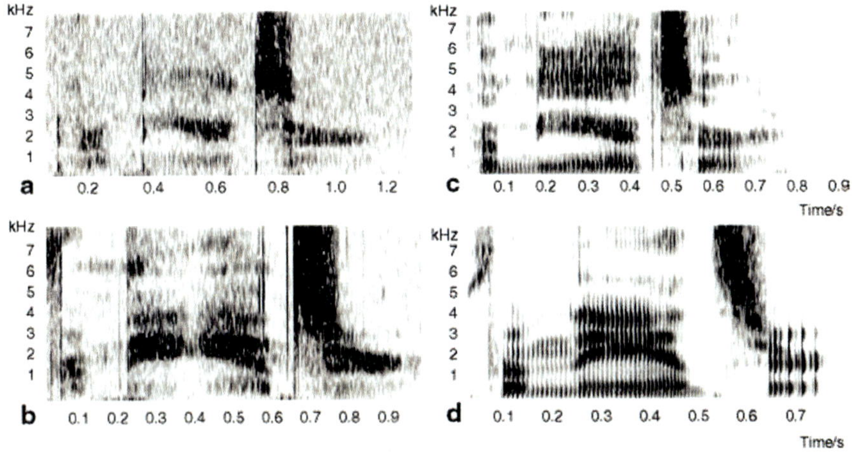

Fig. 10.9 Phrase "from the answer" from two speakers in both neutral and whispered speech mode: **a** lower SNR whisper from speaker I, **b** higher SNR whisper form speaker II, **c** neutral speech from speaker I, **d** neutral speech from speaker II. [6]

are employed, while others fail completely. An example of such "good" and "bad" whispered speech is provided in Fig. 10.9, where the formant structure of the upper whispered speech is buried in the "silent" background noise due to the lower energy. Study in [8] proposed a new approach for confidence space measurement in order to identify whisper speech that can provide good results for neutral trained speaker ID based on spectral analysis. For those whisper that are classified as belonging to "consistent" or "good" speakers, no additional processing will be required, while others are routed to alternative compensation methods.

Table 10.5 Closed-set speaker recognition performance for unvoiced/non-voiced phonemes based on GMMs [16]

Speech mode		Phonemes	Accuracy (%)
Training	Testing		
Neutral	Neutral	Non-voiced	73.76
Neutral	Whisper	Non-voiced	66.49
Neutral	Whisper	Voiced	98.14
Neutral	Whisper	Voiced	57.34

Table 10.6 Closed-set speaker recognition performance for unvoiced based on GMMs [16t]

Speech mode		Warping scale	Accuracy (%)
Training	Testing		
Neutral	Whisper	Mel (MFCC)	67.01
Neutral	Whisper	Linear (LFCC)	71.90
Neutral	Whisper	Exponential (EFCC)	70.55

10.6.2 Phoneme Variability

Besides variability in speaker ID performance, phoneme variability should also be considered. This is due to the fact that differences between whispered and neutral speech focus on the vowels, semi-vowels, nasals, and liquids. While for unvoiced consonants, whispered speech and neutral speech are very similar. Therefore, those phonemes should be treated differently in the system. It is noted that, due to the absence of periodic excitation, the voiced consonants in whispered speech are similar to the unvoiced consonants. This implies that voiced/unvoiced phoneme pairs, such as (z, s), (zh, sh), (t, d), (dzh, tsh), etc., are not separated. In this chapter, all phonemes are mainly divided into two parts: (1) unvoiced phonemes and (2) non-unvoiced phonemes. The recognition results for unvoiced/non-unvoiced phonemes are provided as in Table 10.5 for comparison.

Due to the unique spectral structure of unvoiced phonemes that most information is contained in the higher frequency domain, linear or exponential-scale warping function can take place of Mel-scale. These warped scales were originally investigated by Bou-Ghazale and Hansen [2] for speech under stress and emotion. The recognition results for the corresponding cepstral coefficients: MFCC, LFCC, and EFCC are shown in Table 10.6. LFCC and EFCC differs MFCC in the scale of the frequency warping function, where LFCC is linear (no frequency warping is applied) and EFCC uses an exponential warping function to map linear frequency to the desired frequency scale. It can be observed that features based on linear and exponential scales that emphasis high frequency components provide better performance for speaker ID.

10.6.3 System Combination

Given the presence of speaker and phoneme variability when whisper speech is present, a system framework shown in Fig. 10.10 can be employed. Phoneme variability

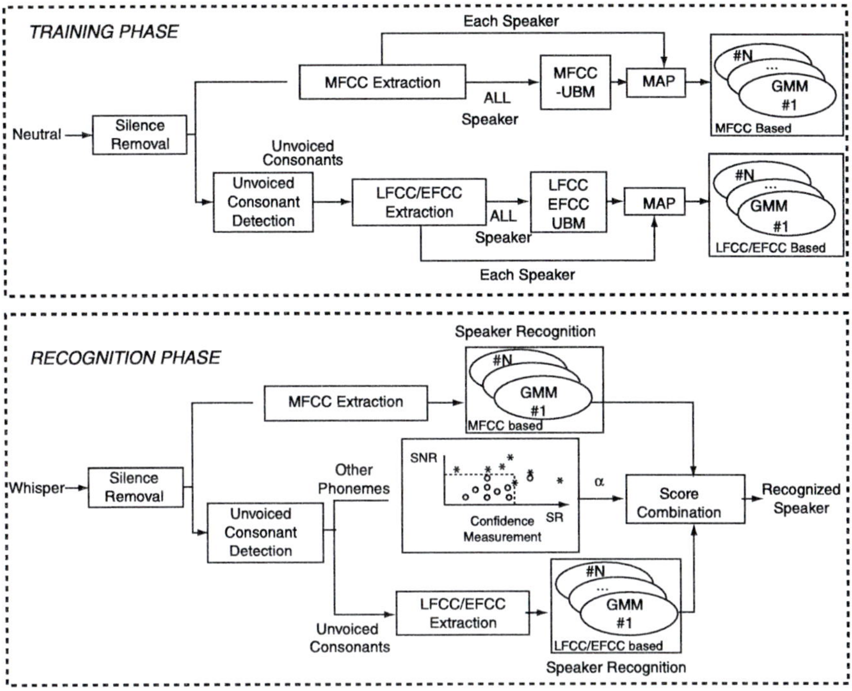

Fig. 10.10 System flow diagram for speaker recognition of whispered speech based on neutral trained GMMs. [9t]

is addressed by introducing additional mixture components within the GMMs for unvoiced phonemes. For example, as discussed in Sect. 6.2, for unvoiced consonants, LFCC and EFCC based features outperformed an MFCC based system by capturing more speaker specific information contained in high frequencies. Thus, unvoiced consonants are separated from other phonemes and processed with LFCC or EFCC feature extraction to enhance overall performance. Only the complete set of utterances from the neutral speech mode is used to train the MFCC-GMM speaker ID system. Next, unvoiced consonants will be separated from other phonemes within the input audio streams and employed to train an LFCC-GMM or EFCC-GMM system. For testing, unvoiced consonants are separated from other phonemes first. Also, speaker variability is incorporated into the system during the decoding phase by introducing a confidence measurement as developed in [8], whose results determine the weight of the scores from the unvoiced and voiced phonemes GMM. If a whispered test utterance is classified as a "good" whisper, more weight will be assigned to scores from voiced phonemes' GMM. Alternatively, if a whispered test utterance is classified as a "bad" whisper, more weight will be given to scores from the unvoiced phonemes' GMM. This decision is made for each speaker within the closed-set speaker ID system. The final results are provided in Table 10.7. The proposed algorithm which performs an unvoiced consonant detection with good/

Table 10.7 Recognition result for closed-set speaker ID [9]

Feature vector	Accuracy (fixed score weighting)	Accuracy (with confidence measurement)
MFCC	79.29%	
MFCC+LFCC	87.30%	88.35%
MFCC+EFCC	87.30%	88.14%

bad whisper partitioning, along with an improved frequency based cepstral feature set and combined GMM classifier structure improves closed-set speaker ID performance from 79.29 to 88.35%, a relative 43.74% reduction in error.

10.7 Future Directions

While the proposed scheme represents one of the first approaches to speaker ID under whisper speech conditions, further research is clearly possible. Additional algorithm development could focus on improving the performance of vowels within whispered speech for speaker ID using neutral trained GMMs. Another direction will be to improve speaker ID for whispered speech in noisy environments. New front-end processing methods, and unsupervised feature compensation methods need to be proposed to address these issues. Note that new front-end processing procedure different from conventional MFCC feature extraction improves the performance, however, additional feature extraction, model training and data transportation will be. Also, methods work well with whispered/neutral mismatched conditions may harm the performance of neutral/neutral matched condition, which required consideration.

References

1. Anderson TW (1958) An introduction to multivariate statistical analysis. Wiley, New York
2. Bou-Ghazale SE, Hansen JHL (2000) A comparative study of traditional and newly proposed features for recognition of speech under stress. IEEE Trans Speech Audio Process 8(4):429–442
3. Chen S, Gopalakrishnan P (1998) Speaker, environment and channel change detection and clustering via the Bayesian information criterion. Proceedings of the Broadcast News Transcription and Understanding Workshop
4. Clark J, Foulkes P (2007) Identification of voices in electronically disguised speech. Int J Speech Lang Law 14(2):195–222
5. Eklund I, Traumuller H (1996) Comparative study of male and female whispered and phonated versions of the long vowels of Swedish. Phonetica 54:1–21
6. Fan X, Hansen JHL (2008) Speaker identification for whispered speech based on frequency warping and score competition. ISCA INTERSPEECH-08, Brisbane, pp 1313–1316
7. Fan X, Hansen JHL (2009) Speaker identification for whispered speech using modified temporal patterns and MFCCs. ISCA INTERSPEECH-09, Brighton, pp 896–899

8. Fan X, Hansen JHL (2010) Acoustic analysis for speaker identification of whispered speech. ICASSP 2010, Dallas, pp 5046–5049
9. Fan X, Hansen JHL (2011) Speaker identification within whispered speech audio streams. IEEE Trans Audio Speech Lang Process 19(5):1408–1421
10. Gao M (2002) Ones in whispered Chinese: articulatory features and perceptual cues. MA thesis, Dept. of Linguist, University of Victoria, British Columbia, Canada
11. Gavidia-Ceballos L (1995) Analysis and modeling of speech for laryngeal pathology assessments. RSPL: Robust Speech Processing Laboratory, Department of Electrical Engineering. Ph.D. thesis, Duke University, Durham, North Carolina
12. Gavidia-Ceballos L, Hansen JHL (1996) Direct speech feature estimation using an iterative EM algorithm for vocal fold pathology detection. IEEE Trans Biomed Eng 43(4):373–383
13. Hansen JHL, Gavidia-Ceballos L, Kaiser JF (1998) A nonlinear based speech feature analysis method with application to vocal fold pathology assessment. IEEE Trans Biomed Eng 45(3):300–313
14. Hansen JHL, Swail C, South AJ, Moore RK, Steeneken H, Cupples EJ, Anderson T, Vloeberghs CRA, Trancoso I, Verlinde P (2000) The impact of speech under 'stress' on military speech technology. NATO Research and Technology Organization RTO-TR-10, AC/323(IST) TP/5 IST/TG-01, March 2000 (ISBN: 92-837-1027-4)
15. Huang R, Hansen JHL (2006) Advances in unsupervised audio classification and segmentation for the broadcast news and NGSW corpora. IEEE Trans Audio Speech Lang Process 14(3):907–919
16. Ito T, Takeda K, Itakura F (2005) Analysis and recognition of whispered speech. Speech Commun 45(2):139–152
17. Jin Q, Jou SS, Schultz T (2007) Whispering speaker identification. IEEE Inter Conf Multimedia Expo, pp 1027–1030
18. Jovicic S, Saric Z (1997) Acoustic analysis of consonants in whispered speech. J Voice 22:263–274
19. Jovicic ST (1998) Formant features differences between whispered and voiced sustained vowels. Acustica-acta 84(4):739–743
20. Kallail K (1984) An acoustic comparison of isolated whispered and phonated vowel samples produced by adult male subjects. J Phon 12:175–186
21. Lass K, Hughes K, Bowyer M, Waters L, Bourne V (1976) Speaker sex identification from voiced, whispered and filtered isolated vowels. J Acoust Soc Am 59:675–678
22. Matsuda M, Kasuya H (1999) Acoustic nature of the whisper. ISCA EUROSPEECH-99, Budapest, pp 133–136
23. Meyer-Eppler W (1957) Realization of prosodic features in whispered speech. J Acoust Soc Am 29(1):104–106
24. Morris RW, Clements MA (2002) Reconstruction of speech from whispers. Med Eng Phys 24(7–8):515–520
25. NIST SRE USA National Institute of Standards and Technology (NIST) speaker recognition evaluation. http://www.itl.nist.gov/iad/mig/tests/sre/. Accessed 25 Jan 2011
26. Patil S, Hansen JHL (2010) The physiological microphone (PMIC): a competitive alternative for speaker assessment in stress detection and speaker verification. Speech Commun: Special Issue Silent Speech Interfaces 52:327–340
27. Schwartz M, Rine H (1968) Identification of speaker sex from isolated whispered vowels. J Acoust Soc Am 44:1736–1737
28. Thomas I (1969) Perceived pitch of whispered vowels. J Acoust Soc Am 46(2B):468–470
29. Zhang C, Hansen JHL (2007) Analysis and classification of speech mode: whispered through shouted. ISCA INTERSPEECH-07, Aug 2007, Antwerp, pp 2289–2292
30. Zhang C, Hansen JHL (2008) Effective segmentation based on vocal effort change point detection. IEEE/ISCA ITRW speech analysis and processing for knowledge discovery, Aalborg, Denmark, 4–6 June, 2008
31. Zhang C, Hansen JHL (2008) An entropy based feature for whisper-island detection within audio streams. ISCA INTERSPEECH-08, Brisbane, pp 2510–2513

32. Zhang C, Hansen JHL (2011) Whisper-island detection based on unsupervised segmentation with entropy based speech feature processing. IEEE Trans Audio Speech Lang Process 19:883–894
33. Zhou B, Hansen JHL (2005) Efficient audio stream segmentation via the combined T^2 statistic and Bayesian information criterion. IEEE Trans Speech Audio Process 13(4):467–474

Part III
Methods and Strategies: Analyzing Features of Speaker Recognition to Optimize Voice Verification System Performance in Legal Settings

Chapter 11
Effects of the Phonological Contents and Transmission Channels on Forensic Speaker Recognition

Kanae Amino, Takashi Osanai, Toshiaki Kamada, Hisanori Makinae and Takayuki Arai

Abstract This chapter introduces experiments on speaker recognition where we focus on two of the factors that affect speaker recognition accuracy: phonological contents of the speech materials used for identifying speakers and the transmission channel difference in automatic speaker verification. Through the experiments, we show that nasal sounds are effective for forensic speaker recognition despite the differences in speaker sets and recording channels. Also we show that performance degradation by the channel difference, in this study air- and bone-conduction, can be improved by devising normalisation methods and acoustic parameters.

11.1 Introduction

As we mentioned in our previous chapter [1], there are many factors that affect speaker recognition performances, in both human- and computer-based methods. Some of them are problematic in forensic speaker recognition; for example, poor transmission quality, existence of noise or voice disguises, non-contemporaneous speech samples are always the cause of worry to the forensics researchers. Other factors such as the phonological contents of the utterances, on the other hand, may be leveraged as a useful clue to the speakers.

This chapter introduces experimental results, where we investigated two of these factors: how the phonological contents of the stimuli affect human speaker recognition and how we alleviate the channel difference problems by elaborating normalisation methods and parameter selection.

K. Amino (✉)
National Research Institute of Police Science,
6-3-1 Kashiwanoha, Kashiwa-shi, Chiba 277-0882, Japan
e-mail: amino@nrips.go.jp

A. Neustein, H. A. Patil (eds.), *Forensic Speaker Recognition*,
DOI 10.1007/978-1-4614-0263-3_11,

11.2 Phonological Contents Effective for Speaker Identification

As mentioned in some studies [2–4], it has been pointed out that the accuracy of speaker identification by listening depends on the phonological contents of the utterances presented to the listeners. In order to confirm what phonological contents are effective for identifying speakers, we conducted perceptual speaker identification experiments. Two factors were assessed in the experiments: (1) effect of the phonological contents; and (2) effect of the speaker familiarity.

Among the earlier studies that examined effective phonological contents for speaker identification in Japanese, Nishio [5] tested various sounds of Japanese for identifying five male and five female familiar speakers and found that (1) people can identify familiar speakers with sentences, words or even a monosyllable or a vowel uttered in isolation, (2) open vowels (e.g. /a/, /ɔ/) were more effective than close vowels (e.g. /i/, /u/), and (3) the identification accuracy ranged from 72 to 99% when the stimulus containing a voiced sound was presented, and from 17 to 34% with the voiceless stimuli. Also, elongated vowels were better than short ones, and the arrangement of the fundamental frequencies degraded the identification rates.

In Nishio's study [5], the selection of the stimuli was not based on phonological grounds. He used only two monosyllables, /pa/ and /ba/, and two repetitions of monosyllables, /sasasa…/ and, /ɾeɾeɾe…/, apart from sentences, words, and isolated vowels. If we were to find the effective sounds for speaker identification and apply them for the practical use, we have to use a variety of sounds; and at the same time, they must be carefully controlled under the experimental conditions. Another thing that was not focused in Nihio's study is the effect of speaker familiarity, which is important for earwitness research [1]; only familiar listeners were involved in his experiment. In order to make them up, we conducted two perceptual speaker identification experiments. In the first experiment, the listeners were familiar with all of the speakers, and in the second, the listeners had not known any of the speakers before. In both experiments we used CV (consonant–vowel) monosyllables as the stimuli. Acoustic analyses of the stimuli and a speaker identification experiment by machine were also conducted.

11.2.1 Experiment 1a: Familiar Speaker Identification

11.2.1.1 Participants and Speech Materials

Fifteen undergraduate students volunteered for participating in the experiments. Ten of them served as the speakers, and five as the listeners. All of the speakers and listeners lived in the same dormitory for at least more than four years; therefore they had known each other very well. Before the experiment, it was confirmed that the listeners were in contact with all of the speakers in daily life. None of them had known hearing impairments. Their native language was Japanese, and their average age at the time of the experiment was 22.9 years old.

Recordings of the speech materials were conducted in a sound-proof room. The speakers were instructed to speak in a normal conversational voice. They uttered Japanese monosyllables in the carrier phrase. The sentences were presented to the speakers on pieces of cardboard randomly, and ten repetitions of each sentence were recorded. The recorded sentences consisted of non-sense words /VCVCVCV/ (V stands for a vowel and C, for a consonant) embedded in a carrier phrase, /… toː o çiʑi çimasɯ/, meaning "I support the 'VCVCVCV' party (a fictional name of the political party)." The names of the fictional political parties were used, because the suffix /–toː/, meaning a political party, forms some compound words that do not have falls in accent in Japanese [6, 7]. Therefore the word "VCVCVCV-to" is uttered with a relatively stable accent pattern after the third mora. The last morae of "VCVCVCV" were excerpted manually to be used as the stimuli, by the method described later.

Speech materials used in the experiments should represent a wide range of speech sounds of Japanese so as to examine what kind of sounds are effective for identifying the speakers, but at the same time, considering the available test time and the burden for the listeners, a limited set of sounds can be used. In this test, only the coronal consonants (/d/, /t/, /n/, /ɲ/, /j/, /z/, /s/, and /ɾ/) and the bilabial nasal (/m/) were selected. The reason for selecting the coronal place of articulation is that this place has the largest number of items in the Japanese phoneme inventory. For the vowel, we selected an open vowel /a/. The reason for using only one vowel /a/ is to make the experiment simple. Notably, /a/ is reported to be the most effective vowel for perceptual speaker recognition of the Japanese five vowels [5, 8, 9] and in other languages [2, 10].

Ten speakers took part in the recording sessions held one by one. All the speech materials were recorded onto digital audiotapes (DAT) using a microphone (SONY ECM-MS957) and a DAT recorder (SONY TCD-D08). The microphone was positioned at approximately 10–15 cm from the speakers' lips. In order to make the situation close to earwitnessing, speech materials were saved with a high quality, at the sampling frequency of 48 kHz with 16-bit resolution.

Stimulus creations were conducted manually by using the computer software Cool Edit [11]. All the excerptions were made based on the waveforms, considering the following two criteria:

1. the mora was cut out to be of its longest possible duration, and
2. the gliding parts should be excluded.

Out of the ten excerpted tokens, five tokens for each syllable were selected on the ground that the sound is in good trim, without any noise on the gliding phases, and that they ultimately sound just as the target phonemes. Screening of the stimuli was made by a native speaker of Japanese other than the experimenter. Furthermore, out of the nine monosyllables excerpted from the sentences, /za/ has some variations when it is in the utterance-initial position; it may be realised as an affricate or a fricative. In this experiment only the samples that were realised as a fricative were

Table 11.1 Percent correct speaker identification for each stimulus monosyllable; $N=250$

Stimuli	Percent correct (%)
/na/	86.0
/ɲa/	85.6
/ma/ /za/	80.8
/sa/	78.8
/ja/	78.4
/da/	78.0
/ɾa/	74.4
/ta/	73.6

used. The total number of the stimuli was 450, that is, corresponding to ten speakers, nine consonants, and five tokens.

Finally, the stimulus array was created; the inter-stimulus interval was 2.0 s, out of which was 0.5 s of white noise. The noise was put central to the inter-stimulus interval, and the remaining space was silence. White noise was inserted because it has the effect of erasing the auditory impression of the preceding stimulus [12, 13].

11.2.1.2 Procedures

The test was carried out in the sound-treated room in a one-by-one manner. The created stimulus array was again recorded onto DAT and presented to the listeners in a random order using the player (SONY TCD-D8) through the headphones (SONY MDR-Z400) at a comfortable sound level.

Before starting the tests, listeners had a chance to hear each speaker's speech samples twice. The sample utterance was /hond͡ʑitsɯ wa seiten naɾi/ (It is fine today). The listeners were informed of the names of the ten speakers, and they were told to write the name of the speaker on the answer sheets for each stimulus. They were instructed to answer intuitively and to leave it blank when they did not know the answer. They took breaks after every 150 trials, and the total test time was about 40 min.

11.2.1.3 Results

The percent correct speaker identification for each monosyllable is shown in Table 11.1. The performances of the five listeners are averaged. The figure shows that the syllables containing the nasal obtained the highest scores, followed by fricatives and oral stops. Also, in the pairs of /da/-/ta/ and /za/-/sa/, the voiced sounds were better than their voiceless counterparts. The advantage of the voiced consonants was also observed in other studies [5, 10, 14].

In order to compare the mean scores of the syllables, analysis of variance (ANOVA) was conducted. The results revealed that the difference among the consonants was not significant ($p=.11$), except that the nasals /n/ and /ɲ/ were significantly better than the tap/flap /ɾ/ and the stop /t/ ($p<.05$). In a t-test, where means of the two groups are compared, nasal and oral sounds were compared and the difference of these two sound classes was significant ($t=2.86, p<.005$).

Voiced consonants gained higher scores than the voiceless counterparts, although the difference was not significant ($t=0.91$). As to the manners of articulation, fricatives (/sa/ and /za/) followed nasals, and oral tap and stops (/ɾa/, /da/ and /ta/) ranked the lowest.

11.2.2 *Experiment 1b: Identification of Previously Unknown Speakers*

11.2.2.1 Participants and Speech Materials

Out of the ten speakers who participated in the above experiment, four speakers were selected and their speech data were used again in this experiment. The selection of the speakers was based on the resemblance of the average fundamental frequency, taking into account the reports that the fundamental frequency has a large effect on unknown speaker identification [15–18], and also that we are focusing on the articulatory properties rather than the phonation properties. The fundamental frequency of the stimuli was analysed through auto-correlation method implemented in Praat [19, 20]. The average fundamental frequency of all the stimuli of the four speakers was 109.9 Hz ($N=160$, i.e., 40 utterances for each speaker, $SD=7.7$) with the range of 102.4–118.5 Hz. Sixteen university students who had never known any of the speakers served as the listeners. They were all native speakers of Japanese with normal hearing.

The stimuli used here were identical to those used above, except that the speech materials of only four speakers described above were used. Nine monosyllables excerpted from carrier sentences were again presented to the listeners in a random order through headphones. The number of the tokens for each speaker was also the same, i.e., five tokens for each monosyllable.

11.2.2.2 Procedures

This experiment was comprised of three sessions: familiarisation, practice and test sessions. The procedure for the experiment is depicted in Fig. 11.1. The listeners went through the familiarisation session at the first stage of the experiment. They listened to the sample sentences of the four speakers, /hond͡ʑit͡sɯ wa seiten naɾi/ ("It is fine today"). The listeners could listen to these sample sentences as many times as they wanted, but these sentences were always presented as one set of the four speakers, i.e., the listeners were not allowed to listen to the utterance of a particular speaker for many times. The speakers were introduced using speaker IDs, from number 1 to number 4, not by their names.

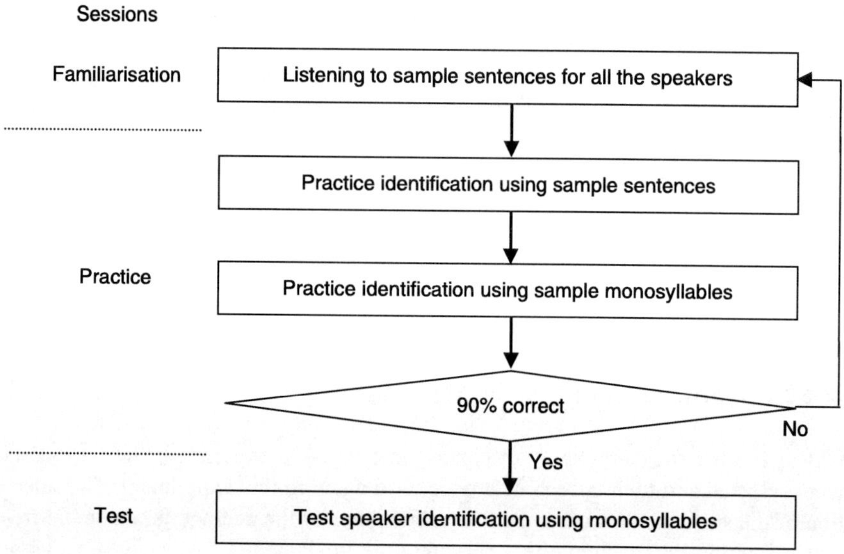

Fig. 11.1 Procedure for Experiment 1b: identification of previously unknown speakers

After the listeners showed some confidence, practice sessions were carried out using these sample sentences. Two types of practice were made; in the first practice, sentence uttered by one of the speakers was presented, and the listeners answered the speaker ID. Feedback was given after each trial. Then we moved on to the second practice session, where they identified the speakers by sample monosyllables. These monosyllables consisted of the consonants that were different from those used in the test session and the vowel /a/. Feedback was given after every trial also in the second practice. We repeated familiarisation and practice sessions until the listeners achieved more than 90% correct identification (out of 10 trials) in both practice sessions. It took the listeners 10–20 min before they reached the desired accuracy.

In the test session that was conducted immediately after the practice session, the listeners answered the speaker ID for each stimulus. The stimuli were presented to the listeners only once using a Praat [19] computer programme; and no feedback was given in this session. The total number of the stimuli was 180 (four speakers, nine syllables and five tokens). The listeners were not allowed to go back to listen to the sample sentences during the test session. The whole experiment was conducted in the same sound-proof room as familiar speaker identification experiment. The participants took a break after every 90 trials.

11.2.2.3 Results

The percentages of the correct speaker identification are shown in Table 11.2. The number of evaluation for each stimulus is 320 (four speakers, five tokens and sixteen listeners). All the stimuli gained higher scores than the chance level (25%).

Table 11.2 Percent correct for each stimulus in identification of previously unknown speakers (the second experiment); $N=320$

Stimuli	Percent correct (%)
/na/	59.0
/ɲa/	53.7
/ja/	52.0
/sa/	50.0
/ma/	49.7
/za/	46.7
/da/	46.3
/ta/	45.7
/ɾa/	44.7

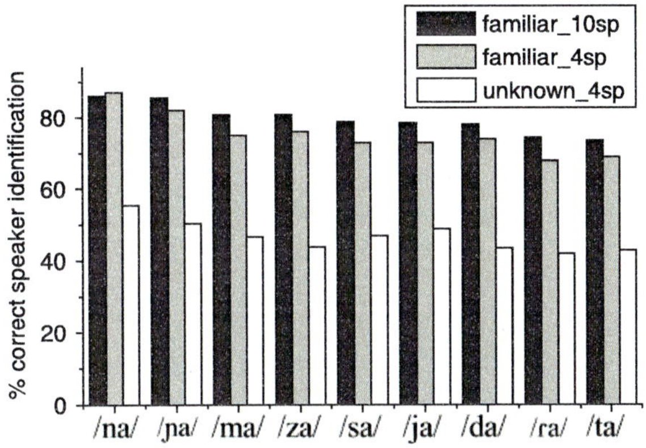

Fig. 11.2 Results of the two experiments: familiar speakers (the first experiment) and unknown speakers (the second experiment). Percent correct for each stimulus syllable in speaker identification tests. The middle (*grey*) bar shows the results of the first experiment, as for the four speakers whose speech materials were used in the second experiment

As was in the above experiment, the nasals /na/ and /ɲa/ ranked the highest, and fricatives and oral stops followed. The nasal /ma/ did not obtain as high score as in familiar speaker identification, on the other hand, /ja/ ranked higher. In one-way ANOVA, the effect of the stimuli was significant ($p<.01$). There were significant differences between the nasal /na/ and the oral stops /da/, /ta/ and /ɾa/.

11.2.3 Discussion: Effects of the Phonological Contents in Relation to Speaker Familiarity

So far we introduced two perceptual speaker identification experiments. Results of these experiments are shown in Fig. 11.2 in comparison to each other. The graph indicates the percent correct for each stimulus. The left bar represents the results for the first experiment (ten speakers, $N=250$); the middle bar, those of the first experiment again, but only the results for the four speakers are shown, whose speech materials were used in the second experiment (four speakers, $N=100$); and the right bar indicates the results for the second experiment (four speakers, $N=320$). Here note that the listeners are different for the central and the right bars, although the speech materials are identical.

As can be seen, the tendencies of the three bars are quite similar, although the identification scores are lower in the second experiment. The effect of the speaker familiarity was significant in ANOVA (F (1, 17)=803.4, p<.001), with familiar speaker identification performance (*Mean*=75.2%, *Standard Error* (*S.E.*)=0.019%) being better than unknown speaker identification performance (*Mean*=46.6%, *S.E.*=0.014%). Also the effect of the stimulus contents was significant (F (8, 17)=10.9, p<.001). The nasal /na/ gained significantly higher score than any other stimuli, and /ɲa/ was significantly better than /da/, /ta/ and /ɾa/.

In our experiments, we can regard familiar speaker identification as a closed-set test, while unknown speaker identification is an open-set test. Poor performance in unknown speaker identification is an expected outcome, just as reported in previous studies [21, 22]. This occurs because familiar and unknown speaker identification tasks undergo different cognitive processes, and the process for unknown speaker identification is a more difficult one [4, 23, 24]. In forensics we have to keep in mind the difference in performances due to speaker familiarity; the victims are often acquainted with perpetrators, while passers-by earwitnesses may not.

Despite the difference in familiarity, the overall tendencies of the two experiments as to the stimulus contents were similar, and no interactions between the stimulus contents and the familiarity to the speakers were seen as for the results of these two experiments. The forensic significance of this result is that whether an earwitness is familiar with the perpetrator or not, utterance containing nasals can be used as more effective speech samples than those consisting of only oral sounds. Since occurrence frequencies of the nasal consonants are relatively high in many languages, including English and Japanese, it is not difficult for us to find nasal portions in the speech samples.

In both experiments, the nasals were more effective for speaker identification than the oral sounds, with the alveolar nasals being better than the bilabial. And when we focus on the manners of articulation, the nasals ranked the highest, then the fricatives and the plosives, or the oral stops, followed them. This ranking of the consonants coincide with the ranking in the sonority scale [25]. The sonority scale is a ranking of segments according to the relative resonance. Vocalic segments are more sonorous than consonantal segments, and voiced segments are more sonorous than voiceless segments. Results of the experiment showed that the more sonorous a consonant is, the more effective it is for perceptual speaker identification. Generally speaking, sonorous consonants tend to be voiced, as with nasals and liquids, thus these sounds contain not only articulatory properties, but also the properties of the sound source created at the vocal folds. As to the individualities contained in the sound source, voiced sounds are reported to be more effective for perceptual speaker identification compared to their voiceless counterparts [5], although speakers' individual differences of course lie in the vocal tract properties, too [26–28].

Apart from sonority and voicing, there were a few differences. The syllable containing an approximant /ja/ ranked higher in unknown speaker identification than in familiar speaker identification. On the contrary, /ma/ did not obtain as high score as in the first experiment.

The effectiveness of the nasals in speaker identification can be explained by the uniqueness of the morphology of the resonators such as the nasal cavity and the paranasal sinuses. Differences in the timing of the velic action may be another factor that differentiates the nasals from oral sounds [29, 30]. Both the morphology and the velic action are something that the speakers cannot intentionally or voluntarily control by themselves. These facts imply that the nasal sounds have relatively stable properties reflecting speaker's physiological and anatomical individualities.

As to the places of articulation, alveolar consonants were better in the scores than bilabials. This tendency is consistent with the results in Amino et al. [31, 32]. Articulatory variations in the nasal production are reported in Fujimura [33]; he suggested that labial /m/ has greater intra-speaker variations than coronal /n/ does. Data in Su et al. [34] also support this claim.

From the perceptual point of view, we do not yet know the reason for the nasal availability. Articulatory and acoustic properties of the nasals have the following three saliencies: first, they contain nasal murmur; secondly, there are transitions to the following vowel; and finally, adjacent vowels are nasalised to some degree. Perceptually, the first two cues, nasal murmur and transitions, contribute to the perception of the place of articulation [35]. Studies on the perception of nasals specific to speaker individuality do not exist as far as the authors know, and perceptual explanations are still awaited.

11.2.4 Acoustic Analyses of the Experimental Stimuli

11.2.4.1 Methodologies of the Spectral Analyses

The results of the perception experiments showed that the stimuli containing a nasal consonant yielded higher speaker identification rates than the stimuli without it. In our experiments above, the stimuli all had the same CV (consonant–vowel) structure where the vowel is controlled to be /a/. Then the differences in the identification results should come from the consonant parts or from the transitions to the following vowel. In order to find out the acoustic characteristics that contribute to the differences among the stimuli, two kinds of acoustic analyses were performed. In the first analysis, cepstral distances were calculated and compared within and among the speakers in order to explain the identification differences among the stimuli. Cepstral distance between two speech samples can be calculated as follows:

$$D(x,y) = \sqrt{\sum_{k=1}^{p} (C_{xk} - C_{yk})^2},$$

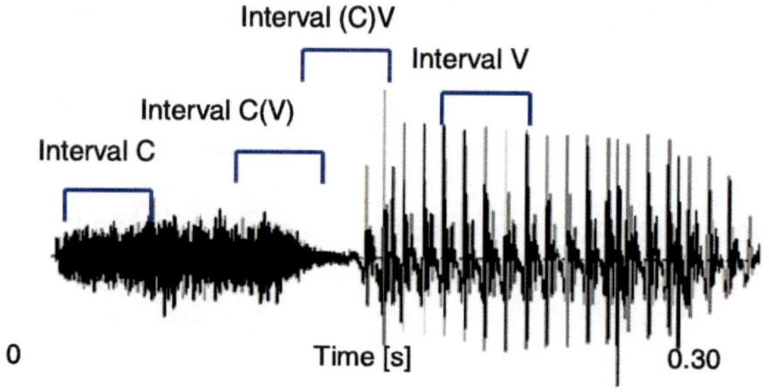

Fig. 11.3 Example of the excerption of the four intervals from a CV (consonant–vowel) syllable

where C_x and C_y are p-th order cepstra for the two speech samples x and y, respectively.

For speaker comparison, we used the ratio of intra- and inter-speaker distances as the measure. This measure is based on the concept of the F-ratio [36, 37], which is used for finding feature discrimination in speaker recognition. When calculating the F-ratio, we usually use the ratio of inter- and intra-speaker variances; however, in this study, we used the ratio of inter- and intra-speaker distances instead. Thus,

$$F = \frac{D_{inter-speaker}}{D_{intra-speaker}},$$

where D represents the cepstral distance. We call this analysis "F-ratio analysis." In the second analysis, or "confusion analysis," our goal was to see the perceptual confusions among the speakers and to determine the portions that are important for perceptual speaker identification.

As the analysis targets for both analyses, four intervals were excerpted from the following six monosyllables uttered by the ten speakers: /ta/, /da/, /sa/, /za/, /ma/ and /na/. Each stimulus had five tokens for each speaker. The interval length was 30 ms; and all the materials were downsampled at 16 kHz before the analyses. The names and criteria for the interval excerption are as follows:

- interval-C: stable consonant part; while /t/ and /d/ omitted,
- interval-C(V): consonant with transition (until the second formant gets stable in the following vowel in spectrograms),
- interval-(C)V: vowel with transition (includes formant transitions), and
- interval-V: stable vowel part.

Example of excerption is shown in Fig. 11.3. All the excerptions were conducted manually based on the waveforms and spectrograms. We did not get interval C for /ta/ and /da/, as they were just silence in some utterances. Several intervals were overlapped with preceding or following interval(s) in some of the samples. As for the remaining syllables, /ja/, /ɾa/ and /ɲa/, we omitted from the analysis targets, since these sounds are realised as momentary or gliding sounds in Japanese, and therefore it is hard to define the boundary of the consonant and vowel.

In both analyses, we used 30th order FFT cepstrum as the analysis parameter. Cepstrum is the inverse Fourier transform of the log power spectrum of a signal [38]. The low-quefrency components, in the representation domain for cepstrum, contain supra-laryngeal information, or filter information of the source-filter model, while higher-quefrency components contain pitch, or source information. We use the former here as a spectral parameter. The zeroth coefficient, which is a measure of energy in the signal, was excluded here. Intra- and inter-speaker cepstral distances were further computed for every possible pair of five tokens of a speaker and of ten speakers, respectively. Thus we obtained fifty-by-fifty square matrices for each monosyllable.

In F-ratio analysis, the ratios of averaged intra-speaker distances to averaged inter-speaker distances were calculated. Larger inter-speaker distance and smaller intra-speaker distance are most probably representing speaker individualities, and greater F-ratio values are desirable for speaker identification purposes.

In confusion analysis, the purpose was to inspect the relationship between perceptual speaker similarities and acoustic properties of the stimuli in more detail. We analysed the perception patterns, that is, confusions among the speakers, observed in the first experiment. In order to examine the perception of speaker identity in relation to the inter- and intra-speaker cepstral distances, confusion patterns were analysed by drawing confusion matrices among the speakers. We made six ten-by-ten square matrices for confusions on each of the six monosyllables and also 24 cepstral-distance matrices, which are corresponding to four intervals for six monosyllables. Then we calculated the correlation coefficients between the confusion matrices and the distance matrices. The correlation coefficients between two variables X and Y can be defined as follows:

$$r = \frac{1}{n-1} \sum_{i=1}^{n} \left[\left(\frac{X_i - \overline{X}}{\sigma_X} \right) \left(\frac{Y_i - \overline{Y}}{\sigma_Y} \right) \right],$$

where n is the sample size, $\overline{X}$ and $\overline{Y}$ are sample means, σ_X and σ_Y are sample standard deviations, respectively. Here we compared matrices only for the same monosyllables.

11.2.4.2 Results of the Spectral Analyses

The average F-ratios for each analysis interval are shown in Table 11.3. We can see that the nasal sounds obtained larger ratio values all through the four intervals. On the other hand, other oral sounds gained relatively higher values only in the vowel part of the stimuli. This means that nasal sounds have longer interval that effectively indicates speaker individuality. As for the interval C, the inter-speaker distances and the ratios of the inter- and intra-speaker distances were the largest in the nasal consonants, and then the fricatives and the oral stops follow them.

Table 11.3 Ratios of inter-speaker to intra-speaker cepstral distances (averaged distances among ten speakers and among five tokens of each speaker, respectively)

Stimuli/intervals		C	C(V)	(C)V	V	Ave.
Nasals	/ma/	2.35	2.26	2.22	2.30	2.28
	/na/	2.08	2.06	2.20	2.21	2.14
Fricatives	/sa/	1.45	1.54	2.05	2.24	1.82
	/za/	1.55	1.55	2.05	1.99	1.79
Stops	/ta/	N/A	1.15	2.06	2.11	1.77
	/da/	N/A	1.15	1.95	1.95	1.68

Table 11.4 Correlation coefficients between the perceptions of the speakers and the *F*-ratios of cepstral distances

Stimuli/intervals		C	C(V)	(C)V	V
Nasals	/ma/	−0.81	−0.79	−0.75	−0.67
	/na/	−0.79	−0.77	−0.62	−0.63
Fricatives	/sa/	−0.38	−0.38	−0.66	−0.69
	/za/	−0.33	−0.33	−0.64	−0.58
Stops	/ta/	N/A	−0.31	−0.60	−0.57
	/da/	N/A	−0.34	−0.64	−0.63

In the statistical analyses, there were significant differences in the ratios among the consonants in ANOVA ($F\,(59, 5) = 19.42$, $p < .001$). In the results of the post-hoc test, where each consonant pair was compared by Tukey's test, the ratios of the two nasal consonants /m/ and /n/ were significantly larger than those of any other consonants. No other pairs or groups were significantly different. We also notice that non-nasal, or oral syllables showed relatively high ratio values in intervals (C)V and V. In fricatives, intervals C and C(V) obtained higher scores in the voiced consonants than in the voiceless counterparts. This is because voiced sounds contain source information as well as resonance information, while voiceless sounds do not, and the source information is known to contain speaker individualities [27, 39].

The results of the confusion analysis are shown in Table 11.4, and the percent correct perceptual identification of the speakers is drawn as the function of the *F*-ratio values in Fig. 11.4.

As can be seen in Table 11.4, all the correlation coefficients were negative, which means that the greater the cepstral distances between two speech samples, the less confused were the listeners. The correlations between the perception and the spectral properties were observed in all the intervals in the nasals, but only in the vowel intervals, (C)V and V, in oral sounds. Figure 11.4 shows the positive correlations between *F*-ratios and percent correct speaker identification. These results lead to the following two implications: first, in stimuli containing a nasal sound, listeners use all four intervals as the cue to identify speakers; and in oral sounds, both stops and fricatives, listeners tend to use only the vowel part as the speaker cue.

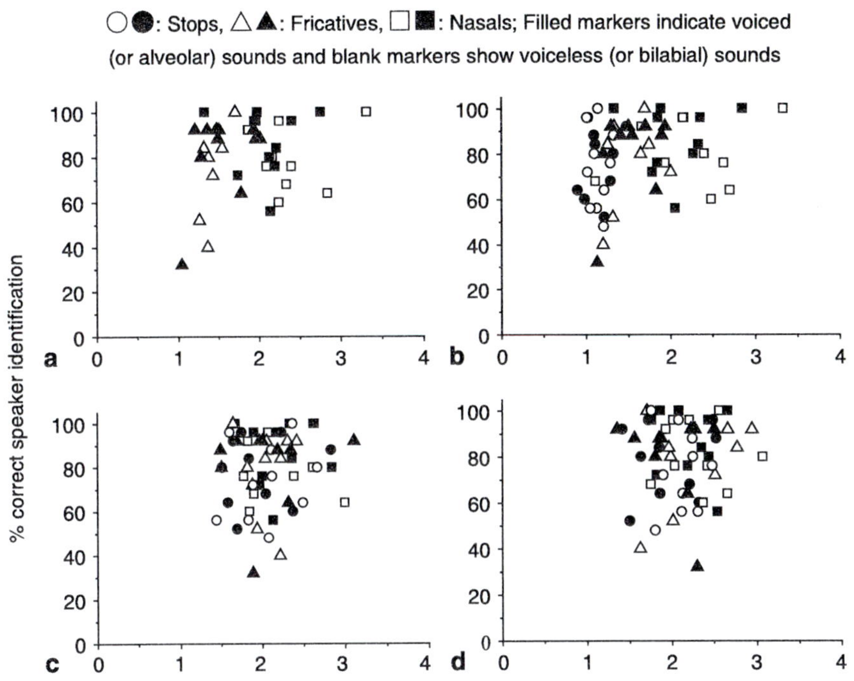

Fig. 11.4 Percent correct perceptual speaker identification as a function of the F-ratios of inter- and intra-speaker cepstral distances for **a** Interval C, **b** Interval C(V), **c** Interval (C)V, and **d** Interval V

11.2.4.3 Duration Analysis

Results of the spectral analyses showed that inter-speaker cepstral distances corresponded to perception of the speaker identity. However, as pointed out by Pollack et al. [40], duration of the presented speech materials may influence the speaker identification accuracy especially when we use short stimuli such as monosyllables. In order to assess the effect of the stimulus lengths, we measured the durations of them.

The average durations for the whole syllable and those for the consonant part (not including the transition to the following vowel) and their standard deviations are shown in Table 11.5.

The consonant duration for the syllable /ja/ was unable to measure, as it was difficult to determine the boundary between the consonant and vowel portions. The total number of tokens for each type is 50, that is, five tokens for each of the ten speakers. The data were submitted to the correlation analysis; consequently the correlation was not significant with the whole duration or the consonant duration. Thus we conclude that the spectral properties of the stimuli are used as the clues to the speaker identity, and not the duration.

Table 11.5 Average durations of the whole syllables and the consonant part (N=50)

Syllables	Whole duration (ms)		Consonant duration (ms)	
	Mean	S.D.	Mean	S.D.
/ma/	274	40	76	14
/na/	273	45	73	18
/ɲa/	277	51	56	17
/ja/	240	34	N/A	N/A
/sa/	301	39	134	19
/za/	267	54	75	23
/ta/	309	35	107	14
/da/	272	52	58	23
/ɾa/	247	37	39	20

Table 11.6 Results of the speaker recognition experiment using inter and intra-speaker cepstral distances. Percent correct verification out of 50 trials is shown

Stimuli/intervals		C	C(V)	(C)V	Ave.
Nasals	/ma/	100	98	100	99.3
	/na/	100	100	100	100.0
Fricatives	/sa/	80	70	98	82.7
	/za/	60	70	98	76.0
Stops	/ta/	N/A	40	98	69.0
	/da/	N/A	48	96	72.0

11.2.4.4 Speaker Identification Experiment Using Inter- and Intra-Speaker Cepstral Distances

In order to evaluate the speaker recognition performances using the cepstral distances, we further conducted a closed-set speaker verification experiment. Inter- and intra-speaker cepstral distances calculated in this section were again used. Three analysis intervals concerned with the consonant, i.e., C, C(V), and (C)V, were tested. There were 50 tokens for each monosyllable and for each analysis interval. One of the 50 tokens was chosen and compared with other 49 tokens, and all of the possible pairs were verified. If the token that showed smallest cepstral distance is of the same speaker as the input token, it is counted as correctly matched. The results were evaluated by percent correct verification where the input token was judged to belong to the same speaker.

The results are summarised in Table 11.6. As the table shows, the nasals yielded almost perfect speaker identification among the ten speakers of this experiment. The advantages of the nasals were observed in all three intervals. On the other hand, oral consonants scored no more than 80% on average. They gained higher scores only in Interval (C)V. The tendencies were similar to those in Table 11.3.

11.2.5 Summary and General Discussion

In this section we introduced two speaker identification experiments by human listening and their results. We assessed the effects of the phonological contents of the stimuli and of the speaker familiarity on speaker identification accuracy. The syllables containing a nasal consonant were the significantly more effective than those without it. Familiarity to the speakers had a large influence on identification performances, but no interaction with the phonological contents was observed.

We compared the spectral properties of the stimuli used in the experiments by calculating intra- and inter-speaker cepstral distances. We found that the inter-speaker distances were greater in nasal sounds than in oral sounds. Furthermore, we analysed the inter-speaker distances for four intervals that temporally ranged from the onset consonant part to the stable vowel part, and we found that all the intervals correlated with the perception of the speaker identity in the stimuli containing a nasal, but not in the stimuli of only oral sounds. The effectiveness of the nasals in speaker identification is also pointed out in other studies, such as Nakagawa and Sakai [41]. However, the correlation between the perceptual confusions among speakers and the inter-speaker cepstral distances in nasal sounds is a new insight.

The speaker-dependency of the nasals can be attributed to their resonance properties; morphologies of the resonance cavities differ considerably among speakers. Especially the shapes of the paranasal sinuses are known to be quite complex and speaker-specific [42, 43]. As to the place of articulation, coronal nasals /n/ and /ɲ/ were more effective for identifying speakers than labial /m/. This difference may be reflecting larger intra-speaker variations in labial articulation [33, 34].

When we look at the rankings of the syllables in the identification test and in the F-ratio analysis of the cepstral distances, which are shown in Table 11.3, we find that they agreed not only in that the nasals had greater inter-speaker variations, but also in the orders of the manners of articulation, i.e., the nasals, the fricatives and the oral stops. This ranking also corresponds to that of the sonority scale [25]. It means that the more sonorous a sound is, the more speaker individualities it contains.

Further analyses of the experimental data showed that speaker identification scores differed among the speakers significantly in the two experiments ($p<.001$). Matsui et al. [44] suggests that the sounds effective for identifying the speakers are different for each speaker. Moreover, Bricker and Pruzansky [45] reported that the identification results may vary according not only to the speakers but also to the speaker ensembles where they are being compared to each other. The coronal nasals /n/ and /ɲ/, the most effective sounds for speaker identification, were not necessarily the most effective sounds for all of the speakers, but we can say that the nasals were relatively effective for most of the speakers.

Differences in the identification performances among listeners were also significant in ANOVA. It is pointed out that the ability to identify speakers is dependent on the individual listener [45], though the listener group of more than twelve people is said to be of typical size to obtain homogeneous data [46, cited in 45].

In our experiments, the average identification rate of the five listeners in familiar speaker identification was 79.6% with the range from 67.1 to 89.1%, whereas that of 16 listeners in unknown speaker identification was 46.6% ranging from 35.0 to 65.0%. Differences in speaker identification performances may also come from the different strategies that the individual applies when identifying speakers or the differences in the priorities of the acoustic cues, as pointed out in [47].

Some researchers [48, 49] propose identifying the phonemes that best indicate speaker individuality through perceptual speaker identification experiments and using them in speaker recognition by machine. Our results suggest that nasal sounds can be a good candidate for this, especially in forensic situations, for the following three reasons. First, they occur frequently in natural speech. This means that we may have more chance to obtain nasal portions in the utterances compared to other consonants. Secondly, the acoustic properties of the nasals are speaker-dependent. This may derive from the individualities in resonant cavity morphologies. Furthermore, the speakers cannot intentionally change the shapes of these cavities; therefore the resonant properties of the nasals may be robust against speaker disguise. Finally, the effectiveness of the nasals for speaker identification did not depend on the degree of speaker familiarity in the perception experiments. Availability of the nasals in perceptual speaker identification may be applied for both familiar and unknown speakers.

11.3 Emerging Standard for Telephone Communication

When a change is brought about in a social system, laws and related fields are always forced to cope with it. Forensics is not an exception. Nowadays, almost everyone has a cell phone. Recent national survey in Japan found that 92.4% of the households own a cell phone at the time of March, 2010 [50]. This new communication infrastructure gave rise to the changes in information technology, in speech communication technologies, and also in forensic sciences. Researchers in forensic phonetics now have to care about new models of cell phones and their sound quality specifications. On the other hand, in order to keep up with the demand, noise-robustness of the cell phones has always been one of the important issues in telephone industry. In line with this trend, bone-conduction cell phones have appeared in the market.

Bone-conduction devices first appeared as hearing-aid in 1812, although the bone conduction itself had been known already in the mid-16c [51]. In bone-conduction hearing, sounds are input through the temporal bone to the inner ear; while in bone-conducted speech production, human speech sounds can be recorded through the temporal bone, zygomatic bone, or frontal bone [52–54]. Also, we can use an earphone type microphone for recording bone-conducted sounds. Bone-conduction microphone is a useful communication tool in noisy speaking environments, especially in a situation where a hands-free and downsized device is desirable; for example, it is used in a construction site, in the military, or on a vehicle.

Bone-conduction microphones do overcome the noise barrier; however, once used in a criminal situation, various problems arise. First of all, the voice quality changes considerably. For example, our speech sounds differently when recorded and played back. This is because people with normal-hearing usually hear their self-voice both through bone and air conduction, whereas the recorded voices are only air-conducted. Forensically, when a perpetrator uses a bone-conduction device while contacting, it may be hard to identify suspects. Otherwise, it becomes difficult to judge whether the air-conducted speech samples later obtained from a suspect belong to the same person or not. Secondly, the advantage of bone conduction that it is noise robust causes trouble in forensic case, as it does not pick up background noise, either. As claimed by many researchers [e.g. 21], background noises are problematic in recognising speakers, but at the same time they often provide us with some important clues about the scenes of the crime. Finally, few studies are available on the acoustic properties of the bone-conducted speech; thus we do not know whether it is possible or not to recognise speakers by using bone-conducted speech.

This section introduces two experiments, speaker verification by machine and by human, in order to investigate the effects of the different conduction channels on the accuracy of speaker verification. We will also describe the outlines of the speech database that we constructed, which contains air-conducted speech (ACS) as well as bone-conducted speech (BCS) samples obtained from the same speakers.

11.3.1 Construction of a Large-Scale Speech Database and Its Pre-Analysis

A large-scale speech database was constructed at National Research Institute of Police Science [55]. The general outline of the database is as follows. The database contains utterances of 632 (313 male and 319 female) speakers. The recordings were conducted in a sound-treated booth; recording sessions were held twice for each speaker. The second recording was conducted three to five months (average 85 days) after the first one. All the speakers lived near Kanto district (the area around Tokyo, Japan) and their age ranged from 18 to 76 years old at the time of the first recording session. They uttered 100 Japanese monosyllables, 66 words, and 64 short sentences. The monosyllables are those used in intelligibility test [56], and the words and sentences included those related to the criminal affairs such as "bomb" and "phone-tracing." Short sentences also included ATR phoneme-balanced sentences. All the utterances were repeated twice.

The speech data were recorded simultaneously through four separate channels using a digital recorder (Roland EDIROL R-4): ACS through a condenser microphone (SONY ECM-23F5), ACS through a cellular phone (NEC, FOMA N902i), BCS through a bone-conduction microphone (TEMCO EML-1-A), and BCS through telephone line (the same as ACS through telephone line). Recorded materials were sampled at 44.1 kHz with 16-bit resolution. The total number of the

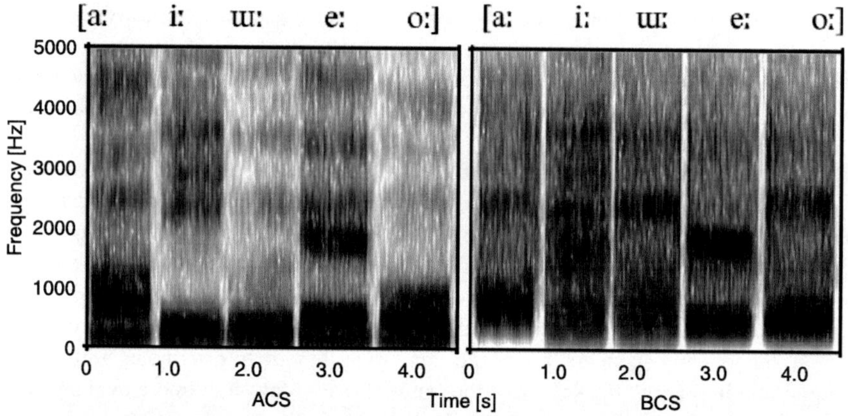

Fig. 11.5 Spectrograms of the Japanese five vowels uttered in isolation; they were recorded simultaneously through air-conduction (*left*) and bone-conduction (*right*) channels

recorded utterances was 2,325,760, that is, corresponding to 632 speakers, 230 utterances repeated twice, two recording sessions, and four channels.

Before conducting speaker verification experiments, we analysed the acoustic differences between ACS and BCS. Figure 11.5 shows the spectrograms of the Japanese five vowels, /a i ɯ e o/ uttered in isolation.

We can see that the spectral energy in the high frequency region is weakened or almost lost in BCS. Also, the figure shows that the vowel formants appeared in different frequency regions at different intensity levels for the identical utterances of ACS and BCS. The spectral differences were remarkable especially in front and mid vowels. These results are consistent with the reports by Ohyama et al. [54].

Spectrograms of Japanese phrase /tsɯɾɯno oŋgaeçi/, which means "The Grateful Crane" (a story title) are shown in Fig. 11.6. Again, the higher frequency components are subject to a drastic damping in BCS. The difference is salient in sibilants, /ts/ and /ç/. When we compare the spectrograms of bone-conducted vowels and phrase, we notice that frequency components in higher regions themselves are present in the vowel spectrogram. This suggests that bone-conducted fricatives do not transmit through body tissue and thus are attenuated more than the vowels are.

11.3.2 Speaker Verification Experiment by Human

11.3.2.1 Procedures

In order to assess the effect of the channel differences on human speaker verification, a perception experiment was conducted. Out of 632 speakers of the database, 60 (30 male and 30 female) speakers were selected, and their sentence utterances recorded through the cell phone (ACS and BCS through telephone line) were used in the experiment. Speech materials were sampled at 44.1 kHz with 16-bit resolution

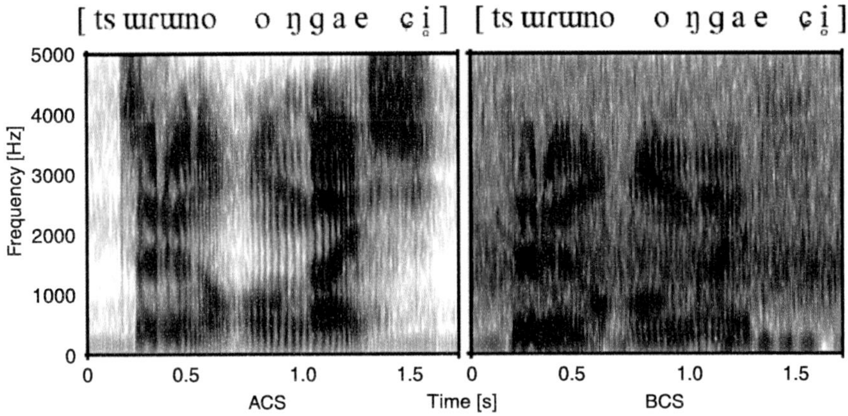

Fig. 11.6 Spectrograms of an utterance recorded simultaneously through air-conduction (*left*) and bone-conduction (*right*) channels

while recordings, but they were all downsampled at 8 kHz before the experiments taking into account the realistic conditions in forensic cases.

One hundred and fifty-five (78 male and the same number of female) listeners participated in the experiment. Their age at the time of the experiment ranged from their 20 to 40 s. The participants were all native speakers of Japanese and had normal hearing. None of them had participated in the recordings of the speech database. We selected two sentences, one long and one short, to present to the listeners: /iç:ɯ:kambakari ɲɯ:jo:kɯo ɕɯzaiɕita/ (4-s, "I visited New York for a week for coverage") and /ke:sat͡sɯɲi jɯ:na/ (2-s, "Do not tell the police").

In order to ease the experimental burden to the listeners, the utterances of 60 speakers were divided into three groups; each group contained 20 (ten male and ten female) speakers. The listeners were also divided into six groups according to their age and gender; and from each listener group, eight or nine listeners were assigned to listen to the utterances of one of the speaker groups.

The stimuli were presented to the listeners in AX form. The number of stimuli presented to each listener was 480, that is, corresponding to 20 speaker pairs (ten same-speaker pairs, nine different-speaker pairs, and one filler pair), two different positions (either A or X), two speaker genders, two sentences, and three channel pairs (ACS–ACS, BCS–BCS, and ACS–BCS). In these stimuli, the sentences presented in A and X positions were always the same one (long or short), and sentences recorded in different sessions were allocated in either position. The results of the experiment were evaluated by percent correct speaker verification.

11.3.2.2 Results

Results of speaker verification by human are shown in Figs. 11.7 and 11.8. The main effect of the presented sentence (long or short) was significant in a paired *t*-test ($p < .001$), long sentence being better than short one. The effect of the channels

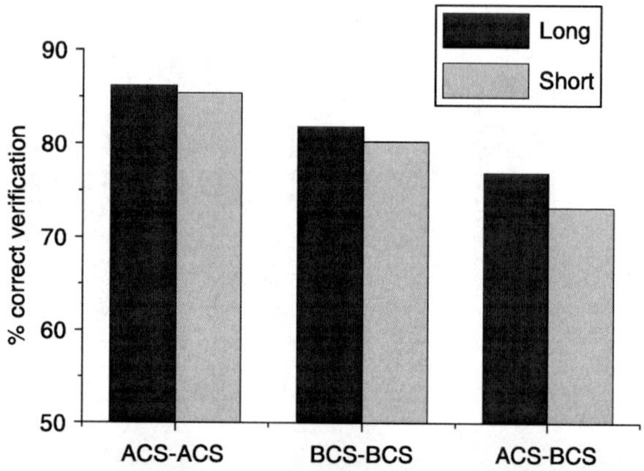

Fig. 11.7 Results of human speaker verification experiment; the effect of the presented sentences (long or short)

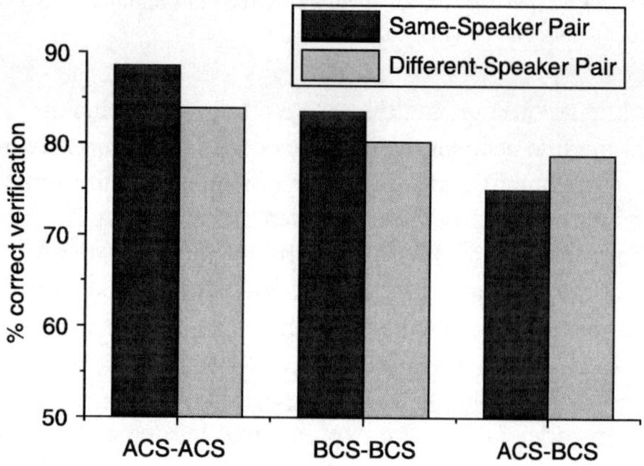

Fig. 11.8 Results of human speaker verification experiment; the effect of the presented speaker pairs (same or different)

was also significant in ANOVA ($p<.05$); ACS–ACS and BCS–BCS pairs obtained higher rates than ACS–BCS pair. The differences in verification rates between same- and different-speaker pairs were not significant in two-way ANOVA, but their interaction with the channel pairs was significant ($p<.001$). This means that the listeners tend to respond positively when the two samples in comparison were recorded through the same channel (ACS–ACS or BCS–BCS pairs), but the other way around when there is a channel difference (ACS–BCS pairs). No significant difference in performances was observed due to the listener's gender.

11.3.3 Speaker Verification Experiment by Machine

11.3.3.1 Procedures

In order to investigate the effects of the channels on speaker verification by machine, we conducted an experiment using words and monosyllables uttered by

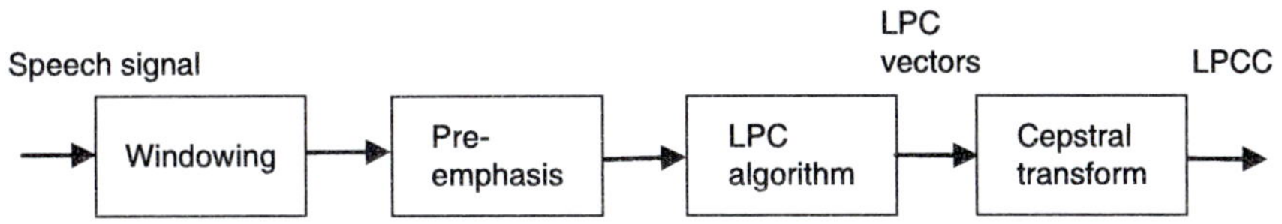

Fig. 11.9 Flowchart of LPC-based cepstrum calculation

200 male and 200 female speakers. Speech materials both recorded directly and through telephone were used. Here again, the directly recorded speech was downsampled at 8 kHz from 44.1 kHz of the recordings before the experiment, taking into account the real forensic situations. They were also adaptively pre-emphasised before the analysis.

For the acoustic parameter, 12th-order linear prediction cepstral coefficients (LPCC) were calculated using a Hamming window of 32 ms long with half-overlapping. In the LPC analysis, a given speech sample is approximated with a linear combination of past samples. It determines the coefficients (LPC coefficients) of a forward linear predictor by minimising the prediction errors. It is possible to calculate cepstral coefficients from these LPC coefficients by using Oppenheim's recursion algorithm (See Oppenheim et al. [57] for detail). A flowchart for LPC-based cepstral analysis is shown in Fig. 11.9.

Then we calculated distances between speech samples of all possible speaker pairs, including same speaker pairs, by applying dynamic time warping (DTW). DTW is used to accommodate differences in timing between two speech samples; it is an algorithm to compute the time axis stretch that optimally maps one time series onto another. The basic principle is to set a range of steps in a space and find the optimal path to proceed through the steps. The optimal path should maximise the local match and minimise the distance between the aligned time series. The path finding in DTW is subject to the constraints on the allowable steps. When there are no differences between two time-series, the warping path would coincide with the diagonal line, but as difference between the two increases, the warping path deviates from the diagonal line. In finding an optimal alignment, DTW sometimes creates unrealistic correspondences, by mismatching long and short features from the two time series. In order to avoid this, slope constraints are applied. See, for example, Senin [58] for detailed review on DTW.

Examples of DTW plots for the two utterances of the word /kɯrɯma/ (car) are shown in Fig. 11.10. We restricted the local slopes between 1/2 and 2. The DTW plot for two samples uttered by the same speaker shows less deviation from the diagonal line compared to that of two different speakers. However, we have to remember that two speech samples uttered by one single speaker may vary as much as, or sometimes more than, two samples uttered by two different speakers, and we cannot simply tell whether two samples were produced by the same speaker from the DTW plots.

Speaker verification was performed either between two ACS samples, two BCS samples and ACS and BCS samples by distance calculation. Minimum distance classifier was used in the verification. Average percent correct verification was used for the evaluation of the results.

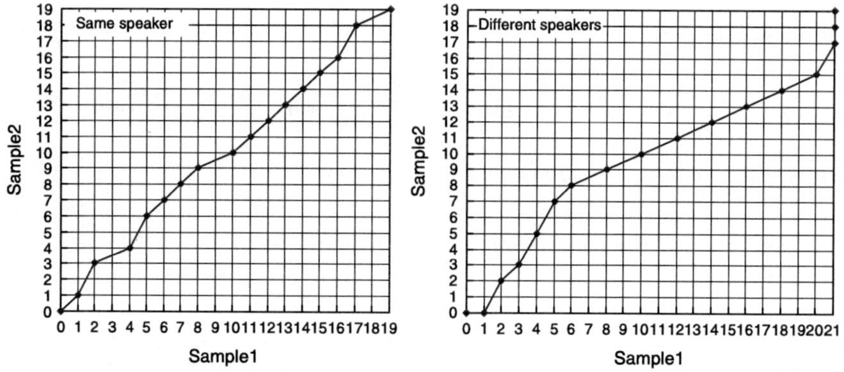

Fig. 11.10 Examples of DTW plots. Time series alignment for the word /kurruma/ (car); two samples uttered by a same speaker (*left*) and by two different speakers (*right*)

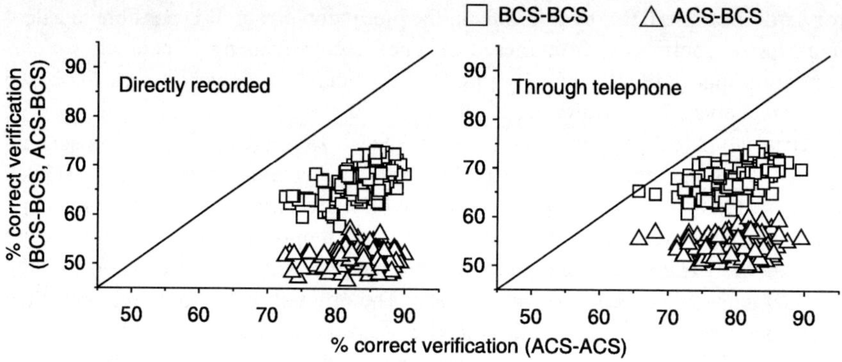

Fig. 11.11 Percentages of correct verification for BCS–BCS and ACS–BCS pairs in comparison with those for ACS–ACS pair; verification using LPCC; the results for directly recorded speech (*left*) and for speech recorded through telephone (*right*)

11.3.3.2 Results

The verification rates of the word utterances were compared between all possible pairs of the recording channels. In total, we had six combinations: ACS–ACS, BCS–BCS, ACS–BCS, and their telephone counterparts. Verification results for BCS–BCS and ACS–BCS pairs are plotted in Fig. 11.11 in comparison with those for ACS–ACS pairs. The average verification rates for each channel pair are shown in Fig. 11.12. The verification score was degraded by almost 17% in BCS–BCS pair, compared to the scores of ACS–ACS pair. Moreover, the verification rates deteriorated significantly when there is a difference in the recorded channels. The differences in verification rates were significant between all channel pairs in ANOVA (p<.001).

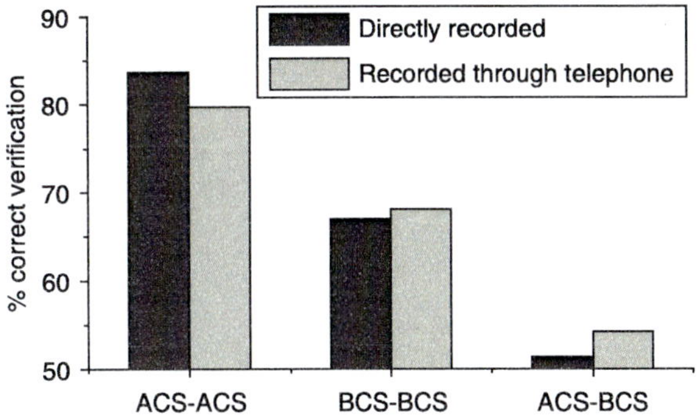

Fig. 11.12 Verification rates for each channel-pair; LPCC was used as the parameter

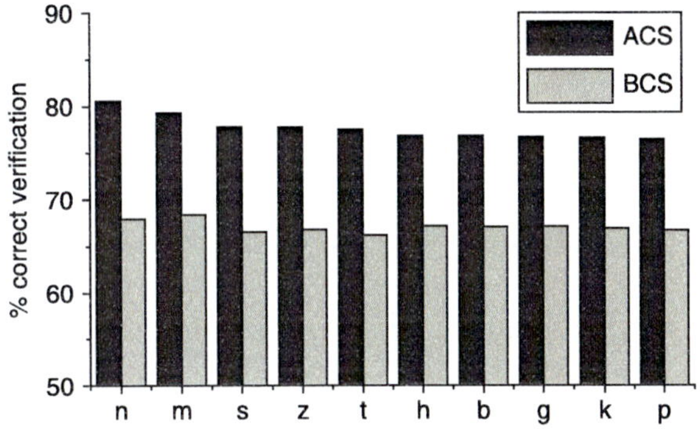

Fig. 11.13 Results of speaker verification using monosyllables ordered according to the consonant rankings; ACS (*left*) and BCS (*right*); the parameter used was LPCC

The results for monosyllables are summarised according to the onset consonants and shown in Fig. 11.13. With both ACS and BCS, the nasals, /m/ and /n/, obtained the highest verification accuracy. Results of ANOVA showed a significant tendency that the nasal consonants are better than oral consonants ($p = .052$). In ACS results, the rankings of the consonants broadly coincide with those in familiar speaker identification experiment (in Sect. 11.2.1); the nasals are followed by fricatives, voiced stops and voiceless stops. On the other hand, in BCS, the rankings of the fricatives and stops are reversed. This is presumably because the acoustic characteristics of BCS accompany damping of higher-frequency components as we have seen in the examples of the sibilants in Fig. 11.6 above.

11.3.3.3 Discussion

We have seen that the speaker verification accuracy using BCS was far less than that of ACS. Also, the difference in the recording channels in the speech data being compared has a great influence on the verification performances. Practically, it is quite possible that the speech samples we obtain from a suspect are recorded through different channel than the perpetrator's samples that are already recorded. Neverthe-

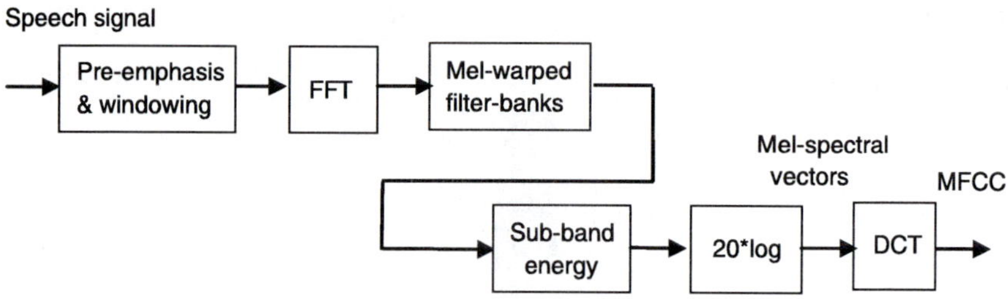

Fig. 11.14 Flow chart of mel-frequency cepstrum calculation

less, the percent correct verification that we gained in this experiment for samples recorded through different channels is about 51%, i.e., just above the chance level.

There are several ways to improve the speaker verification results when there is a difference in the recording channels. Among them, we examined the effects of the analysis parameters and normalisations of the transmission systems in our further experiments. Procedural details are given in the following section.

11.3.4 Improving Speaker Verification Performances When Speech Materials are Recorded Through Different Channels

11.3.4.1 Acoustic Parameters

In this experiment, we used the identical speech materials and methodology as those used in the above experiment (in Sect. 11.3.3) except that we used 12th-order mel-frequency cepstral coefficients (MFCC) and local peaks of the speech spectra (PEAK) as the acoustic parameters instead of 12th-order LPC cepstral coefficients. Both parameters are commonly used in speech and speaker recognition [59–61, among others].

The mel-frequency cepstrum is based on calculating the cepstrum from the log spectra obtained from a filter-bank with filters spaced on mel-warped frequency [60]. Mel-scale takes human perception with respect to frequencies into consideration. The analysis flow chart and the mel-frequency filter-bank are shown in Figs. 11.14 and 11.15, respectively.

We obtained PEAKs based on spectral envelopes extracted through LPC analysis. The PEAK values were represented by the differences of formant frequencies between two speech samples.

Results of the experiment are shown in Figs. 11.16 and 11.17 in comparison with the verification rates using LPC cepstrum. Here, too, minimum distance classifier was used for speaker verification. The verification rates improved slightly with BCS–BCS and ACS–BCS pairs when we used MFCCs as the parameter, as we can see in Fig. 11.16 that triangles and circles are above the diagonal line in

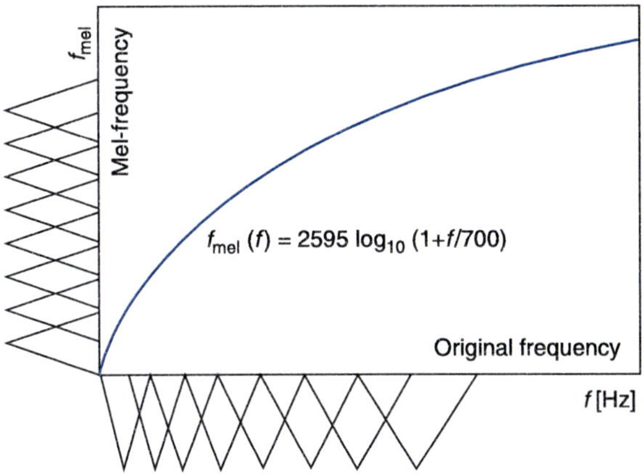

Fig. 11.15 Mel-frequency filter-bank. Triangular filters are uniformly distributed at the mel-warped frequency scale, but not so at the original frequency scale

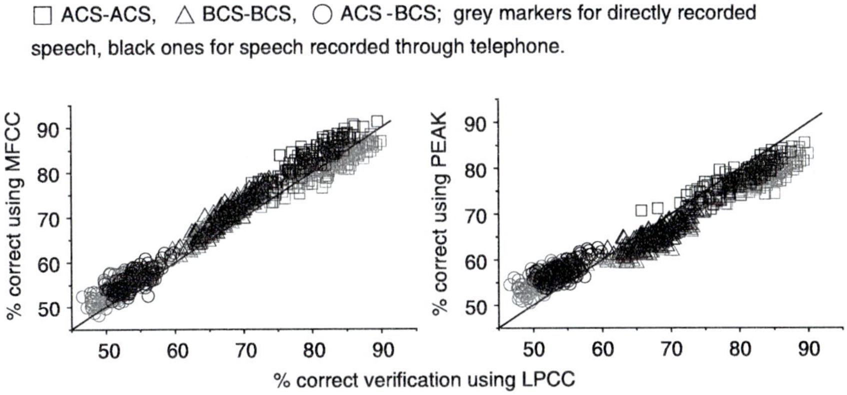

Fig. 11.16 Percentages of correct verification when using MFCC (*left*) and PEAK (*right*) in comparison with LPCC as the parameters

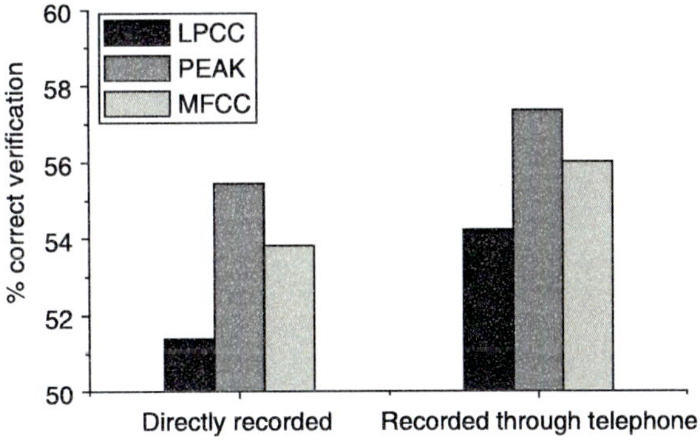

Fig. 11.17 Results of speaker verification using LPCC, MFCC, and PEAK as the parameters; speech recorded directly (*left*) and through telephone line (*right*)

the left graph; on the other hand, when PEAKs were used, performance improved in only ACS–BCS pairs; only the circles are above the diagonal line in the right graph. Figures 11.16 and 11.17 together show that PEAKs were more effective for speaker verification including channel difference, especially when the speech data are recorded through telephones.

11.3.4.2 Normalisation of the Transmission Systems

Another way to improve speaker verification accuracy against channel differences is to apply some kind of normalisation. Speaker individuality conveyed by the speech samples is relatively enhanced through channel normalization. We used two pre-existing methods that are used in order to make up for the channel differences: cepstral mean normalisation (CMN) [62] and standardisation-normalisation transformation (SNT) [63]. CMN is the method used for minimising the effect of the channel differences by subtracting the cepstral mean calculated across the utterance from the cepstra of each frame. It is a useful method for reducing multiplicative noise. The normalised vector can be expressed as in the following equation:

$$\tilde{\mathbf{C}}_j^T = (\mathbf{C}_j - \mu)^T, \tag{11.1}$$

where $\mathbf{C}_j{}^T = [C_{jl}, \ldots, C_{ji}, \ldots, C_{jp}], (j = 1, \ldots, N)$, is the p-th order cepstrum vector, N is the total number of frames in an utterance, T represents transposition, and μ is the mean vector, hence,

$$\mu = \frac{1}{N}\sum_{j=1}^{N} \mathbf{C}_j.$$

On the other hand, SNT is the method we proposed [63], where the mean and standard deviation of an acoustic parameter are used for standardisation and normalisation so that the norm is set to unity. SNT is similar to mean and variance normalisation (MVN) [64], which is a combination of CMN and cepstral variance normalisation (CVN). The difference between SNT and MVN is that the former uses the mean value that is calculated from a population samples, whereas the latter uses the values that rely on particular utterances or particular speakers.

In SNT, the transformed vectors $\tilde{\mathbf{C}}_j^T$ can be obtained by standardisation and normalisation of the parameter vectors, $\mathbf{C}_j{}^T = [C_{jl}, \ldots, C_{ji}, \ldots, C_{jp}], (j = 1, \ldots, N)$, where N denotes the total number of utterances produced by a population. Standardisation is performed by using the mean vector, μ, and inverse covariance matrix, $\mathbf{\Lambda}^{-1}$, of a population, while normalisation is achieved by dividing the standardised vectors by their Euclidean norm. Thus SNT is expressed as in the following expression,

$$\tilde{\mathbf{C}}_j^T = \frac{(\mathbf{C}_j - \mu)^T \Lambda^{-1}}{\left\|(\mathbf{C}_j - \mu)^T \Lambda^{-1}\right\|}. \tag{11.2}$$

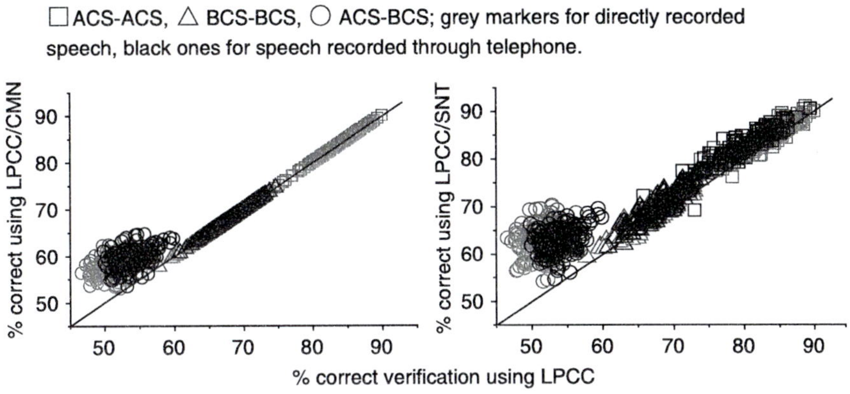

Fig. 11.18 Percentages of correct verification when applying CMN (*left*) and SNT (*right*) compared to those without any normalisations; LPCC was used as the parameter

The mean μ is calculated from a population; and the matrix $\mathbf{\Lambda}^{-1}$ is a diagonal matrix where each of the diagonal entries is the standard deviation of a population for each order. The i-th and ii-th elements of μ and $\mathbf{\Lambda}^{-1}$, respectively, are calculated as follows:

$$\mu_i = \frac{1}{N}\sum_{j=1}^{N} C_{ji},$$

$$\Lambda^{-1}{}_{ii} = \sqrt{\frac{1}{N-1}\sum_{j=1}^{N}(C_{ji}-\mu_i)^2}\,.$$

$\|\mathbf{C}_j\|$ is the Euclidian norm of $\mathbf{C}_j$ and can be calculated by the following equation:

$$\|\mathbf{C}_j\| = \sqrt{\sum_{i=1}^{p} C_{ji}^2}\,.$$

As to the population size needed for obtaining reliable statistics, see Noda and Osanai [65].

The results of speaker verification after applying either of these two normalisations are shown in Figs. 11.18 and 11.19, when using LPCC and MFCC as the parameters, respectively. Both applying CMN and SNT improved verification performances of ACS–BCS pairs significantly ($p<.001$), as can be seen in the graphs that circles are above the diagonal line. Figure 11.20 shows that the improvement was greater with telephone speech compared to directly recorded speech. The improvement scores were better with SNT than CMN by approximately three points on average. We conclude that both CMN and SNT improved verification performances for speech materials with channel differences.

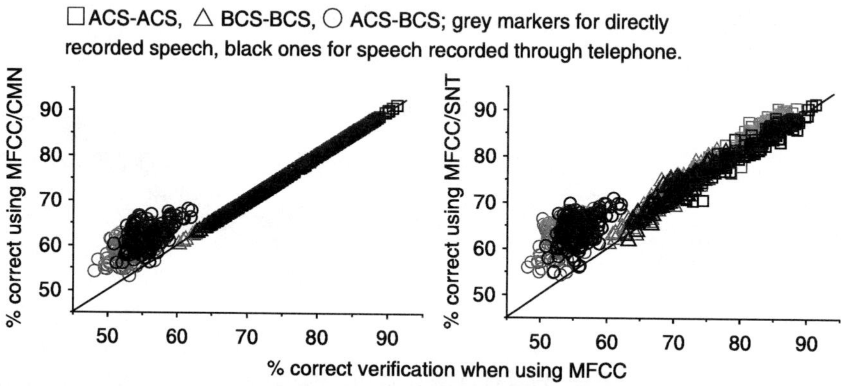

Fig. 11.19 Percentages of correct verification when applying CMN (*left*) and SNT (*right*) compared to those without any normalisations; MFCC was used as the parameter

Fig. 11.20 Summary of the verification results when applying or not applying normalisations, and using LPCC or MFCC as the parameters

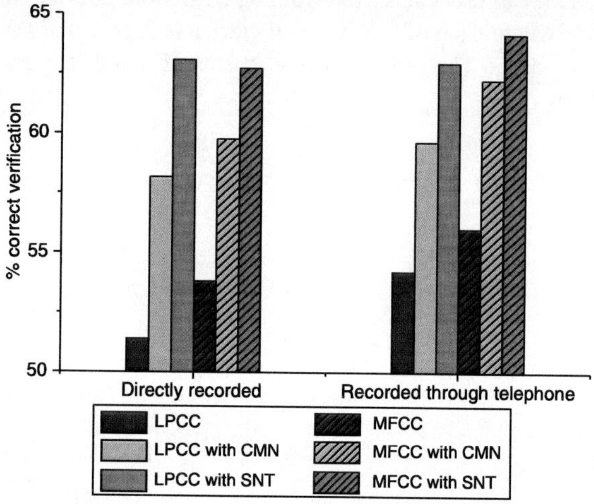

11.3.5 General Discussion

Through the experiments, we confirmed that we could compensate the performance degradations due to the channel difference by using suitable acoustic parameters and normalisation methods. We obtained slightly better results when we used PEAKs and MFCCs as the parameter and SNT for normalisation compared to using LPCC and CMN. Current state-of-the-art systems used in forensic automatic speaker recognition employ MFCC as the parameter and CMN for channel compensation [1, 66]. The reason why SNT was more advantageous than CMN is that SNT uses the normalised norm for an acoustic parameter. Our previous experiments showed

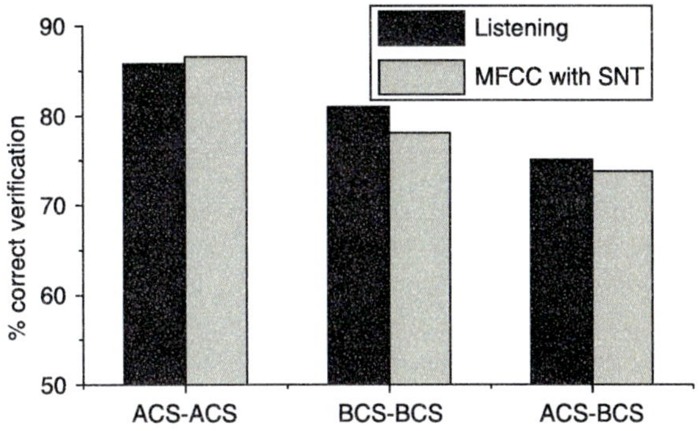

Fig. 11.21 Comparison of human and machine speaker verification performances as a function of channel pairs; machine recognition using MFCC as the parameter and SNT as the normalisation method

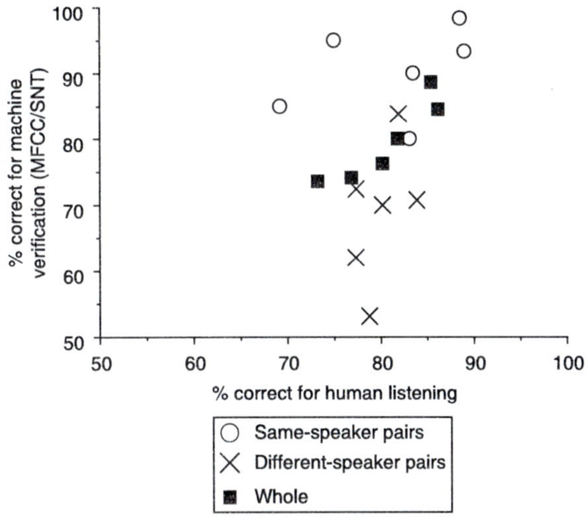

Fig. 11.22 Comparison of human and machine speaker verification performances; the effect of the presented speaker pairs (same or different) together with the whole results

that the speaker verification rate was improved significantly when we eliminated the radial components of the acoustic parameters [63], which means setting the norm to be unity has an influence on the verification performance.

Results of the human and machine verification experiments are compared in Figs. 11.21 and 11.22. The plotted data for machine recognition are the results using MFCC as the parameter with SNT normalisation. The verification performance of humans was slightly better when BCS samples were involved, although the difference was not significant. Figure 11.22 shows that the verification accuracy was higher in machine recognition when same-speaker pairs were presented. However, when the compared speech samples were of different two persons, the behaviour was different; performances of machine recognition ranged from 50 to 85% according to the input samples, whereas human beings could verify these samples at 80% correct almost always. We may obtain still better results when we recruit trained forensic phoneticians as the human listeners [21].

In interpreting the results, we should consider two limitations. One is that the speech materials used for two verification experiments were different. We used words and monosyllables for experiment by human and sentences for that by machine. The other is that not all acoustic properties were accessible in machine verification. For instance, temporal properties were relativised by applying DTW in machine recognition, whereas human listeners could exploit them as well as other properties. In machine recognition, we can only use the properties that are explicitly submitted for evaluation. We should redesign an additional experiment if we are to compare human and machine performance in a strict sense; but no matter how we do it, machines can use only limited and discrete speech information.

11.4 Summary

In this chapter, we introduced experimental findings on two factors that affect speaker recognition performances. In the first experiment, we investigated the effect of the phonological contents on speaker identification accuracy and found that the listeners could identify speakers when the stimuli containing a nasal sound were presented compared to the stimuli without it, regardless of the familiarity to the speakers. The rankings of the consonants according to the speaker identification accuracy coincided with the sonority scale. Similar tendency was confirmed in the second experiment, where speaker verification experiment by machine was conducted using a different, and larger, speaker set.

The effectiveness of nasals may derive from the morphological differences among speakers; the cepstral distances among speakers were also greater in nasal consonants than in oral consonants. The shapes of the speech organs responsible for nasal articulation, such as nasal cavity and paranasal sinuses, are difficult to change at speakers' will; therefore it is hard for them to change the resonant properties of the nasal sounds. Moreover, nasals appear relatively more frequently in natural speech in many languages besides Japanese. Taking these together, using nasals in forensic speaker recognition seems to be more effective than using other consonants.

In the second experiment, we investigated the effects of using speech materials recorded through different channels and bone-conducted speech samples in speaker verification. Verification accuracy was degraded significantly for both human and machine verification when we used bone-conducted speech samples; it was further degraded when using air-conducted and bone-conducted speech samples together. Another interesting finding of this experiment is that the nasal sounds were more effective for speaker verification than other consonants even with bone-conducted speech, although the difference was not as salient as air-conducted speech.

In order to improve the performances in machine speaker verification, we tried two other acoustic parameters, MFCC and local spectral peaks (PEAKs), and two channel normalisations, CMN and SNT. All of them increased the verification accuracy; PEAKs being slightly better than MFCCs, and SNT being better than

CMN. Comparison between human and machine verification revealed that the performance of human listeners was slightly better than machine when bone-conducted speech samples were involved, although we cannot simply compare them as the speech materials (words or sentences) used for verification were not the same.

For future tasks, we will find a way to exploit the availability of nasals in forensic speaker recognition. For that we will need to look in detail for which aspects or what acoustic properties of the nasals are in charge of speaker individuality. Effects of the nasal diseases, e.g. empyema and allergic rhinitis, and change of the acoustic properties of the nasals over time should be investigated, too. Furthermore, it may be more practical when we test our results with a framework that is actually used in forensic situations; for example, by using Bayesian method [67]. For the comparison between verifications by human and machine, we need another experiment to show the difference, especially in the forensic context. There are several studies that investigated the difference between forensic phoneticians and naive listeners in speaker recognition performances [reviewed in 21], but the study on trained phoneticians and automatic methods is still missing. In order to develop a better solution to forensic speaker recognition, collaboration between forensic phoneticians and engineers will be crucial [1], and this will require understanding the strong points and limitations of both human- and machine-based methods.

Acknowledgement This work was supported by KAKENHI (21510185, 20700177, 21710174, and 22·3118) and Sophia University Open Research Centre from MEXT.

References

1. Amino K, Osanai T, Kamada T, Makinae H, Arai T (2011) Historical and procedural overview of forensic speaker recognition. In: Neustein A, Patil HA (eds) Advances in forensic speaker recognition. Springer
2. Bricker P, Pruzansky S (1966) Effects of stimulus content and duration on talker identification. J Acoust Soc Am 40:1441–1450
3. Hollien H, Schwartz R (2000) Aural-perceptual speaker identification: problems with noncontemporary samples. Forensic Linguist 7:199–211
4. Nygaard L (2005) Perceptual integration of linguistic and nonlinguistic properties of speech. In: Pisoni DB, Remez RE (eds) The handbook of speech perception. Blackwell, pp 390–413
5. Nishio T (1964) Can we recognise people by their voices? Gengo Seikatsu 158:36–42
6. Hirayama T (ed) (1960). Dictionary of the Japanese accents. Tokyodo Publishing Company
7. Kindaichi H, Akinaga K (1981) Dictionary of Japanese accents, 2nd edn. Sanseido Publishing Company
8. Kitamura T, Akagi M (1995) Speaker individualities in speech spectral envelopes. J Acoust Soc Jpn(E) 16:283–289
9. Amino K (2003) The characteristics of the Japanese phonemes in speaker identification. Proc Sophia Univ Linguist Soc 18:32–43
10. Ramishvilli GS (1966) Automatic voice recognition. Eng Cybernet 5:36–42
11. Syntrillium Software Corporation (1996). Cool Edit Ver. 96. A computer software
12. Pisoni DB (1975). Auditory short-term memory and vowel perception. Mem Cognit 3:7–18
13. Repp BH, Healy AF, Crowder RG (1979) Categories and context in the perception of isolated steady-state vowels. J Exp Psychol Human Percept Perform 5:129–145

14. Sambur MR (1975) Selection of acoustic features for speaker identification. IEEE Trans Acoust Speech Sig Process 23:176–182
15. Clarke FR, Becker RW (1969) Comparison of techniques for discriminating among talkers. J Speech Hear Res 12:747–761
16. Kitamura T, Mokhtari P (2005) Proceedings of the autumn-2005 meeting of the acoustical society of Japan, Sendai, Japan, Paper 2-Q-29, pp 525–526 (in Japanese)
17. Kitamura T, Saito T (2006). Effects of acoustic modifications on perception of speaker characteristics for sustained vowels. Tech Rept IEICE 106:43–48
18. Kitamura T, Saito T (2007) Effects of acoustic modification on perception of speaker characteristics for sustained vowels Acoust Sci Tech 28:434–437
19. Boersma P, Weenink D (2010) Praat: doing phonetics by computer. http://www.praat.org/ (Computer programme)
20. Boersma P (1993) Accurate short-term analysis of the fundamental frequency and the harmonics-to-noise ratio of a sampled speech. Proc Inst Phonetic Sci 17:97–110 (University of Amsterdam)
21. Hollien H (2002) Forensic voice identification. Academic Press, San Diego
22. Hollien H, Majewski W, Doherty TE (1982) Perceptual identification of voices under normal, stress, and disguise speaking conditions. J Phonetics 10:139–148
23. Van Lacker D, Kreiman J, Emmorey K (1985) Familiar voice recognition: patterns and parameters part 1: recognition of backward voices. J Phonetics 13:19–38
24. Van Lacker D, Kreiman J (1985) Familiar voice recognition: patterns and parameters part 2: recognition of rate-altered voices. J Phonetics 13:39–52
25. Selkirk E (1984) Phonology and syntax: the relation between sound and structure. MIT Press, Cambridge
26. Coleman RO (1973) Speaker identification in the absence of inter-subject differences in glottal source characteristics. J Acoust Soc Sm 53:1741–1743
27. Bachorowski JA, Owren MJ (1999) Acoustic correlates of talker sex and individual talker identity are present in a short vowel segment produces in running speech. J Acoust Soc Am 106:1054–1063
28. Kitamura T, Honda K, Takemoto H (2005) Individual variation of the hypopharyngeal cavities and its acoustic effects. Acoust Sci Tech 26:16–26
29. Engwall O, Delvaux V, Metens T (2006) Interspeaker variation in the articulation of nasal vowels. Proc Int'l Seminar on Speech Production, pp 3–10
30. Amino K, Arai T (2009) Speaker-dependent characteristics of the nasals. Forensic Sci Int'l 185:21–28
31. Amino K, Sugawara T, Arai T (2006) Effects of the syllable structure on perceptual speaker identification. Tech Rept IEICE 105:109–114
32. Amino K, Arai T, Sugawara T (2007) Effects of the phonological contents on perceptual speaker identification. In: Mueller, C, Schoetz, S (eds) Speaker classification 2. Springer, Berlin, pp 83–92
33. Fujimura O (1962) Analysis of nasal consonants. J Acoust Soc Am 34:1865–1875
34. Su LS, Li KP, Fu KS (1974) Identification of speakers by use of nasal coarticulation. J Acoust Soc Am 56:1876–1882
35. Kurowski K, Blumstein SE (1984) Perceptual integration of the murmur and formant transitions for place of articulation in nasal consonants. J Acoust Soc Am 76:383–390
36. Pruzansky S, Mathews MV (1964) Talker-recognition procedure based on analysis of variance. J Acoust Soc Am 36:2041–2047
37. Wolf JJ (1972) Efficient acoustic parameters for speaker recognition. J Acoust Soc Am 51:2044–2056
38. Bogert BP, Healy MJR, Tukey JW (1963) The quefrency analysis of time series for echoes: cepstrum, pseudo autocovariance, cross-cepstrum and saphe cracking. In: Rosenblatt M (Ed.). Proc Sympo Time Series Analysis. Wiley, pp 209–243 (Chapter 15)
39. Matsui T, Furui S (1990) Text-independent speaker recognition using vocal tract and pitch information. Proc 1st conf spoken lang proc, pp 137–140

40. Pollack I, Pickett JM, Sumby WH (1954) On the identification of speaker by voice. J Acoust Soc Am 26:403–406
41. Nakagawa S, Sakai T (1979) Feature analysis of Japanese phonetic spectra and considerations on speech recognition and speaker idetification. J Acoust Soc Jpn 35:111–117
42. Miura T (1980). New edition of hearing and speech. Corona Publishing Company
43. Dang JW, Honda K (1996) Acoustic characteristics of the human paranasal sinuses derived from transmission characteristics measurement and morphological observation. J Acoust Soc Am 100:3374–3383
44. Matsui T, Pollack I, Furui S (1993) Perception of voice individuality using syllables in continuous speech. Proc aut meet acoust soc Jpn, pp 379–380
45. Bricker P, Pruzansky S (1976) Speaker recognition. In: Lass N (ed) Contemporary issues in experimental phonetics. Academic Press, New York, pp 295–326
46. Williams CE (1964) The effects of selected factors on the aural identification of speakers. Sect 3, Rept ESD-TDR-65-153, electronics systems division, air force systems command
47. Philippon AC, Cherryman J, Bull R, Vrij A (2007) Earwitness identification performances: the effect of language, target, deliberate strategies and indirect measures. Appl Cogn Psychol 21:539–550
48. O'Shaughnessy D (2001) Speech communication—human and machine, 2nd edn. Addison-Wesley Publishing Company
49. Bonastre JF, Bimbot F, Boe LJ, Campbell JP, Reynolds DA, Magrin-Chagnolleau I (2003) Person authentication by voice: a need for caution. Proc Eurospeech, pp 1–4
50. Economic and Social Research Institute, Japan Cabinet Office (2010) Monthly consumer confidence survey covering all of Japan. http://www.esri.cao.go.jp/en/stat/shouhi/1004shouhi-e.html
51. Berger KW (1976) Early bone conduction hearing aid devices. Arch Otolaryngol 102:315–318
52. Watson NA (1937) Hearing of speech by bone conduction. J Acoust Soc Am 9:99–106
53. Kirikae I, Kawamura S, Muto J, Oshima H (1954) A contribution to bone conduction audiometry. J Oto-Rhino-Laryngol Jpn 58:226–234
54. Ohyama M, Miyoshi Y, Shoji K, Yamamoto S, Taniguchi C (1976) Acoustic analysis of speech signals (bone-conducted sounds) picked up at the head. J Oto-Rhino-Laryngol Jpn 79:963–972
55. Makinae H, Osanai T, Kamada T, Tanimoto M (2007) Construction and preliminary analysis of a large-scale bone-conducted speech database. IEICE Tech Rep 107, SP2007-40, pp 97–102
56. Kondo T, Sakamoto S, Amano S, Suzuki Y (2010) Speech recognition threshold in noise for 100 Japanese monosyllables in "Familiarity-controlled word lists 2003 (FW03)". J Acoust Soc Jpn 66:105–111
57. Oppenheim AV, Schafer RW, Stockham TG (1968) Nonlinear filtering of multiplied and convolved signals. IEEE Trans Audi Electroacoust 16:437–466
58. Senin P (2008) Dynamic time warping algorithm review. http://seninp.googlepages.com/699fall08report.pdf
59. Mermelstein P (1976) Distance measures for speech recognition: psychological and instrumental. Status report on speech research, Haskins Lab., SR-47, pp 91–103
60. Davis S, Mermelstein P (1980) Comparison of parametric representations for monosyllabic word recognition in continuously spoken sentences. IEEE Trans Acoust Speech Sig Process ASSP-28:357–366
61. Campbell JP Jr (1997) Speaker recognition: a tutorial. Proc IEEE 85:1437–1462
62. Liu FH, Stern RM, Huang XD, Acero A (1993) Efficient cepstral normalization for robust speech recognition. Proc ARPA human lang tech workshop, pp 69–74
63. Osanai T, Ozeki K, Tanimoto M (2006). Feature parameter transformation in speaker verification using vowels uttered in isolation. J Acoust Soc Jpn 62:848–855
64. Jain P, Hermansky H (2001) Improved mean and variance normalisation for robust speech recognition. Proc int conf acoust speech sig process

65. Noda H, Osanai T (1990) On the relation between the number of speakers and the reliability of recognition rate in speaker recognition. J IEICE J73-A:717–724
66. Nakasone H, Beck SD (2001) Forensic automatic speaker recognition. Proc a speaker Odyssey—the speaker recognition workshop, pp 139–142
67. Gonzalez-Rodriguez J, Ortega-Garcia J, Lucena-Molina JJ (2001) On the application of the Bayesian approach to real forensic conditions with GMM-based systems. Proc a speaker Odyssey—the speaker recognition workshop, pp 135–138

Chapter 12
Aerodynamic and Acoustic Theory of Voice Production

T. V. Ananthapadmanabha

Abstract A theory of voice production for vowels has to deal with two related problems; the problem of biomechanical modeling of vocal fold vibrations and the problem of calculating volume-velocity airflow through the glottis or the glottal airflow. This report is a tutorial on the second problem. We call this the aerodynamic and acoustic theory of voice production. Calculation of glottal airflow is difficult since it depends on an interaction between (1) the nonlinear time varying glottal impedance specified in the *time domain* and (2) the subglottal and vocal tract input impedances specified in the *frequency domain*. The effect of glottal geometry on the glottal impedance and the role of glottal impedance elements like kinetic resistance, viscous resistance and glottal inductance in determining glottal airflow are discussed. Methods to calculate vocal tract or subglottal input impedance based on a transmission line analog model and a formant network model are presented. Equations to find glottal airflow with source-filter interaction are derived. A digital pole-zero modeling of input impedance is proposed for an efficient and accurate computation of glottal airflow. The role of various factors in determining the so called residue, ripple and superposition components of glottal airflow is discussed with examples. The time domain response of a vowel is calculated using the glottal airflow with source-filter interaction. The instantaneous frequency and instantaneous bandwidth of an interactive vowel response are computed and interpreted. Further research is needed to extend the theory to the case of breathy vowels, vowel onsets, consonant to vowel and vowel to consonant transitions where the acoustic waves are superposed on a large dynamically changing mean airflow. A good understanding of the theory guides one in appropriate modeling and interpretation of voice source. The relevant features in voice source for a specific application such as forensic speaker identification can thus be identified. The author believes that habitually formed relative dynamic variations in voice source parameters are of greater significance in forensic speaker recognition.

T. V. Ananthapadmanabha (✉)
Voice and Speech Systems, 53, "Girinivas", Temple Road, 13th Cross, Malleswaram, Bangalore 560003, India
e-mail: tva_vss@yahoo.com, tva.blr@gmail.com

A. Neustein, H. A. Patil (eds.), *Forensic Speaker Recognition*,
DOI 10.1007/978-1-4614-0263-3_12, © Springer Science+Business Media, LLC 2012

12.1 Introduction

'Voice' conveys a speaker's individuality, his/her emotional state, socio-cultural background, local accent etc. in addition to the linguistic content. Further, an individual can voluntarily change his/her voice. The term 'voice' is used to represent the *abstract perceptual image* formed by a listener, which is *uniquely* associated with a speaker or a singer. It is still an open and challenging problem to determine the objective correlates of voice that can uniquely determine the identity of a speaker from the spoken material.

Study of voice is a multi-disciplinary subject covering neurology, anatomy and physiology, acoustics, system theory, signal processing, aesthetics etc. An understanding of voice is of interest to a wide cross section of professionals; actors, singers, speech therapists, voice consultants, acousticians, speech scientists, ENT specialists, phono-surgeons, speech language pathologists, signal processing engineers, communication engineers, computer scientists, forensic experts etc.

In literature related to speech production the terms 'voice' and 'voice source' are often used interchangeably as though the 'attributes of a voice' can be determined from the 'voice source' *alone*. Further, in a restricted sense the term 'voice source' refers only to the *pulse shape of glottal airflow* and its dynamic characteristics. This report uses the term 'voice source' in this restricted sense. In a broad sense, the term 'voice source' may also include F0-level and intonation. Additionally, supra-segmental features such as duration of segments, pauses, intensity level etc. may also determine the 'perceived voice quality'. One of the research topics is to determine the extent to which 'voice source' determines uniquely the identity of a speaker especially in the context of forensics.

Broadly, there are two streams of research related to 'voice source'. Firstly, those based on the underlying physiological mechanism of speech production. Secondly, those based on linear prediction residual which has very little or no reference to the physiological mechanism but relies heavily on signal characteristics. This report on 'voice source' is concerned with the former approach.

On the *theoretical* side, speech research involves the development of a speech production model. Such a development is based on the acoustic theory of vocal tract, the theory of biomechanics of vocal fold vibrations and the theory of aerodynamics and acoustics governing glottal airflow. These theories are inter-related. On the *practical* side, speech research involves the problem of deriving the control parameters of a production model from a speech signal. This is a speech analysis problem. Control parameters are required for the development of practical applications like forensic speaker identification, text-to-speech synthesis, speech coding speech recognition etc. This report is a tutorial related to the aerodynamic and acoustic *theory of voice production*. Speech analysis techniques for deriving the control parameters are not covered here except for a brief mention in Sects. 12.9.3 and 12.9.4.

The acoustic theory of speech production [1–5] is a well developed topic primarily dealing with the relationship between the vocal tract shape and its transfer function. The theory of voice production received much less attention in the early days of speech research (1950s and 1960s). A fixed shape for the glottal pulse for a given speaker was assumed to be adequate for synthesis. Probably this is the reason for using the terms 'voice' and 'voice source' interchangeably. In his early work, Fant refers to the characteristics of voice source as a 'constant factor' [3] represented spectrally by a second order lowpass filter with cut-off frequency of about 100 Hz. Experiments on static models of glottis provided data on glottal impedance [6]. In his early work, Flanagan remarked that the glottal impedance is very much higher than the vocal tract input impedance and hence the voice source and vocal tract filter are assumed to be linearly separable. He computed the volume-velocity airflow through the glottis and oral pressure with such an assumption [4, 7]. Accordingly, the theory was restricted to the computation of volume velocity airflow, ignoring the influence of subglottal and vocal tract systems. In other words, the interaction between the source and filter was assumed to be negligible. That theory showed the shape of volume velocity airflow to be nearly the same as that of the shape of glottal area function. The acoustic effect of coupling of vocal tract to sublgottal system via the glottal impedance was represented by an increase in the bandwidths of the lower formants.

During late 1970s and early 1980s, research related to voice source gained greater importance with an aim to produce natural sounding text to speech synthesis and high quality speech coders. One of the earliest theories on voice is on the self-oscillations of vocal folds due to an interaction between the myoelastic and aerodynamic forces [8]. This led to a series of sophisticated mechanical models of vocal fold vibrations; one-mass, two-mass, multi-mass etc. [9–18]. An integrated model for speech synthesis emerged combining biomechanics of phonation, aerodynamics of glottal airflow and acoustics of vocal tract [9–11, 19, 20]. Despite technological advances in digital computing, the integrated models have remained as software simulation tools on supercomputers due to their numerical complexity. Deriving control parameters of an integrated model from a speech signal is still an open challenge because the number of control parameters far exceeds the number of parameters than can reliably be measured from a speech signal.

Meanwhile, direct measurements of oral pressure, subglottal pressure, glottal area and glottal airflow via inverse filtering became available [21–23]. Unlike early theories, it came to be realized that the voice source, viz., the volume velocity airflow though the glottis or the glottal pulse shape, is dependent on the vocal tract filter, a phenomenon called source-filter interaction [24–33]. Experiments on static models of glottis [34–37] provided more accurate measurements on glottal impedance.

The effect of source-filter interaction on formants is not explicitly seen in an integrated model. Our analytical approach to study the source-filter interaction not only led to an efficient method for computing glottal airflow but also to an interpretation of the factors that determine the glottal airflow and its components [32]. It also enabled us to compute the temporal and spectral characteristics of the interactive vowel response [33] and to study the perceptual importance of the interaction effect [38, 39].

This chapter is a **tutorial** on the aerodynamic and acoustic theory of voice production. Although this theory has been in the literature since 1980s, the author has taken this opportunity to consolidate his previous research works with his colleagues by providing the full technical details, including the derivation of equations; to correct, clarify and amplify certain points in the previous works; and to present some new results. Only vowel production model is considered here. It involves the computation of glottal airflow and its components as well as the corresponding vowel response. The aim is to understand the concept of acoustic source and factors that influence source characteristics. Some important related topics like spectral characteristics of voice source, spectral characteristics of interactive vowel response, perceptual importance of interactive response are not covered here due to limitations of space.

How valid is the assumption of the early theories of a 'constant' or fixed glottal pulse shape for a given speaker? This assumption led early researchers to believe that the uniqueness of voice is related to the fixed pulse shape and hence the terms 'voice' and 'voice source' began to be used interchangeably. To address this question we have to consider speech analysis, inverse filtering, pulse shape modeling, robust estimation of voice source model parameters, voice source dynamics etc. As mentioned earlier, these topics are not covered here except for a brief mention in Sects. 12.9.3 and 12.9.4.

This chapter is **not** an exhaustive review of literature on voice source. Only some common references which are relevant to source-filter interaction and associated topics are cited here. Apologies to those researchers who have contributed to our improved understanding of voice source but whose works might have been left out in the list of references.

12.2 Candidates for the Voice Source

12.2.1 Vowel Production

A deep inhalation precedes vowel production. An excess lung pressure is built up. During vowel production, air is released slowly through a pair of vibrating vocal folds by a controlled expiratory phase of respiration. Over a short duration of the vowel, the excess lung pressure can be represented by an analogous *DC* voltage source, P_L. Through out this chapter we will use analogous electrical and acoustic terms, current for volume airflow and voltage for differential pressure.

The vocal folds are set into self-oscillations by a complex interplay of aerodynamic and acoustic factors on one hand and the mass and elastic forces of intrinsic laryngeal muscles on the other [8, 18]. Glottis is the orifice between the vocal folds. The quasi-periodic cycles of opening and closing gestures of vocal folds control the area and geometry of the glottis. The changing area and geometry of glottis interrupts the expiratory airflow from the lungs to produce quasi-periodic puffs of airflow

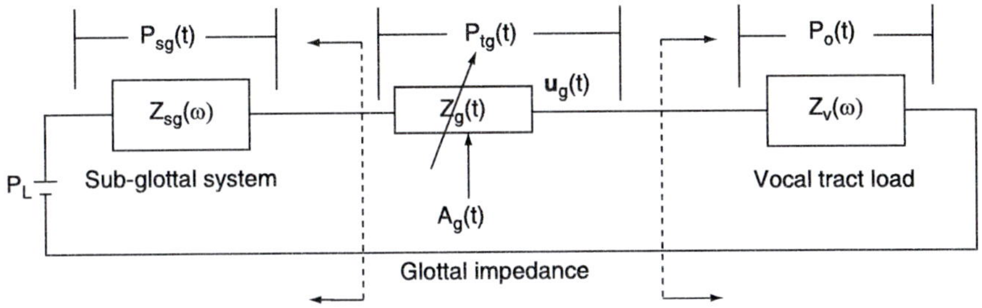

Fig. 12.1 Block diagram of an electrical equivalent circuit of vowel production

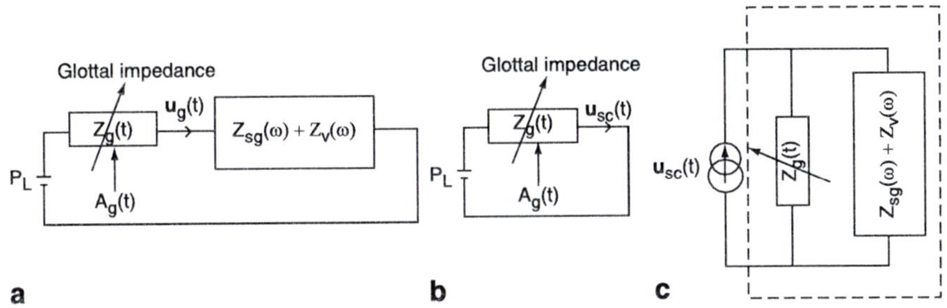

Fig. 12.2 **a** Thevinin's equivalent circuit. **b** Short-circuit current or no load flow **c** Norton's equivalent circuit of vowel production

called variously as volume velocity airflow, glottal volume velocity, glottal airflow, voice source, glottal wave, glottal pulses, source, excitation, true glottal flow etc.

A detailed three-dimensional modeling of mechanical vocal fold vibrations is not discussed here. Instead, it is assumed that the glottal area function for each cycle, $A_g(t)$, is directly given as an input variable. The electrical equivalent circuit diagram of vowel production is shown in Fig. 12.1. The time varying glottal area function, $A_g(t)$, determines the time varying nonlinear aerodynamic glottal impedance, $Z_g(t)$. A discussion on glottal impedance is presented in Sect. 12.3. Although P_L is a constant, the time varying glottal impedance, $Z_g(t)$, produces a time varying glottal airflow, $u_g(t)$ at the input terminals of vocal tract. In other words, the larynx acts as a *DC* (mean flow) to *AC* converter thus producing audible frequencies.

The glottal airflow $u_g(t)$ is not determined by the aerodynamic glottal impedance alone. The input impedance of the subglottal system, $Z_{sg}(\omega)$, and that of the vocal tract, $Z_v(\omega)$, act as acoustic load in series with the time varying glottal impedance to determine the glottal airflow signal $u_g(t)$ as shown in Fig. 12.2a. Note that the input impedance, also referred to as driving point impedance, is a complex function of frequency.

The acoustic input impedance of subglottal and vocal tract loads can be computed given the geometry of the trachea and the vocal tract respectively. This is discussed in Sect. 12.4. Input impedance of vocal tract, $Z_v(\omega)$ is for the closed glottis condition as it excludes the effect of glottal impedance and subglottal system,

Fig. 12.1. Accordingly, the vocal tract transfer function to be used for vowel synthesis also corresponds to the closed glottis condition. There are two candidates for defining the source: (a) short circuit current or no load airflow and (b) actual glottal airflow which we refer to as true glottal flow.

12.2.2 Short Circuit Current or No Load Airflow

Short circuit current or the no load airflow $u_{sc}(t)$ corresponds to the current when the combined effect of input impedances $Z_{sg}(\omega)$ and $Z_v(\omega)$ is replaced by a short circuit or zero input impedance in Fig. 12.2b. The short circuit current $u_{sc}(t)$ is determined only by the aerodynamic impedance $Z_g(t)$ and is given by

$$u_{sc}(t)Z_g(t) = P_L \tag{12.1}$$

The glottal impedance $Z_g(t)$ is time varying, nonlinear and dependent on P_L as will be discussed in greater detail in Sect. 12.3.

An advantage of using short circuit current or no load airflow is that it is independent of vocal tract input impedance. That is, the airflow is independent of the positions of articulators. But, short circuit current or no load airflow is only a theoretical concept since vocal tract is invariably present. In fact, according to some theories of phonation, vocal tract load has to be necessarily present for vocal folds to vibrate [18]. Even if one considers the case of an excised larynx, the radiation load is still present.

As per Norton's equivalent circuit, the short circuit current source acts as an input to a parallel combination of glottal impedance $Z_g(t)$ and acoustic loads (box shown by dashed lines in Fig. 12.2c). In other words, we can use short circuit current or no load airflow as source, provided we modify the vocal tract transfer function to include the effect of time varying nonlinear glottal impedance (box shown by dashed lines in Fig. 12.2c). This complicates the definition of vocal tract transfer function. The well established early acoustic theories of speech production [1–4] compute the transfer function of vocal tract ignoring the influence of the source and still use $u_{sc}(t)$ as the source since these models assume the magnitude of input impedance to be negligible compared to glottal impedance. However, these early theories [3, 4] account for the parallel combination of vocal tract load and glottal impedance by a suitable increase of bandwidths of formants.

12.2.3 Actual Glottal Airflow or the True Glottal Flow

Applying Kirchoff's loop equation to the equivalent circuit shown in Fig. 12.1, we get

$$u_g(t)Z_g(t) = P_{tg}(t) = P_L - P_O(t) - P_{sg}(t) \tag{12.2a}$$

where $P_{tg}(t)$ is the transglottal pressure, $P_o(t)$ is the oral pressure just above the glottis and $P_{sg}(t)$ is the subglottal pressure. Comparing Eq. (12.1) with Eq. (12.2a) it is clear that early theories assumed $P_o(t)$ and $P_{sg}(t)$ to be negligible compared to P_L. Direct measurements of pressure just above and below the glottis [21–23] have clearly shown that $P_o(t)$ and $P_{sg}(t)$ pressure signals show transients of significant amplitude (20–30%) comparable to P_L.

We refer to the airflow, $u_g(t)$, as 'true glottal flow'. Note from Eq. (12.2a) that $u_g(t)$ is influenced by $P_o(t)$ and $P_{sg}(t)$ which in turn depend on the input current $u_g(t)$. The vocal tract and subglottal loads have a significant effect on the true glottal flow. The source is dependent on the load (or the vocal tract filter). This phenomenon is referred to as source-filter interaction.

The Problem: Given the lung pressure, P_L, the glottal area function $A_g(t)$, and the input impedances, the problem is to determine the true glottal flow $u_g(t)$ by solving Eq. (12.2a). The solution for $u_g(t)$ is difficult. Firstly, the glottal impedance $Z_g(t)$ is time-varying and nonlinear.

Secondly, the input impedance consists of a distributed network of analog electrical elements dependent on frequency and hence the input impedances, $Z_v(\omega)$, $Z_{sg}(\omega)$ are specified in the frequency domain. But, the terms $P_o(t)$ and $P_{sg}(t)$ in Eq. (12.2a) are in the time domain.

The pressure (voltage) drop $P_o(t)$ **cannot** be written simply as $u_g(t)Z_v(\omega)$ since $u_g(t)$ is in the time domain and $Z_v(\omega)$ is a complex function of frequency. We **cannot** use inverse Fourier transform of the frequency domain equations

$$P_o(\omega) = U_g(\omega)Z_v(\omega) \text{ and} \tag{12.2b}$$

$$P_{sg}(\omega) = U_g(\omega)Z_{sg}(\omega) \tag{12.2c}$$

since the Fourier transform $U_g(\omega)$ assumes that $u_g(t)$ is already known for all 't' determination of which is the problem. Because of this inherent difficulty some researchers have used a *single lumped element* (resistance or inductance) to represent the distributed load of $Z_v(\omega)$ and/or chosen a single frequency, ω_0 (typically the fundamental frequency) [21, 25, 27–29] to compute the airflow.

Ideally, in order to solve Eq. (12.2a) we need to know the following:

1. Factors that determine the glottal impedance, $Z_g(t)$. See Sect. 12.3.
2. A method for finding input impedances $Z_v(\omega)$, $Z_{sg}(\omega)$. See Sect. 12.4.
3. An explicit equation for $P_o(t)$ and $P_{sg}(t)$ in the time domain. See Sect. 12.5.

12.3 Static Aerodynamic Glottal Impedance

Initially the glottal impedance for a static model of the glottis is discussed. This is later extended to a dynamic case.

12.3.1 Kinetic Resistance of a Rectangular Glottis

The excess subglottal pressure (potential energy) separates out the vocal folds which are held together (adducted) by elastic forces in the laryngeal muscles. When the vocal folds separate, an air jet at a high velocity (in kinetic energy form) escapes through the glottis. The relation between the potential and kinetic energy is given by the Bernoulli (conservation of energy) equation

$$\Delta P = ½\,\rho\, v^2 = ½\,\rho\,[U_0/A_g]^2 \tag{12.3a}$$

Where ΔP is the differential pressure across the static glottis (transglottal pressure), ρ is the density of air, v is the air particle velocity, U_0 is the volume velocity airflow and A_g is the glottal area. Eq. (12.3a) can be re-written as

$$U_0^2 = A_g^2\,[2\,\Delta P/\rho] \tag{12.3b}$$

The relationship between the current U_0 and voltage difference ΔP is non-linear.

Glottis is a three-dimensional slit whose area changes with the depth and its geometry changes with time. In that case what is the glottal area A_g to be used in Eq. (12.3b)? This aspect is discussed later. For the present assume a static rectangular glottis with area A_g constant with depth.

Van den Berg et al. [6] proposed a modification to Eq. (12.3a) as

$$\Delta P = ½\,k\,\rho\, v^2 = ½\,k\,\rho\,[U_0/A_g]^2 \quad \text{or} \tag{12.3c}$$

$$U_0^2 = A_g^2[2\,\Delta P/(k\rho)] = [2/(k\rho)]A_g^2\,[\Delta P] \tag{12.3d}$$

where a factor '*k*' has been introduced. A value of $k=k_1-k_2=1.375-0.5=0.875$ was proposed by van den Berg et al. The factor $k_1=1.375$, called entry coefficient, arises for an area of the jet different from the physical area of the glottis. The factor $k_2=0.5$, called the exit recovery coefficient, arises since exit pressure is assumed to be lower than the atmospheric pressure there by increasing the pressure differential. Later experiments on static realistic models of glottis [34] have shown that the exit recovery coefficient k_2 is not very significant and that k_1 is much lower than the value suggested by van den Berg et al. [6].

In keeping with the electrical analogy, the factor $\Delta P/U_0$ is referred to as the kinetic resistance of the glottis denoted by R_k and given by

$$R_k = [1/A_g][0.5k\rho\Delta P]^{½} \tag{12.3e}$$

Note that the kinetic resistance depends on the input pressure (voltage) and is non-linear.

12.3.2 Effect of Glottal Geometry on Kinetic Resistance

Glottis is a three-dimensional slit formed between two vibrating vocal folds. For chest register type of phonation, vocal fold vibratory pattern has a vertical phase difference in the movement of the lower and upper margins [40]. The movement of lower margins leads the movement of upper margins. During the opening phase, the glottis has a convergent shape and during the closing phase the glottis has a divergent shape. Two issues arise: (a) What is the effect of convergent/divergent geometry on glottal impedance? (b) Which is the glottal area to be used since the glottal area changes with depth? Several researchers have addressed the effect of glottal geometry [11, 12, 15, 34–37]. It is generally known that a convergent glottis has a lower kinetic resistance compared to a divergent shape. However, we have observed empirically that, when the glottal inlet is stream lined rather than abrupt then (1) there is not a very significant difference between the kinetic resistances for convergent and divergent glottal shapes, (2) the glottal area to be used is the minimum area irrespective of the change with respect to depth and (3) a value of 1.1 for k can be used irrespective of glottal geometry [35, 36].

12.3.3 Viscous Resistance of the Glottis

Viscous loss arises due to friction between airflow and inner surfaces of vocal folds. The pressure drop due to viscous resistance for a rectangular glottis is given by the formula [6]

$$\Delta P_v = R_v U_0, \; R_v = [12\mu D l^2 / A_g^3] \qquad (12.3f)$$

Where μ=the coefficient of viscosity of air, D is the depth of the glottis, l=length of the glottis. Note: Instead of the factor 'l^2' in Eq. 12.3f, the factor 'l' has been shown in the equations for the viscous resistance in [32, 36]. This correction may please be noted. However, the appropriate factor 'l^2' has been used in the numerical calculations of airflow reported in those references.

During vocal fold vibrations, the glottis assumes convergent and divergent shapes as mentioned earlier. The area of glottis varies with the depth. Hence a correction has to be applied to the above equation as shown in Appendix 1. The equation for viscous resistance becomes

$$R_v = [k_a 12\mu D l^2 / A_g^3] \qquad (12.3g)$$

where the factor k_a has been introduced to take care of the geometry of the glottis. For example, for a ratio of widths of 4, k_a=5/32. This means that for a convergent or divergent glottis with the above ratio of widths the viscous resistance is about 5/32 the value of that of a rectangular glottis of area equal to the minimum glottal area.

Several issues are to be noted with respect to viscous resistance. Strictly speaking viscous resistance is applicable for a laminar flow with a parabolic velocity profile. This happens for very low pressure difference of less than 1 cm water pressure for airflow through an orifice [41]. A minimum pressure difference of about 4 cm water is required to initiate vocal fold vibrations [18]. In experiments with a static glottis it has been observed that an air jet is formed with a separation of flow especially for a divergent glottal shape [34, 37]. The inclusion of viscous resistance term for large pressure difference and for a divergent glottis with separation of airflow is hence questionable. Even if viscous resistance is to be included, then its value is reduced by a factor k_a in case of a convergent or divergent glottis as shown in Appendix 1.

12.3.4 Quasi-Static Glottal Flow Equation

Extending the static glottal impedance to the case of time varying glottal area function, $A_g(t)$, the glottal airflow under no load condition including the kinetic and viscous resistances is given by modifying Eq. (12.3d)

$$U_0^2(t) = [2/(k\rho)]A_g^2(t)[\Delta P - R_v U_0(t)] \tag{12.3h}$$

The above equation implies that the variation with respect to time is represented by a series of quasi-static states of the glottis. The effect of dynamic conditions and the effect of acoustic loads are ignored. The quasi-static equation has been used in early studies on glottal flow [3, 4, 7].

12.3.5 Glottal Inductance

The term glottal inertance is sometimes used in place of glottal inductance. The mass of air column (or air plug) within the glottis is modeled by glottal inductance and is given by

$$L_g(t) = \rho D/A_g(t) \tag{12.3i}$$

Since the glottal inductance is time varying, the pressure drop ΔP_L due to the glottal inductance for no load condition is given by

$$\begin{aligned} \Delta P_L &= [d/dt][L_g(t)U_0(t)] = [d/dt][\rho D U_0(t)/A_g(t)] \\ &= [\rho D][d/dt][v] \end{aligned} \tag{12.3j}$$

where $U_o(t)$ is the no load airflow and v is the air particle velocity. Assuming only the kinetic resistance, the air particle velocity for no load condition is given by

$$v = U_0(t)/A_g(t) = [2\Delta P/(k\rho)]^{1/2} \tag{12.3k}$$

Air particle velocity v is a constant for a constant pressure difference ΔP across the glottis, i.e., under no load condition. Hence the pressure drop due to glottal inductance is zero. However, in the presence of viscous resistance and acoustic load, the particle velocity is not strictly a constant. Flanagan also remarks that the effect of glottal inductance is small [4]. Direct measurements on models of glottis have shown that the effect of glottal inductance is small [42]. We will study the effect of glottal inductance in Sect. 12.8.

12.3.6 Glottal Impedance and Acoustic Loads

In the presence of subglottal and vocal tract loads, the transglottal pressure P_{tg} is given by the equation

$$P_{tg} = P_{tg}(t) = P_L - \Delta P_v - \Delta P_{Lg} - P_o(t) - P_{sg}(t) \qquad (12.31)$$

Including the viscous resistance drop $\Delta P_v = R_v u_g(t)$, but omitting the effect of glottal inductance, using the Eq. (12.3d), true glottal flow can be written as

$$u_g^2(t) = A_g^2(t)\,[2/(k\rho)]\,[P_L - R_v u_g(t) - P_o(t) - P_{sg}(t)] \qquad (12.3m)$$

Generally the computations are made using discrete signals. Let the sampling interval be ΔT and sampling frequency be $F_s = 1/\Delta T$. For a discrete signal simulation and for ease of numerical computation the above equation is re-written as

$$k_k u_g^2(n) + R_v u_g(n) = [P_L - P_o(n) - P_{sg}(n)] \qquad (12.3n)$$

where $k_k = [(k\rho/2)/A_g^2(n)]$ (12.3o)

Equation 12.3n is a quadratic equation of the form $Ax^2 + Bx = C$. In order to solve the above quadratic equation for the *n-th* sample of true glottal flow, $u_g(n)$, the terms $P_o(n)$ and $P_{sg}(n)$ have to be known which depend the input impedances of vocal tract and subglottal loads. This leads to the issue of input impedance calculation.

12.4 Input Impedance Calculation

We will now discuss the modeling of input or driving point impedance of vocal tract. Similar method can be used for modeling input impedance of subglottal system. Input impedance has to be calculated for a given vocal tract geometry. For an intended vowel sound, there is a specific configuration of articulators that determines vocal tract geometry or shape. Vocal tract geometry is obtained either by x-ray measurements [3], MRI data [43] or by means of an articulatory model [44, 45].

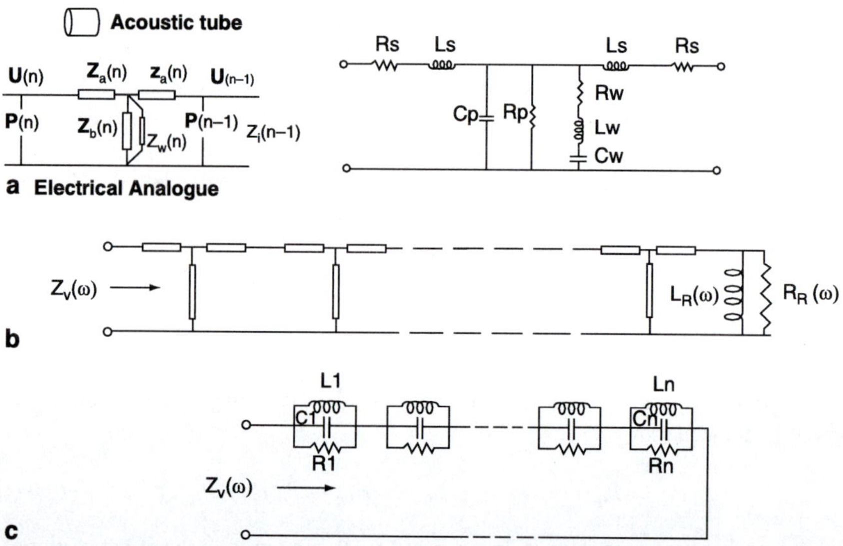

Fig. 12.3 **a** Electrical equivalent circuit of a small cylindrical section of vocal tract with yielding wall and **b** Distributed chain of T-networks terminating in radiation impedance **c** Series of formant or RLC networks

Direct measurement and modeling of input impedance of subglottal system has also been reported [46, 47].

For a steady vowel sound, input impedance needs to be calculated only once. For connected speech, input impedance needs to be computed whenever there is a significant change in vocal tract shape. However, a sudden change of input impedance or transfer function would cause undesirable transients at the instant of change. Hence input impedance and transfer function are computed for every sampling instant.

There are two models used for computation of vocal tract input impedance: (1) Transmission line analog (TL) model to be discussed in Sect. 12.4.1. (2) Series of *RLC* or formant networks model to be discussed in Sect. 12.4.2 Both these models are in the analog domain. Suitable transformation has to be used in order to obtain the digital domain representation which will be discussed in Sect. 12.4.3.

12.4.1 Transmission Line Analog (TL) Model

Vocal tract is divided along its axis from glottis to lips into a large number of sections. Assuming one-dimensional acoustic wave propagation, each section of vocal tract is replaced by a cylindrical section of average sectional area.

An equivalent electrical 'T-network' represents each section, Fig. 12.3a. [1–4]. The distributed elements Z_a and Z_b are related to the sectional geometry via hyperbolic functions; $Z_a = Z_0 tanh(\gamma l/2)$, $Z_b = Z_0 csch(\gamma l)$, where Z_0 is the characteristic impedance, γ is the propagation constant and l is the section length. $Z0$ and γ are determined by *R, L, C* and *G* elements of the section. When distributed elements are used, section lengths need not be the same for all sections. When section lengths are equal and small, for low frequencies, retaining only the first term in the expansion of the hyperbolic functions leads to lumped R_S, L_S, C_P, and G_P elements for each cylindrical section. Use of lumped elements reduces the number of computations and is generally used only in time domain computation. The '*R*' and '*G*' elements are frequency dependent. Yielding wall effect is taken into account by an additional series R_W, L_W, C_W elements which shunts the T-network. The 'T-network' is also referred to as four-pole or two-port network. For the entire vocal tract, the T-networks are connected in series which is referred to as ladder network, Fig. 12.3b. The ladder network terminates in the radiation impedance which is frequency dependent [4].

Input impedance of vocal tract, $Z_v(\omega)$, is computed in frequency domain over a desired frequency range using transmission parameters and chain matrix multiplication [19, 20, 48–50]. Let us say we are interested in determining input impedance at an angular frequency, ω. The amplitude of pressure and flow phasors of the chosen frequency 'ω' at an *i-th* section, $P(i, \omega)$, $U(i, \omega)$, are related to pressure and flow at $(i-1)$-*th* section, $P(i-1, \omega)$, $U(i-1, \omega)$, through a matrix. The '*ABCD*' elements of the matrix are called the transmission parameters.

The equivalent electrical impedances of the *i-th* T-network, with either the distributed or the lumped elements, determine the transmission parameters. A product of chain matrices is written for $i=0$ to L where $i=0$ is the glottal end and $i=L$ is the lip end. The input impedance $P(0, \omega)/U(0, \omega)$ at the glottis at the chosen frequency 'ω' can thus be obtained. Similarly the transfer function $U(L, \omega)/U(0, \omega)$ can be obtained where $U(L, \omega)$ is the flow at the lip end. The calculation is carried out over the desired frequency range of ω.

Instead of using variables $P(i, \omega)$ and $U(i, \omega)$, one can write a relation between two successive sections $(i, i-1)$ using the ratio of flows $U(i, \omega)/U(i-1, \omega)$ and input impedance $Z(i-1, \omega)=P(i-1)/U(i-1)$ as variables [51]. Here the computation begins at the lip end and proceeds towards the glottal end. At the lip end, input impedance is the radiation impedance. The computation is carried out over the desired frequency range of ω.

Similar procedure may be used for computing input impedance of subglottal system except that the yielding wall effect is ignored and proper boundary condition is used for lungs (short circuit or loss less termination instead of radiation impedance).

Time domain methods to compute input impedance also exist [52–56]. However, these time domain methods make certain assumptions. The axis of vocal tract is divided into a number of small *equal length* sections, typically about 0.5 cm. This imposes a constraint on the total length of vocal tract since number of sections is an integer. There is a constraint relating the frequency range that can be used and section length. Typically for a 20 section vocal tract, the maximum frequency ($Fs/2$) is about 20 kHz. Lumped equivalent electrical impedance elements are used instead of hyperbolic functions. Usually the time domain method ignores the frequency dependency of 'R_S' and 'G_P' elements and yielding wall effects. It approximates the radiation impedance. When these approximations are not made, a digital filter model represents the influence of each factor. Consequently, the resulting computations are more involved.

12.4.2 Formant Network Model

Input impedance $Z_v(\omega)$ can be represented by a series of *RLC* or formant networks, Fig. 12.3c [24, 25, 32, 46]. This is also called Foster reactance network. This representation is an approximation to a distributed parameter TL model for the following reasons (a) Internal losses are assumed to be frequency independent (b) The effect of yielding wall which is distributed through out the vocal tract is represented as lumped elements (c) Frequency dependency of radiation impedance is approximated and its effect at the lips is distributed in all *RLC* networks. (d) Higher pole correction factor [3, 57, 58] is ignored; theoretically there are a large number of formants. Each formant has a frequency response extending from 0 to infinity in the analog domain. By using only a finite number of low frequency formants, the effect of upper formants (higher poles) on the shaping of low frequency spectrum, called higher pole correction, is ignored. In a formant network model, the term '*R*' does not include losses due to glottal coupling since input impedance $Z_v(\omega)$ corresponds to the closed glottis condition.

Any reference to spatial location within the vocal tract is lost in a formant network model. Hence, pressure or flow at a given location within vocal tract cannot be known. In a distributed parameter TL model, spatial reference to vocal tract is preserved. This is especially important for synthesis of sounds where the excitation is within the vocal tract.

In order to determine the *R, L, and C* elements of the networks, input impedance of a distributed model is computed in the frequency domain initially with a large frequency spacing, say, 50 or 100 Hz over the desired range. The resonant frequencies are located as peaks in the real part of $Z_v(\omega)$. Subsequently, input impedance of the same distributed model is recomputed with narrower frequency spacing, say 1 or 0.5 Hz, only around the estimated resonant frequency. If required the peak location can be obtained to a higher accuracy with interpolation. Magnitude of the peak in real part of $Z_v(\omega)$ at the resonant frequency gives '*R*'. Values of '*L*' and '*C*' elements are computed using the peak resonant frequency (damped resonant frequency) and the 3-dB down bandwidth around the resonant frequency from the log magnitude function of $Z_v(\omega)$. Distinction between the damped and undamped resonant frequencies has to be kept in mind. This procedure is repeated for all the resonances (formants) within the frequency range of interest.

12.5 Computation of True Glottal Flow

12.5.1 TL Model

The following approach of finding true glottal flow for a distributed parameter TL model has been presented in [19, 50]. Vocal tract input impedance, $Z_v(k\Delta\omega)$, is computed at equally spaced frequencies, $\Delta\omega$, in the frequency domain over the desired frequency range, i.e., for $k=0$–$0.5Fs/\Delta\omega$. Fourier inverse of $Z_v(k\Delta\omega)$ is computed to obtain the discrete time domain response $z_v(n)$. The oral pressure $P_o(n)$ is then obtained as

$$P_o(n) = u_g(n) * z_v(n) = u_g(n)z_v(0) + \sum u_g(n-j)z_v(j) \qquad (12.4a)$$

where '*' denotes the convolution operation and the summation is for $j=1$ *to* L_v, L_v is the number of samples required to represent $z_v(n)$. Typically FFT of size 2,048 (or 4,096) may have to be used resulting in L_v of 1,024 (or 2,048) samples. Similarly for the subglottal system

$$P_{sg}(n) = u_g(n) * z_{sg}(n) = u_g(n)z_{sg}(0) + \sum u_g(n-l)z_{sg}(l) \qquad (12.4b)$$

where $z_{sg}(n)$ is Fourier inverse of input impedance of subglottal system $Z_{sg}(k\Delta\omega)$, summation is for $l=1$ *to* L_{sg}, L_{sg} is the number of samples required to represent $z_{sg}(n)$.

Substituting Eq. (12.4a) and Eq. (12.4b) in Eq. 12.3n we get,

$$\begin{aligned} k_k u_g^2(n) + [R_v + z_v(0) + z_{sg}(0)]u_g(n) = P_L - \sum u_g(n-j)z_v(j) \\ - \sum u_g(n-l)z_{sg}(l) \end{aligned} \qquad (12.4c)$$

Equation (12.4c) is a quadratic equation in $u_g(n)$ with all known coefficients and hence can be easily solved. Only the positive solution is meaningful. If true glottal flow is required for more than one pitch period, then the samples of true glottal flow of previous cycle(s) have to be retained. The time domain response $z_v(n)$ has to be updated whenever there is a change of vocal tract shape. Care must be taken not to introduce any transients during updating.

Vocal tract for the closed glottis condition has very low internal loss. Hence the time domain signal $z_v(n)$ exists for a substantial duration there by increasing the number of computations required in the convolution of Eq. (12.4a). Typically, for every sample of $u_g(n)$, 2,048 operations may be required. Using input impedance of a TL model is more realistic but is computationally very intensive. Also, it does not provide any insight into the effect of time varying glottal impedance on the temporal or spectral characteristics of the response of a formant.

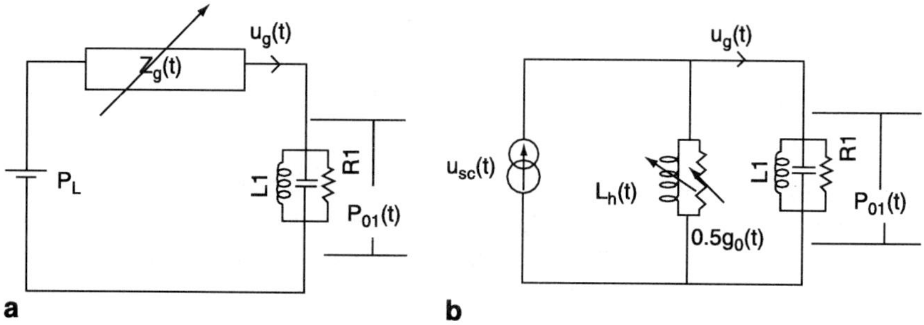

Fig. 12.4 Computation of true glottal flow for one formant load **a** Thevinin's equivalent circuit **b** Norton's equivalent circuit

12.5.2 Formant Network Model

The use of *RLC* or formant networks simplifies the computation of true glottal flow. The approach of using lumped *RLC* networks for computing true glottal flow has been outlined in [25, 31, 32]. Consider the first formant load for the sake of simplicity as shown in Fig. 12.4a. The oral pressure for the first formant load $P_o(t)=P_{o1}(t)$. True glottal flow is the sum of three branch currents:

$$u_g(t) = [1/L_1] \int P_{o1}(t)dt + C_1[d/dt]P_{o1}(t) + P_{o1}(t)/R_1 \qquad (12.5a)$$

The discrete signal representation of Eq. (12.5a) based on bilinear transformation leads to the equation

$$P_{o1}(n) = D_{01}u_g(n) - D_{11}\ P_{o1}(n-1) - D_{21}\ S_{o1}(n-1) \qquad (12.5b)$$

See Appendix 2 for derivation. Assuming only the first formant network for vocal tract input impedance and neglecting subglottal system in Eq. 12.3n and substituting $P_{o1}(n)$ in place of $P_o(n)$ we get

$$k_k u_g^2(n) + [R_v + D_{01}]u_g(n) = [P_L + D_{11}P_{o1}(n-1) + D_{21}S_{o1}(n-1)] \quad (12.5c)$$

Equation 12.5c is a quadratic equation in $u_g(n)$ with all known coefficients and hence can be easily solved. Only the positive solution is meaningful. $P_{o1}(n)$ and $S_1(n)$ have to be updated for every sample. If true glottal flow is required for more than one pitch period, these values have to be updated even during the closed phase interval when $u_g(n)$ may be zero.

The above equation can easily be generalized for '*N*' number of *RLC* networks in series as

$$P_o(n) = \sum P_{oi}(n) = \sum [D_{0i}u_g(n) - D_{1i}P_{oi}(n-1) - D_{2i}S_{oi}(n-1)] \qquad (12.5d)$$

where the summation is for $i=1$ *to N*.

Similarly for subglottal system with '*M*' number of *RLC* networks,

$$P_{sg}(n) = \sum P_{sj}(n) = \sum [D_{0sj} u_g(n) - D_{1sj} P_{oi}(n-1) - D_{2sj} S_{sj}(n-1)] \tag{12.5e}$$

where the summation is for $j=1$ *to* M. Substituting (12.5d) and (12.5e) in (12.3n) we get

$$k_k u_g^2(n) + [R_v + D_0]u_g(n) = P_L + \sum D_{1i} P_{oi}(n-1) + \sum D_{2i} S_{oi}(n-1) + \sum D_{1sj} P_{sj}(n-1) + \sum D_{2sj} S_{sj}(n-1) \tag{12.5f}$$

where $D_0 = \Sigma D_{0i} + \Sigma D_{0sj}$. Equation (12.5f) is a quadratic equation in $u_g(n)$ with all known coefficients and hence can be easily solved. Only the positive solution is meaningful. $P_{oi}(n)$, $S_{oi}(n)$, for $i=1$ to N and $P_{sj}(n)$ and $S_{sj}(n)$ for $j=1$ to M have to be updated for every sample. If true glottal flow is required for more than one pitch period, these values have to be updated even during the closed phase interval. Note that the order of computation for every sample of $u_g(n)$ is $2(N+M)$ instead of 2,048 samples as in the case of a TL model. Also, oral and subglottal pressure signals are available as by products.

12.5.3 Digital Pole-Zero Model

In case of a TL model, use of discrete convolution involving inverse Fourier transform of input impedance for computing true glottal flow is computationally intense. The author proposes a computationally more efficient digital pole-zero model to represent input impedance of vocal tract. See Sect. 12.5.3.1.

In case of a formant network model, use of trapezoidal rule for integration in Eq. (12.5a) to derive Eq. (12.5b) implicitly means that a bilinear transformation is used for mapping analog input impedance on to digital domain. In bilinear transformation, the range of analog frequency axis (0, ∞) is mapped nonlinearly onto the range (0, $F_s/2$) in digital domain. The author's experiments have shown that this nonlinear warping of frequency axis of bilinear transformation introduces distortions in the computed airflow. The author proposes impulse invariance technique for mapping the input impedance of a formant network on to digital domain, which is more accurate compared to bilinear transformation. See Sect. 12.5.3.2. A bilinear transformation may be implicitly used in some of the integrated source-filter models, mechanical models of vocal fold vibrations etc leading to distortions.

12.5.3.1 Digital Pole-Zero Filter for TL Model

The frequency response of input impedance $Z_v(\omega)$ has alternate poles and zeros. It can be shown that the z-transform of inverse Fourier transform of input impedance

$z_v(n)$ obtained using a transmission line analog model can be represented by a digital pole-zero filter of the form

$$P_o(z)/U_g(z) = Z_v(z) = [1 - z^{-1}][b_0 + b_1 z^{-1} + b_2 z^{-2} + \ldots b_q z^{-q}] \\ /[1 + a_1 z^{-1} + a_2 z^{-2} + \ldots a_p z^{-p}] \tag{12.6a}$$

Where p is the number of poles equal to the number of vocal tract sections and q is the number of zeros equal to $(p-1)$. The zero at DC or $\omega=0$ has been brought out explicitly. The denominator coefficients $\{a_1, a_2, \ldots a_p\}$ can be determined using the covariance method of linear prediction applied over the derivative of $z_v(n)$ after $(q+p)$ samples. The numerator coefficients $\{b_0, b_1, b_2, \ldots b_q\}$ may subsequently be solved writing direct equations for the derivative of $z_v(n)$, for $n=1$ *to* q and using $\{a_1, a_2, \ldots a_p\}$ already determined. From Eq. 12.6a we write

$$P_o(z) = U_g(z)[1 - z^{-1}][b_0 + b_1 z^{-1} + b_2 z^{-2} + \ldots b_q z^{-q}]/[1 + a_1 z^{-1} \\ + a_2 z^{-2} + \ldots a_p z^{-p}] \tag{12.6b}$$

It may be noted from RHS of Eq. 12.6b that the variable $P_o(n)$ depends on the *derivative* of true glottal flow. The importance of the derivate of true glottal flow will be frequently encountered in further discussion. The inverse z-transform of Eq. (12.6b) gives

$$P_o(n) = b_0 u_g(n) + \sum [b_j - b_{j-1}] u_g(n-j) - \sum a_i P_o(n-i) \tag{12.6c}$$

where the summation is for $j=1$ *to* q and for $i=1$ *to* p. The oral pressure $P_o(n)$ in Eq. (12.6c) can be substituted in to Eq. 12.3n to get

$$k_k u_g^2(n) + [R_v + b_0] u_g(n) = P_L - \sum [b_j - b_{j-1}] u_g(n-j) \\ + \sum a_i P_o(n-i) \tag{12.6d}$$

where the summation is for $j=1$ *to* q and for $i=1$ *to* p.

Equation 12.6d is a quadratic equation in $u_g(n)$ with all known terms and hence can be easily solved. Only the positive solution is meaningful. If true glottal flow is required for more than one pitch period, $P_o(n)$ has to be updated using Eq. (12.6c) for every sample even during the closed phase interval.

In a transmission line analog model, the number of computations for every sample of $u_g(n)$ using Eq. (12.4b) or (12.4c) is of the order of 2,048 where as using a digital pole-zero filter model the number of computations as per Eq. (12.6d) is of the order (p+q), typically about 40 for a 20 section vocal tract. A pole-zero digital filter model implementation is more efficient and has the additional advantage of representing the relatively more accurate transmission line analog model.

Similar procedure may be followed to include the input impedance of the subglottal system.

12.5.3.2 Digital Filter for Formant Network Model

Instead of using bilinear transformation as discussed earlier in Sect. 12.5.2, the author recommends the use of impulse invariance technique for mapping the analog input impedance of an *RLC* network on to the digital domain. Differentiating Eq. (12.5a) and multiplying throughout by L_1 gives

$$\begin{aligned} P_{o1}(t) + L_1 C_1 [d^2/dt^2] P_{o1}(t) \\ + [L_1/R_1][d/dt] P_{o1}(t) = [L_1][d/dt] u_g(t) \end{aligned} \qquad (12.6e)$$

Note from RHS of Eq. (12.6e) that the variable $P_{o1}(t)$ depends on the *derivative* of true glottal flow. Applying Laplace transform to the differential equation Eq. (12.6e), dividing throughout by $L_1 C_1$ and re-arranging the terms we get

$$P_{o1}(s) = [L_1 s U_g(s)]\ [\omega_1^2/(s^2 + 2\alpha_1 s + \omega_1^2)] \qquad (12.6f)$$

where the undamped resonant frequency $\omega_1{}^2 = 1/(L_1 C_1)$ and $2\alpha_1 = 1/(C_1 R_1)$. The second factor within square brackets on RHS of Eq. (12.6f) is the transfer function of first formant network with *DC* gain (gain at $s=0$) equal to unity. Writing the equivalent digital representation of the second factor of Eq. (12.6f) on RHS based on impulse invariance technique and noting that '*s*' in the first factor of RHS represents differentiation gives

$$\begin{aligned} P_{o1}(z) = [L_1 F_s\ (1 - z^{-1})\ U_g(z)]\ [(1 + a_{11} + a_{21})/(1 + a_{11} z^{-1} \\ + a_{21} z^{-2})] \text{ or} \end{aligned} \qquad (12.6g)$$

$$P_{o1}(z) = G_{o1} U_g(z)\ (1 - z^{-1}) - a_{11} z^{-1} P_{o1}(z) - a_{21} z^{-2} P_{o1}(z) \qquad (12.6h)$$

where $G_{o1} = F_s L_1 (1 + a_{11} + a_{21})$, $a_{11} = -2\ cos(2\pi F_1 \Delta T) exp(-\pi B_1 \Delta T)$ and $a_{21} = exp(-2\pi B_1 \Delta T)$. F_1 is the damped or peak resonant frequency. The inverse z-transform of the Eq. 12.6h gives

$$P_{o1}(n) = G_{01}[u_g(n) - u_g(n-1)] - a_{11} P_{o1}(n-1) - a_{21} P_{o1}(n-2) \quad (12.6i)$$

Note that the *n-th* sample of $P_{o1}(n)$ depends only on one past sample of $u_g(n)$ and two past samples of $P_{o1}(n)$ assuming only the first formant network for vocal tract input impedance. Neglecting the subglottal system in Eq. 12.3n and substituting $P_{o1}(n)$ in place of $P_o(n)$ as in Eq. 12.6i we get

$$\begin{aligned} k_k u_g^2(n) + [R_v + G_{01}] u_g(n) = [P_L + a_{11} P_{o1}(n-1) + a_{21} P_{o1}(n-2)] \\ + [G_{01}] u_g(n-1) \end{aligned} \qquad (12.6j)$$

Equation 12.6j is a quadratic equation in $u_g(n)$ with all known terms and hence can be easily solved. Only the positive solution is meaningful. If true glottal flow is required for more than one pitch period, $P_{o1}(n)$ has to be updated for every sample even during the closed phase interval.

The above equation can be extended to include N_v number of vocal tract *RLC* networks in series:

$$P_o(n) = \sum P_{oi}(n) = \sum [G_{0i} u_g(n) - a_{1i} P_{oi}(n-1) - a_{2i} P_{oi}(n-2)] \quad (12.6k)$$

where the summation is for $i=1$ *to* N_v. Substituting Eq. 12.6k in (12.3n) we get

$$\begin{aligned} k_k u_g^2(n) + [R_v + G_0]u_g(n) = [P_L + \sum a_{1i} P_{oi}(n-1) \\ + a_{2i} P_{oi}(n-2)] + [G_0]u_g(n-1) \end{aligned} \quad (12.6l)$$

where $G_0 = \Sigma G_{oi}$ and the summation is for $i=1$ *to* N_v.

A similar procedure may be followed to include the effect of subglottal system.

In case of a formant network model, only $2N_v$ computations are required to find the *n-th* sample of $u_g(n)$. Further, the number of *RLC* networks used for computing true glottal flow may be different from the number of networks used in the transfer function. Thus true glottal flow may be computed using only the first formant load where as vowel output can be computed with four formants. This is not possible in TL model. Although use of *RLC* networks is less accurate compared to a TL model, it has the computational advantage and the results can be easily interpreted in terms of the response of formants. We will be using the digital filter representation of formant network model to present the results in this chapter.

12.5.3.3 Relationship Between TL Model and Formant Network Model

The digital filter of a formant network model can be derived from the digital filter of a TL model. Using partial fraction expansion of complex variable theory, $Z_v(z)$ of Eq. (12.6a), can be written as

$$\begin{aligned} P_o(z)/U_g(z) &= Z_v(z) = [1 - z^{-1}]B(z)/A(z) \\ &= [1 - z^{-1}][B_1(z)/A_1(z) + B_2(z)/A_2(z) + \ldots + B_N(z)/A_N(z)] \end{aligned} \quad (12.6m)$$

Where $A_i(z)$ represents either the *i-th* real root or *i-th* conjugate root pair of $A(z)$ and $B_i(z)$ is its residue. Assume only conjugate root pairs. Sum of terms $B_i(z)/A_i(z)$ on RHS of Eq. (12.6m) represents a series connection of '*RLC*' formant networks. Compare input impedance of first formant network derived from Eq. (12.6g) to the first term in expansion of Eq. (12.6m). Formant network model assumes that there are no real roots, $B_i(z)$ is a constant, that the roots are arranged according to the ascending order of the resonant frequency, and that only the first N_v terms are retained ($N_v \le N$).

12.6 Components of True Glottal Flow

12.6.1 Dynamic Glottal Impedance

Considering one formant load with only the kinetic resistance and combining Eq. (12.31) and Eq. (12.5a) we get

$$\begin{aligned} u_g(t) &= [1/L_1] \int P_{o1}(t)dt + C_1[d/dt]P_{o1}(t) + P_{o1}(t)/R_1 \\ &= A_g(t)[2/(k\rho)]^{1/2}[P_L - P_{o1}(t)]^{1/2} \\ &= [A_g(t)\{2/(k\rho)P_L\}]^{1/2}[1 - P_{o1}(t)/P_L]^{1/2} \\ &= u_{sc}(t)[1 - 0.5P_{o1}(t)/P_L] = u_{sc}(t) - 0.5g_o(t)P_{o1}(t)] \end{aligned} \tag{12.7a}$$

where $g_o(t)=u_{sc}(t)/P_L$ is the no load glottal conductance. It is assumed that $[P_{o1}(t)/P_L]<1$ so that $(1-x)^{1/2}=(1-0.5x)$. Example: If we assume the maximum amplitude of oral pressure to be about 0.2 of lung pressure, then $x=0.2$ and the first three terms in the series expansion of $(1-x)^{1/2}$ will be 1, 0.1, 0.005. Hence retaining only the first two terms in the series expansion seems justified. Then

$$\begin{aligned} [1/L_1] \int P_{o1}(t)dt + C_1[d/dt]P_{o1}(t) + P_{o1}(t)/R_1 \\ + 0.5\, g_o(t)P_{o1}(t) = u_{sc}(t) \end{aligned} \tag{12.7b}$$

Differentiating Eq. 12.7b with respect to 't' and using the notation $g_0'(t)$ for the derivative of $g_0(t)$ we get

$$\begin{aligned} [1/L_1 + 0.5g_0'(t)]P_{o1}(t) + C_1[d^2/dt^2]P_{o1}(t) \\ + [1/R_1 + 0.5g_0(t)][d/dt]P_{o1}(t) = [d/dt]u_{sc}(t) \end{aligned} \tag{12.7c}$$

The first coefficient in Eq. (12.7c) indicates a parallel combination of two inductances L_1 and $L_h(t)=[2/g_0'(t)]$. The third term indicates a parallel combination of two resistances, R_1 and $R_g(t)=2/g_0(t)$. This shows that the dynamic glottal impedance in the Norton's equivalent circuit (Fig. 12.4b.) consists of a hypothetical dynamic glottal inductance, $L_h(t)$, and a dynamic glottal resistance $R_g(t)$. We refer to $L_h(t)$ as 'hypothetical' since it is different from the acoustic glottal inductance due to the mass of air or air plug in the glottis as discussed in Sect. 12.3.5. The dynamic glottal conductance is given by $0.5g_0(t)$. In literature the concept of dynamic glottal conductance has been well recognized [3, 4] but not the concept of hypothetical dynamic glottal inductance.

12.6.2 Relation Between True Glottal Flow and No Load Airflow

A concept called pseudo Laplace transform has been used in [32]. The author would like to clarify that the concept of pseudo Laplace transform is required *only for the sake of an interpretation* of the components of true glottal flow. The assumption of pseudo Laplace transform has neither been used in the numerical computation of the true glottal flow already presented in Sect. 12.5 nor in deriving Eq. 12.7c showing the presence of a hypothetical glottal inductance.

Dividing Eq. 12.7c throughout by C_1, we get

$$[1/L_1C_1][1+0.5L_1g_0'(t)]P_{o1}(t)+[d^2/dt^2]P_{o1}(t)+[1/C_1][1/R_1 + 0.5g_0(t)]\,[d/dt]P_{o1}(t)=[1/C_1]\,[d/dt]u_{sc}(t) \quad (12.7d)$$

Note that $P_{o1}(t)$ depends on the *derivative* of the no load airflow.

The oral pressure $P_{o1}(t)$ can be interpreted as the output of a linear time-varying one formant network for no load airflow as input (Fig. 12.4b). The same oral pressure $P_{o1}(t)$ is produced as the output of a time-invariant one formant network for true glottal flow as input (Fig. 12.4a). The problem is to determine the characteristics of the time-varying formant network.

Equation 12.7d is a linear differential equation with time varying coefficients. Strictly speaking Laplace transform is not applicable to Eq. 12.7d. We define a *pseudo* Lapalace transform ignoring the fact that $g_o(t)$ and $g_0'(t)$ are time varying. Replacing $g_0'(t)$ by $g_0'(\tau)$ and $g_0(t)$ by $g_0(\tau)$, taking the Laplace transform of Eq. 12.7d and re-arranging the terms we get

$$[s^2+2\alpha_{1t}s+\omega_{1t}^2]\;P_{o1}(s)=L_1\omega_1^2\;sU_{sc}(s) \quad (12.7e)$$

where $2\alpha_{1t}=2\alpha_1[1+0.5R_1g_0(\tau)]$ and $\omega_{1t}^2=\omega_1^2[1+0.5L_1g_0'(\tau)]$, and as before $2\alpha_1=[1/R_1C_1]$, $\omega_1^2=[1/L_1C_1]$. The terms α_1 and ω_1 are time invariant as in Eq. 12.6f where as α_{1t} and ω_{1t} are time varying. Re-writing Eq. 12.7e

$$P_{o1}(s)=[L_1sU_{sc}(s)][\omega_1^2/(s^2+2\alpha_{1t}s+\omega_{1t}^2)] \quad (12.7f)$$

The second factor is the transfer function of time-varying one formant network. From Eq. (12.6f) and Eq. (12.7f) we get the relation

$$\begin{aligned}U_g(s)&=U_{sc}(s)\;[(s^2+2\alpha_1s+\omega_1^2)/(s^2+2\alpha_{1t}s+\omega_{1t}^2)]\\&=U_{sc}(s)\;[H_{1t}(s)]\end{aligned} \quad (12.7g)$$

The above equation based on pseudo Laplace transform gives a relationship between the true glottal flow $u_g(t)$ and the no load airflow $u_{sc}(t)$ in the Laplace domain.

Let us interpret $u_g(t)$ as a dependent variable or the response, $u_{sc}(t)$ as an independent variable or input excitation and $H_{1t}(s)$ as system transfer function. Instead of interpreting the response as a convolution of input and impulse response of system transfer function we use partial fraction expansion. Using partial fraction expansion theorem of complex variable theory, Eq. 12.7g for $U_g(s)$ can be written as a sum of two components:

$$U_g(s)=[\mathit{Residue1}][\mathit{poles\ of}\ U_{sc}(s)]+[\mathit{Residue2}][\mathit{poles\ of}\ H_{1t}(s)] \quad (12.7h)$$

The Laplace inverse of the first term on RHS of Eq. 12.7h gives a time domain component which we have referred to as 'residue component' of true glottal flow [32]. Since this term depends only on poles of input, $U_{sc}(s)$, it is analogous to 'particular integral component' as per the circuit theory [48].

We have referred to the Laplace inverse of the second term on RHS of Eq. 12.7h as the 'ripple component' of true glottal flow [32]. This component depends on the poles of system transfer function, $H_{1t}(s)$, and it corresponds to the 'complimentary solution' as per the circuit theory [48]. The poles of $H_{1t}(s)$ are excited at the instants (epochs) where the input $u_{sc}(t)$ or one of its derivatives has a discontinuity. This gives rise to a transient response. Note that the system $H_{1t}(s)$ is time varying.

An implicit term not seen in Eq. 12.7h arises due to initial conditions and is called the 'superposition component'. We will explain these three components in some more detail in subsequent sections.

Separation of the derivative of true glottal flow obtained by inverse filtering into a residue and a ripple component has been used in a practical application of speaker identification [59]. See author's cautious remarks in Sect. 12.9.2.

12.6.3 Residue Component of True Glottal Flow

The need for computing residue component of airflow arises for the purpose of modeling only the gross features of the source pulse shape.

Using Taylor series expansion, $u_{sc}(t)$ can be written as a series of polynomial terms in powers of 't' around $t=0$ over a segment where it is continuously differentiable. See Appendix in [32]. Hence, $U_{sc}(s)$ can be written as a series of polynomial terms in $(1/s)$ with poles at $s=0$. The residue of the first term for $s=0$ is given as $\omega_1^2/\omega_{1t}^2 = [1/\{1+0.5L_1 g_0'(\tau)\}]$. Since $g_0'(\tau)$ is positive during the opening phase and negative during the closing phase, the residue component rises relatively more gradually and falls more steeply compared to $u_{sc}(t)$. Of the three load elements, only the load inductance element L_1 appears in the residue term for the first term in the expansion. Similarly it can be shown that only the load inductance appears in the residue of other terms in the expansion, i.e, for higher powers of $(1/s)$ [32].

This suggests a method for computing residue component. The residue component $u_r(t)$ of airflow can be estimated by retaining only the inductance element(s) of formant network(s). Since inductances of all formant networks are in series, the effective inductance, L_T, is the sum of inductances in *RLC* networks. If a single formant load is used then the residue component has to be computed using only the inductance of that formant load. On the other hand if several *RLC* networks are used then L_T corresponds to the sum of inductances in these networks. The oral pressure $P_o(n)$ is given by

$$P_o(n) = Fs\ L_T\ [u_r(n) - u_r(n-1)] \tag{12.7i}$$

where $u_r(n)$ is the residue component of airflow. Note that the oral pressure is determined by the *derivative* of residue component. The residue flow is given by rewriting the Eq. 12.3n with input impedance represented only by L_T as follows:

$$k_k u_r^2(n) + [R_v + F_s L_T]u_r(n) = P_L + F_s L_T u_r(n-1) \qquad (12.7\text{j})$$

The positive solution to the above quadratic equation gives an *estimate* of residue component of airflow.

It needs to be clarified that this method of estimating residue component of airflow is a good *approximation* and not an exact solution. An equation similar to Eq. 12.7g can be derived for the relationship between $u_r(t)$ and $u_{sc}(t)$ using the inductance as the only load element. Then $u_r(t)$ itself consists of two components. The first term gives the desired residue component of airflow. The second term represents transients excited at instants of discontinuity (onset and/or peak) in the source or its derivative(s) due to an inductive load. Using only the inductance element to find residue component of airflow assumes that the transients due to inductive load are negligible.

12.6.4 Ripple Component of Airflow

Computationally, ripple component of airflow is obtained as the difference between true glottal flow and estimated residue component. It may be noted from Fig. 12.4b. that the current $u_g(t)$ excites the time invariant closed glottis condition formant network. But, the oral pressure is the response of a time-varying formant network due to glottal coupling. It is the ripple component in $u_g(t)$ that induces the time varying bandwidth and time varying formant frequency during the open phase of the present glottal cycle.

In literature the presence of ripple component in glottal flow has been noted [4]. But, the use of pseudo Laplace transform gives a theoretical basis for the ripple component. It may be noted that ripple component does not imply fine movement of vocal folds since it is not present in the glottal area function but only in the airflow. The ripple component needs to be isolated from true glottal flow in order to study non-exponential decay and dispersion of a formant frequency within the open phase of the present glottal cycle.

12.6.5 Superposition Component

As per the circuit theory [48], one of the components in the solution to a differential equation of a linear network depends on the *initial condition of the circuit elements at the instant of switching on the circuit.* For glottal flow computation, the initial condition corresponds to the oral pressure at the instant of glottal onset. This gives rise to the superposition component in the airflow.

At the onset of the very first glottal cycle (t=0), the oral pressure is zero. Some energy of the transient response of a formant due to the excitation at the glottal closure of the first glottal cycle still remains at the onset of the second glottal cycle. This implies that at the onset of the second glottal cycle, the oral pressure is non-zero (*initial condition*). This remaining energy gets carried over to the second glottal cycle. The carry over response gets *superposed* with the formant response due to the excitation of second glottal cycle. See Sect. 12.8.8, Fig. 12.16. The superposed component of formant response undergoes a non-exponential damping and dispersion over open phase of the second cycle. This initial condition in the oral pressure introduces a component in true glottal flow called 'superposition component'. The superposition component of true glottal flow induces the time varying bandwidth and time varying formant frequency of the superposed component over the open phase of the second (and subsequent) glottal cycle(s).

It may be noted that there is no superposition component during the very first glottal cycle. The superposition component of glottal flow during the open phase of the second glottal cycle is isolated from true glottal flow in the following manner. True glottal flow is calculated for successive glottal cycles. The onset of the very first glottal cycle is shifted to coincide with the onset of the second glottal cycle and subtracted from the glottal flow of the second glottal cycle to determine a difference component which gives the 'superposition component' of flow during the open phase of the second cycle. If superposition extends beyond two cycles, the onset of glottal flow of the first glottal cycle is shifted to coincide with the onset of the third glottal cycle and is subtracted from glottal flow of the third glottal cycle to determine the superposition component during the open phase of the third cycle and so on. The superposition component needs to be isolated from true glottal flow in order to study instantaneous formant frequency and instantaneous bandwidth of the response within the open phase of the subsequent glottal cycles. See Sect. 12.8.8.

12.7 Vowel Synthesis

12.7.1 Derivative of True Glottal Flow as Voice Source

The transfer function of TL model is computed in a manner similar to the input impedance calculation described in Sect. 12.4.1. In TL model, radiation at lips is represented by a parallel circuit of radiation inductance and radiation resistance [4]. The electrical current through the radiation resistance corresponds to the time varying airflow at the lips, $u_L(t)$. The relationship between the output response $u_L(t)$ and the input source $u_g(t)$ is defined via the transfer function in the frequency domain as

$$H(\omega) = U_L(\omega)/U_g(\omega) \tag{12.8a}$$

The transfer function $H(\omega)$ corresponds to the closed glottis condition. The acoustic wave propagation within the vocal tract assumes the conservation of mass and hence the gain of the transfer function at DC or $\omega=0$ is unity.

The radiated acoustic pressure far away from the lips, *p(t)* is the time derivative of $u_L(t)$, except for a scale factor. The scale factor is determined by the inverse square law which states that the intensity of sound wave decreases as the inverse of the square of the distance from the source. Ignoring the scale factor, we get

$$p(t) = [d/dt][u_L(t)] \text{ or} \tag{12.8b}$$

$$P(\omega) = [j\omega][U_L(\omega)] = [j\omega][U_g(\omega)H(\omega)] = [j\omega U_g(\omega)][H(\omega)] \tag{12.8c}$$

$$p(t) = [d/dt]u_g(t) * h(t) \tag{12.8d}$$

where '*' denotes the convolution operation in the time domain. Computing free field pressure of the response, *p(t)*, given the source and the vocal tract transfer function is called synthesis. From Eq. 12.8d, the source is the *derivative* of true glottal flow, $[d/dt]u_g(t)$. We have also noted the role of the derivative of airflow in determining the oral pressure.

In case of a formant network model, there is no specific representation of lips or radiation load resistance. The transfer function $H(\omega)$ of a TL model is *approximated* by a *cascade* of formant networks. This involves approximations as already mentioned for the case of input impedance. The formant network model is considered to be practically adequate. We use the impulse invariance method to map the analog transfer function on to the digital domain. Accordingly, the z-transform of *H(z)* is given by

$$H(z) = H_1(z)\, H_2(z)\, H_3(z) \ldots \tag{12.8e}$$

where $H_i(z)$ is the z-transform of the *i-th* formant with unity gain at $\omega=0$ given by

$$H_i(z) = [1 + a_{1i} + a_{2i}]/[1 + a_{1i}z^{-1} + a_{2i}z^{-2}] \tag{12.8h}$$

with $a_{1i} = -2cos(2\pi F_i \Delta T)exp(-\pi B_i \Delta T)$, $a_{2i} = exp(-2\pi B_i \Delta T)$, where F_i and B_i are the *i-th* time invariant (damped) formant frequency and bandwidth respectively. The bandwidth B_i is for the closed glottis condition. The time domain response $h_i(n)$ is of the form $exp(-\pi B_i n)sin(2\pi F_i n)$.

It is possible to use partial fraction expansion and express the transfer function *H(z)* as a sum of responses of $H_1(z)$, $H_2(z)$ etc. which gives rise to a parallel formant synthesizer [4]. The poles of $H_i(z)$ are the same as the poles of $A_i(z)$ of input impedance in Eq. 12.6m. One may wonder as to what constitutes the difference between input impedance and transfer function of a parallel formant synthesizer, since in both cases the result is *a sum of responses for the same poles*. The difference arises due to the residue terms. For example, for a cylindrical tube of uniform sectional area in the analog domain, the residues for a parallel formant synthesizer are in proportion 1:−1/3:1/5:−1/7 where as for the input impedance the residues are in proportion 1:1/3:1/5:1/7 [60]. Input impedance is a pole-zero function where as the transfer function is all-pole.

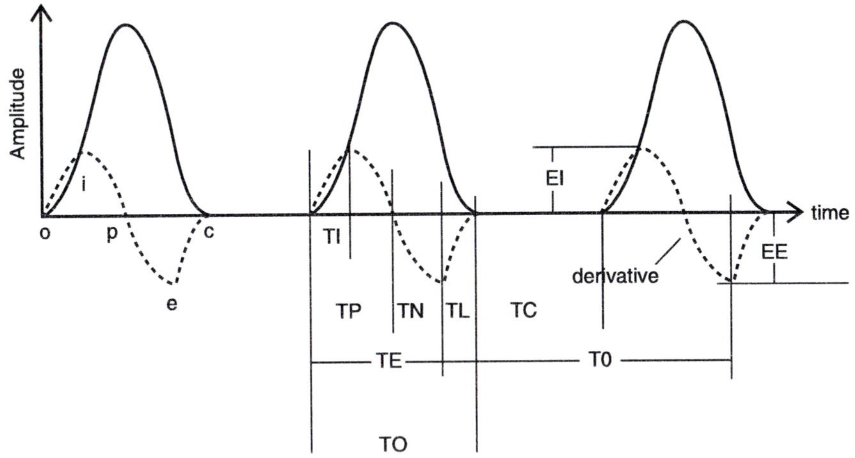

Fig. 12.5 Typical glottal area function and its derivative (not to scale)

12.7.2 Interactive Response of Formants

The derivative of true glottal flow of first glottal cycle appended with superposition components of subsequent cycles is used as an excitation to the first formant network to obtain an interactive response of the first formant. This interactive response consists of an exponential decay with constant formant frequency during the closed phase(s) and a frequency modulated wave with non-exponential damping during the open phase(s). One can study the instantaneous formant frequency and bandwidth during the open phase. See Sect. 12.8.8. In contrast when residue flow is used as an excitation to a time invariant formant network it results in a non-interactive response.

12.8 Results

We have presented the general theoretical background on computation of true glottal flow, its components and vowel response. Some specific results will be presented and discussed in this section.

12.8.1 Typical Glottal Area Function

Describing glottal area function or volume velocity airflow by mathematical equations is called modeling. A common terminology is used to describe important instants, amplitudes and parameters of both glottal area and glottal pulse (volume velocity airflow) models. These are shown in Fig. 12.5 and are also described below.

12.8.1.1 Important Instants and Intervals

o: onset, i: inflexion, p: peak, e: epoch, c: closure

TI: Onset interval, *TP*: Opening interval, or Opening Phase *TN*: Closing interval or closing phase, *TL*: Leakage interval, *TO*=Open phase interval=*TP*+*TN*+*TL*, *TC*: Closed phase interval

12.8.1.2 Important Parameters

Pitch Period T0: is the interval between two successive epochs.

Open Quotient, OQ, is the ratio of total interval for which the glottis is open, *TO*, to pitch period, *T*0. *OQ*=*TO*/*T*0=(*TP*+*TN*+*TL*)/*T*0. A high open quotient means that the glottis is open for a large part of the pitch period. For glottal (or airflow) pulse typical value is about 60%. A value of 100% means that the glottis never closes implying a breathy voice. For a pressed voice, the open quotient is about 30%.

Speed Quotient, SQ, is the ratio of relative opening interval to closing interval [*TP*/(*TN*+*TL*)]. For example *SQ*=1 means that opening and closing intervals are of equal duration. Usually, for a glottal area function the opening interval is shorter compared to closing interval (*SQ*<1). The opening and closing intervals are misnomers when used to describe the glottal airflow since the peak of glottal flow is delayed with respect to glottal area. See Sect. 8.4. See Fig. 12.15.

Leakage Quotient, LQ, is the interval *TL* relative to pitch period *T*0. *LQ*=*TL*/*T*0. The glottal area function is said to have an abrupt closure when *TL*=0. When *TL* is non-zero *in the glottal area function*, the closure is said to be non-abrupt. Whenever there is a glottal leakage or glottal chink or a nodule or when voice is breathy, then non-abrupt glottal closure is seen both in glottal area and in airflow. The interval *TL* may be zero for glottal area but finite for glottal flow due to various reasons. See Sects. 12.8.2 and 12.9.2. Hence, leakage interval in case of glottal flow has to be carefully interpreted.

12.8.1.3 A Model for Glottal Area Function

Different models differ in terms of mathematical functions used to describe pulse shape between the important instants. Some of the early models can be seen in Ref. [4]. In this work, we use the following equation to represent a typical glottal area function [61, 62] for the case *TL*=0:

$$\begin{aligned} A_g(t) &= A_{gmax}[0.5 - 0.5cos(\pi t/TP)] \quad for\ 0 \le t \le TP \\ &= A_{gmax}cos[(\pi/2)(t - TP)/(TN)] \quad for\ TP < t \le TE \end{aligned} \tag{12.9a}$$

Note that when $TL=0$, $TO=TE$. When TL is non-zero, a parabolic function [62, 63] is used to model the glottal area over the interval $TE<t\leq TO$.

LF model [64] has become very popular amongst researchers on studies related to voice source. In LF model, the segment TE is modeled as a product of two functions; an exponential and a sinusoid, with the exponential function having a positive exponent. See also [65]. In LF model, the segment TL is modeled by an exponential function [64]. See also [66]. LF model is suitable only for cases with SQ>>1. Since the exponential function is a monotonically increasing (or decreasing) positive function, the zero-crossing in the flow (or area) at $t=TP$ is ensured only when the frequency of the sinusoidal function is (1/2TP). When $SQ=1$ ($TP=TN$), one full sine wave cycle is seen in the derivative of flow (or area) signal. When $SQ<1$ ($TN>TP$), more than one sine wave cycle is seen in the derivative of flow (or area) signal. The model is hence unrealistic for $SQ\leq 1$. An exponential with a positive exponent can arise only from an unstable system with active element(s) in the process of the generation of the signal [48]. For these reasons the author prefers the model given in [61, 62]. See also comments in Sect. 12.9.2 on aliasing in voice source spectrum.

12.8.1.4 Typical Constants and Variables

It is a standard practice to use MKS system. However, for the dimensions involved with glottal area and airflow it is more convenient to use the CGS system. Typical values of some of the constants and variables are:

$\rho=1.14\times10^{-3}$ g/cm^3, $\mu=1.84\times10^{-4}$ g/cm s. k=kinetic resistance coefficient$=1.1$. k_a=Viscous resistance coefficient$=1$ for a rectangular glottis; else depends on the ratio of widths for convergent/divergent glottis. For a ratio of widths of 4, $k_a=5/32$. Depth of the glottis, $D=0.3$ cm. Length of the glottis, $l=1.8$ cm. $P_L=8$ cm water or 8 g=8*980 dynes/cm^2. Typical maximum glottal area $A_{gmax}=20$ mm^2$=0.2$ cm^2. Assume $T0=10$ ms. $OQ=0.5$, $SQ=1$, $TL=0$ unless otherwise specified.

12.8.2 *Relative Importance of Viscous and Kinetic Resistances Under No-Load Condition*

As per Eq. (12.3h), when R_v is assumed to be zero, the no load airflow $U_0(t)$ is simply a scale factor times the glottal area; $U_0(t)=3{,}536\,[A_g(t)]$, for $P_L=8$ cm water, $A_g(t)$ in sq cm, and $U_0(t)$ is in cc/s. Consider the case where both R_k and R_v (with $k_a=1$) are present. Substituting the typical values we find $R_k=2.21/A_g(t)$ and $R_v=0.0021/A_g^{\,3}(t)$. The viscous resistance dominates only for a very small glottal area. The viscous resistance is equal to kinetic resistance ($R_v=R_k$) for an area of the glottis equal to 3 mm^2 for $P_L=8$ cm water. As the glottal area exceeds 3 mm^2, the viscous resistance sharply decreases. For example for a value of $A_g(t)=0.5$ mm^2, $R_v=17170\ \Omega$ and $R_k=442\ \Omega$ where as for $A_g(t)=20$ mm^2, $R_v=0.26\ \Omega$ and $R_k=11\ \Omega$.

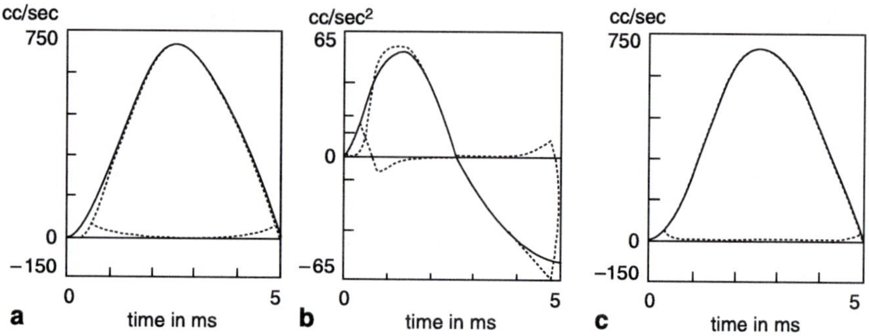

Fig. 12.6 No load condition. **a** Glottal airflow. *Solid*: R_k only. *Dashed*: R_k+R_v, $k_a=1$. **b** Derivative for the airflow shown in (a). The y-axis scale for the derivative has to be multiplied by Fs. **c** Glottal airflow, *Solid*: R_k only. *Dashed*: R_k+R_v, with time varying k_a. Difference signal is also shown for all cases

The no load airflow for the typical glottal area function defined in Eq. (12.9a) is shown in Fig. 12.6a for two different conditions; (1) with only the kinetic resistance and (2) with both kinetic and viscous resistances. The maximum difference in the flow for the two cases is about 7% of the maximum glottal flow of case (1). The derivative of flow is shown in Fig. 12.6b. The main effect of viscous resistance is to smooth out airflow near the glottal onset and closure [4, 32]. In the frequency domain the effect of viscous resistance manifests as a lowpass filtering beyond 3 kHz [32].

Although glottal area has an abrupt closure ($TL=0$), the derivative of airflow indicates a non-abrupt termination arising due to high viscous resistance for low glottal area. Also note the increase in the parameters '*EI*' and '*EE*' in the presence of viscous resistance. It is unfortunate that there is a common terminology to describe both the glottal area and glottal airflow signals. In this example, the leakage quotient for the glottal airflow is about 3%. Caution must be exercised when interpreting the leakage coefficient as measured from the airflow. Although the leakage coefficient for the airflow is about 3%, it does not imply a non-abrupt closure of vocal fold movement since in this example the glottal area function has an abrupt closure. Similarly, it may be noted the terms 'opening phase' and 'closing phase' used to describe the airflow do not exactly correspond to intervals of separation or approximation (actual physical movement) of vocal folds.

In the above example a value of $k_a=1$ has been used which is applicable for a rectangular glottis. We can include the effect of glottal geometry on R_v by suitably varying k_a with respect to time assuming a convergent glottis at onset and a divergent glottis at closure. One such example of flow computation with varying k_a is shown in Fig. 12.6c. A ratio of widths of 4 at glottal onset is changed to ¼ at closure gradually over the glottal cycle. In this case, the effect of viscous resistance is negligible.

The glottal airflow obtained by means of inverse filtering for natural vowels often shows an abrupt closure and sometimes even an abrupt onset. This indirectly

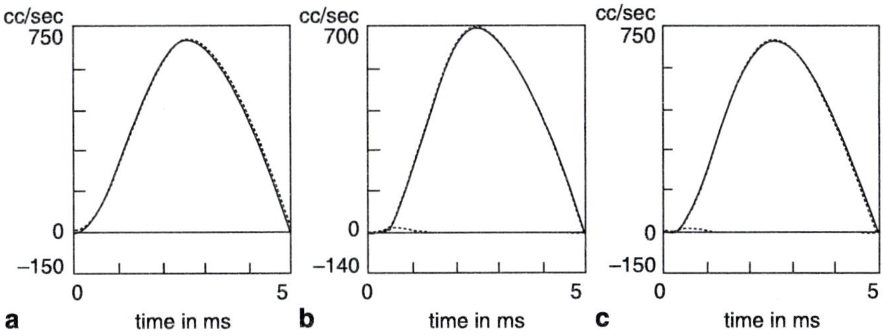

Fig. 12.7 No load condition. **a** Glottal airflow. *Solid*: R_k only. *Dashed*: R_k+L_g. The two curves are overlapping. **b** Glottal airflow. *Solid*: R_k+R_v. *Dashed*: $R_k+R_v+L_g$, $k_a=1$ and **c** Glottal airflow. *Solid*: R_k+R_v. *Dashed*: $R_k+R_v+L_g$., time varying k_a. Difference signal is also shown for all cases

implies either a very low prominence or even an absence of viscous resistance. We have already mentioned that viscous resistance dominates only for transglottal pressure less than 1 cm water [41] and that a minimum of 4 cm water pressure is required to initiate vocal fold vibrations [18]. Its inclusion in glottal impedance for phonation is hence questionable.

12.8.3 *Effect of Glottal Inductance*

Figure 12.7a shows glottal airflow computed for two different conditions; (1) with only the kinetic resistance and (2) with kinetic resistance and glottal inductance. It may be seen that the two airflow graphs are almost identical. The particle velocity is a constant in the absence of viscous resistance and for the no load condition. The pressure drop due to effective glottal inductance depends on the derivative of particle velocity and hence is zero in this case. Results of glottal airflow calculation in the presence of viscous resistance is shown in Figs. 12.7b, c. In the presence of viscous resistance, the air particle velocity is not a constant and hence the glottal inductance has a finite but small effect on the true glottal flow [4, 42].

12.8.4 *Factors Influencing the Residue Component of Airflow*

Viscous resistance appears to produce unrealistic leakage in an airflow signal. Its inclusion needs justification. The effect of glottal inductance is very small. Hence for further illustrations to be presented, only the kinetic resistance of glottal impedance is considered.

The residue component of airflow (henceforth called residue airflow) is computed using only the load inductance element as per Eq. (12.7b). Glottal airflow

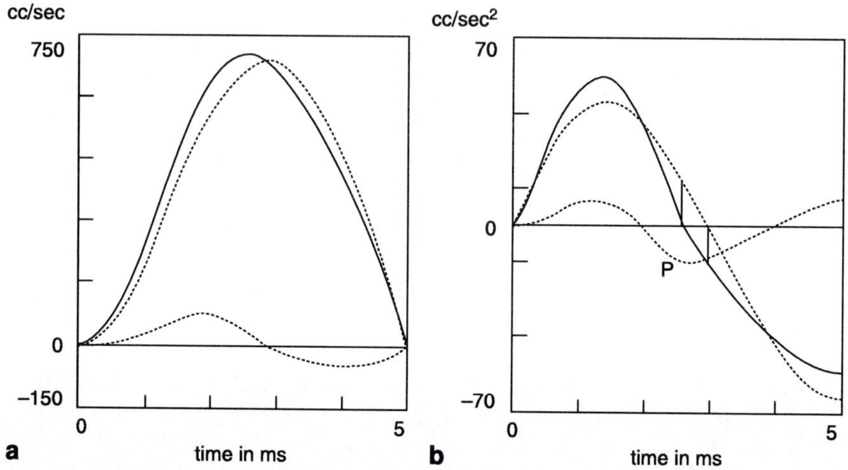

Fig. 12.8 **a** *Solid*: no load glottal airflow with only the kinetic resistance. *Dashed*: residue airflow for an inductance load of 6 mH along with the kinetic resistance. **b** The derivative of airflow shown in (a). Instants of zero-crossings are shown by vertical lines. The difference signal is also shown

computed using only the kinetic resistance is compared with the residue airflow computed for the case of an inductance load, L_T, of 6 mH in Fig. 12.8a. The derivative of airflow is shown in Fig. 12.8b.

The following differences between no load glottal airflow and residue airflow may be noted. (a) The instant of the peak in residue airflow occurs later than the instant of the peak in no load airflow. The instant of the peak in residue airflow is delayed with respect to the instant of peak in the no load airflow (or glottal area). This delay is also referred to as latency, say T_d. The residue airflow pulse is skewed relatively more to the right. This can be noted by a shift in the zero-crossing in the derivative of the residue airflow towards right. (b) Residue airflow shows a lower peak amplitude (or the maximum) compared to the peak amplitude (or the maximum) in no load airflow. (c) In the derivative of residue airflow, the positive peak amplitude parameter *EI* is lower and the negative peak amplitude *EE* is greater compared to the respective parameter values in the derivative of no load airflow. We discuss each of these differences in some detail.

12.8.4.1 Delay or Latency

Direct glottal area measurements show a pulse with an abrupt rise and a gradual fall [67, 68]. On the other hand inverse filtering experiments have shown the opposite trend in the airflow pulse shape, i.e., a gradual rise and an abrupt fall [27, 28]. This mismatch between the shape of glottal area and shape of glottal flow has been intriguing. See Fig. 12.15. In early studies on the computation of glottal flow the effect of source-filter interaction was ignored [3, 4]. Now it is well known that the

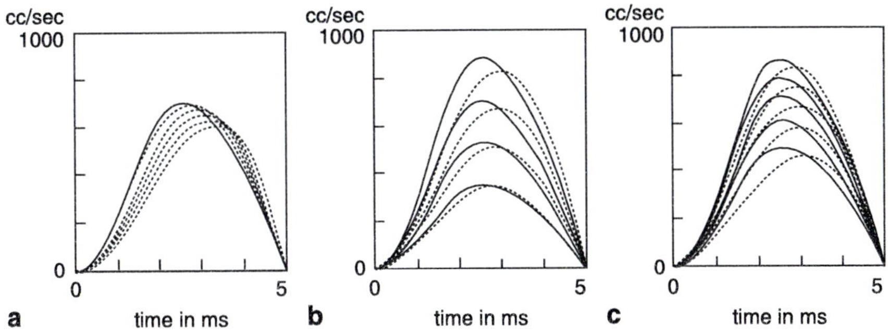

Fig. 12.9 **a** No load airflow (*solid*) and residue airflow (*dashed*) for different inductances of the load. **b** No load airflow (*solid*) and residue airflow (*dashed*) for different peak glottal areas. **c** No load airflow (*solid*) and residue airflow (*dashed*) for different lung pressures

inductive load of vocal tract is the main cause for the skewing of the airflow pulse to the right [27, 28, 32]. Intuitively, this could have been anticipated since from an electrical circuit theory point of view, a lag in the current with respect to the voltage is analogous to a phase shift due to an inductive load. This result can also be interpreted as the effect of Lenz's law which states that an inductive (inertia) element opposes any sudden change in the current thereby introducing a delay or phase shift in the current with respect to voltage in the transient response of an electrical circuit. In the above discussion residue airflow is to be compared to current (or response) and no load airflow is to be compared to the voltage (or input excitation).

What are the factors that determine the latency or delay? As seen above, one obvious factor is the inductance of acoustic load. But there are other factors as well. According to transient analysis of a simple electrical circuit with resistance and inductance elements, the phase shift in current relative to voltage is related to the time constant given by '*L/R*' where '*L*' is the inductance and '*R*' is the resistance. In case of residue airflow, the term '*L*' can be identified with the inductance of the acoustic load, L_T. The term '1/*R*' can be identified with the dynamic glottal conductance which is a time varying function. Since the delay or latency occurs near the flow peak, the maximum glottal conductance of no load airflow, g_{0max}, can be used in place of '1/*R*'. The maximum value of glottal conductance for no load flow is given by $A_{gmax}/[0.5k\rho P_L]^{1/2}$. Hence we expect the delay, T_d, to be directly proportional to L_T and A_{gmax} and inversely proportional to square root of P_L. See also [27].

The residue airflow computed for various conditions is shown in Figs. 12.9 and 12.10. Residue airflow computed for varying inductance (4, 8, 12, 16 and 20 mH) with a fixed maximum glottal area of 0.2 cm^2 and P_L of 8 cm water pressure is shown in Fig. 12.9a. As expected, the delay increases with increasing value of inductance. A decrease in the maximum of residue airflow with increasing inductance may also be noted. The measured slope of delay Vs inductance is about 0.05 ms/mH up to about 15 mH and for higher values of inductance the slope is less (about 0.04 ms/mH), Fig. 12.10a. The discrete jumps in the measured delay arise since the delay is measured up to an accuracy of one sampling interval.

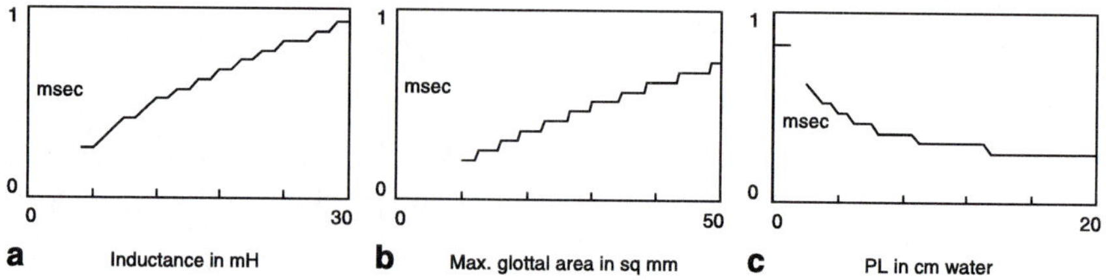

Fig. 12.10 Delay between the peak locations in residue airflow and no load airflow as a function of **a** load inductance, **b** maximum glottal area and **c** lung pressure

Residue airflow computed for varying maximum glottal area (0.05, 0.1, 0.15, 0.2 and 0.25 cm^2) with a fixed inductance of 8 mH and P_L of 8 cm water pressure is shown in Fig. 12.9b. It may be noted that the delay increases with increasing value of maximum glottal area. The measured slope of delay Vs maximum glottal area is about 0.02 ms/mm^2 over the range 10–50 mm^2, Fig. 12.10b.

Residue airflow computed for varying lung pressure (4, 6, 8, 10 and 12 cm water) with a fixed inductance of 8 mH and maximum glottal area of 0.2 cm^2 is shown in Fig. 12.9c. The delay is decreasing with increasing lung pressure. The delay decreases as the square root of lung pressure at a rate of 0.1 ms/cm change in pressure from 2 cm water pressure with the delay being 0.75 ms for 2 cm water pressure, Fig. 12.10c.

As predicted the delay is directly proportional to L_T and A_{gmax} but inversely proportional to the square root of lung pressure, P_L. The measured maximum delay is of the order of 1 ms for the entire range considered here. For a 10 ms pitch period with $OQ=0.5$, $SQ=1$ this implies a change of 1 ms in 2.5 ms or about 40% change relative to the interval *TP*. The maximum change in *SQ* for residue airflow is by a factor of about 2.3.

12.8.4.2 Maximum Residue Airflow

It is important to note that there is a decrease in the maximum residue airflow with increasing inductance of load. We explain this result with reference to Fig. 12.8. During the opening phase (interval *TP*), the derivative of residue airflow is positive causing a positive drop across the inductance and hence a *decreased* transglottal pressure. Hence the residue airflow is lower compared to no load airflow over the interval *TP*. At the instant, '*p*' when the glottal area and hence the no load airflow reaches the maximum value, the derivative of residue airflow is still positive due to the delay (Fig. 12.8b). Longer the delay, greater is the pressure drop. Hence at instant '*p*', the transglottal pressure is lower than the lung pressure resulting in a lower residue airflow compared to the maximum no load airflow. When the derivative of residue airflow reaches zero, the transglottal pressure is equal to lung pressure but by now the value of glottal area is less than the maximum. Hence the maximum residue airflow is less than the maximum no load airflow. Subsequently, the derivative of the residue airflow is increasingly negative resulting in an increased transglottal pressure which causes the residue airflow to be greater in magnitude compared to no load airflow despite the decreasing glottal area.

12.8.4.3 Maximum Negative Amplitude in the Derivative of Flow, EE

The maximum residue airflow is lower compared to the maximum no load airflow and hence one expects the maximum negative amplitude in the derivative also to be lower in the residue airflow, Fig. 12.8b. But, the slope at closure (*EE* in the derivative) of residue airflow is higher compared to that in no load flow. This is because the rate of increase of transglottal pressure for $t > TP$ is much steeper than the rate of decrease of glottal area and the counter acting the effect of lower peak airflow.

12.8.4.4 Modeling the Residue Airflow

Equations for residue airflow can be derived from the equations for the no load airflow (or short circuit current) by introducing a suitable time varying phase function. The no load airflow is merely a scale factor times the glottal area function. Let the maximum value of no load airflow be U_{0max}. Using Eq. 12.9a, the no load airflow can be written as

$$\begin{aligned} U_0(t) &= U_{0max}[0.5 - 0.5cos(\pi t/TP)] \quad for\ 0 \leq t \leq TP \\ &= U_{0max}\ cos[(\pi x/2TN),\ x = t - TP, \quad for\ 0 < x \leq (TE - TP) \\ &= 0 \quad for\ TE < t \leq T0 \end{aligned} \tag{12.9b}$$

Assume the delay between residue airflow and no load airflow T_d to be known. Also assume the maximum value of residue flow to be U_{rmax}. We show later that U_{rmax} can be estimated from U_{0max} and the delay T_d. Define the interval $TP' = TP + T_d$. The interval TP' is the opening phase of the residue airflow. We retain the same form of equations for the residue airflow as those used for no load airflow but introduce a time varying phase shift term. The residue flow $U_r(t)$ is written as

$$\begin{aligned} U_r(t) &= U_{rmax}\ [0.5 - 0.5cos\{(\pi t/TP) - \phi_P(t)\}] \quad for\ 0 \leq t \leq TP' \\ &= U_{rmax} cos\{(\pi x/2TN) - \phi_N(x)\}, \\ x &= (t - TP'), \quad for\ 0 < x \leq (TE - TP') \\ &= 0 \quad for\ TE\ < t \leq T0 \end{aligned} \tag{12.9c}$$

For the residue airflow, the frequency of the cosine segments during the opening (1/2*TP*) and closing (1/4*TN*) phases are the same as those in no load airflow but only the phase term is different. The argument of cosine term in $U_r(t)$ for the interval 0 to TP' has to reach a phase angle of π radians at $t = TP'$. The phase term $\phi_P(t)$ is assumed to vary linearly over the interval 0 to TP' such that its value is $\phi_P(0) = 0$ at $t = 0$ and $\phi_P(TP') = \pi T_d/TP$ at $t = TP'$. Hence

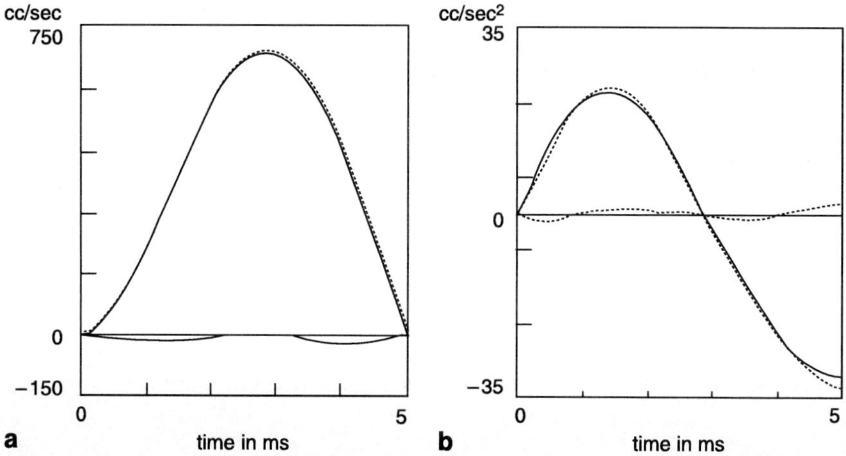

Fig. 12.11 **a** Computed residue airflow (*solid*) and predicted residue airflow (*dashed*) **b** Derivative of computed residue airflow (*solid*) and predicted residue airflow (*dashed*)

$$\phi_P(t) = [\pi T_d/TP]\,[t/TP'] \quad (12.9d)$$

Define $TN' = TE - TP' = TE - TP - T_d$. The argument of cosine term in $U_r(t)$ for the interval TP' to TE has to reach a phase angle of $\pi/2$ radians at $x = TN'$. The phase term $\phi_N(x)$ is assumed to vary linearly over the interval TP' to TE such that its value is $\phi_N(0) = 0$ at $x = 0$ and $\phi_N(TN') = (\pi/2)[(TN' - TN)/TN]$

$$\phi_N(x) = (\pi/2)\,[(TN' - TN)/TN]\,[(x/TN')] \quad (12.9e)$$

Since $TN' < TN$, the phase term $\phi_N(x)$ is negative for $t > TP'$.

A very good match is seen between the computed residue airflow and the predicted residue airflow using equations, (12.9c), Fig. 12.11. The results are for the case with $L_T = 6.14$ mH, peak glottal area of 20 mm^2 and $P_L = 8$ cm water. Similar results with good match has been observed over a wide range of the parameter values of load inductance, peak glottal area and lung pressure.

The maximum residue flow U_{rmax} can be estimated to a reasonable accuracy from the no load airflow maximum U_{0max} using the relation

$$U_{rmax} = U_{0max}\,[0.5 - 0.5cos\{\pi t/TP - \phi_P(TP)\}] \quad (12.9f)$$

The equations for phase, (12.9d) and (12.9e) are valid for $SQ = 1$. The transient component in computed residue airflow at $t = TP$ is of very small amplitude for $SQ = 1$. For the assumed glottal area model, for SQ away from unity, a transient is excited at $t = TP$ due to discontinuity in the derivative. For $SQ \neq 1$ a correction has to be applied to the phase term as given in Appendix 3. An alternate interpretation is that Eq. (12.9c) with phase terms as in Eqs. (12.9d) and (12.9e) models the residue component without transient terms, the Laplace inverse of the first term in partial fraction expansion of Eq. (12.7h). This in fact may be the required solution since the numerically computed residue airflow with inductance load is only a good approximation as it includes transients. Confirmation of this result requires further research.

In practice, one is interested in the inverse solution of finding the no load airflow (glottal area) given the residue airflow. By means of inverse filtering of vowel sounds, one estimates (but for a scale factor) the gross pulse shape of glottal flow which corresponds to residue airflow. By including the delay T_d as one of the parameters of the model and using Eqs. (12.9c)–(12.9f), the no load airflow can be estimated given the residue airflow. Further refinement can be achieved by an analysis-by-synthesis procedure.

The difference signal between residue airflow and no load airflow (Fig. 12.8) has a frequency component which is nearly twice of those used in Eq. (12.9a) for the segments during opening and closing phases of glottal area. The presence of a component with twice the frequency has been derived in [32]. It is possible to model residue airflow by adding a component of twice the frequency of the opening and closing phases of the glottal area. This model also gives good results for $SQ=1$ but requires more parameters and hence is not illustrated here.

12.8.5 Factors Influencing the Ripple Component of Airflow

For illustrations to be presented in this section only the kinetic resistance of glottal impedance has been considered. Consider the first formant load of vowel /a/ with $L_1=6.14$ mH, $F_1=660$ Hz and $B_1=32$ Hz. The ripple component is simply the difference between true glottal flow and computed residue airflow. The true glottal flow and residue airflow and their derivatives are shown in Fig. 12.12. The gross shape of the pulse for both cases is almost the same. The magnitude of ripple component is very small. The ripple component has zero amplitude both at glottal onset and closure.

The first formant load is determined by three parameters; inductance, first formant frequency and bandwidth. The effect of these factors individually on the ripple component is now presented. Each of these factors is varied while keeping the other two factors fixed. The difference between true glottal flow and residue airflow is determined for each of these cases. The results are shown in Fig. 12.13.

The magnitude of the ripple component increases with L_1, Fig. 12.13a, but not in the same proportion. L_1 increases successively by a factor of five. The magnitude of the ripple component increases at a much lower rate. The measured absolute maximum value of the ripple component for the three cases is 9, 14 and 21 cc/s respectively relative to a maximum flow of about 700 cc/s.

The magnitude of the ripple component decreases as the formant frequency increases, Fig. 12.13b. This general trend is to be expected since the magnitude of ripple component is related to magnitude of residue of $U_{sc}(s)$ at the formant frequency. The magnitude of this residue decreases with increasing formant frequency since

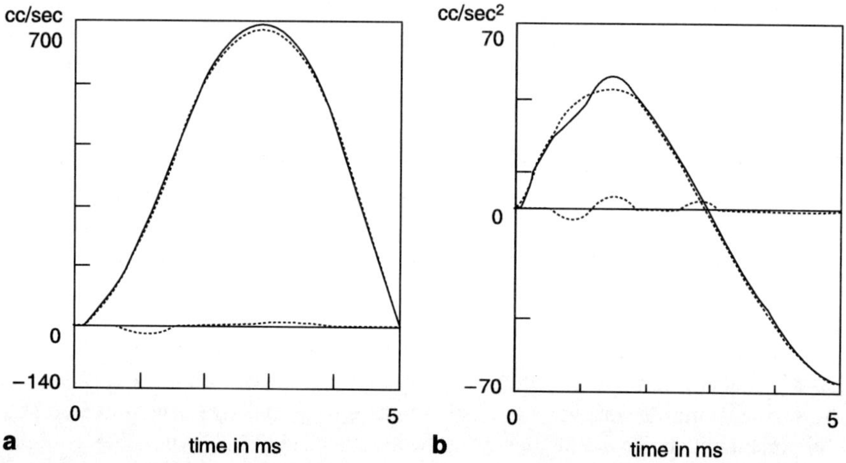

Fig. 12.12 **a** True glottal airflow (*solid*) and residue airflow (*dashed*) for first formant load. Difference signal is the ripple component. **b** Derivative of flow shown in (a). Difference signal is also shown

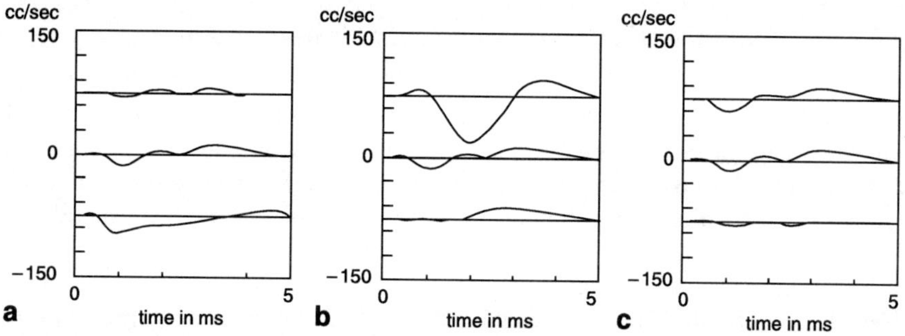

Fig. 12.13 Ripple component—the difference between true glottal and residue airflow for different conditions. *Top* to *bottom*: **a** For Inductance $L1=0.2$, 1 and 5 times 6.14 mH, $F1=660$ Hz, $B1=32$ Hz **b** For $F1=330$, 660 and 1320 Hz, $L1=6.14$ mH and $B1=32$ Hz. **c** For $B1=3.2$, 32 and 320 Hz, $F1=660$ Hz, $L1=6.14$ mH

$U_{sc}(s)$ has pole(s) at $s=0$. The measured absolute maximum of ripple component for the three cases is 56, 14 and 13 cc/s respectively relative to a maximum flow of about 700 cc/s.

The frequency of ringing seen in ripple component appears to be much lower than the formant frequency. This may arise because of the time varying hypothetical inductance which increases the inductance of the formant network by a maximum factor as high as thirty there by brings down the time varying formant frequency during the open phase. See Fig. 12.17.

The ripple component with changes in bandwidth is illustrated in Fig. 12.13c. There is not a significant difference in the magnitude of ripple component between

Table 12.1 Input impedance parameters of subglottal system (Sub) and vocal tract for different vowels

	*F*1	*B*1	*L*1	*F*2	*B*2	*L*2	*F*3	*B*3	*L*3
	Hz	Hz	mH	Hz	Hz	mH	Hz	Hz	mH
Sub	615	246	3.80	1,355	155	0.72	2,110	140	0.27
/a/	660	32	6.14	1,060	31	1.80	2,418	56	0.12
/o/	524	35	5.73	847	24	1.46	2,357	28	0.06
/u/	277	69	7.97	611	23	1.39	2,374	20	0.03
/i/	269	63	7.31	2,257	22	0.38	2,876	26	0.07
/e/	444	31	2.77	1,940	49	0.41	2,716	192	0.43

the two cases $B1=3.2$ and 32 Hz. The magnitude of ripple component is very small for $B1=320$ Hz. This is to be expected since the bandwidth of first formant coupled with glottal conductance is very much high. For this example, the maximum bandwidth due to the dynamic glottal conductance is about 720 Hz. See Fig. 12.17. The measured absolute maximum of the ripple component for the three cases is 15, 14, and 7 cc/s respectively relative to a maximum flow of about 700 cc/s.

12.8.6 Relative Importance of Vocal Tract and Subglottal Formants

The parameters of input impedance for subglottal system [46] and vocal tract [69] for different vowels are given in Table 12.1. Input impedance for vocal tract has been computed using x-ray data of Russian vowels [3]. A clarification needs to be made here. There are two entries for the vocal tract area at $x=0$ and $x=0.5$ cm in Table 2.33–1 of Ref. [3] and both these areas refer to the same (first) section and the duplicated entries must not be interpreted as if there are two sections [70].

The true glottal flow is computed for vowel /a/ by including successively more number of formants of vocal tract and subglottal system. The results are shown in Fig. 12.14 One can notice that the first formant of vocal tract and the first formant of subglottal system have the most significant effect. The effect of second and third formants of vocal tract and subglottal system is negligible. The value of load inductance decreases for successively higher formants.

Based on the results presented earlier it follows that the effect on the flow decreases with decreasing inductance. Also, we have seen that the magnitude of ripple component decreases with increasing formant frequency. Hence the relative effect of upper formants is much less significant. Similar trend is seen for other vowels but not illustrated here.

One possible simplification to the representation of input impedance of vocal tract for computation of glottal airflow is to include only the first formant network of vocal tract in series with the sum of inductances of the remaining formant networks. Similarly, one may include only the first formant network of subglottal system in series with the sum of inductances of the remaining formant networks.

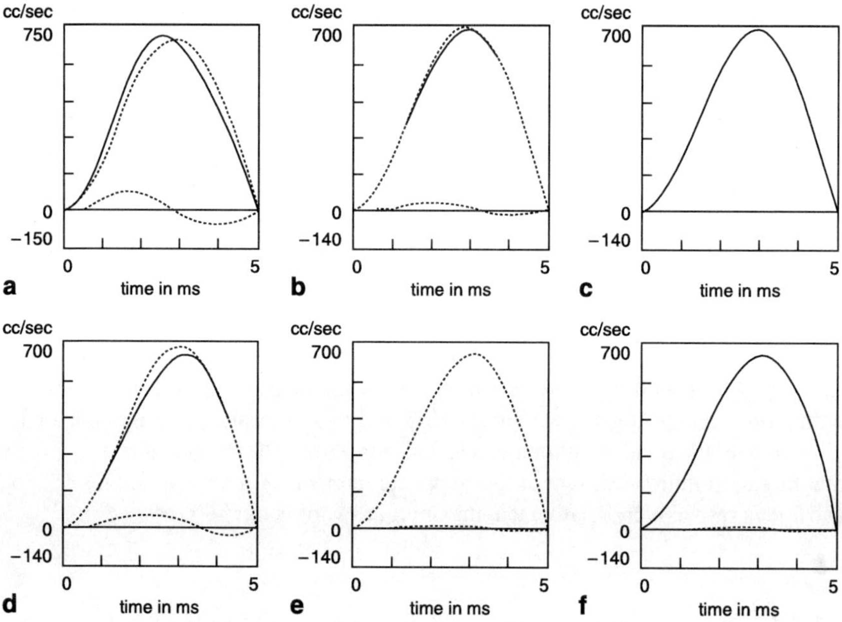

Fig. 12.14 **a** No load airflow (*solid*) and true glottal flow with F_{1V} load (*dash*) True glottal flow: **b** F_{1V} (*solid*) and F_{1V} and F_{2V} loads (*dashed*) **c** F_{1V}, F_{2V} (*solid*) and F_{1V}, F_{2V} and F_{3V} loads (*dashed*) **d** F_{1V}, F_{2V} and F_{3V} (*solid*) and F_{1s} load (*dashed*) **e** F_{1V}, F_{2V} and F_{3V} (*solid*) and F_{1s} and F_{2s} loads (*dashed*) **f** F_{1V}, F_{2V} and F_{3V} (*solid*) and F_{1s}, F_{2s} and F_{3s} loads (*dashed*). Difference signal is also shown for all cases

12.8.7 Validation of Glottal Flow Computation

Validation of true glottal airflow computation has been made using simultaneously measured glottal area and airflow [71]. Glottal area of an adult male speaker phonating vowel /a/ has been recorded using trans-illumination and photoglottography. Simultaneously, the vowel response at the lips has been recorded using pneumo-tachograph mask. The measured glottal area corresponds to the projected (minimum) area of the non-uniform glottis. Absolute scale factor is not available. Hence the measured glottal area signal is scaled such that the peak value corresponds to an arbitrarily chosen 20 mm^2. Using the measured glottal area, assuming a lung pressure of 10 cm water, the glottal airflow has been computed which is referred to as the predicted true glottal airflow assuming the input impedance of vocal tract of vowel /a/. The recorded vowel is inverse filtered to obtain the measured airflow. Two different criteria have been used in inverse filtering [72] (1) a flat closed phase criterion and (2) a smooth spectrum of the source criterion. The measured glottal area, the predicted true glottal airflow and the measured airflow are shown in Fig. 12.15 [36]. It may be seen that there is a good agreement between the measured glottal

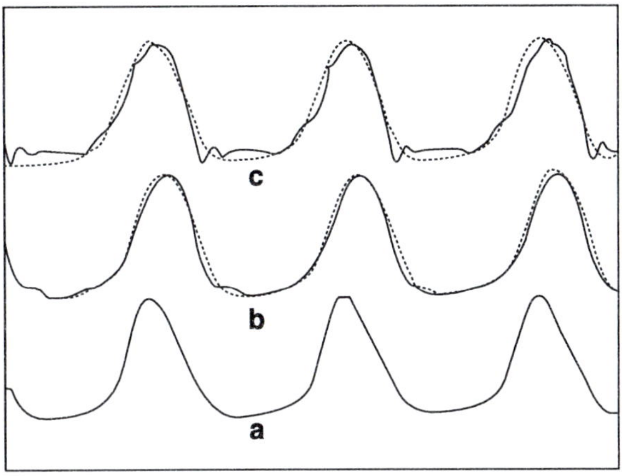

Fig. 12.15 **a** Measured glottal area suitably scaled. **b** Predicted true glottal airflow (*dashed*) and measured glottal airflow (*solid*) based on smooth source spectrum criterion **c** Predicted true glottal airflow (*dashed*) and measured glottal airflow (*solid*) based on flat closed phase criterion

airflow and the predicted glottal airflow thus validating the method of computing the true glottal flow. See also [73].

12.8.8 Interactive Formant Response

12.8.8.1 First Formant Response of a Synthetic Vowel

Initially residue airflow is computed using only the inductance load element. The derivative of residue flow of first glottal cycle is used as an excitation to the first formant network of vowel /a/ to obtain non-interactive vowel response. This is a hypothetical situation since residue airflow is computed using only the inductance load but vowel response is computed using the first formant network. The bandwidth of formant network corresponds to the closed glottis condition. The non-interactive response along with the excitation signal is shown in Fig. 12.16a. The initial part of the response over the interval $t=0$ to $t=TP$ resembles the excitation signal itself. The transient response of the first formant with significant ripples due to excitation at glottal onset are seen over the interval 0 to TO. The amplitude of transient component is large since the (closed glottis) bandwidth of the transfer function is low and in the case of residue airflow there is no ripple component in airflow to cause damping due to glottal coupling. Note the significant peak amplitude of non-interactive response near glottal onset of the second cycle, $t=T0$ (10 msec). In case of residue airflow there is no superposition component of airflow during the second glottal cycle since only an inductance load is used to compute the flow. There is no change in the formant frequency and bandwidth over the open phase interval of the second cycle. Even at $t=2T0$, the response has not completely decayed.

The true glottal flow is computed for the case of first formant input impedance of vowel /a/ as per Table 12.1. The derivative of true glottal airflow of first glottal

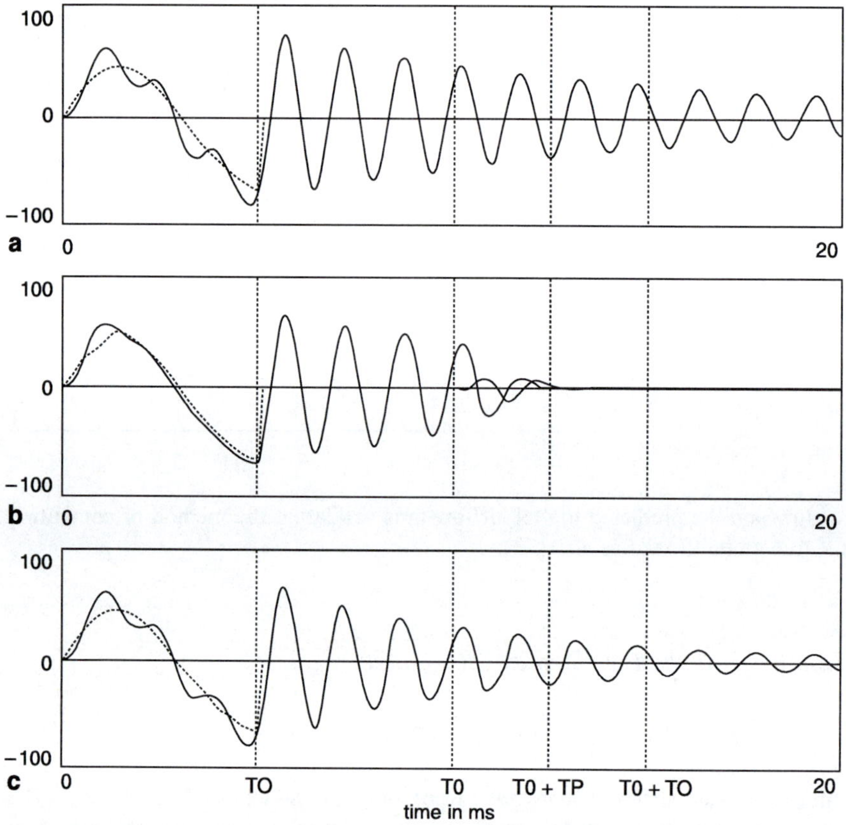

Fig. 12.16 **a** Response of first formant network for residue flow **b** Response of first formant network for the true glottal flow of the first cycle and superposition component of the second glottal cycle. **c** Response of first formant network with effective bandwidth for residue flow. The derivative of the airflow and superposition component are shown by *dashed curves*

cycle and superposition component of second glottal cycle are used as an excitation to the first formant network to obtain interactive vowel response. The interactive response along with the excitation signal is shown in Fig. 12.16b.

The initial part of the response over the interval $t=0$ to $t=TP$ resembles the excitation component. The transient response of first formant due to an excitation at glottal onset has a very low amplitude as seen over the interval 0 to TO. The ripple component in true glottal flow causes a significant damping due to glottal coupling. During closed phase interval extending from TO to $T0$, the formant ringing with an exponential envelope and constant F_1 can be noted. Note the significant peak amplitude near the glottal onset of the second cycle, $t=T0$. During the open phase interval of the second cycle $t=T0$ to $t=TO+T0$, there is a very rapid decrease in the amplitude of the interactive response. The superposition component in the open phase of second glottal cycle (shown by the dashed curve) has caused this rapid decay. Negligible energy is carried beyond one half of the open phase interval of the second cycle.

When residue airflow is used as input, the first formant bandwidth corresponds to the closed glottis condition whereas when true glottal flow is used as input, there is an increased damping due to glottal coupling during the open phase. The spectral level at F_1 for the non-interactive response is higher compared to the spectral level at F_1 for the interactive response. When residue airflow is used as excitation, an effective bandwidth greater than the closed phase bandwidth has to be used to ensure the same spectral level at F_1 for both the cases [39]. Alternately, energy in non-interactive and interactive responses over the interval $t = TO$ *to* $2T0$ may be equalized. Non-interactive response obtained using a bandwidth of 53 Hz is shown in Fig. 12.16c. The damping during the closed phase is relatively more rapid compared to the case shown in Fig. 12.16a.

12.8.8.2 Instantaneous Resonant Frequency and Instantaneous Bandwidth

According to psuedo Laplace transform theory, Sect. 6.3.1, during the open phase, first formant response has a non-exponential damping due to glottal conductance and a time varying resonant frequency due to hypothetical glottal inductance. In literature the non-exponential damping due to glottal conductance has been noted [26]. However, the effect of hypothetical inductance on time varying resonant frequency has not been reported.

The instantaneous resonant frequency and bandwidth may be estimated from the interactive response excluding the interval of the first glottal cycle. The interactive response $p(n)$ for a second order all-pole (one formant) filter can be written as

$$p(n) = -b_1 p(n-1) - b_2 p(n-2) \textit{ and}$$
$$p(n-1) = -b_1 p(n-2) - b_2 p(n-3) \qquad (12.8a)$$

For an interactive response, the coefficients $(b1, b2)$ are time varying but assumed to be constant over an interval of four samples. The coefficients $(b1, b2)$ can be estimated using the known samples of the interactive response $\{p(n-3), p(n-2), p(n-1), p(n)\}$ by solving the simultaneous Eq. 12.8a. This is called the direct method of solving for the all-pole coefficients. It is preferable to use double precision arithmetic for a better accuracy. Direct method can be used in case of a synthesized interactive response over the open phase interval of second cycle since it is due only to the superposition component. During this interval, there are no other components, such as excitation signal or the transients, in the interactive response. The estimated instantaneous resonant frequency, F_{1e}, and bandwidth, B_{1e}, are given by the inverse relations to Eq. 12.8a:

$$B_{1e} = (Fs/2\pi)\, ln\, (b_2);\; F_{1e} = (Fs/2\pi)\, cos^{-1}(0.5b_1/b_2{}^{-1/2}) \qquad (12.8b)$$

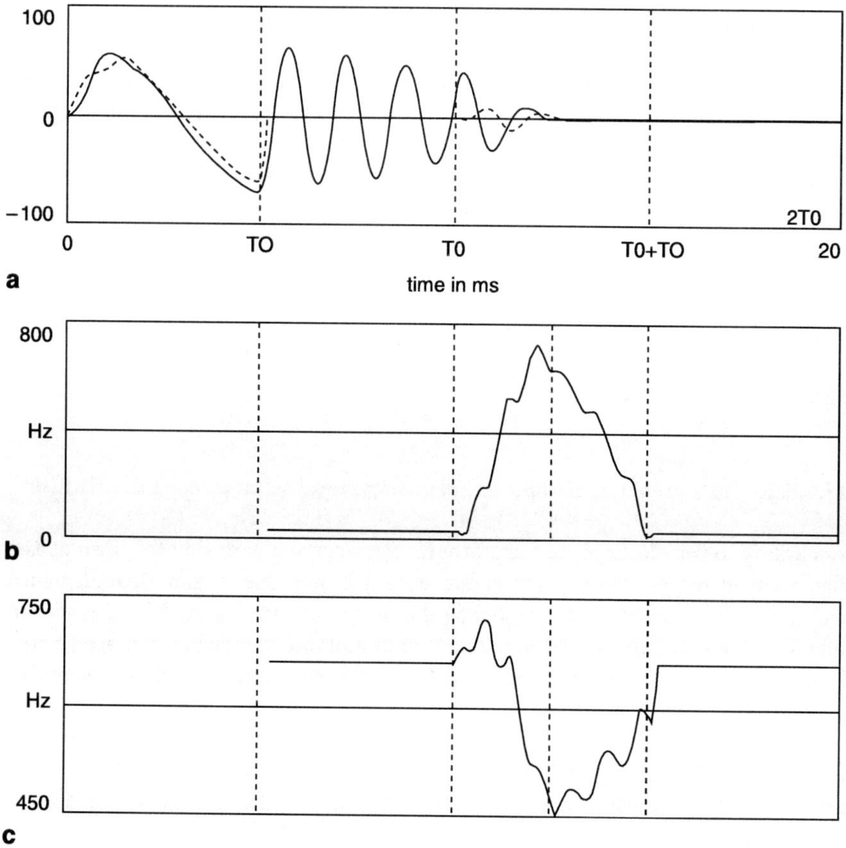

Fig. 12.17 **a** Interactive response of first formant network for the true glottal flow of the first cycle and superposition component of the second glottal cycle. **b** Estimated instantaneous bandwidth. **c** Estimated instantaneous resonant frequency

The estimated F_{1e} and B_{1e} for an interval immediately after TO and up to $2T0$ are shown in Fig. 12.17. As expected, the estimated bandwidth and resonant frequency are constant during the closed phase and correspond correctly to the values specified for synthesis.

During open phase of second glottal cycle, B_{1e} increases till about the middle of open phase interval and then decreases. This change in B_{1e} is brought about by the no load conductance, $g_0(t)$ which is directly proportional to $A_g(t)$. Hence the shape of instantaneous bandwidth is similar to that of $A_g(t)$. For this example, the maximum value of the instantaneous bandwidth is about 720 Hz.

During open phase of second glottal cycle, F_{1e} is initially greater than the closed glottis formant frequency (660 Hz) and reaches a value of about 720 Hz and then decreases to a minimum of 450 Hz and rises again to reach the closed glottis formant frequency by the end of open phase. This change in F_{1e} is brought about by the

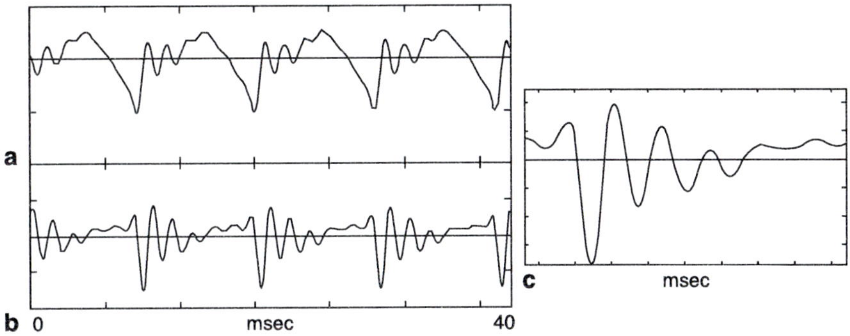

Fig. 12.18 **a** Selective inverse filter output of F1 of a natural vowel, **b** its pre-emphasized signal and **c** a zoomed segment of pre-emphasized signal

hypothetical inductance, $g_0'(t)$. The shape of the instantaneous resonant frequency approximates that of the derivative of $A_g(t)$.

The variation in F_{1e} and B_{1e} is not smooth and the expected shape of variation in B_{1e} differs from that of $A_g(t)$ and the expected shape of variation in F_{1e} differs from that of derivative of $A_g(t)$. Expected shape of variation is based on pesudo Lapalce transform theory which is only an approximation to the exact solution. Also, we have assumed the coefficients $(b1, b2)$ to be constant over four samples.

12.8.8.3 Formant Response of Natural Vowels

The theoretical prediction of non-exponential damping of a formant and dispersion in resonant frequency can also be seen in natural vowels. A standard inverse filter uses a cascade of anti-formant networks for all the formants over the frequency range of the signal [74]. In selective inverse filtering [75], the anti-formant network of the selected formant is left out in the cascade of anti-formant networks. It is equivalent to using the output of the standard inverse filter as an excitation signal to synthesize the response of the selected formant.

An example of selective inverse filter output of a natural vowel for first formant is shown in Fig. 12.18a. The presence of gross pulse shape in the response makes it difficult to notice the non-exponential damping. Use of pre-emphasis suppresses the gross pulse shape and enhances the formant ringing, Fig. 12.18b. The sudden decrease of the amplitude and changes in zero-crossing interval of the formant ringing over the open phase can be noted in the zoomed segment, Fig. 12.18c.

Unlike a synthesized vowel, superposition component of airflow cannot be isolated for a natural vowel. It is difficult to estimate the instantaneous formant frequency and bandwidth because of the presence of gross pulse shape component and transients due to excitation at glottal onset and/or glottal peak. Generally covariance method of LP is used for speech analysis over closed phase and for short intervals [76–78]. But, covariance method of linear prediction does not give reliable results during the open phase. The author tested the covariance method of LP on pre-emphasized non-interactive response of the first formant. In case of covariance method of LP no windowing is to be used. The author has used an analysis interval of 1 ms and analysis order of 2 as well as 3. The author finds that during the glottal open phase, due to the presence of other signal components (gross pulse shape of excitation signal, transients excited at onset and/or peak), the estimated formant frequency and bandwidth show fluctuations though a non-interactive (time invariant) response has been analyzed. Hence covariance method of linear prediction is not suitable to estimate the instantaneous formant frequency and bandwidth of natural vowels. Despite this fact, some researchers [79, 80] have used the covariance method of LP over the open phase and doubted the presence of dispersion effect. In this work, presence of dispersion effect has been derived theoretically and demonstrated practically for an interactive response, Fig. 12.17c. Also, it may be noted that pseudo Laplace transform theory has not been used in deriving the presence of hypothetical glottal inductance.

12.9 Conclusion

12.9.1 Summary of Work Presented

Studies on voice source fall broadly into two groups:

(a) Theory of voice production. This has two parts (1) Mechanical modeling of vocal fold vibrations. A good review on mechanical modeling of vocal fold vibrations as can be seen in [18]. (2) Aerodynamic and acoustic theory of voice production covered in this work.
(b) Estimation of voice source dynamics from natural speech samples. Important aspects of part (b) will be covered briefly in Sect. 12.9.3.

In this chapter we have discussed the various factors that influence the true glottal flow and its components. The relative role of static glottal impedance elements; kinetic resistance, viscous resistance and glottal inductance has been discussed including the effect of glottal geometry. It has been noted that the viscous resistance is significant only for very low transglottal pressure of the order of 1 cm water and that the viscous resistance is much lower than usually reported for convergent and divergent glottal shapes. Its inclusion produces unrealistic glottal flow closure. Hence it may not be present or its effect may be negligible. The effect of glottal inductance has been shown to be very small. Input impedance calculation for distributed TL and formant network models has been presented. A digital filter has been proposed to model the input impedance. Equations to the solution to true glottal flow in the presence of source-filter interaction have been derived. It is shown that the dynamic glottal impedance has both glottal conductance and hypothetical glottal inductance. Using pseudo Laplace transform, components such as residue, ripple and superposition have been identified in the true glottal flow. Residue airflow has been modeled using the same form equations as used for no load airflow but for an additional time varying phase term. Effect of various factors in determining these components and the role of these components in determining the vowel response have been illustrated. The importance of the derivative of glottal flow in determining oral pressure and vowel response has been noted. Instantaneous bandwidth and instantaneous resonant frequency of the formant response over the open phase interval has been illustrated for a synthetic interactive response. Selective inverse filter output of a natural vowel sample has been illustrated. Difficulties in the estimation of instantaneous formant frequency and bandwidth of a natural vowel have been highlighted.

Due to limitation of space some issues related to source-filter interaction have not been covered in this report but are available in the literature; Spectral characteristics of true glottal flow and its components, spectral characteristics of interactive response [32, 33], perceptual importance of interactive response [38, 39, 81], use of modulated jet noise generated at the glottis for synthesis [82].

12.9.2 Future Research Directions on the Theoretical Aspects

There is a need to conduct basic research to further our understanding on voice source. Models of glottal impedance can be improved by experiments on static models of glottis with *AC* flow superposed on a mean flow as reported in [41]. A rigorous comparison of input impedance and transfer function of a distributed parameter TL model and formant network models *including phase response* has to be made in order to understand the approximations involved. Possibility of analytically solving nonlinear time varying glottal flow equation instead of pseudo Laplace transform has to be explored. Decomposition instead of de-convolution to obtain voice source may be attempted. Detailed study of selective inverse filtered formant responses has to be undertaken. Estimation of instantaneous first formant frequency and bandwidth of natural vowels during open phase and its utility in estimating glottal conductance and inversion from the glottal flow to glottal area have to be explored. Source-tract interaction for a breathy voice has been reported assuming a *DC* shift in the glottal area function [83]. But, it has been shown that the return phase or the leakage interval changes dynamically for CV and VC transitions [62]. A dynamic modeling of glottal area for voice initiation, CV and VC transitions, needs to be made. Some measurements have been reported on oral and glottal flow for VCV utterances [84]. But a rigorous theoretical analysis as well as numerical computation of glottal flow needs to be undertaken in the presence of a dynamic abduction component in the glottal area and for cases where the mean transglottal pressure varies within a glottal cycle.

12.9.3 Voice Source Dynamics

Various topics come under the study of voice source dynamics. To mention a few: (1) Identification of the closed glottis interval (2) Speech analysis to estimate formant frequencies and bandwidths during the closed glottis interval (3) Effect of glottal leakage on the estimated formant data (4) Techniques of inverse filtering and their implementation for steady vowels and connected speech including semi-automatic methods (5) Modeling of voice source (6) Estimation of model parameters or the study of voice source dynamics (7) Relative merits of different voice source models (8) Applications using the dynamics of estimated voice source parameters. Etc. The author has made several research contributions related to the above topics. But, a critical review of the above topics is beyond the scope of the present report. However, a couple of points will be noted.

Importance of the Derivative Flow: We have seen in this report, the importance of the derivative of true glottal flow in determining the oral and subglottal pressure, residue flow and the vowel response. The author has referred to the derivative of true glottal flow as voice source. The author has shown that many significant features are highlighted in the derivate signal, the voice source [62]. It is important to note that the short-time spectrum of vowel response reaching a listener contains the spectral component of the *derivative* of the source. Hence the author has emphasized modeling of voice source, the derivative of glottal flow [62]. Prior to this work [62], research on voice reported temporal and spectral properties of volume velocity airflow [4, 74, 85]. Subsequent research has been restricted to study of voice source or the derivative of airflow. Some samples of such works are [59, 64, 86].

Importance of Spectral Matching: For speech signal recorded with poor low frequency response or poor phase response, matching voice source parameters in the spectral domain is a better choice [87, 88]. Voice source pulse is a finite duration signal and hence theoretically speaking not a band limited signal. This implies that there is an inevitable aliasing in the spectrum of voice source. Voice source signal of LF model seems to possess a greater aliasing effect relative to other models [64]. Simultaneous matching between the measured and modeled voice source both in the time and spectral domains leads to a selection of an appropriate model to capture the voice source dynamics. Separation of estimated glottal flow obtained using inverse filtering to residue and ripple components has been reported [59]. But one has to be cautious not to label any mismatch between gross shape of the model and the measured voice source pulse as a ripple component. Simultaneous temporal and spectral matching has to be attempted.

Importance of Analytic Signal Modeling: It has been shown that an analytic signal model captures the phase characteristics of speech signal to a better accuracy [78]. Hence an analytic signal model for voice source pulse shape is a better choice for natural speech samples [87]. It can resolve the temporal characteristics into voice source and phase effect. Recently there has been a growing number of studies on matching voice source models to the inverse filter output (derivative of glottal flow) signal, examples, [59, 89]. In these studies the effect of phase response on the inverse filter output seems to have been overlooked.

12.9.4 Forensic Speaker Identification

Perceived voice quality is determined not only by the 'voice source' but also by vocal and nasal tract characteristics [90]. Thus a number of studies on speaker recognition use phonetic features [91–93]. Conversely, it may be noted that 'voice source' alone carries significant phonetic intelligibility [94]. We have already remarked in the introduction that 'voice source' in a restricted sense implies the glottal pulse shape. But, in a broad sense, 'voice source' includes F0-level, intonation as well as the supra-segmental features such as duration of segments, intensity level etc. Thus there have been many studies on speaker identification using not only pulse shape features but also other features as well. Example, [95, 96]. We mentioned in the introduction that there are two broad streams of research on voice source; those inspired by physiological model of speech production and those using linear prediction residual. The author has related the physiological model and the LP residual in his thesis [97]. An interpretation of LP residual via-a-vis glottal pulse shape and effect of phase can be seen in [78]. Interestingly, there are a number of papers on speaker recognition based on LP residual, for example [98].

There are notably two major challenges in forensic speaker recognition. The signal recorded is often over a noisy band-limited channel. Secondly, a speaker can voluntarily change his/her voice. To address the first problem, synthesized speech or vowel can be generated using model pulse shape and played over a channel that is similar to the one used for recording the speaker's voice. The synthesized speech or vowel may be recorded and analyzed. The distortions introduced in the model pulse shape may thus be noted. This may guide one in deriving the model parameters to a better accuracy. In this context, simultaneous spectral and time domain matching has to be used [87].

Although a speaker can voluntarily change his/her voice, certain dynamic habitual patterns are relatively more difficult to change. An analogy is to 'style of walking or gait'. When one attempts to change one's gait, it would look so unnatural or slow. The relative dynamics in source parameters may thus prove useful in recognizing a speaker or in identifying unnaturalness in rendition.

Research on voice source is both theoretically challenging and intellectually stimulating. Voice source research, one day, may lead us to the possibility of reproducing high quality natural sounding synthetic speech of a specific speaker's voice and in capturing that elusive unique attribute called 'voice' associated with a speaker for speaker recognition.

12.9.5 Dedication

The preparation of this article nearly coincides with two events; the golden jubilee of the publication of the famous '*Acoustic theory of speech production*' by Profes-

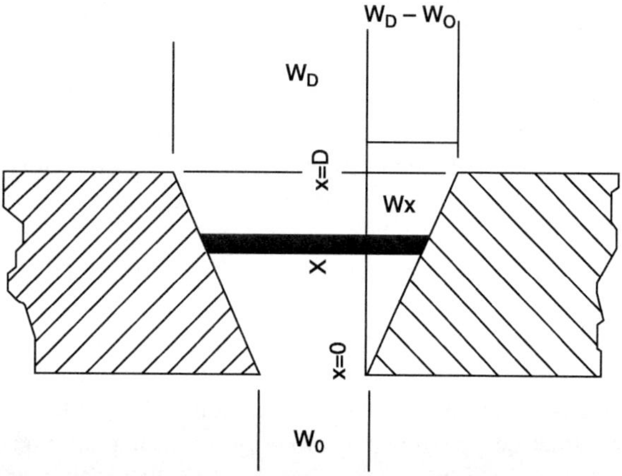

Fig. 12.19 A model of divergent glottis to compute viscous resistance

sor Gunnar Fant and the first death anniversary of Prof. Fant. The author would like to dedicate this chapter to the memory Prof. Fant. The author cherishes the memory of exciting research collaboration he had with Prof. Fant during 1981–1985 at the Department of Speech Communication and Music Acoustics, Royal institute of Technology, Stockholm, Sweden.

Appendix 1: Viscous Resistance of a Non-Uniform Glottis

Consider a glottis with a geometry as shown in Fig. 12.19.

The area of glottis at depth x is given by $A_g(x)=l\,W(x)$. The viscous resistance for a small section of width $W(x)$ at x of depth dx is given by the formula

$$R_v(x) = [12\mu dx l^2 / A_g^3(x)] = [12\mu dx / l W^3(x)] \tag{12.9}$$

The width is varying with the depth given by the equation

$$W(x) = W_0 + [W_D - W_0][x/D] \tag{12.10}$$

Integrating over the range $x=0$ to $x=D$, the total viscous resistance is given by

$$R_v = [12\mu D/l][W_D + W_0]/[2W_D^2 W_0^2] \tag{12.11}$$

Let $W_D=\beta W_0$, then

$$R_v = [12\mu D/l][1+\beta]/[2\beta^2 W_0^3] = [k_a][12\mu D l^2/A_0^3] \tag{12.12}$$

where the minimum area of the glottis $A_0=lW_0$, and $k_a=[1+\beta]/(2\beta^2)$. For a ratio of $W_D=4W_0$, $k_a=(5/32)$.

Appendix 2: Computation of True Glottal Flow for One Formant Network

Consider discrete signal representation with sampling interval=ΔT and sampling frequency $Fs=1/\Delta T$. Using the trapezoidal rule for the integration, the first term on RHS of Eq. (12.5a) can be written as

$$\int P_{o1}(t)dt = S_{o1}(n-1) + 0.5[P_{o1}(n) + P_{o1}(n)]\Delta T \tag{12.13}$$

where $S_{o1}(n-1) = \sum P_{o1}(j) \quad for \quad j = 0 \; to \; n-1.$ (12.14)

Substituting the above in Eq. 12.5a and multiplying throughout by L_1 we get

$$\begin{aligned} L_1 u_g(n) = {} & S_{o1}(n-1) + 0.5[P_{o1}(n) + P_{o1}(n-1)]\Delta T \\ & + [L_1 C_1 Fs][P_{o1}(n) - P_{o1}(n-1)] + [L_1/R_1]P_{o1}(n) \end{aligned} \tag{12.15}$$

Dividing the above equation throughout by $1/L_1C_1$, substituting $\omega_1{}^2=1/(L_1C_1)$ and $2\alpha_1=1/(R_1C_1)$, we get

$$\begin{aligned} \omega_1^2 L_1 u_g(n) = {} & \omega_1^2 S_{o1}(n-1) + P_{o1}(n)[0.5\omega_1^2\Delta T + Fs + 2\alpha_1] \\ & + P_{o1}(n-1)[0.5\omega_1^2\Delta T - Fs] \end{aligned} \tag{12.16}$$

$$P_{o1}(n) = D_{01}u_g(n) - D_{11}P_{o1}(n-1) - D_{21}S_{o1}(n-1) \tag{12.17}$$

Where

$$D_{01} = \omega_1^2 L_1/[0.5\omega_1^2\Delta T + Fs + 2a_1],$$
$$D_{11} = [0.5\omega_1^2\Delta T - Fs]/[0.5\omega_1^2\Delta T + Fs + 2a_1] \text{ and}$$
$$D_{21} = \omega_1^2/[0.5\omega_1^2\Delta T + Fs + 2\alpha_1]$$

Appendix 3: Correction to Phase Term When SQ Is Not Equal to One

The following corrections have been arrived at by trial and error in order to obtain the best match between the predicted residue airflow and the computed residue airflow. It is an empirical result for which a theoretical justification is wanting.

$$\phi_P(t) = PI * ((TP' - TP)/TP) * [(t/TP')]^{SQ1}, \; SQ1 = SQ^{-0.5} \qquad (12.18)$$

$$\phi_N(t) = PI * ((TN' - TN)/TN) * [(t/TN')]^{SQ} \qquad (12.19)$$

References

1. Dunn HK (1950) The calculation of vowel resonances and an electrical vocal tract. J Acoust Soc Am 22(6):740–753 (Reproduced in speech analysis, Schafer RW, Markel JD (eds), IEEE Press)
2. Stevens KN, Kaowski S, Fant G (1953) An electrical analog of the vocal tract. J Acoust Soc Am 25(4):734–742
3. Fant G (1960) Acoustic theory of speech production. Mouton, Hague
4. Flanagan JL (1965) Speech analysis, synthesis and perception, 1st edn. Springer, New York (2nd edn 1972)
5. Sondhi MM (1974) Model for wave propagation in a lossy vocal tract. J Acoust Soc Am 55(5):1070–1075
6. Van Den Berg Jw, Zantema JT, Doornebal P (1957) On the air response and the Bernoulli effect of the human larynx. J Acoust Soc Am 29:626–631
7. Flanagan JL (1959) Estimates of intraglottal pressure during phonation. J Speech Hear Res 2:168–172
8. Van Den Berg Jw, Zantema JT, Doornebal P (1959) Myoelastc-aerodynamic theory of voice production. J Speech Hear Res 1:227–243
9. Flanagan JL, Landgraf LL(1968) Self-oscillating source for vocal tract synthesizers. IEEE Trans Audio Electroacoust AU-16:57–64
10. Flanagan JL, Cherry L (1969) Excitation of vocal-tract synthesizers. J Acoust Soc Am 45(3):764–769
11. Ishizaka K, Flanagan JL (1972) Synthesis of voiced sounds from the two mass model of the vocal cords. BSTJ 51:1233–1267
12. Ishizaka K, Matsudiara M (1972) Fluid mechanical consideration of vocal cord vibrations. SCRL Monograph No. 8, Santa Barbara, California
13. Ishizaka K, Matsudiara M (1972) Theory of vocal cord vibrations. Rep Univ Electro Commn 23:107–136
14. Flanagan JL, Ishizaka K, Shipley KL (1975) Synthesis of speech from a dynamic model of the vocal cords and vocal tract. BSTJ 54:485–506
15. Broad DJ (1979) The new theories of vocal fold vibration. Speech and language: advances in basic research and practice, vol 2. Academic, New York, pp 203–256
16. Titze IR (1980) Comments on the myoelastic-aerodynamic theory of phonation. J Speech Hear Res 23:495–510
17. McGowan R (1991) Phonation from a continuum mechanics points of view. In: Gauffin J, Hammarberg B (eds) Vocal fold physiology: acoustics, perceptual and physiological aspects of voice mechanisms. Singular publishing, San Diego, pp 65–72

18. Titze IR (1994) Principles of voice production. Prentice Hall, Englewood Cliff
19. Sondhi MM, Schroeter J (1987) A hybrid time-frequency domain articulatory speech synthesizer. IEEE Trans ASSP 35(7):955–967
20. Schroeter J, Sondhi MM (1992) Speech coding based on physiological models of speech production. In: Furui S, Sondhi MM (eds) Advances in speech signal processing. Marcel Dekker, New York, pp 231–267
21. Hiki S, Koike Y, Takahashi H (1970) Simultaneous measurement of subglottal and supraglottal pressure variation. 79th meeting, ASA, paper DD4, April 1970
22. Kitzing P, Lofquist A (1975) Subglottal and oral air pressures during phonation—preliminary investigation using a miniature transducer system. Med Bio Eng 13:644–648 (Sept 1975)
23. Cranen B, Boves L (1985) Pressure measurement during speech production. J Acoust Soc Am 77:1543–1551
24. Mrayati M, Guerin B, Boe LJ (1976) Etude de l'impedeance d'entrée du conduit vocal—couplage source-conduit vocal. Acoustica 35:330–340
25. Guerin B, Mrayati M, Carre R (1976) A voice source taking into account of coupling with the supraglottal cavities. ICASSP 1:47–50
26. Fant G, Liljencrants J (1979) Perception of vowels and truncated intraperiod decay envelopes. STL-QPSR 1:79–84
27. Rothenberg M (1981) An interactive model for the voice source. STL-QPSR 4:1–7
28. Rothenberg M (1983) Acoustic interaction between the glottal source and the vocal tract. In: Bless DM, Abbs JH (eds) Proc. Conf. Vocal Fold Physiology, Kurume, Japan, 1980, pp 305–323, College Hill, San Diego,
29. Al-Ansari A, Guerin B, Degryse D (1981) Subglottal impedance effects on the vocal source signal. IV FASE Symp April:21–24 (Venezia)
30. Fant G (1981) The source-filter concept in voice production. STL-QPSR 1:21–37
31. Ananthapadmanabha TV, Fant G (1981) Glottal flow calculations. Paper JJ3, 102nd ASA meeting, Miami Beach, Florida
32. Ananthapadmanabha TV, Fant G (1982) Calculation of true glottal flow and its components. STL-QPSR 1:1–30 (Also Speech Commun 1:167–184)
33. Fant G, Ananthapadmanabha TV (1982) Truncation and superposition. STL-QPSR 2–3:1–17
34. Scherer RW (1981) Laryngeal fluid mechanics: steady flow considerations using static models. PhD thesis, University of Iowa
35. Gauffin J, Binh N, Ananthapadmanabha TV, Fant G (1981) Glottal geometry of volume-velocity waveform. Proc. research conf. voice physiology, Madison, WI, USA, May 31,Jun 4, 1981
36. Ananthapadmanabha TV, Gauffin J (1983) Some results on the aerodynamic and acoustic factors in phonation. STL-QPSR 1, 1983(and also in Titze IR, Scherer R (eds) Proc. vocal fold physiology conf., Denver Center for the performing Arts, Denver, Colarado, 1983)
37. Scherer R, Titze IR (1983) Pressure-flow relationships in a model of the laryngeal airway with diverging glottis. In: Bless DM, Abbs JH (eds) Vocal fold physiology: contemporary research and clinical issues. College-Hill Press, San Diego
38. Ananthapadmanabha TV, Nord L, Fant G (1982) Perceptual discriminability of nonexponential/exponential damping of the first formant of vowel sounds. In: Carlson R, Granstrom B (eds) The representation of speech in the peripheral auditory system. North-Holland, Amsterdam, pp 217–222
39. Nord L, Ananthapadmanabha TV, Fant G (1984) Signal analysis and perceptual tests of vowel responses with an interactive source filter model. STL-QPSR 25(2–3):25–52
40. Hirano M (1981) Clinical examination of voice. Springer, New York
41. Ingard U, Ising H (1967) Acoustic nonlinearity of an orifice. J Acoust Soc Am 42(1):6–17
42. Laine U, Karjalainen M (1986) Measurements on the effects of glottal opening and flow on the glottal impedance. ICASSP , paper 31.6.1, pp 1621–1625
43. Badin P, Bailly G, Raybaudi M, Segebarth C (1998) A three-dimensional linear acoustic articulatory model based on MRI data. Proc 5th ICSLP, vol 2, pp 417–420
44. Coker CH (1976) A model of articulatory dynamics and control. Proc IEEE 64(4):452–460

45. Mermelstein P (1973) Articulatory model for the study of speech production. J Acoust Soc Am 53(4):1070–1082
46. Ishizaka K, Masudiara M, Kaneko T (1976) Input acoustic impedance measurement of the subglottal system. J Acoust Soc Am 60:910–917
47. Koike V, Hirano M (1973) Glottal area time function and subglottal pressure variation. J Acoust Soc Am 54:1618–1672
48. Van Valkenberg ME (1976) Network analysis. Prentice-Hall, New Delhi
49. Wakita H, Fant G (1978) Toward a better vocal tract model. STL-QPSR 1:9–29
50. Sondhi MM, Sinder DJ (2004) Articulatory modeling: a role in concatenative text to speech synthesis. In: Narayanan S, Alwan A (eds) Text to Speech Synthesis: New Paradigms and Advances. Pearson education, India (205, Chapter 4), pp 85–109
51. Badin P, Fant G (1984) Notes on vocal tract computation. STL-QPSR 2–3:53–107
52. Kelly JL, Lochbaum CC (1962) Speech synthesis. In ICA 4, paper G42 (Also in Speech synthesis, Rabiner LR, Flanagan JL (eds), Dowden Wiley, 1973)
53. Maeda S (1982) A digital simulation method of the vocal tract system. Speech Commun 1:199–229
54. Strube HW (1982) Time varying wave digital filters and vocal tract models. ICASSP 1982:923–926
55. Laine U (1982) Modeling lip radiation in the z-domain. ICASSP 1982:1992–1995
56. Liljencrants J (1985) Speech synthesis with a reflection-type line analog. DSc dissertation, Department of speech communication and music acoustics, RIT, Stockholm
57. Gold B, Rabiner LR (1968) Analysis of digital and analog formant synthesizers. RLE Tech rep.465, June 1968
58. Laine U (1988) Higher pole correction in vocal tract models and terminal analogs. Speech Commun 7:21–40
59. Plumpe MD, Quatieri TF, Reynolds DA (1999) Modeling of the glottal flow derivative waveform with application to speaker identification. IEEE Trans Acoust Speech Signal Process 7(5):569–585
60. Fant G Private communication
61. Rosenberg AE (1971) Effect of glottal pulse shape on the quality of natural vowel. J Acoust Soc Am 49:583–590
62. Ananthapadmanabha TV (1984) Acoustic analysis of voice source dynamics. STL-QPSR 2–3:1–24
63. Ananthapadmanabha TV (1993) Working papers, MIT speech group report, vol ix
64. Fant G, Liljencrants J, Lin Q (1985) A four parameter model of glottal flow. STL-QPSR 4:1–13
65. Matasuek MR, Batalev VS (1980) A new approach to the determination of glottal waveform. IEEE Trans ASSP 28:616–622
66. Rice DL (1974) Articulatory tracking of the acoustic speech signal. Proc. speech commn seminar, Stockholm pp 21–26
67. Kitzing P, Lofquist A (1979) Evaluation of voice therapy by means of photoglottography. Folia Phoniatrica 31:103–109
68. Titze IR (1984) Parameterization of the glottal area, glottal flow and vocal fold contact area. J Acoust Soc Am 75(2):57–580
69. Laine U Input impedance data were computed and provided by Laine.
70. Fant G Private communication clarifying the x-ray data
71. Lofquist A, Ananthapadmanabha TV Unpublished. Simultaneous Measurement of glottal area using trans-illumination photo-glottographic equipment and airflow using a mask
72. Liljencrants J INA program for inverse filtering. KTH, Stockholm
73. Cranen B, Boves L (1983) Pressure 77
74. Miller RL (1959) Nature of vocal cord wave. J Acoust Soc Am 31:667–677
75. Holmes JN (1962) An investigation of the volume velocity waveform at the larynx during speech by means of an inverse filter. Proc IV Intnl Cong on Ac, pp 1–4, August 1962

76. Atal BS, Hanauer SL (1971) Speech analysis and synthesis by linear prediction of the speech wave. J Acoust Soc Am 50:637–655 (Aug 1971)
77. Makhoul J (1975) Linear prediction: a tutorial review. Proc. IEEE, vol 63, April 1975, pp 561–580 (Also in Schafer RW, Markel JD (eds) Speech Analysis, IEEE Press)
78. Ananthapadmanabha TV, Yegnanarayana B (1979) Epoch extraction from linear prediction residual. IEEE Trans ASSP 27:309–319
79. Cheng Y, Guerin B (1987) A study of the source-filter interactive concept and its application to male and female speech synthesis. Bulletin du Lab de la Commun. Parlee, No. 1A, pp 29–66, INPG-ENSERG, Grenoble, France
80. Lin Q (1990) Speech production theory and articulatory speech synthesis. PhD thesis, KTH
81. Childers DG, Lee CK (1991) Voice quality factors: analysis, synthesis and perception. J Acoust Soc Am 90(5):2394–2410
82. Ananthapadmanabha TV, Prasad MG (1989) A note on jet noise component in phonation. Tech Mem, AT&T Bell Labs
83. Rothenberg M (1983) Source-tract interaction in breathy voice. In: Titze IR, Scherer RC (eds) Vocal fold physiology—biomechanics, acoustics and phonatory control. Denver center for the performing arts, Colarado
84. Lofquist A, Koenig L, McGowan RS (1995) Vocal tract aerodynamics in/aCa/utterances: measurements. Speech Commun 16:49–66
85. Sundberg J, Gauffin J (1981) Waveform and spectrum of glottal voice source. STL-QPSR 2–3
86. Fujisaki H, abd Ljungquist M (1986) Proposal and evaluation of models for the glottal source waveform. ICASSP 1986, paper 31.2.1, pp 1605–1608
87. Ananthapadmanabha TV(1991) Spectral parameters of a voice source model. 122nd meeting ASA, 1991. (Also in Speech technology for man-machine interaction Rao PVS, Kalia BB (eds) Tata McGraw Hill, New Delhi, 1993)
88. Ananthapadmanabha TV (1995) See discussion section on "waveforms and spectrum envelopes". In: Fujimura O, Hirano M (eds) Vocal fold physiology: voice quality control. Singular Publishing, San Diego, pp 347–353 (Fujimura O, Chap 22)
89. Jankowski CR (1996) Fine structure features for speaker identification. PhD thesis, MIT, USA
90. Laver J (1980) Phonetic description of voice quality. Cambridge University Press, UK
91. Nolan JF (1983) The phonetic bases of speaker recognition. Cambridge University Press, Cambridge
92. Koenig BE (1986) Spectrographic voice identification: a forensic survey. J Acoust Soc Am 79:2088–2090
93. Amino,K, Arai T (2009) Speaker-dependent characteristics of the nasals. Forensic Sci Int 185(1–3):21–28 (Mar 2009)
94. Ananthapadmanabha TV (1982) Intelligibility carried by speech source functions. STL-QPSR 4:49–64
95. Quatieri TF, Jankowski CR, Reynolds DA (1994) Energy onset times for speaker identification. IEEE Sig Process Lett 1(11):160–162 (Nov 1994)
96. Jankowski CR, Quatieri TF, Reynolds DA (1996) Fine structure features for speaker identification. IEEE international Conference on acoustics, speech and signal processing, pp II-689–692, May 1996
97. Ananthapadmanabha TV(1978) Epoch extraction of voice speech. PhD thesis, Indian institute of science, Bangalore
98. Mahadeva Prasanna SR, Gupta CS, Yegnanarayana B (2006) Extraction of speaker-specific excitation information from linear prediction residual of speech. Speech Commun 48(10):1243–1261 (Oct 2006)

Chapter 13
Prosodic Features for Speaker Recognition

Leena Mary

Abstract In this chapter the effectiveness of syllable-based prosodic features for speaker recognition is discussed. The term prosody represents a collection of characteristics such as intonation, stress and timing, primarily expressed using variations in pitch, energy and duration at various levels of speech. Prosody reflects the learned/acquired speaking habits of a person and hence contributes for speaker recognition. Because prosodic features are less affected by channel mismatch and noise, they are particularly well suited for speaker forensics, a field that demands accurate identification of suspects with as few mitigating conditions as possible. In this chapter, the author describes a method for extracting prosodic features directly from speech signal. Applying this method, speech is segmented into syllable-like regions using vowel onset points (VOP). The locations of VOPs serve as reference for extraction and representation of prosodic features. The effectiveness of the prosodic features for speaker recognition is demonstrated for extended task of NIST speaker recognition evaluation 2003. Combining evidence from spectral features with that of the proposed prosodic features helps to improve overall speaker recognition accuracy.

13.1 Introduction

Human beings use several levels of perceptual cues for speaker recognition, ranging from high-level cues such as semantics, pronunciations, idiosyncrasies and prosody to low-level cues such as acoustic aspects of speech [1]. The high-level features such as prosody and idiolect are the behavioral attributes of the speaker, different from physiological characteristics of the speech production system. Human beings derive evidence regarding the identity of a speaker from certain prosodic cues such as pitch gestures, accents, and speech rate. It is generally recognized that human listeners can better recognize those speakers who are familiar to them, than those who are relatively less familiar. This increased ability is due to speaker-specific prosody and idiosyncrasies that are recognized by the listener, either consciously or

L. Mary (✉)
Rajiv Gandhi Institute of Technology, Kottayam 686501, Kerala, India
e-mail: leena.mary@rit.ac.in

A. Neustein, H. A. Patil (eds.), *Forensic Speaker Recognition*,
DOI 10.1007/978-1-4614-0263-3_13, © Springer Science+Business Media, LLC 2012

otherwise [2]. “Familiar-speaker” differences, surely relate to the features such as intonation, stress and timing, which are collectively referred as prosody.

Speaker characteristics are manifested in speech signal as a result of anatomical differences inherent in speech production organs and differences in acquired speaking habits of individuals [3]. Physiological differences include the differences in the shape and size of oral tract, nasal tract, vocal folds and trachea. This can lead to difference in the vocal tract dynamics and excitation characteristics. The acquired habits are characteristics that are learned over a period of time, mostly influenced by the social environment and also by the characteristics of the first or native language in the “critical period” (lasting roughly from infancy until puberty) of learning. The way prosodic characteristics are manifested in speech give important information regarding the identity of a speaker. Idiosyncrasies of a speaker are reflected in the usage of certain words and phrases and it is present even at the semantic level.

Differences in speaker characteristics may be summarized as follows:

1. Vocal tract size and shape
2. Excitation characteristics
3. Prosody
4. Idiolect
5. Semantic

In order to represent differences among speakers in terms of characteristics of excitation source, vocal tract system and prosody, features should be extracted from levels of speech at which these characteristics are manifested [4]. The approximated levels of representation of speaker-specific features are the following:

1. Subsegmental: The main source of excitation for production of speech is the glottal vibration. In each glottal cycle, the instant of glottal closure is the instant at which significant excitation of vocal tract takes place. Hence a small region around the instant of glottal closure contains significant information about the speaker. In order to represent the excitation source characteristics of a speaker, excitation sequence corresponding to duration of 1–5 ms, which is less than one pitch period is considered.
2. Segmental: Vocal tract system characteristics are extracted from a window of speech signal that contains a few pitch periods (10–30 ms). Time varying nature of the vocal tract system is taken into account by sliding the window by about 5–10 ms.
3. Suprasegmental: Features corresponding to a larger span of speech (>100 ms) which go beyond segments are referred as suprasegmental features. The suprasegmental features mostly represent the habitual attributes of a speaker such as prosody and idiolect. Prosodic characteristics such as intonation, duration and stress are visible only for a large span of speech. One important aspect of prosody is that it spans over long segments like syllables, words, and utterances and reflects differences in speaking style, language background, sentence type, and emotions to mention a few [5].

Conventional text independent speaker recognition systems rely mostly on spectral features derived through short-term spectral analysis, to represent vocal tract characteristics. This approach does not attempt to model the long-term speaker-specific characteristics such as prosody present in the speech signal. Moreover it is understood that the spectral features are affected by channel characteristics and noise. The long-term features are relatively less affected by channel mismatch and noise. In order to incorporate long-term features, system generally requires significantly more data for training. Hence in 2001, National Institute of Standards and Technology (NIST) introduced the extended data task of speaker recognition evaluation (SRE), which provides multiple conversation sides, for speaker training [6]. This helped in the study of long-term features for speaker recognition. A workshop was conducted at the John Hopkins University to explore a wide range of features for speaker verification using NIST 2001 extended data task as its test bed [7].

Incorporation of long-term features into speaker recognition systems provides complementary evidence to the spectral-based systems. To make use of the speaker-specific information present at larger span of speech, most of the existing speaker recognition systems use segment boundaries and text labels obtained using automatic speech recognizers. For building a speech recognizer, several man-hours are needed for preparing manually labeled corpora. In this chapter, a method has been described for extracting prosodic features directly from the acoustic speech signal.

Remaining part of the chapter is organized as follows: Sect. 13.2 defines the term prosody and Sect. 13.3 explains the speaker-specific aspect of prosody. In Sect. 13.4, robustness of prosodic features against channel mismatch and noise is illustrated. The relevance of prosodic features for forensic speaker recognition is discussed in Sect. 13.5. The discussions in Sects. 13.2–13.5 are general pertaining to the use prosody for speaker recognition. In Sect. 13.6, a method for automatic extraction and representation of prosodic features for speaker recognition is suggested by the author, which is supported by the results of the experimental studies in Sect. 13.7. Finally, Sect. 13.8 gives a summary and limitations of the material presented in this chapter.

13.2 What Is Prosody?

Speech is primarily intended to convey some message. It is conveyed through a sequence of legal sound units in a language. However speech cannot be merely characterized as a sequence of sound units. There are some characteristics that lend naturalness to speech. The variation of pitch provides some recognizable melodic properties to speech. This controlled modulation of pitch is referred as *intonation*. The sound units are shortened or lengthened in accordance to some underlying pattern giving *rhythmic* properties to speech. Some syllables or words may be made more prominent than others, resulting in linguistic *stress*. The information gleaned from melody, timing and stress in speech increases the intelligibility of spoken message, enabling the listener to segment continuous speech into phrases and words with ease [8]. It is also capable of conveying many more lexical and nonlexical information such as lexical tone, prominence, accent and emotion. The characteristics that make us perceive these effects are collectively referred to as prosody. Prosodic cues include stress, rhythm and intonation. Each cue is a complex perceptual entity, expressed primarily using three acoustic parameters: pitch, energy and duration.

Prosodic characteristics convey some important information regarding the identity of the speaker. Since each speaker has unique physiological characteristics of speech production and speaking style, speaker-specific characteristics are reflected in prosody. However it is very difficult even for a listener to describe the nature of speaker-specific prosodic features that he or she will be using for recognition. Therefore it is a challenging task to identify, extract and represent prosodic features for recognizing a speaker.

13.3 Speaker-Specific Aspect of Prosody

The prosodic characteristics as manifested in speech give important information regarding the speaking habit of a person. Pitch is a perceptual attribute of sound. The physical correlate of pitch is the fundamental frequency (F_0) of vibration of vocal folds. It is speaker-specific due to differences in the physical structure of the vocal folds among speakers. The average value of F_0 is generally higher for children and females, due to smaller size of the vocal folds. The distribution of fundamental frequency (F_0) values varies among speakers as illustrated in Fig. 13.1. These histograms are prepared using text independent speech corresponding to four different speakers in OGI database. Researchers have attempted to capture the global statistics of F_0 values of a speaker using appropriate distributions for speaker verification task [9].

The F_0 value is controlled either by varying the subglottal pressure or laryngeal tension or a combination of both [10], which is speaker-specific. The dynamics of F_0 contour is influenced by several factors such as the identity of the sound unit spoken, position with respect to phrase or words, context (the units that precede and follow), speaking style of a particular speaker, intonation rules of the language, type of sentence (interrogative or declarative) etc. The dynamics of F_0 contour and energy contour can be different among speakers due to different speaking style and accent. The dynamics of F_0 contour will be different for two speakers, even when they utter the same text in the same context as illustrated in Fig. 13.2. However when a given speaker repeats the same text, the characteristics of F_0 contour are consistent and this is true across speakers as illustrated in Fig. 13.3. The presence of speaker-specific information in temporal dynamics of F_0 contour may be used for characterizing a speaker. This property has been used in text-dependent speaker verification, using dynamic time warping (DTW) [11]. It has been shown that the dynamics of F_0 contour can also contribute to text-independent speaker verification [12, 13]. Other prosodic features useful for speaker recognition are duration (e.g. pause statistics, phone or syllable duration), speaking rate, and energy distribution among others [5].

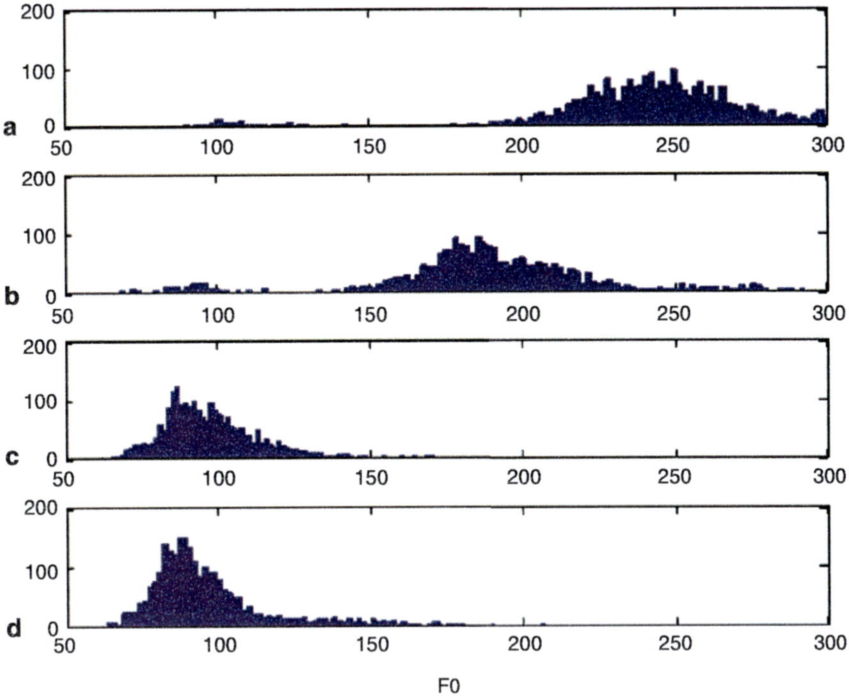

Fig. 13.1 Variation in histogram of F_0 for **a**, **b** two female, **c**, **d** two male speakers (taken from spontaneous speech in OGI database)

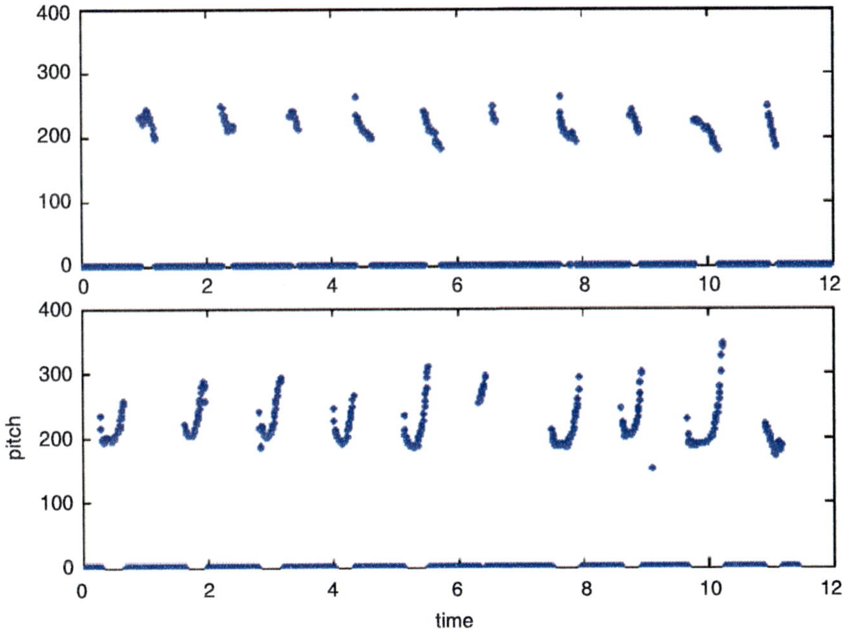

Fig. 13.2 Variation in dynamics of F_0 contour of two different female speakers while uttering *One, Two, Three, Four, Five, Six, Seven, Eight, Nine, Ten*

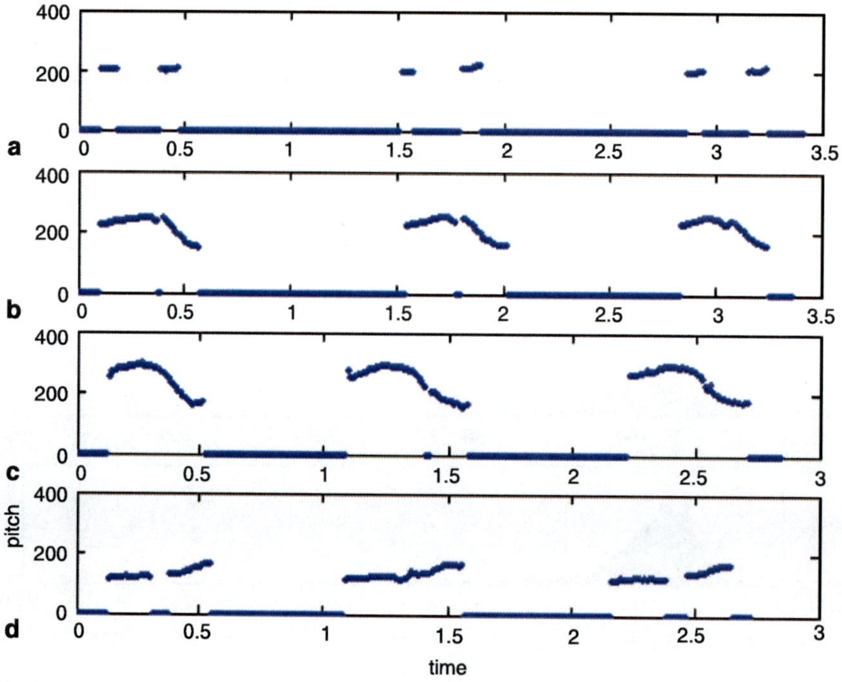

Fig. 13.3 Variation in F_0 contour dynamics of four different speakers: **a, b, c** Three different female voices. **d** Male voice. All repeating the same text *Monday, Monday, Monday*

13.4 Robustness of Prosodic Features

Most of the current speaker recognition systems rely on the spectral features derived through short time spectral analysis of the speech signal. The magnitude of the short-time spectrum encodes information about vocal tract shape of the speaker [14–16]. Therefore spectral features are widely used for speaker modeling. Speaker recognition systems based on spectral features perform well in favorable acoustic conditions, however the performance degrades due to noise and unmatched acoustic conditions [17]. For example, a speaker model trained using speech collected over a microphone may not give correct result for a genuine test utterance collected over land line or cellular environment. Prosodic features derived from pitch, energy and duration are relatively less affected by channel variations and noise [18]. Though the systems based on spectral features outperform the prosody-based systems, their combined performance may provide the needed robustness to recognition systems.

The effect of channel variations on spectral feature vectors and F_0 contour are illustrated in Figs 13.4 and 13.5, respectively [19]. The same utterance *Don't carry an oily rag like that* recorded through three different channels, available in Texas Instruments and Massachusetts Institute of Technology (TIMIT) database, is used

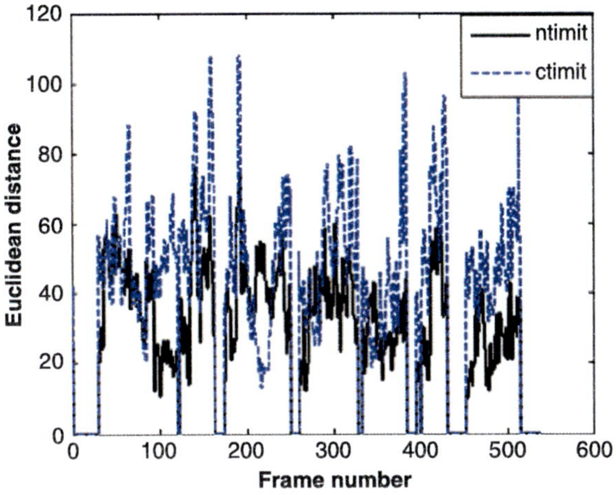

Fig. 13.4 Euclidean distance of LPCC feature vectors on a frame-to-frame basis for the same speaker and text *Don't carry an oily rag like that.* The *solid line* corresponds to the distance of NTIMIT data and *dashed line* corresponds to CTIMIT data with reference to TIMIT data [19]

for comparing the effect of channel variability and noise. Channels correspond to TIMIT, NTIMIT and CTIMIT represent speech collected over close-speaking microphone, noisy channel and cellular environment, respectively. Figure 13.4 shows the difference in Euclidean distance of LPCC features of NTMIT and CTIMIT sentence with reference to the corresponding TIMIT sentence. This distance would have been ideally zero if LPCC features were unaffected by channel variability and noise. Figure 13.5 illustrates the robustness of F_0 contour characteristics against channel variations [19]. In Fig. 13.5, the F_0 contours remain the same for all the cases except some durational variation of voiced region in (b) and (c) compared to (a).

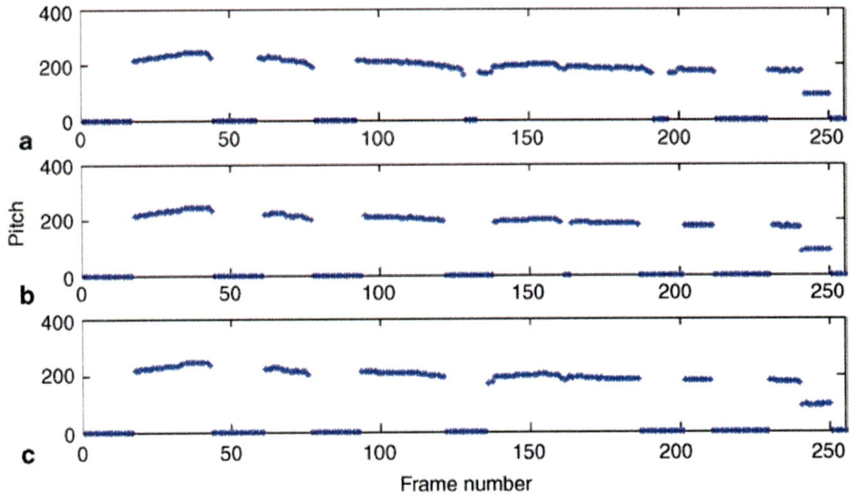

Fig. 13.5 F_0 contours of **a** TIMIT **b** NTIMIT and **c** CTIMIT sentence of the same speaker for the same sentence *Don't carry an oily rag like that* [19]

13.5 Relevance of Prosodic Features in Forensic Speaker Recognition

The widespread use of telephones has resulted in an increased use of human voice as an instrument in the commission of crimes [20]. As a result, forensic automatic speaker recognition (FASR) has become an important tool in forensic sciences. When using FASR, the goal is to identify whether an unknown voice of a questioned recording belongs to a particular known speaker. The interpretation of recorded speech as evidence in the forensic context presents particular challenges such as difference in the background noise, phone handset, transmission channel and recording devices used in investigative activities. The comparison of traditional forensic speaker recognition with automatic speaker recognition systems indicates that prosodic features may be acceptable to legal and forensic practitioners, because of their overlap with traditional methods.

The conventional automatic speaker recognition, which relies on spectral features, suffers from the problem of lack of robustness (as illustrated in Fig. 13.4) and interpretability for forensic applications [21]. Due to the higher dimensionality of spectral features, it is hard to visualize them. The prosodic features described in terms of pitch contour, duration and amplitude are based on perceptual cues and hence more interpretable to a non-expert or jury [22]. Figures 13.2 and 13.3 indicate that dynamics of pitch contour can give a clue about true identity of the speaker in the forensic contexts. Robustness of prosodic features in cases of acoustic mismatch (as illustrated in Fig. 13.5) may provide an edge for them in forensic applications. As the samples of training and testing (known and questioned recordings) may belong to different acoustic conditions in terms of background noise and channel, use of spectral features alone may not be a reliable option for performing FASR. The use of prosodic features offers a better choice for FASR in case of channel mismatch and noise.

13.6 Extraction of Prosodic Features for Speaker Recognition

In general, there are two broad approaches for extracting prosodic features from speech, the automatic speech recognizer (ASR)-based approach and ASR-free approach. The ASR-based approach uses the explicit subword boundaries obtained from ASR for extracting the prosodic features in terms of duration, pitch and energy [23].

However for speaker recognition, the use of ASR may not be needed. In most of the ASR-free approaches, pitch contour dynamics are represented using parameters derived from linear stylized pitch segments [24–26]. In another approach, inflection points and start or end of voicing of pitch and energy trajectories are used to segment the speech signal, and features are derived from linear stylized segments of pitch and energy contour [25]. The segmented trajectories are then quantized and labeled into a small set of classes that describe the dynamics of F0 contour and energy contour. N-grams of these labels are used to model the characteristics of a speaker or a language [21]. Recent approaches to automatic syllable-like segmentation include the use of vowel detection [27] and group delay function of minimum phase signal [28]. These approaches have the advantage that features are derived directly from the speech signal. In another approach, valley points of energy contour is aligned with pitch contour for segmenting long pitch segments into shorter ones [29].

In this section, a technique based on vowel onset points (VOP) for extraction and representation of prosodic features is discussed [4, 19, 30]. The technique combines salient features of both approaches mentioned above, namely, association with the syllabic pattern as in the ASR-based approach, and extraction of features without explicit speech recognition as in the ASR-free approach.

13.6.1 Choice of Syllable as the Basic Unit

All spoken utterances can be considered as sequence of syllables that constitute a continual rhythmic alternation between opening and closing of mouth while speaking [31]. Syllable of CV type provides an articulatory pattern beginning with a tight restriction and ending with an open vocal tract, resulting some rhythm that is especially suited both to the production and the perception mechanisms [32]. It is demonstrated that the tonal events are aligned to the segmental events such as onset and /or offset of a syllable [33]. Therefore syllable appears to be a natural choice for the basic unit for representing prosody [4].

For representing syllable-based rhythm, intonation, and stress, the speech signal should be segmented. Segmenting speech into syllables is typically a language-specific mechanism, and thus it is difficult to develop a language independent algorithm for this. Segmentation into syllable-like units may be accomplished with the knowledge of vowel onset points (VOPs) as illustrated in Fig. 13.6a, where VOP refers to the instant at which the onset of vowel takes place in a syllable [34]. The VOP detection algorithm described in the next section relies on strength of excitation and does not use any language level knowledge. There may be limitations in this approach, however since it provides a language-independent solution to the segmentation problem, it is adopted [4]. The speech between two successive VOPs corresponds to syllable-like regions as illustrated Fig. 13.6a. The F_0 contour of syllable-like regions may be then associated with the locations of VOPs as shown in Fig. 13.6b, for feature extraction. Sections 13.6.2 and 13.6.3 describe the methods used for detection of VOPs and extraction of pitch respectively. Both methods use the strength excitation derived from excitation source characteristics.

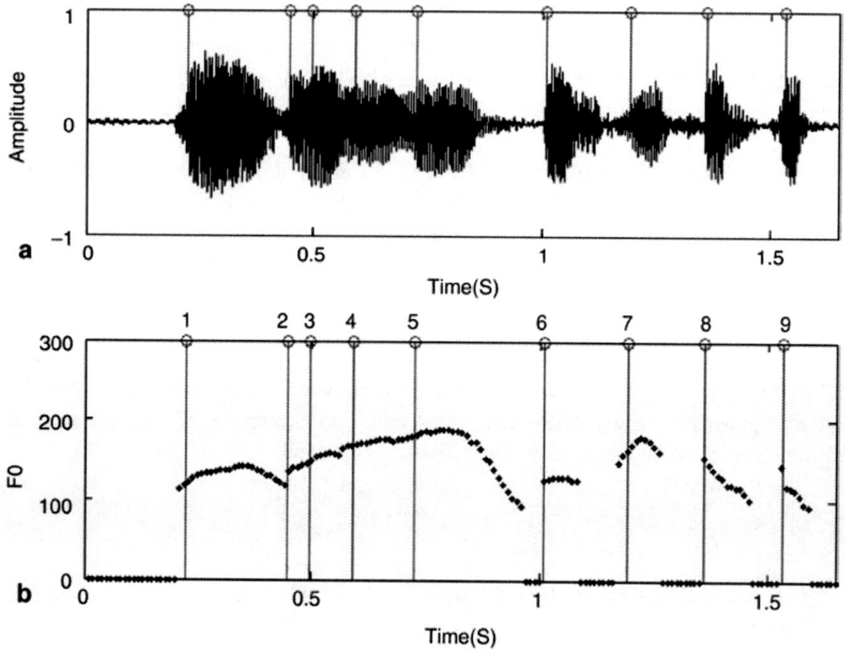

Fig. 13.6 **a** segmentation of speech into syllable-like units using automatically detected VOPs, **b** F_0 contour associated with VOPs (marked '1'–'9')

13.6.2 Detection of Vowel Onset Points

Vowel onset point is an important event in speech production, which may be described in terms of changes in the vocal tract and excitation source characteristics. A technique that relies on excitation source characteristics is described here for the extraction of VOPs from continuous speech [35, 36]. It uses the Hilbert envelope $h_e(n)$ of LP residual $e(n)$ which is defined as

$$h_e(n) = \sqrt{e^2(n) + e_h^2(n)} \tag{13.1}$$

where $e_h(n)$ is the Hilbert transform of $e(n)$, and is given by

$$e_h(n) = \begin{Bmatrix} IDFT\,[-jE(\omega)], & 0 < \omega < \pi \\ IDFT\,[jE(\omega)], & 0 > \omega > -\pi \\ 0 & \omega = 0,\ \pi \end{Bmatrix} \tag{13.2}$$

where IDFT denotes the inverse discrete Fourier transform, and $E(\omega)$ is the Fourier transform of $e(n)$. The Hilbert envelope approximately represents the strength of excitation as shown in Fig. 13.7 [37, 38].

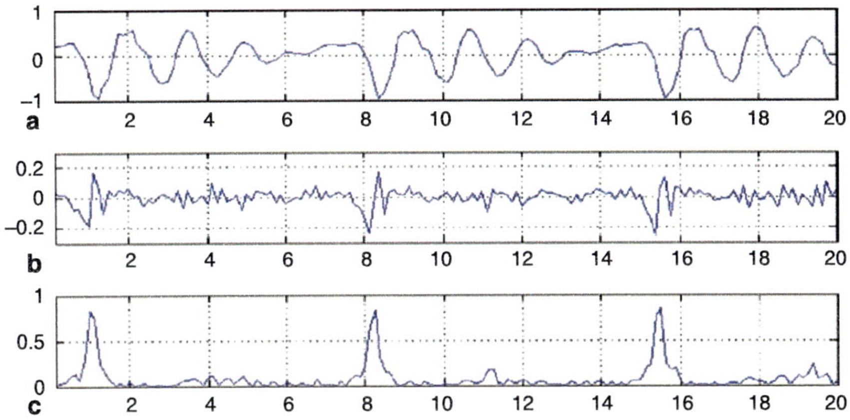

Fig. 13.7 **a** A segment of voiced speech and its, **b** LP residual, **c** Hilbert envelope of the LP residual

The strength of excitation at the instants of glottal closure for voiced sounds is generally higher compared to the strength at random instants present in the unvoiced sound. Also, the strength of excitation at the instants of glottal closure for vowels is higher compared to the strength of the voiced consonants. Therefore, the strength of excitation represented by the Hilbert envelope shows a significant change at the transition from consonant to vowel, and hence can be used as a cue for detecting VOP. The instant with maximum excitation within a pitch period corresponds to the instant of glottal closure. The places with significant change in the strength of excitation give the evidence for the detection of VOPs.

Figure 13.8 shows the speech waveform with manual marked VOPs, the Hilbert envelope of the LP residual, the VOP evidence, output of peak picking algorithm, and the hypothesized VOPs. The VOP evidence is obtained from the Hilbert envelope of the LP residual by multiplying it with a Gabor window. A Gabor window in this case is a modulated Gaussian pulse (as illustrated in Fig. 13.9) that is defined as

$$g(n) = \frac{1}{\sqrt{2\pi}\sigma} e^{-\frac{n^2}{2\sigma^2}} \cos(\omega n) \tag{13.3}$$

where σ is the spatial spread and ω is the modulating frequency.

By taking the sum of the product for every sample shift, the VOP evidence plot is obtained as shown in Fig. 13.8c, from which peaks are located using a peak-picking algorithm. Spurious peaks represented using downward arrows in Fig. 13.8d are eliminated using the characteristics of the VOP evidence plot. It is done by checking the VOP evidence plot between two peaks for the presence of a negative region of sufficient strength. If peaks are two true VOPs, then there will be a negative region of sufficient strength between them due to the vowel region. If there is no negative region, then the first peak is eliminated, as it is spurious. The algorithm for automatic detection of VOP is given in Table 13.1.

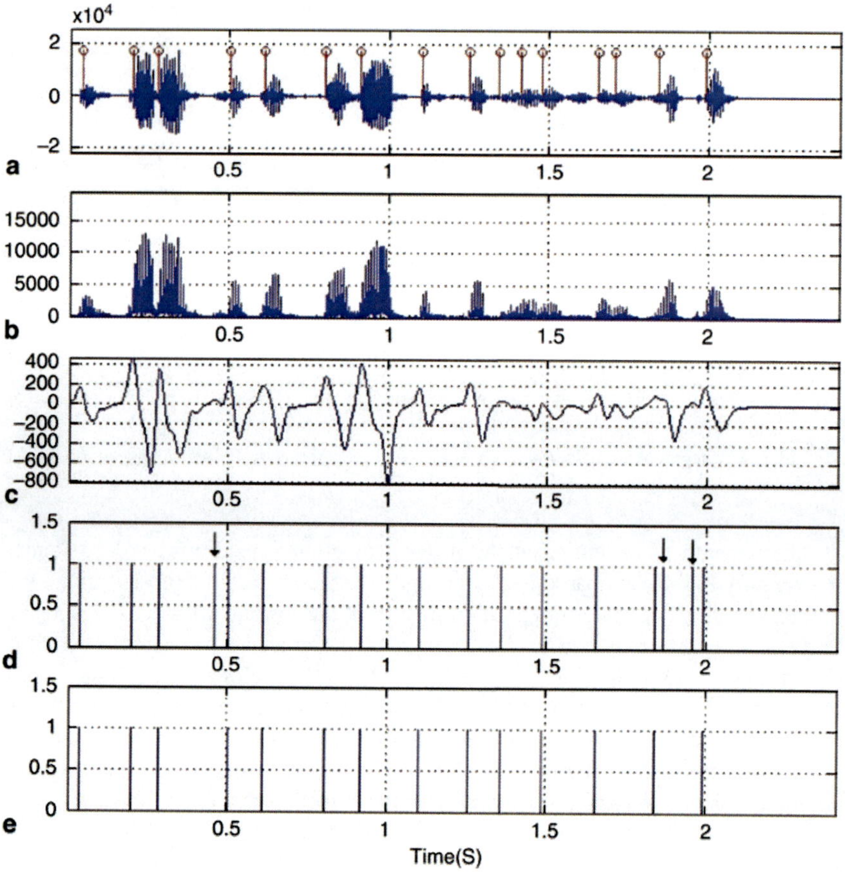

Fig. 13.8 **a** Speech waveform with manual marked VOPs, **b** Hilbert envelope of LP residual, **c** VOP evidence plot, **d** Output of peak picking algorithm, **e** Hypothesized VOP after eliminating few spurious peaks [35]

Table 13.1 Algorithm for automatic detection of vowel onset points [33]

1.	Preemphasize input speech by differencing
2.	Compute LP residual using 10th order LP analysis using frame size 20 ms and frame shift 5 ms (frame overlap 75%)
3.	Compute Hilbert envelope of the LP residual
4.	Obtain the VOP evidence plot from the Hilbert envelope by convolving it with the Gabor filter (σ=100, ω=0.0114 and analysis window size 100 ms)
5.	Identify peaks in the VOP evidence plot
6.	For each peak, if there is no negative region with reference to the next peak, then eliminate such a peak as it is spurious
7.	Hypothesize the remaining peaks as the VOPs

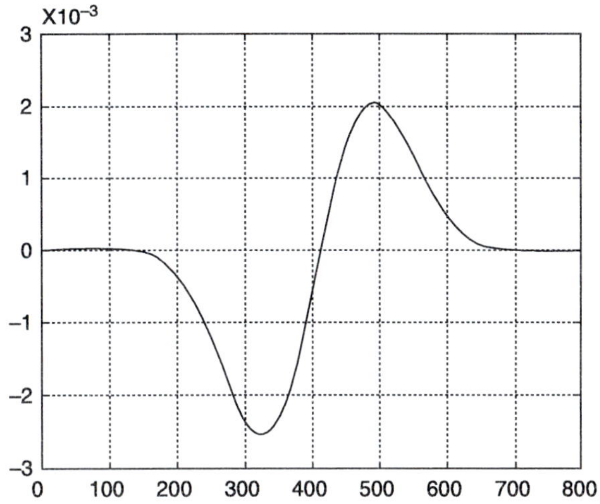

Fig. 13.9 Gabor window for $\sigma=100$, $\omega=0.0114$ and $n=800$

13.6.3 Extraction of Pitch

There are several algorithms proposed in the literature for the extraction of pitch [39]. These algorithms may be broadly classified into three categories, namely, algorithms using time domain properties, algorithms using frequency domain properties and algorithms using both time and frequency domain properties of speech signals. A method proposed in the literature based on the glottal closure instants is described in this section [35, 38, 40, 41]. This approach employs autocorrelation analysis of Hilbert envelope of the LP residual.

Figure 13.10 shows segment of voiced speech, its LP residual, Hilbert envelope of LP residual and corresponding autocorrelation sequences. In the autocorrelation sequence, the distance of the first major peak from the center peak is marked as the pitch period. Comparing to the pitch period of the neighboring frames validates this pitch value. If the variation of pitch period is within a threshold, its value is retained for next stage of validation, else it is set to zero. Second stage of validation uses similarity of the samples around the first major peak in autocorrelation sequence of the adjacent frames for voiced speech. This similarity can be measured by comparing the small segment (2.5 ms on either side of the first major peak) of present frame with that of previous frame, computed using the correlation coefficient defined as follows.

$$c = \frac{\sum |(x-\bar{x})|\,|(y-\bar{y})|}{\sqrt{\sum (x-\bar{x})^2}\sqrt{\sum (y-\bar{y})^2}} \tag{13.4}$$

where x and y represents samples around the first major peak in the current frame and the previous frame respectively and $\bar{x}$ and $\bar{y}$ represents their mean. The sequence of correlation coefficients obtained as per Eq. (13.4) is smoothed using

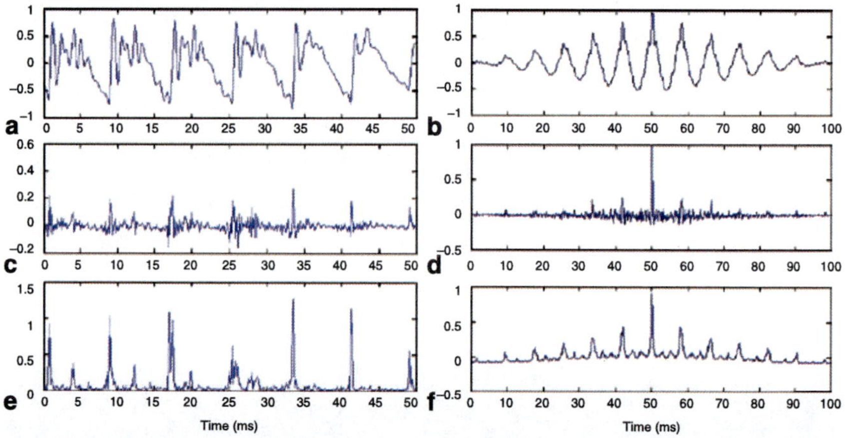

Fig. 13.10 **a** Segment of voiced speech and its, **b** autocorrelation sequence, **c** Segment of LP residual and its, **d** autocorrelation sequence, **e** Segment of Hilbert envelope of the LP residual and its, **f** autocorrelation sequence [35]

Table 13.2 Algorithm for the extraction of pitch

1.	Preemphasize input speech
2.	Compute LP residual using 10th order LP analysis using frame size of 20 ms and frame shift of 5 ms
3.	Compute Hilbert envelope of the LP residual
4.	Perform autocorrelation of the Hilbert envelope
5.	In the autocorrelation sequence, find the first major peak after the center peak and find its distance from the center peak
6.	Find the similarity between the small segment of the present frame with the corresponding segment from the previous or next frame using correlation coefficient given in Eq. (13.4)
7.	Median smooth the correlation coefficient values by five points
8.	If correlation coefficient is greater than 0.7, then declare the distance as pitch value, else set pitch value as zero

median filtering. If the correlation coefficient c value is higher than a threshold, then corresponding pitch value is taken and otherwise it is set to zero. Table 13.2 describes the algorithm for the extraction of pitch using this method.

13.6.4 Feature Parameterization

The association of F_0 contour with locations of VOP [4, 18] for extraction of prosodic features is shown in Fig. 13.11

The nature of F_0 variations for a segment of pitch contour between two consecutive VOPs may be a rise, a fall, or a rise followed by a fall in most of the cases. It is

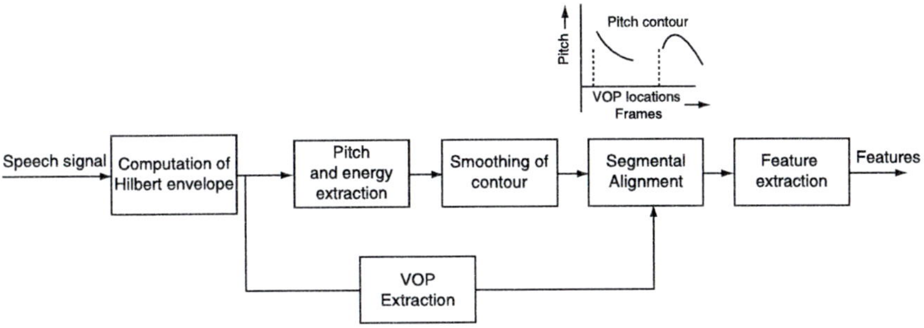

Fig. 13.11 Association of F_0 contour with locations of VOP for prosodic feature extraction

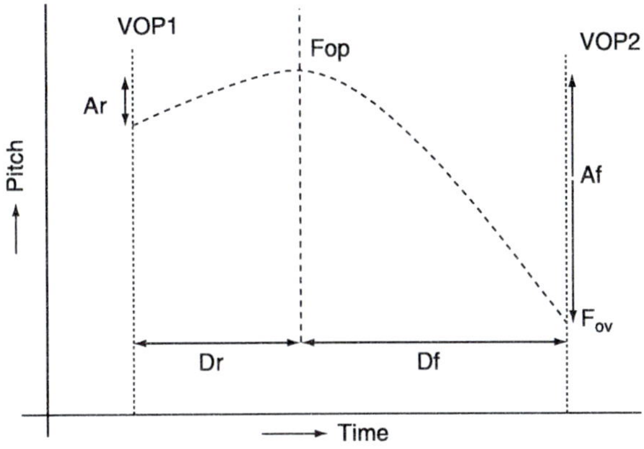

Fig. 13.12 Segmental alignments of F_0 contour with locations of VOP. Tilt parameters A_t and D_t defined in terms of A_r, A_f, D_r and D_f represent the dynamics of a segment of F_0 contour

assumed that more complex F_0 variations are unlikely within a segment. With reference to Fig. 13.12, tilt parameters [42], namely amplitude tilt (A_t) and duration tilt (D_t) for a segment of F_0 contour are defined as follows:

$$A_t = \frac{A_r - A_f}{A_r + A_f} \tag{13.5}$$

$$D_t = \frac{D_r - D_f}{D_r + D_f} \tag{13.6}$$

where A_r and A_f represent the rise and fall in F_0 amplitude, respectively, with respect to peak value of fundamental frequency F_{0p}. Similarly D_r and D_f represent the duration taken for rise and fall respectively. Figure 13.13a–f show F_0 contours with different values of tilt. From Fig. 13.13e, f, it is clear that amplitude tilt do not show the change in F_0. Therefore the difference between F_{0p} and F_{0v} may be included in the feature set.

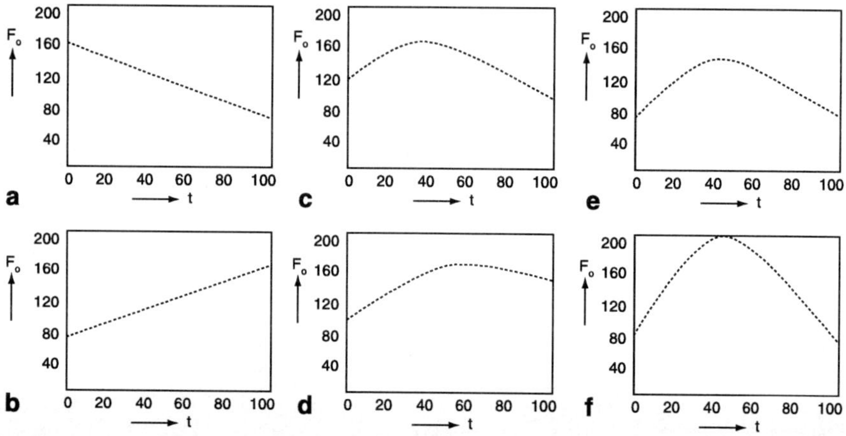

Fig. 13.13 Illustration of F_0 contours represented using various tilt parameters. **a** $A_t=(0-80)/(0+80)=-1$, $D_t=(0-100)/(0+100)=-1$; **b** $A_t=(80-0)/(80+0)=1$, $D_t=(100-0)/(100+0)=1$; **c** $A_t=(40-60)/(40+60)=-0.2$, $D_t=(40-60)/(40+60)=-0.2$; **d** $A_t=(60-20)/(60+20)=0.5$, $D_t=(60-40)/(60+40)=0.2$; **e** $A_t=(80-80)/(80+80)=0$, $D_t=(50-50)/(50+50)=0$; **f** $A_t=(120-120)/(120+120)=0$, $D_t=(50-50)/(50+50)=0$

The use of tilt parameters help to represent F_0 patterns quantitatively, instead of quantizing and labelling of F_0 patterns as in other approaches [13].

Studies have shown that, speakers can vary the prominence of pitch accents by varying the height of the fundamental frequency peak, to express different degrees of emphasis. Likewise, the listener's judgment of prominence reflects the role of F_0 variation in relation to variation in prominence [43]. To express the height of the F_0 peak, the difference between peak and valley fundamental frequency ($\Delta F_0=F_{0p}-F_{0v}$) is used in this study. It has been observed that the length of the F_0 peak (length of onset) has a role in the perceptual prominence [43]. This may be represented using the distance of F_0 peak location with respect to VOP (D_p).

From Fig. 13.13e, f, it is clear that the tilt parameters do not represent the height of F_0 peak. But Studies have indicated that listeners are more sensitive to variations in F_{0p} [35]. Hence peak value of F_0 (F_{0p}) for each segment of F_0 contour may be useful for speaker recognition. An increase in F_0 may be obtained by increasing the vocal fold tension, by increasing the subglottal pressure, or a combination of them. Therefore F_{0p} and F_{0mean} obtained for each segment of F_0 contour may reflect some physiological as well as habitual aspect of a speaker. The change in log energy (ΔE) along with F_0 change gives a quantitative measure of stress characteristics, therefore may be specific to a particular speaker. Thus the parameters identified for characterizing the speaker-specific aspect of prosody are the following:

1. $F_{0\ mean}$
2. F_0 peak (F_{0p})
3. Change in F_0 (ΔF_0)
4. Distance of F_0 peak with respect to VOP (D_p)

Table 13.3 Procedure for the extraction of prosodic features for a speech region between two consecutive VOPs as illustrated in Fig. 13.11

1.	Preemphasize input speech
2.	Compute LP residual using 10th order LP analysis using frame size of 20 ms and frame shift of 5 ms
3.	Compute Hilbert envelope of the LP residual
4.	Perform pitch extraction algorithm as described in step 4–8 of Table 13.2 and energy to obtain smoothed pitch and energy contour
5.	Perform VOP detection algorithm as described in step 4–7 of Table 13.1
6.	Take pitch values corresponding to the speech region between consecutive VOPs as illustrated in Fig. 13.12, find out maximum pitch F_{0p}, minimum pitch F_{0v}, mean pitch F_{0mean}, Change in pitch ΔF_0, distance of F_{0p} from VOP1 D_p, A_r, A_f, D_r and D_f
7.	Compute amplitude tilt A_t and duration tilt D_t using Eqs. (13.5) and (13.6)
8.	Compute change in energy ΔE
9.	Store the computed 7-dimesional feature vectors [F_{0mean}, F_{0p}, ΔF_0, D_p, A_t, D_t, ΔE]

5. Amplitude tilt (A_t)
6. Duration tilt (D_t)
7. Change of log energy (ΔE)

Table 13.3 summarizes the procedure followed for the extraction of prosodic feature vectors.

13.7 Results from Experimental Studies

The effectiveness of the proposed prosodic features is demonstrated using the first subset of NIST 2003 extended data task. Unlike the traditional speaker recognition tasks, the extended data task provides more speech data for training (4-side/8-side/16-side, where each conversation side contains approximately 2.5 min of speech). Each target model is tested with a set of 1-side test utterances, where the task is to find out whether the particular test utterance belongs to the target speaker or not. The first split in NIST 2003 extended data task is selected to carry out the experiments, which consists of 137, 54 and 74 speaker models for the 16-side, 8-side, and 4-side cases, respectively. The speaker models are evaluated using 1076, 1238 and 1258 test utterances for the 16-side, 8-side and 4-side cases, respectively.

It is hypothesized that the distribution of syllable level prosodic feature vectors are speaker-specific. To capture the distribution of the feature vectors, autoassociative neural network (AANN) models or alternatively conventional Gaussian mixture models (GMM) can be used. The AANN is a feedforward neural network which tries to map an input vector onto itself, and hence the name autoassociation or identity mapping [44]. It consists of an input layer, an output layer and one or more hidden layers. To capture the distribution of the feature vectors, examples are presented in a random order to the AANN and the network is trained using standard backpropagation algorithm. It has been demonstrated that the AANN has the ability

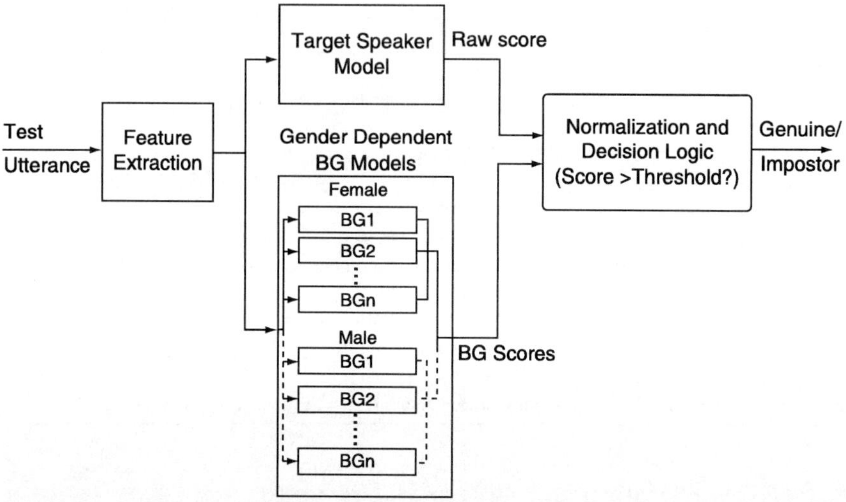

Fig. 13.14 Block diagram of prosody-based speaker verification system, showing the testing of an unknown utterance against target speaker model and a set of background models

to capture the distribution of input data [45]. During testing, for each prosodic feature vector in the test utterance, the error between the output and the input of AANN is noted. This error is converted into confidence value using $C_i = \exp(-E_i)$, where E_i is the squared error for the ith syllable [45]. The average confidence is computed as

$$C = \frac{1}{N} \sum\nolimits_{i=1}^{N} C_i \qquad (13.7)$$

where C_i is the confidence value for the ith syllable, and N is the number of syllables in the test utterance.

For each target speaker, one AANN model is trained for 500 epochs to capture the distribution of prosodic feature vectors. The structure of the AANN model used for capturing the distribution of the speaker-specific prosodic features is *7L 28N 2N 28N 7L*, where *L* represent units with linear activation function, *N* represent units with nonlinear activation function, and the numerals represent the number of units in the layers. A set of background models built from a known set of impostor speakers (taken from another split of the same database) helps to fix a global threshold for verification, to decide whether the test utterance belongs to the target speaker or not. The background models consist of a set of male and female models as illustrated in Fig. 13.14.

Score normalization is used for scaling the likelihood scores, which helps to find a global speaker independent threshold for the decision making. Feature vectors obtained from the test utterance is presented to the target speaker model as well as to a set of background models as shown in Fig. 13.14 For each test utterance,

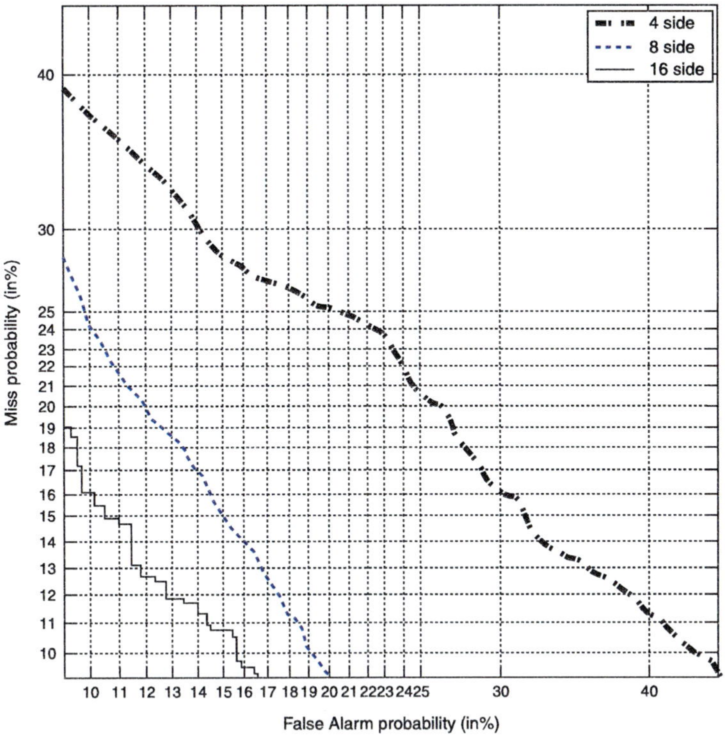

Fig. 13.15 DET *curve* showing the performance of prosody-based system for 16-side, 8-side and 4-side conversational cases [30]

the decision on the gender is made based on the average score of male or female background model set. The raw score obtained from target speaker model is test normalized using scores of the background (BG) models. The normalized score C_n is computed from raw score C as:

$$C_n = \frac{C - \mu_g}{\sigma_g} \tag{13.8}$$

where μ_g and σ_g represent mean and standard deviation of BG scores corresponding to the hypothesized gender of test utterance.

The prosody-based speaker verification system resulted in an equal error rate (EER) of 12.4, 15 and 23% for 16-side, 8-side and 4-side conversational cases of the particular data set, respectively. Performance is shown using the DET curves in Fig. 13.15. Better performance in case of 16-side and 8-side training cases show that more training speech is required for capturing the prosodic characteristics well.

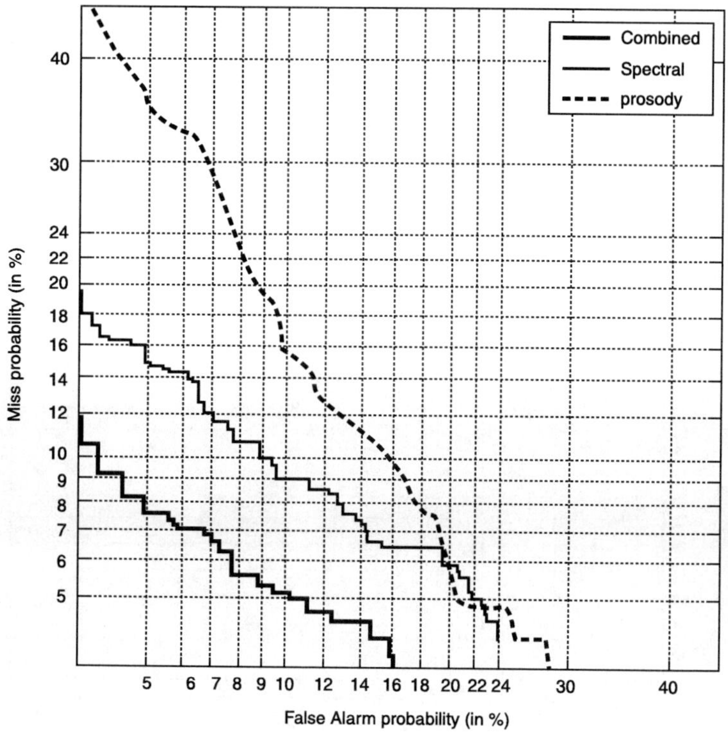

Fig. 13.16 DET *curve* showing the performance of spectral based system, prosody based system and combined system for 16-side conversational case [30]

13.7.1 Combining Evidence from Prosody and Spectral-Based Systems

The evidence about the speaker from different features may be combined in several ways to achieve better performance. One simple approach is the weighted addition of evidences from different systems. The spectral-based baseline speaker verification system [45] using weighted linear prediction cepstral coefficients (WLPCC) gives an EER of 9.5% for the same 16-side data set. Even though this baseline system is not the best performing one, it is used to illustrate the complementary information provided by the prosodic features. Combining by simple addition of the evidences in 16-side cases results in an EER of 6.9 as illustrated in Fig .13.16. Combining evidence for 8-side and 4-side result in an EER of 9.3 and 11 respectively as illustrated in Figs. 13.17 and 13.18. Even though the performance of prosodic features is inferior to spectral features in all the cases the combination of evidence significantly improve the performance, illustrating their complementary nature for speaker verification task.

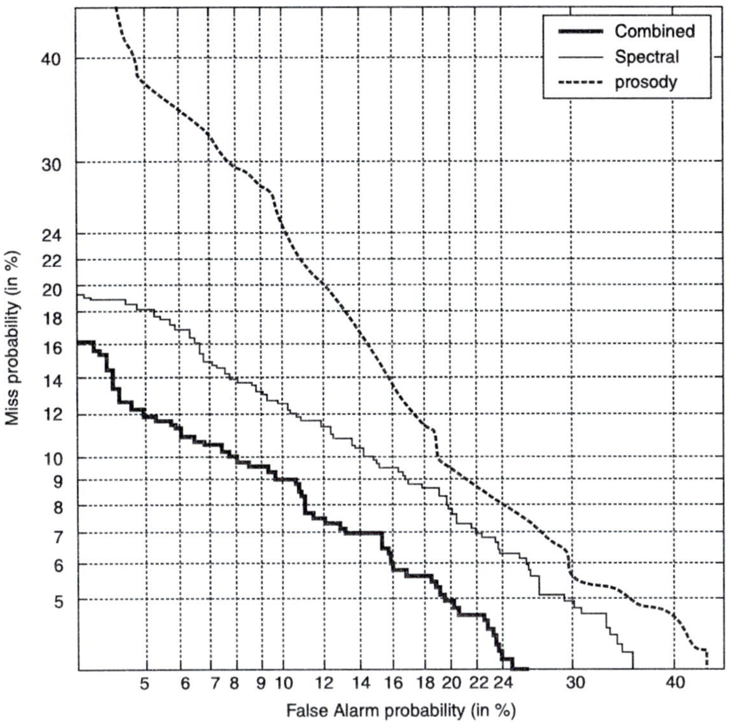

Fig. 13.17 DET *curve* showing the performance of spectral based system, prosody based system and combined system for 8-side conversational case [19]

13.8 Summary

In this chapter, the usefulness of prosodic features for text independent speaker recognition is discussed. An approach for extraction and representation of prosodic features is described. This approach eliminates the requirement of automatic speech recognizer for prosodic feature extraction, but still gives a meaningful association of prosodic features with the corresponding syllable sequence. This is done with the knowledge of VOPs, detected automatically from the Hilbert envelope of the LP residual of the speech signal. The region between two successive VOPs is considered as a syllable-like region, and parameters are derived to represent duration, dynamics of F_0 contour and energy variations corresponding to each region. For evaluating the potential of prosodic features for speaker verification, experimental results are reported for NIST SRE 2003 extended data. The performance seems to be significant, especially for cases where more speech data was available for training the models.

Prosody spans over longer segments of speech such as syllables, words and sentences. Throughout this chapter, prosodic features discussed are with reference to

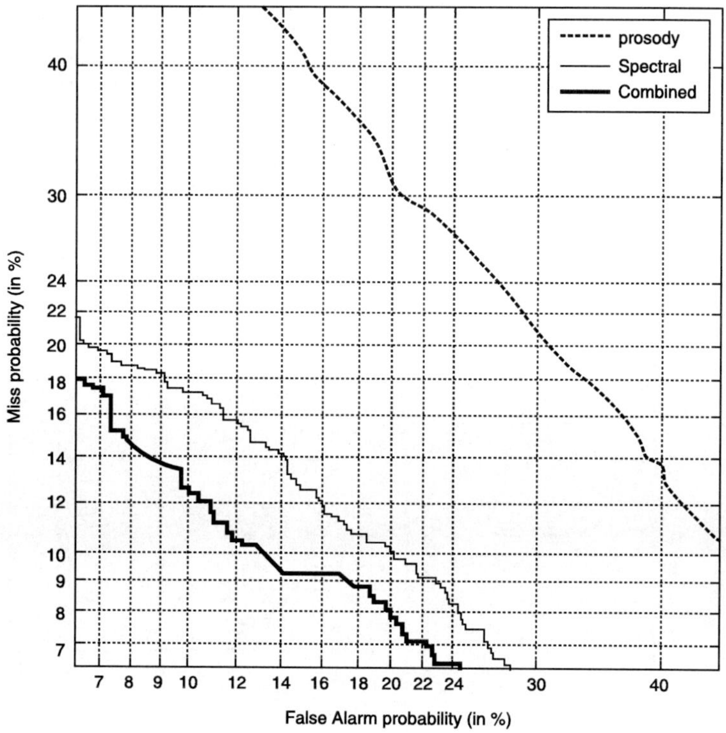

Fig. 13.18 DET *curve* showing the performance of spectral based system, prosody based system and combined system for 4-side conversational case [4]

the syllabic sequence. Prosody at the word and sentence level is also speaker-specific, and there is a possibility of using them for speaker recognition applications. Prosody corresponding to a particular word or phrase also may be useful in the context of forensic applications.

Acknowledgement The author would like to thank Prof. B. Yegnanarayana and members of Speech and Vision Laboratory of IIT Madras, India during 2002–2006 for their support to carry out the study described in this chapter.

References

1. Heck LP (2002) Integrating high-level information for robust speaker recognition in John Hopkins University workshop on SuperSID, Baltimore, Maryland. http:\\www.cslp.jhu.edu/ws2002/groups/supersid
2. Doddington GG (2001) Speaker recognition based on idiolectic differences between speakers. Proc. EUROSPEECH, Aalborg, Denmark, pp 2521–2524
3. Campbell JP (1997) Speaker recognition: a tutorial. Proc IEEE 85(9):1437–1462

4. Mary L (2006) Multilevel implicit features for language and speaker recognition. Ph. D. Thesis, Indian Institute of Technology, Madras
5. Kinnunen T, Li H (2010) An overview of text-independent speaker recognition: from features to supervectors. Speech Commun 52:12–40
6. NIST (2001) Speaker recognition evaluation website: http://www.nist.gov/speech/tests/spk/2001
7. Reynolds D, Andrews W, Campbell J, Navratil J, Peskin B, Adami A, Jin Q, Klusacek D, Abramson J, Mihaescu R, Godfrey J, Jones D, Xiang B (2003) The superSID project: exploiting high-level information for high-accuracy speaker recognition Proc. IEEE Int. Conf. Acoust., Speech and Signal Processing, Hong Kong, China, 4, pp 784–787
8. Shriberg E, Stolcke A, Hakkani-Tur D, Tur G (2000) Prosody-based automatic segmentation of speech into sentences and topics. Speech Commun 32:127–154
9. Sonmez MK, Heck L, Weintraub M, Shriberg E (1997) A lognormal tied mixture model of pitch for prosody-based speaker recognition. Proc. EUROSPEECH, Rhodes, Greece. 3, pp 1391–1394
10. Atkinson JE (1978) Correlation analysis of the physiological factors controlling fundamental voice frequency. J Acoust Soc Am 63(1):211–222
11. Yegnanarayana B, Prasanna SRM, Zachariah JM, Gupta CS (2005) Combining evidence from source, suprasegmental and spectral features for a fixed-text speaker verification system. IEEE Trans Speech Audio Process 13(4):575–582
12. Atal B (1972) Automatic speaker recognition based on pitch contours. J Acous Soc Am 52(3):1687–1697
13. Adami AG, Mihaescu R, Reynolds DA, Godfrey JJ (2003) Modeling prosodic dynamics for speaker recognition. Proc. ICASSP, Hong Kong, China, 4, pp 788–791
14. Makhoul J (1975) Linear prediction: a tutorial review. Proc IEEE 63:561–580
15. Furui S (1981) Cepstral analysis technique for automatic speaker verification. IEEE Trans Speech Audio Process 29:254–272
16. Reynolds DA, Rose R (1995) Robust text-independent speaker identification using Gaussian mixture speaker models. IEEE Trans Speech Audio Process 3:72–83
17. Reynolds DA (1996) The effect of handset variability on speaker recognition performance: Experiments on the switchboard corpus. Proc. ICASSP, Atlanta, GA, USA, 1, pp 113–116
18. Thyme-Gobbel AE, Hutchins SE (1996) On using prosodic cues in automatic language identification. Proc. Int. Conf. Spoken Language Processing, Philadelphia, PA, USA, 3, pp 1768–1772
19. Mary L, Yegnanarayana B (2008) Extraction and representation of prosodic features for language and speaker recognition. Speech Commun 50:782–796
20. Drygajlo A (2007) Forensic automatic speaker recognition. IEEE Signal Process Mag 132–135
21. Shriberg E, Stolcke A (2008) The case for automatic higher level features in forensic speaker recognition. Proc. Interspeech, Brisbane, Australia, pp 1509–1512
22. Rose P (2006) Technical speaker recognition: evaluation, types and testing of evidence. Comp Speech Lang 20:159–1914
23. Shriberg E, Ferrer L, Kajarekar S, Venkataraman A, Stolcke A (2005) Modeling prosodic feature sequences for speaker recognition. Speech Commun 46:455–472
24. Sonmez MK, Shriberg E, Heck L, Weintraub M (1998) Modeling dynamic prosodic variation for speaker variation. Proc. ICSLP, Sydney, Australia, 7, pp 3189–3192
25. Adami AG, Mihaescu R, Reynolds DA, Godfrey JJ (2003) Modeling prosodic dynamics for speaker recognition. Proc. ICASSP, Hong kong, China, 4, pp 788–791
26. Peskin B, Navratil J, Abramson J, Jones D, Klusacek D, Reynolds D, Xiang B (2003) Using prosodic and conversational features for high-performance speaker recognition: report from JHU WS'02. Proc. ICASSP, Hong kong, China, 4, pp 792–795
27. Rouas J, Farinas J, Pellegrino F, Andre-Obrecht R (2005) Rhythmic unit extraction and modelling for automatic language identification. Speech Commun 47:436–456

28. Nagarajan T, Murthy HA (2006) Language identification using acoustic log-likelihoods of syllable-like units. Speech Commun 48:913–926
29. Dehak N, Kenny P, Dumouchel P (2007) Continuous prosodic features and formant modeling with joint factor analysis for speaker verification. Proc. of Interspeech, pp 1234–1237
30. Mary L, Yegnanarayana B (2006) Prosodic features for speaker verification. Proc. of Interspeech, Pittsburgh, Pennsylvania, pp 917–920
31. MacNeilage PF (1998) The frame/content theory of evolution of speech production. Behav Brain Sci 21:499–546
32. Krakow RA (1999) Physiological organization of syllables: a review. J Phonetics 27:23–54
33. Atterer M, Ladd DR (2004) On the phonetics and phonology of "segmental anchoring" of F0: evidence from German. J Phonetics 32:177–197
34. Prasanna SRM, Gangashetty SV, Yegnanarayana B (2001) Significance of vowel onset point for speech analysis. Proc. Signal Proc. Com, Indian Institute of Science, pp. 81–88
35. Prasanna SRM (2004) Event-based analysis of speech. Ph D Thesis, Indian Institute of Technology, Madras
36. Prasanna SRM, Yegnanarayana B (2005) Detection of vowel onset point events using excitation source information. Proc. of Interspeech, pp 1133–1136
37. Prasanna SRM, Zachariah JM (2002) Detection of vowel onset point in speech. Proc. IEEE Int Conf Acoust Speech, Signal Processing, Orlando, Fl, USA 4:4159
38. Ananthapadmanabha TV (1978) Epoch extraction of voice speech. Ph. D. Thesis, Indian institute of Science, Bangalore
39. Hess W (1983) Pitch determination of speech signals. Springer, Berlin
40. Ananthapadmanabha TV, Yegnanarayana B (1979) Epoch extraction fromlinear prediction residual for identification of closed glottis interval. IEEE Trans ASSP 27:309–319
41. Ananthapadmanabha TV, Yegnanarayana B (1975) Epoch extraction of voice speech. IEEE Trans ASSP 23:562–570
42. Taylor P (2000) Analysis and synthesis of intonation using the tilt model. J Acoust Soc Am 107(3):1697–1714
43. Gussenhoven C, Reepp BH, Rietveld A, Rump HH, Terken J (1997) The perceptual prominence of fundamental frequency peaks. J Acoust Soc Am 102(5):3009–3022
44. Yegnanarayana B (1999) Artificial neural network. Prentice Hall of India, New Delhi
45. Yegnanarayana B, Kishore SP (2002) AANN-An alternative for GMM for pattern recognition. Neural Netw 15(3):459–469

Chapter 14
Speaker Identification Using Intermediate Matching Kernel-Based Support Vector Machines

A. D. Dileep and C. Chandra Sekhar

Abstract Gaussian mixture model (GMM) based approaches have been commonly used for speaker recognition tasks. Methods for estimation of parameters of GMMs include the expectation-maximization method which is a non-discriminative learning based method and the large margin method which is a discriminative learning based method. Discriminative classifier based approaches to speaker recognition include support vector machine (SVM) based classifiers using dynamic kernels such as generalized linear discriminant sequence kernel, probabilistic sequence kernel, GMM supervector kernel and Bhattacharyya distance based kernel. Recently, the intermediate matching kernel (IMK) has been proposed as a dynamic kernel for recognition of objects in an image represented using a set of local feature vectors. The IMK-based SVMs give a better performance than the state-of-the-art GMM-based approaches for speaker identification tasks, because they are well suited for meeting the basic challenge of providing reliable scores of intra-speaker variation of suspects and scores of inter-speaker variation of the potential population which is crucial to law enforcement and counter terrorism agencies in evaluating the strength of the evidence at hand. Thus, the IMK-based SVMs can be used to build the speaker recognition models in the FSR (forensic speaker recognition) systems. However, it is necessary to develop techniques to determine the strength of evidence from the outputs of SVM-based models. The SVM-based models are trained using discriminative methods and their generalization ability is good. We propose to use the IMK-based SVM classifier for speaker identification from the speech signal of an utterance represented as a set of local feature vectors. The main issue in building the IMK-based SVM classifier is selection of the virtual feature vectors using which the local feature vectors from the representations of two different utterances are matched. We explore the use of components of universal background GMM as the set of virtual feature vectors. We compare the performance of the GMM-based approaches and the dynamic kernel SVM-based approaches to speaker identification. The 2002 and 2003 NIST speaker recognition corpora are used in evaluation

C. C. Sekhar (✉)
Speech and Vision Laboratory, Department of Computer Science and Engineering,
Indian Institute of Technology Madras, Chennai 600036, Tamilnadu, India
e-mail: chandra@cse.iitm.ac.in

A. Neustein, H. A. Patil (eds.), *Forensic Speaker Recognition,*
DOI 10.1007/978-1-4614-0263-3_14, © Springer Science+Business Media, LLC 2012

of different approaches to speaker identification. Results of our studies show that the dynamic kernel SVM-based approaches give a significantly better performance than the GMM-based approaches. For speaker identification task, the IMK-based SVM gives a performance that is comparable to that of SVMs using any of the other dynamic kernels. The storage requirements and the computational complexity of the IMK-based SVMs are less than of SVMs using any of the other dynamic kernels.

14.1 Introduction

Speaker recognition tasks include speaker identification and speaker verification [1, 2]. Speaker identification involves identifying a speaker among a known set of speakers using a speech utterance produced by the speaker. Speaker verification involves accepting or rejecting the claim of a speaker based on a speech utterance and is used in a voice based authentication system. The speaker recognition tasks involve processing continuous feature vectors extracted from the speech signal of an utterance. A feature vector typically consists of features obtained by the spectral analysis of a frame of the speech signal. The Mel-frequency cepstral coefficients (MFCC) are the commonly used features. For speaker recognition tasks, the sequence information is not considered to be important. Therefore, each utterance is represented by a set of feature vectors. The size of the set is dependent on the duration of the utterance. In the pattern classification based approaches to speaker recognition tasks, the varying length patterns corresponding to the training and test examples represented as set of feature vectors are given as input to a classification model.

Two major categories of approaches to pattern classification are the generative model based approaches and the discriminative model based approaches. In the generative model based approaches, a statistical model is built for each class to capture the distribution of the training data of the class. The commonly used generative models are the Gaussian mixture model (GMM) and the hidden Markov model (HMM). The GMM is used for static pattern classification tasks in which a pattern is represented by a feature vector, and for varying length pattern classification tasks in which a pattern is represented by a set of feature vectors. The HMM is used for sequential pattern classification tasks in which a pattern is represented by a sequence of feature vectors. The Bayes classifier is used to classify a test pattern based on the posterior probabilities of the classes for the test pattern. The posterior probability of a class is computed using the likelihood of the test pattern being generated by the generative model of the class and the prior probability of the class. The maximum likelihood (ML) based method is commonly used for estimation of parameters of the generative model for each class. The ML based method is a non-discriminative training based method because the estimation of parameters for each class is done independently, i.e., the data of other classes is not used in estimation of parameters of the model for that class. Recently, the discriminative training based large margin method has been proposed for estimation of parameters of generative models [3]. In this method, the parameters of the generative models of all the classes are estimated simultaneously by solving an optimization problem to maximize the distance between the boundaries of the classes. When the number of classes is large, the optimization problem solving in the large margin method for GMMs or HMMs is computationally highly intensive. It may not be practical to implement the large margin GMM-based speaker recognition systems for a large number of speakers.

The discriminative model based approaches to pattern classification, such as the support vector machine (SVM) based approaches [4], construct the decision boundaries between the classes without having to capture the distributions of the data of the classes. The discriminative models are trained by using the data of all the classes simultaneously. The discriminative training based approaches are expected to give a better generalization performance than the non-discriminative training based approaches, especially when the classes are confusable. The SVM-based approaches have been shown to give a better classification performance in comparison to the statistical model based approaches (Bayes classifier) for several static pattern classification tasks. The choice of the kernel function used is important for the performance of SVM-based approaches, and several kernel functions have been proposed for static patterns. The kernel functions designed for static patterns are called static kernels. Recently, the SVM-based approaches have been proposed for varying length pattern classification tasks in which a varying length pattern is represented by a set of feature vectors. These approaches are suitable for speaker recognition tasks in which the speech signal of an utterance from a speaker is represented by a set of feature vectors. The main issue in building an SVM-based classifier for varying length patterns represented by sets of feature vectors is the design of a suitable kernel function that gives a measure of similarity between a pair of sequential patterns of different lengths. Kernel functions designed for varying length patterns are referred to as dynamic kernels. Recently, different types of dynamic kernels have been proposed and the SVM-based classifiers using these dynamic kernels have been developed for speaker recognition tasks [5–8]. In this chapter, we present the different types of dynamic kernels. We also study the performance of SVM-based classifiers using a dynamic kernel, namely the intermediate matching kernel (IMK), for speaker recognition.

The GMM-based classifying MFCC features based representation is the state-of-the art speaker recognition system and is adopted by the law enforcement agencies as well [9, 10]. In forensic speaker recognition (FSR), an analyst is required to determine the source (suspected speaker) for the unknown voice of the questioned recording (trace) [11, 12]. One of the difficulties in using speech as an identifying characteristic is the variabilities present in the speech. There is within-speaker (within-source) variability as well as between-speakers (between-sources) variability. Consequently, forensic speaker recognition methods should provide a statistical (probabilistic) evaluation, which attempts to give an indication of the strength of the evidence, given the estimated within-source variability and the between-sources variability. Hence, conventional speaker recognition techniques are not directly applicable in forensic analysis, and it is necessary to adapt them to the requirements of the judicial system. The core of the automatic FSR system is building a model for the features of each speaker, testing the features of utterance(s) from the questioned recording against this model and obtaining the likelihood that this utterance could have come from this speaker [9]. Statistical models such as GMMs are commonly used to build the speaker models. The estimated likelihood scores given by the speaker recognition system are then interpreted and evaluated to know the strength of the evidence [12]. In [11, 13], the automatic FSR system is built based on a Bayesian interpretation, having a two-stage modeling approach. In the first stage, the speaker models are built using GMMs. In the second stage, the scores of the within-source (intra-speaker variation of suspect) and scores of the between-sources (inter-speaker variation of the potential population) are used to model the within-source and between-sources distributions [14]. The likelihood ratio between them is considered to evaluate the strength of the evidence. The IMK-based SVMs give a better performance than the state-of-the-art GMM-based approaches for speaker identification tasks. Therefore, the IMK-based SVMs can be used to build the speaker recognition models in the FSR systems. However, it is necessary to develop techniques to determine the strength of evidence from the outputs of SVM-based models. The SVM-based models are trained using discriminative methods and their generalization ability is good. Therefore, the SVM-based models can be helpful in developing the FSR systems with an improved performance.

The organization of the rest of the chapter is as follows: Sect. 14.2 describes the SVM-based approach to pattern classification. The GMM-based approaches to speaker recognition are presented in Sect. 14.3. The dynamic kernels and the SVM-based approaches to speaker recognition are presented in Sect. 14.4. The intermediate matching kernel based SVM is described in Sect. 14.5. Our studies on speaker recognition using the GMM-based approaches and the SVM-based approaches are presented in Sect. 14.6. The summary and conclusions are presented in Sect. 14.7.

14.2 Support Vector Machines for Pattern Classification

In this section we describe the support vector machines (SVMs) for pattern classification. The SVM [4, 15, 16] is a linear two-class classifier. An SVM constructs the maximum margin hyperplane (optimal separating hyperplane) as a decision surface to separate the data points of two classes. The margin of a hyperplane is defined as the minimum distance of training data points from the hyperplane. We first present the construction of an optimal separating hyperplane for linearly separable classes. Then we present the construction of an optimal separating hyperplane for linearly non-separable classes that have some training examples which cannot be classified correctly. Finally, we discuss building an SVM for nonlinearly separable classes by constructing an optimal separating hyperplane in a high dimensional feature space induced by kernel function.

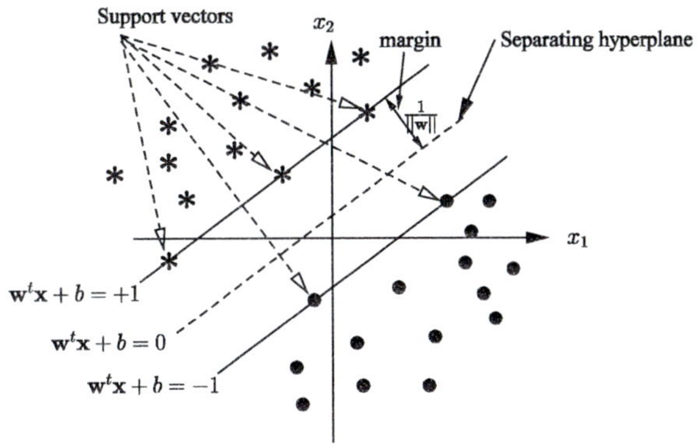

Fig. 14.1 Illustration of constructing the optimal hyperplane for linearly separable classes

14.2.1 Optimal Separating Hyperplane for Linearly Separable Classes [17]

Suppose the training data set of the two classes consists of L examples, $\{\mathbf{x}_i, y_i\}_{i=1}^{L}$, $\mathbf{x}_i \in \mathbb{R}^d$ and $y_i \in \{+1, -1\}$, where $\mathbf{x}_i$ is ith training example and y_i is the corresponding class label. Figure 14.1 illustrates the construction of an optimal separating hyperplane for linearly separable classes in the two-dimensional input space of $\mathbf{x}$. A hyperplane is specified as $\mathbf{w}^t\mathbf{x}+b=0$, where $\mathbf{w}$ is the parameter vector and b is the bias. The examples with class label $y_i=+1$ are the data points lying on the positive side of the hyperplane and the examples with class label $y_i=-1$ are the data points lying on the negative side of the hyperplane. Now, a separating hyperplane that separates the data points of two linearly separable classes satisfies the following constraints:

$$y_i(\mathbf{w}^t\mathbf{x}_i + b) > 0 \quad \text{for} \quad i = 1, 2, \ldots, L \tag{14.1}$$

The distance between the nearest example and the separating hyperplane, called the margin, is given by $1/\|\mathbf{w}\|$. The problem of finding the optimal separating hyperplane that maximizes the margin is the same as the problem of minimizing the Euclidean norm of the parameter vector $\mathbf{w}$. For reducing the search space of $\mathbf{w}$, the constraints that the optimal separating hyperplane must satisfy are specified as follows:

$$y_i(\mathbf{w}^t\mathbf{x}_i + b) \geq 1 \quad \text{for} \quad i = 1, 2, \ldots, L \tag{14.2}$$

The learning problem of finding the optimal separating hyperplane is a constrained optimization problem stated as follows: Given the training data set of two classes, find the values of $\mathbf{w}$ and b such that they satisfy the constraints in (14.2) and the parameter vector $\mathbf{w}$ minimizes the following cost function:

$$J(\mathbf{w}) = \frac{1}{2}\|\mathbf{w}\|^2 \tag{14.3}$$

The constrained optimization problem is solved using the method of Lagrangian multipliers. The primal form of the Lagrangian objective function is given by

$$\mathcal{L}_p(\mathbf{w},\ b,\ \boldsymbol{\alpha}) = \frac{1}{2}\|\mathbf{w}\|^2 + \sum_{i=1}^{L} \alpha_i[y_i(\mathbf{w}^t\mathbf{x}_i + b) - 1] \tag{14.4}$$

where the non-negative variables α_i are called Lagrange multipliers. The saddle point of the Lagrangian objective function provides the solution for the optimization problem. The solution is determined by first minimizing the Lagrangian objective function with respect to $\mathbf{w}$ and b, and then maximizing with respect to $\boldsymbol{\alpha}$. The two conditions of optimality due to minimization are as follows:

$$\frac{\partial \mathcal{L}_p(\mathbf{w}, b, \boldsymbol{\alpha})}{\partial \mathbf{w}} = \mathbf{0} \tag{14.5}$$

$$\frac{\partial \mathcal{L}_p(\mathbf{w}, b, \boldsymbol{\alpha})}{\partial b} = 0 \tag{14.6}$$

Application of optimality conditions gives

$$\mathbf{w} = \sum_{i=1}^{L} \alpha_i y_i \mathbf{x}_i \tag{14.7}$$

$$\sum_{i=1}^{L} \alpha_i y_i = 0 \tag{14.8}$$

Substituting the expression for $\mathbf{w}$ from (14.7) in (14.4) we get the dual form of the Lagrangian objective function.

$$\begin{aligned}\mathcal{L}_p(\mathbf{w}, b, \boldsymbol{\alpha}) = {} & \frac{1}{2}\left(\sum_{i=1}^{L} \alpha_i y_i \mathbf{x}_i\right)^t \left(\sum_{j=1}^{L} \alpha_j y_j \mathbf{x}_j\right) \\ & + \sum_{i=1}^{L} \alpha_i \left[y_i \left(\sum_{j=1}^{L} \alpha_j y_j \mathbf{x}_j^t\right) \mathbf{x}_i + b - 1 \right]\end{aligned} \tag{14.9}$$

$$\begin{aligned}\mathcal{L}_p(\mathbf{w}, b, \boldsymbol{\alpha}) = {} & \frac{1}{2}\left(\sum_{i=1}^{L} \alpha_i y_i \mathbf{x}_i\right)^t \left(\sum_{j=1}^{L} \alpha_j y_j \mathbf{x}_j\right) \\ & + \sum_{i=1}^{L}\sum_{j=1}^{L} \alpha_i \alpha_j y_i y_j \mathbf{x}_i^t \mathbf{x}_j - \sum_{i=1}^{L} \alpha_i y_i b + \sum_{i=1}^{L} \alpha_i\end{aligned} \tag{14.10}$$

Using the condition in (14.8) and simplifying (14.10), the dual form can be derived as a function of Lagrangian multipliers $\boldsymbol{\alpha}$, as follows:

$$\mathcal{L}_d(\boldsymbol{\alpha}) = \sum_{i=1}^{L} \alpha_i - \frac{1}{2} \sum_{i=1}^{L} \sum_{j=1}^{L} \alpha_i \alpha_j y_i y_j \mathbf{x}_i^t \mathbf{x}_j \tag{14.11}$$

The optimal values of Lagrangian multipliers are determined by maximizing the dual form of the objective function $\mathcal{L}_d(\boldsymbol{\alpha})$ subject to the following constraints:

$$\sum_{i=1}^{L} \alpha_i y_i = 0 \tag{14.12}$$

$$\alpha_i \geq 0 \quad \text{for} \quad i = 1,\ 2,\ \ldots, L \tag{14.13}$$

This optimization problem is solved using quadratic programming methods [18]. The data points for which the values of the optimal values of Lagrange multipliers are not zero are the support vectors. For these data points the distance to the optimal separating hyperplane is minimum. Hence, the support vectors are the training data points that lie on the margin, as illustrated in Fig. 14.1. Support vectors are those data points that lie closest to the decision surface and define the optimal parameter vector $\mathbf{w}^*$. For the optimal values of Lagrange multipliers $\{\alpha_j^*\}_{j=1}^{L_s}$, the optimal parameter vector $\mathbf{w}^*$ is given by

$$\mathbf{w}^* = \sum_{j=1}^{L_s} \alpha_j^* y_j \mathbf{x}_j \tag{14.14}$$

where L_s is the number of support vectors. The discriminant function of the optimal separating hyperplane in terms of support vectors is given by

$$D(\mathbf{x}) = \mathbf{w}^{*t}\mathbf{x} + b^* = \sum_{j=1}^{L_s} \alpha_j^* y_j \mathbf{x}^t \mathbf{x}_j + b^* \tag{14.15}$$

where b^* is the optimal value of bias.

Next we present a method to construct an optimal separating hyperplane for linearly non-separable classes.

14.2.2 Optimal Separating Hyperplane for Linearly Non-Separable Classes

The training data points of two linearly non-separable classes cannot be separated by a hyperplane without classification error. In such cases, it is desirable to find an optimal hyperplane that minimizes the probability of classification error for the

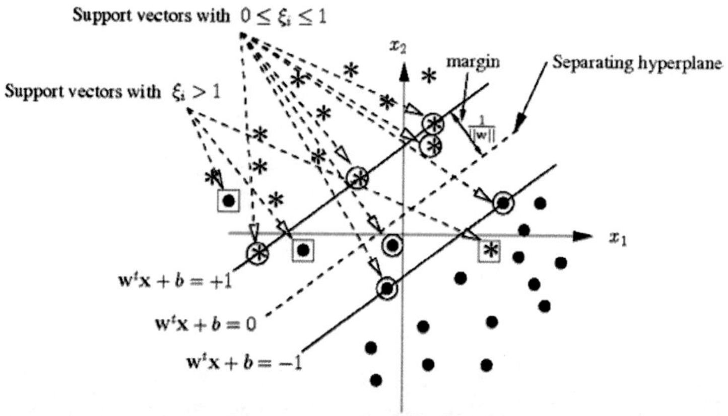

Fig. 14.2 Illustration of constructing an optimal separating hyperplane for linearly non-separable classes. Data points enclosed with circles correspond to support vectors with $0 \le \xi_i \le 1$ and data points enclosed in squares correspond to support vectors with $\xi_i > 1$

training data set. A data point is misclassified by a separating hyperplane when the data points do not satisfy the constraint in (14.2). This corresponds to a data point that falls either within the margin region or on the wrong side of the separating hyperplane as illustrated in Fig. 14.2.

For the construction of an optimal separating hyperplane for linearly non-separable classes, the constraints in (14.2) are modified by introducing the nonnegative slack variables ξ_i as follows:

$$y_i(\mathbf{w}^t\mathbf{x}_i + b) \ge 1 - \xi_i \quad \text{for} \quad i = 1, 2, \ldots, L \tag{14.16}$$

The slack variable ξ_i is a measure of the deviation of a data point $\mathbf{x}_i$ from the ideal condition of separability given in (14.2). For $0 \le \xi_i \le 1$, the data point falls inside the region of separation, but on the correct side of the separating hyperplane. For $\xi_i > 1$, the data point falls on the wrong side of the separating hyperplane. The support vectors are those particular data points that satisfy the constraint in (14.16) with equality sign. The cost function for linearly non-separable classes is given as

$$J(\mathbf{w}, \xi) = \frac{1}{2}\|\mathbf{w}\|^2 + C\sum_{i=1}^{L}\xi_i \tag{14.17}$$

where C is a user-specified positive parameter that controls the trade-off between the complexity of the classifier and the number of non-separable data points. Using the method of Lagrange multipliers to solve the constrained optimization problem as in the case of linearly separable classes, the dual form of the Lagrangian objective function can be obtained as follows [17]:

$$\mathcal{L}_d(\boldsymbol{\alpha}) = \sum_{i=1}^{L}\alpha_i - \frac{1}{2}\sum_{i=1}^{L}\sum_{j=1}^{L}\alpha_i\alpha_j y_i y_j \mathbf{x}_i^t \mathbf{x}_j \tag{14.18}$$

subject to the constraints:

$$\sum_{i=1}^{L} \alpha_i y_i = 0 \tag{14.19}$$

$$0 \leq \alpha_i \leq C \quad \text{for} \quad i = 1, 2, \ldots, L \tag{14.20}$$

It may be noted that the maximum value that the Lagrangian multipliers α_i can take is C for the linearly non-separable classes. For the optimal values of Lagrange multipliers $\{\alpha_j^*\}_{j=1}^{L_s}$, the optimal parameter vector $\mathbf{w}^*$ is given by

$$\mathbf{w}^* = \sum_{j=1}^{L_s} \alpha_j^* y_j \mathbf{x}_j \tag{14.21}$$

where L_s is the number of support vectors. The discriminant function of the optimal separating hyperplane for an input vector $\mathbf{x}$ is given by

$$D(\mathbf{x}) = \mathbf{w}^{*t}\mathbf{x} + b^* = \sum_{j=1}^{L_s} \alpha_j^* y_j \mathbf{x}^t \mathbf{x}_j + b^* \tag{14.22}$$

where b^* is the optimum bias.

14.2.3 Support Vector Machine for Nonlinearly Separable Classes

For nonlinearly separable classes, an SVM is built by mapping the input vector $\mathbf{x}_i$, $i=1, 2,\ldots, L$ into a high dimensional feature vector $\mathbf{\Phi}(\mathbf{x}_i)$ using a nonlinear transformation $\mathbf{\Phi}$, and constructing an optimal separating hyperplane defined by $\mathbf{w}^t\mathbf{\Phi}(\mathbf{x}) + b = 0$ to separate the examples of two classes in the feature space $\mathbf{\Phi}(\mathbf{x})$. This is based on the Cover's theorem [19] which states that an input space where the patterns are nonlinearly separable may be transformed into a feature space where the patterns are linearly separable with a high probability, provided two conditions are satisfied [17]. The first condition is that the transformation is nonlinear and the second condition is that the dimensionality of the feature space is high enough. Figure 14.3 illustrates the conversion of a nonlinearly separable pattern classification task in a 2-dimensional input space of $\mathbf{x}$ into a linearly separable pattern classification task in a 3-dimensional feature space of $\mathbf{\Phi}(\mathbf{x})$. It is seen that the nonlinearly separable data points $\mathbf{x}_i = [x_{i1}, x_{i2}]^t$, $i = 1, 2,\ldots, L$ in a two-dimensional input space are mapped onto three-dimensional feature vectors $\mathbf{\Phi}(\mathbf{x}_i) = \left[x_{i1}^2,\ x_{i2}^2,\ \sqrt{2}x_{i1}x_{i2}\right]^t$, $i=1, 2,\ldots, L$ where they are linearly separable.

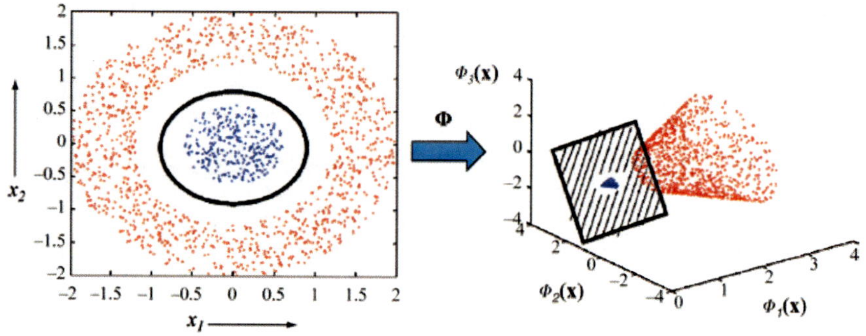

Fig. 14.3 Illustration of conversion of a nonlinearly separable pattern classification in the input space of **x** to a linearly separable pattern classification task in the feature space of **Φ(x)**

For the construction of the optimal hyperplane in the high dimensional feature space **Φ(x)**, the dual form of the Lagrangian objective function in (14.18) takes the following form:

$$\mathcal{L}_d(\boldsymbol{\alpha}) = \sum_{i=1}^{L} \alpha_i - \frac{1}{2}\sum_{i=1}^{L}\sum_{j=1}^{L} \alpha_i \alpha_j y_i y_j \boldsymbol{\Phi}(\mathbf{x}_i)^t \boldsymbol{\Phi}(\mathbf{x}_j) \tag{14.23}$$

subject to the constraints:

$$\sum_{i=1}^{L} \alpha_i y_i = 0 \tag{14.24}$$

$$0 \le \alpha_i \le C \quad \text{for} \quad i = 1, 2, \ldots, L \tag{14.25}$$

For the optimal $\boldsymbol{\alpha}^*$, the optimal parameter vector $\mathbf{w}^*$ is given by

$$\mathbf{w}^* = \sum_{j=1}^{L_s} \alpha_j^* y_j \boldsymbol{\Phi}(\mathbf{x}_j) \tag{14.26}$$

where L_s is the number of support vectors. The discriminant function of the optimal hyperplane for an input vector **x** is given by

$$D(\mathbf{x}) = \mathbf{w}^{*t} \boldsymbol{\Phi}(\mathbf{x}) + b^* = \sum_{j=1}^{L_s} \alpha_j^* y_j \boldsymbol{\Phi}(\mathbf{x})^t \boldsymbol{\Phi}(\mathbf{x}_j) + b^* \tag{14.27}$$

It is seen that the term in (14.23) and (14.27) involve computation of the innerproduct $\boldsymbol{\Phi}(\mathbf{x}_i)^t\boldsymbol{\Phi}(\mathbf{x}_j)$. Computation of innerproducts in a high dimensional feature space is avoided by using an innerproduct kernel, $K(\mathbf{x}_i, \mathbf{x}_j)$, defined as $K(\mathbf{x}_i, \mathbf{x}_j) = \boldsymbol{\Phi}(\mathbf{x}_i)^t\boldsymbol{\Phi}(\mathbf{x}_j)$ [20]. A valid innerproduct kernel $K(\mathbf{x}_i, \mathbf{x}_j)$ for two pattern vectors $\mathbf{x}_i$ and $\mathbf{x}_j$ is a symmetric function for which the Mercer's theorem holds good. The Mercer's theorem can be stated as follows [17]: Let $K(\mathbf{x}_i, \mathbf{x}_j)$ be a continuous symmetric kernel that is defined in the closed interval $\boldsymbol{u} \leq \mathbf{x}_i \leq \boldsymbol{v}$ and likewise for $\mathbf{x}_j$. The kernel $K(\mathbf{x}_i, \mathbf{x}_j)$ can be expanded in the series

$$K(\mathbf{x}_i, \mathbf{x}_j) = \sum_{l=1}^{\infty} \lambda_l \, \varphi_l(\mathbf{x}_i) \, \varphi_l(\mathbf{x}_j) \tag{14.28}$$

with positive coefficients, $\lambda_l > 0$ for all l. For this expansion to be valid and for it to converge absolutely and uniformly, it is necessary and sufficient that the condition

$$\int_{\boldsymbol{u}}^{\boldsymbol{v}}\int_{\boldsymbol{u}}^{\boldsymbol{v}} K(\mathbf{x}_i, \mathbf{x}_j) g(\mathbf{x}_i) g(\mathbf{x}_j) d\mathbf{x}_i \, d\mathbf{x}_j \tag{14.29}$$

holds for all $g(\mathbf{x}_i)$ such that

$$\int_{\boldsymbol{u}}^{\boldsymbol{v}} g^2(\mathbf{x}_i) d\mathbf{x}_i < \infty \tag{14.30}$$

The functions $\varphi_l(\mathbf{x}_i)$ are called eigenfunctions of the expansion and the coefficients λ_l are called eigenvalues of the expansion. The fact that all of the eigenvalues are positive means that the kernel $K(\mathbf{x}_i, \mathbf{x}_j)$ is positive definite. Thus, satisfying the Mercer's property is the necessary condition for all the kernel functions.

The objective function in (14.23) and the discriminant function of the optimal hyperplane in (14.27) can now be specified using the kernel function as follows:

$$\mathcal{L}_d(\boldsymbol{\alpha}) = \sum_{i=1}^{L} \alpha_i - \frac{1}{2} \sum_{i=1}^{L} \sum_{j=1}^{L} \alpha_i \alpha_j y_i y_j K(\mathbf{x}_i, \mathbf{x}_j) \tag{14.31}$$

$$D(\mathbf{x}) = \mathbf{w}^{*t} \boldsymbol{\Phi}(\mathbf{x}) + b^* = \sum_{j=1}^{L_s} \alpha_j^* y_j K(\mathbf{x}, \mathbf{x}_j) + b^* \tag{14.32}$$

The kernel gram matrix is an $L \times L$ matrix whose elements are the values of kernel function for all the pairs of examples in the training. The kernel gram matrix must be symmetric and positive semidefinite. The convergence of the convex optimization techniques used for maximizing the objective function in (14.31) needs the kernel gram matrix to be positive semidefinite.

The architecture of a support vector machine for two-class pattern classification that implements the discriminant function of the hyperplane in (14.32) is given in Fig. 14.4.

The number of hidden nodes corresponds to the number of support vectors. The training examples corresponding to the support vectors are determined by the number of hidden nodes corresponding to the number of support vectors, and the training examples corresponding to the support vectors are determined by maximizing

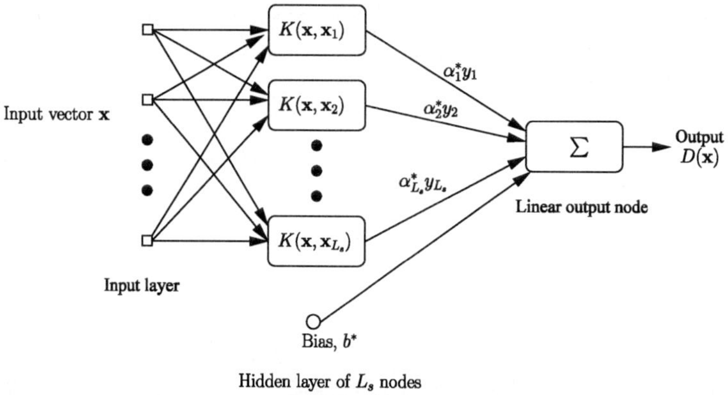

Fig. 14.4 Architecture of a support vector machine for two-class pattern classification. The class of the input pattern **x** is given by the sign of the discriminant function $D(\mathbf{x})$. The number of hidden nodes corresponds to the number of support vectors L_s. Each hidden node computes the innerproduct kernel function $K(\mathbf{x}, \mathbf{x}_i)$ on the input pattern **x** and a support vector $\mathbf{x}_i$

the objective function in (14.28) using a given training data set and for a chosen kernel function.

Some commonly used innerproduct kernel functions are as follows:

$$\text{Polynomial kernel: } K(\mathbf{x}_i, \mathbf{x}_j) = \left(a\mathbf{x}_i{}^t\mathbf{x}_j + c\right)^p$$
$$\text{Sigmoidal kernel: } K(\mathbf{x}_i, \mathbf{x}_j) = \tanh\left(a\mathbf{x}_i{}^t\mathbf{x}_j + c\right)$$
$$\text{Gaussian kernel: } K(\mathbf{x}_i, \mathbf{x}_j) = \exp\left(-\delta||\mathbf{x}_i - \mathbf{x}_j||^2\right)$$

Here, $\mathbf{x}_i$ and $\mathbf{x}_j$ are vectors in the d-dimensional input pattern space, a and c are constants, p is the degree of the polynomial and δ is a nonnegative constant used for numerical stability in Gaussian kernel function. The dimensionality of the feature space is $\binom{p+d}{d}$ for the polynomial kernel [15]. The feature spaces for the sigmoidal and Gaussian kernels are of infinite dimension. The kernel functions involve computations in the d-dimensional input space and avoid the innerproduct computation in the high dimensional feature space. The above mentioned kernel functions are kernels for vectorial representation of data. The kernels for non-vectorial representation of data include kernels on graphs, kernels on sets, kernels on text, kernels on structured data like string, trees, etc [17]. The graph kernels [21] such as random walk kernels, shortest path kernels, and all paths kernels are used in bioinformatics for protein function prediction, protein structure comparison, RNA structure comparison, and biological network comparison. The kernels on sets include intersection kernel, union complement kernel, and agreement kernel. Kernels on text include vector space kernels, semantic kernels, and latent semantic kernels [22]. These kernels are used for text categorization and document classification. Spectrum kernel [23], mismatch kernel [24, 25], String subsequence kernel [26], and marginalised kernels [27] are the examples of string kernels. These kernels are used in various applications in bioinformatics such as protein classification, protein function prediction, and protein structure prediction. The best choice of the kernel function for a given pattern classification problem is still a research issue [9]. The suitable kernel function and its parameters are chosen empirically.

The complexity of a two-class support vector machine is a function of the number of support vectors (L_s) determined during its training. Multiclass pattern classification problems are generally solved using a combination of two-class SVMs. Therefore, the complexity of a multiclass pattern classification system depends on the number of SVMs and the complexity of each SVM used. In the next subsection, we present the commonly used approaches to multiclass pattern classification using SVMs.

14.2.4 Multiclass Pattern Classification Using SVMs

Support vector machines are originally designed for two-class pattern classification. Multiclass pattern classification problems are commonly solved using a combination of two-class SVMs and a decision strategy to decide the class of the input pattern [28]. Each SVM has the architecture given in Fig. 14.4 and is trained independently. Now we present the two approaches to decomposition of the learning problem in multiclass pattern classification into several two-class learning problems so that a combination of SVMs can be used. The training data set $\{\mathbf{x}_i, c_i\}_{i=1}^{L}$ consists of L examples belonging to T classes. The class label is $c_i \in \{1, 2, \ldots, T\}$. For the sake of simplicity, we assume that the number of examples for each class is the same, *i.e.*, $L_t = L/T$.

14.2.4.1 One-Against-the-Rest Approach

In this approach, an SVM is constructed for each class by discriminating that class against the remaining $(T-1)$ classes. The classification system based on this approach consists of T SVMs. All the L training examples are used in constructing an SVM for each class. In constructing the SVM for the class t the desired output y_i for a training example $\mathbf{x}_i$ is specified as follows:

$$y_i = \begin{cases} +1, & \text{if} \quad c_i = t \\ -1, & \text{if} \quad c_i \neq t \end{cases} \tag{14.33}$$

The examples with the desired output $y_i = +1$ are called *positive* examples. The examples with the desired output $y_i = -1$ are called *negative* examples. An optimal separating hyperplane is constructed to separate L_t positive examples from $L(T-1)/T$ negative examples. The much larger number of negative examples leads to an imbalance, resulting in the dominance of negative examples in determining the decision boundary [29]. The extent of imbalance increases with the number of classes and is significantly high when the number of classes is large. A test pattern

$\mathbf{x}$ is classified by using the *winner-takes-all* strategy that uses the following decision rule:

$$\text{Class label for } \mathbf{x} = \arg\max_t D_t(\mathbf{x}) \qquad (14.34)$$

where $D_t(\mathbf{x})$ is the discriminant function of the SVM constructed for the class t.

14.2.4.2 One-Against-One Approach

In this approach, an SVM is constructed for every pair of classes by training it to discriminate the two classes. The number of SVMs used in this approach is $T(T-1)/2$. An SVM for a pair of classes s and t is constructed using $2L_t$ training examples belonging to the two classes only. The desired output y_i for a training example $\mathbf{x}_i$ is specified as follows:

$$y_i = \begin{cases} +1, & \text{if} \quad c_i = s \\ -1, & \text{if} \quad c_i = t \end{cases} \qquad (14.35)$$

The small size of the set of training examples and the balance between the number of positive and negative examples lead to a simple optimization problem to be solved in constructing an SVM for a pair of classes. When the number of classes is large, the proliferation of SVMs leads to a complex classification system.

The *maxwins* strategy is commonly used to determine the class of a test pattern $\mathbf{x}$ in this approach. In this strategy, a majority voting scheme is used. If $D_{st}(\mathbf{x})$, the value of the discriminant function of the SVM for the pair of classes s and t, is positive, then the class s wins a vote. Otherwise, the class t wins a vote. Outputs of SVMs are used to determine the number of votes won by each class. The class with the maximum number of votes is assigned to the test pattern. When there are multiple classes with the same maximum number of votes, the class with the maximum value of the total magnitude of discriminant functions (TMDF) is assigned. The total magnitude of discriminant functions for the class s is defined as follows:

$$\text{TMDF} = \sum_t |D_{st}(\mathbf{x})| \qquad (14.36)$$

where the summation is over all t with which the class s is paired. The maxwins strategy needs evaluation of discriminant functions of all the SVMs in deciding the class of a test pattern.

The SVM-based classifiers have been successfully used in various applications like image categorization, object categorization [30], text classification [26], handwritten character recognition [31], speech recognition [16], speaker recognition and verification [5, 7], and speech emotion recognition [32]. In this paper, we use the SVM-based classifier for the speaker recognition task. However, Gaussian mixture models (GMMs) are commonly used for speaker recognition. In the next section, we describe the GMM-based approaches to speaker recognition.

14.3 GMM-Based Approaches to Speaker Recognition

Gaussian mixture models are commonly used for speaker recognition and verification tasks. Generative approaches like GMMs focus on fitting a statistical model for a given data by estimating the density of the data. The methods used for estimating the parameters of a GMM are based on non-discriminative learning as the model for each class is trained independently.

In this section, we present the different approaches based on GMMs that are commonly used for speaker recognition.

14.3.1 *Gaussian Mixture Model*

Let $\mathbf{D}=\{\mathbf{x}_1, \mathbf{x}_2, \ldots, \mathbf{x}_{Nc}\}$ be the set of feature vectors extracted from the training utterances of a speaker [1]. The number of feature vectors in the set $\mathbf{D}$ is N_c. The likelihood of a feature vector $\mathbf{x}$ being generated by a GMM that represents a linear superposition of Q Gaussian components, and specified by a set of parameters $\boldsymbol{\theta}$ is given by

$$p(\mathbf{x}|\boldsymbol{\theta}) = \sum_{q=1}^{Q} \pi_q \mathcal{N}(\mathbf{x}\,|\boldsymbol{\mu}_q, \boldsymbol{\Sigma}_q) \tag{14.37}$$

where π_q, $\boldsymbol{\mu}_q$, and $\boldsymbol{\Sigma}_q$ are the mixture weight, mean vector, and covariance matrix of the qth component respectively, and $\mathcal{N}(\mathbf{x}|\boldsymbol{\mu}_q, \Sigma_q)$. The set of parameters, $\boldsymbol{\theta}=\{\pi_q, \boldsymbol{\mu}_q, \boldsymbol{\Sigma}_q\}$, $q = 1, 2, \ldots, Q$ is estimated from the training data $\mathbf{D}$ [32]. Maximum likelihood (ML) method is commonly used for estimating the set of parameters.

In the ML method, the parameters $\boldsymbol{\theta}$ of a GMM are estimated using the following criterion:

$$\begin{aligned}
\boldsymbol{\theta}^{\text{ML}} &= \arg\max_{\boldsymbol{\theta}} \ \ln p(\mathbf{D}|\boldsymbol{\theta}) \\
\boldsymbol{\theta}^{\text{ML}} &= \arg\max_{\boldsymbol{\theta}} \sum_{n=1}^{Nc} \ln p(\mathbf{x_n}|\boldsymbol{\theta}) \\
\boldsymbol{\theta}^{\text{ML}} &= \arg\max_{\boldsymbol{\theta}} \sum_{n=1}^{Nc} \ln \sum_{q=1}^{Q} \pi_q \mathcal{N}(\mathbf{x}_n\,|\boldsymbol{\mu}_q, \boldsymbol{\Sigma}_q)
\end{aligned} \tag{14.38}$$

As a closed form expression for solution of (14.42) does not exists, it is required to use an iterative algorithm such as expectation-maximization (EM). After a suitable initialization of parameters $\boldsymbol{\theta}$, the following two steps in the EM method are repeated until convergence to obtain an ML estimate of $\boldsymbol{\theta}$ as $\boldsymbol{\theta}^{\text{ML}}$ [33]:

Expectation (E)-Step: Compute the responsibility terms, γ_{nq}, $n=1, 2, \ldots, N_c$ and $q=1, 2, \ldots, Q$, using the current values of $\boldsymbol{\theta}$ as

$$\gamma_{nq} = \frac{\pi_q \mathcal{N}(\mathbf{x}_n \,|\boldsymbol{\mu}_q, \boldsymbol{\Sigma}_q)}{\sum_{j=1}^{Q} \pi_j \mathcal{N}(\mathbf{x}_j \,|\boldsymbol{\mu}_j, \boldsymbol{\Sigma}_j)} \tag{14.39}$$

The responsibility term γ_{nq} corresponds to the posterior probability of the qth component for $\mathbf{x}_n$.

Maximization (M)-Step: Re-estimate the parameters using γ_{nq} to obtain new estimates of parameters, $\boldsymbol{\theta}^{new}$ as follows:

$$\begin{aligned}
\boldsymbol{\mu}_q^{new} &= \frac{1}{N_q} \sum_{n=1}^{N_c} \gamma_{nq} \mathbf{x}_n \\
\boldsymbol{\Sigma}_q^{new} &= \frac{1}{N_q} \sum_{n=1}^{N_c} \gamma_{nq} \left(\mathbf{x}_n - \boldsymbol{\mu}_q^{new}\right)\left(\mathbf{x}_n - \boldsymbol{\mu}_q^{new}\right)^t \\
\pi_q^{new} &= \frac{N_q}{N_c}
\end{aligned} \tag{14.40}$$

where $N_q = \sum_{n=1}^{N_c} \gamma_{nq}$

Given a test utterance, represented by a set of feature vectors as, $T=\{\mathbf{t}_1, \mathbf{t}_2, \ldots, \mathbf{t}_M\}$, where $\mathbf{t}_l \in \mathbb{R}^d$, the class label y for T is assigned according to the decision rule given below:

$$y = \arg\max_c \sum_{l=1}^{M} \ln p(\mathbf{t}_l | \boldsymbol{\theta}_c^{\mathrm{ML}}) \tag{14.41}$$

where $\boldsymbol{\theta}_c^{\mathrm{ML}}$ is the set of parameters estimated using the ML method from the training data of the class c.

As the objective function in (14.41) is a nonlinear function of $\boldsymbol{\theta}$, it may have several maxima. Depending on the initial values of $\boldsymbol{\theta}$, the EM method gives an estimate corresponding to a local maximum. This point estimate is used in the computation of the likelihood for the test data. The number of components in the GMM, i.e, the value of Q is empirically chosen. A large value of Q may lead to overfitting. Another limitation of the ML method is that, it may not provide a good estimate of parameters when sufficient amount of training data is not available. In such situations, an alternate method of obtaining robust parameter estimates is through model adaptation. In the next subsection we briefly describe the process of obtaining an estimate of parameters using model adaptation.

14.3.2 GMM Adaptation

Estimation of GMM parameters using the ML method yields robust estimates only when sufficient training data is available. For many tasks like speaker identification, the amount of training data available per speaker is limited. In such cases, robust estimates can be obtained through adaptation of the universal background model (UBM) [34]. The adaptation provides a tight coupling between the speaker model and UBM. The UBM is a large GMM trained using the training data of all the speakers to represent the speaker independent distribution of data. A speaker dependent GMM is obtained by adapting the UBM to the data of that speaker. Maximum a posteriori (MAP) adaptation method is commonly used for the adaptation. The adaptation is carried out using the EM method. The first step of the EM method estimates the sufficient statistics such as mixture weight, mean, and variance of the training data of a class for each component in the UBM. In the second step of the EM method, these new estimates of sufficient statistics are then combined with the old sufficient statistics from the UBM parameters using a data dependent mixing coefficient. The data-dependent mixing coefficient is designed so that the components with high counts of data from the class rely more on the new sufficient statistics for final parameter estimation, and the components with low counts of data from the class rely more on the old sufficient statistics for final parameter estimation.

Given a UBM and training data of a class, $\mathbf{D}=\{\mathbf{x}_1, \mathbf{x}_2, \ldots, \mathbf{x}_{Nc}\}$, the probabilistic alignment of the vectors onto the UBM components is determined. That is, for the component q in the UBM, the probability of $\mathbf{x}_n$ being generated by the component q, γ_{nq} is computed as

$$\gamma_{nq} = \frac{\pi_q \mathcal{N}(\mathbf{x}_n \,|\boldsymbol{\mu}_q, \boldsymbol{\Sigma}_q)}{\sum_{j=1}^{Q} \pi_j \mathcal{N}(\mathbf{x}_j \,|\boldsymbol{\mu}_j, \boldsymbol{\Sigma}_j)} \tag{14.42}$$

The effective number of examples belonging to the component q is given as

$$N_q = \sum_{n=1}^{N_c} \gamma_{nq} \tag{14.43}$$

The weighted mean of the examples belonging to the qth component is obtained as

$$E_q(\mathbf{x}_n) = \frac{1}{N_q} \sum_{n=1}^{N_c} \gamma_{nq} \mathbf{x}_n \tag{14.44}$$

The variance of the effective examples belonging to the qth component is obtained as

$$E_q(\mathbf{x}_n^2) = \frac{1}{N_q} \sum_{n=1}^{N_c} \gamma_{nq} \mathbf{x}_n^2 \tag{14.45}$$

where $\mathbf{x}_n^2$ is the short-hand notation for diag$(\mathbf{x}_n\mathbf{x}_n^t)$ [34].

These new sufficient statistics obtained from the training data are used to update the old UBM sufficient statistics for the component q and obtain the adapted parameters for the component q as follows:

$$\widehat{\pi}_q = \left[\lambda_q^{(\pi)}\frac{N_q}{N_c} + \left(1 - \lambda_q^{(\pi)}\right)\pi_q\right]\tau \tag{14.46}$$

$$\hat{\boldsymbol{\mu}}_q = \lambda_q^{(\mu)}\frac{E_q(\mathbf{x}_n)}{N_c} + (1 - \lambda_q^{(\mu)})\boldsymbol{\mu}_q \tag{14.47}$$

$$\widehat{\boldsymbol{\sigma}}_q^2 = \lambda_q^{(v)}\frac{E_q(\mathbf{x}^2)}{N_c} + \left(1 - \lambda_q^{(v)}\right)\left(\boldsymbol{\sigma}_q^2 + \boldsymbol{\mu}_q^2\right) - \widehat{\boldsymbol{\mu}}_q^2 \tag{14.48}$$

The adaptation coefficients $\left\{\lambda_q^{(\pi)}, \lambda_q^{(\mu)}, \lambda_q^{(v)}\right\}$ control the balance between the old and new estimates. The scale factor, τ, in (14.46) is used to ensure that the sum of all the mixture weights is unity. The data-dependent adaptation coefficient $\lambda_q^{(\rho)}, \rho \in \{\pi, \mu, v\}$, used in the above equations is defined as

$$\lambda_q^{(\rho)} = \frac{N_q}{N_q + r^{(\rho)}} \tag{14.49}$$

where $r^{(\rho)}$ is a is a fixed relevance factor for parameter ρ.

The ML method for estimation of parameters of a GMM is based on non-discriminative learning, as the model for each class is trained independently. For confusable classes, the discriminative learning based approaches may give a better performance compared to the non-discriminative learning based approaches. The large margin method has been proposed for discriminative learning of GMMs [3]. Recently, the SVM-based approaches to speaker identification and verification have been proposed. These approaches use the dynamic kernels for handling the varying length patterns extracted from speech utterances [5–8]. In the next section, we describe the SVM-based approaches to speaker recognition.

14.4 SVM-Based Approaches to Speaker Recognition

In the speaker recognition task, the speech signal of an utterance processed using a short-time analysis technique is represented as a set of feature vectors. The size of the set of feature vectors depends on the duration of the utterance. The SVM with standard kernels like Gaussian kernel and polynomial kernel cannot handle such varying length patterns. The kernels designed for varying length patterns are referred to as dynamic kernels [35]. Different approaches for designing dynamic kernels are as follows: (1) Explicit mapping based approaches [5, 6], where a set of feature vectors is mapped onto a fixed dimensional representation and a kernel function is defined in the space of that representation, (2) probabilistic distribution based approaches [7, 8], where a suitable distance measure for two sets of feature vectors is kernelized, and (3) Matching based approaches [36, 37], where a kernel function is defined by matching the feature vectors in the pair of examples. Some of the dynamic kernels commonly used for speaker recognition and verification tasks are the generalized linear discriminant sequence kernel [5], probabilistic sequence kernel [6], GMM supervector kernel [7], and Bhattacharyya distance based kernel [8]. The generalized linear discriminant sequence kernel and probabilistic sequence kernel are designed using the first approach. The GMM supervector kernel and Bhattacharyya distance based kernel are designed using the second approach. In this section, we describe these dynamic kernels.

14.4.1 Generalized Linear Discriminant Sequence Kernel

Generalized linear discriminant sequence (GLDS) kernel [5] uses an explicit expansion into a kernel feature space defined by the polynomials of degree p. The GLDS kernel is derived from the generalized linear discriminant method used in [38]. In [38] polynomials are considered as the generalized linear discriminant functions. For speaker verification task in [38], the polynomial classifier estimates the parameters of the speaker model, $\mathbf{w}_{spk}$ based on the following criteria:

$$\mathbf{w}_{spk} = \arg\min_{\mathbf{w}} \left[\sum_{i=1}^{T_{tgt}} \left| \mathbf{w}^t \boldsymbol{\Psi}(\mathbf{x}_i) - 1 \right|^2 + \sum_{j=1}^{T_{imp}} \left| \mathbf{w}^t \boldsymbol{\Psi}(\mathbf{y}_j) \right|^2 \right] \tag{14.50}$$

where T_{tgt} is the total number of feature vectors from the training data of the target speaker, T_{imp} is the total number of feature vectors from the training data of imposters, $\mathbf{x}_i$ are feature vectors from the training data of the target speaker, $\mathbf{y}_i$ are feature vectors from the training data of imposters, $\boldsymbol{\Psi}(\mathbf{x}_i)$ and $\boldsymbol{\Psi}(\mathbf{y}_i)$ are the polynomial expansions of the of $\mathbf{x}_i$ and $\mathbf{y}_i$ respectively.

The GLDS kernel is derived using the polynomial expansions of feature vectors in the examples represented as sets of feature vectors. Let $\mathbf{X}=\mathbf{x}_1, \mathbf{x}_2 \ldots \mathbf{x}_m \ldots \mathbf{x}_M$, where $\mathbf{x}_m \in \mathbb{R}^d$, be a set of M feature vectors. A feature vector $\mathbf{x}_m$ is represented in a higher dimensional feature space $\boldsymbol{\Psi}$ as a polynomial expansion $\boldsymbol{\Psi}(\mathbf{x}_m)=[\psi_1(\mathbf{x}_m), \psi_2(\mathbf{x}_m), \ldots, \psi_r(\mathbf{x}_m)]^t$, where r is the number of monomials of elements of $\mathbf{x}_m$. The expansion $\boldsymbol{\Psi}(\mathbf{x}_m)$ includes all monomials of elements of $\mathbf{x}_m$ upto and including degree p. The set of feature vectors $\mathbf{X}$ is represented as a fixed dimensional vector $\boldsymbol{\Phi}(\mathbf{X})$ which is obtained as follows:

$$\boldsymbol{\Phi}^{\mathrm{GLDS}}(\mathbf{X}) = \frac{1}{M} \sum_{m=1}^{M} \boldsymbol{\Psi}(\mathbf{x}_m) \tag{14.51}$$

The GLDS kernel between two examples $\mathbf{X} = \{\mathbf{x}_1, \mathbf{x}_2, \ldots, \mathbf{x}_M\}$ and $\mathbf{Y} = \{\mathbf{y}_1, \mathbf{y}_2, \ldots, \mathbf{y}_N\}$ is given as

$$K^{\text{GLDS}}(\mathbf{X},\mathbf{Y}) = \left(\frac{1}{M}\sum_{m=1}^{M}\boldsymbol{\Psi}(\mathbf{x}_m)\right)^t \mathbf{S}^{-1}\left(\frac{1}{N}\sum_{n=1}^{N}\boldsymbol{\Psi}(\mathbf{y}_n)\right) \tag{14.52}$$

Let T be the total number of feature vectors from all the examples in the training data set which includes the data belonging to two classes. The correlation matrix $\mathbf{S}$ is defined as follows:

$$\mathbf{S} = \frac{1}{T}\mathbf{R}^t\mathbf{R} \tag{14.53}$$

where $\mathbf{R}$ is the matrix whose rows are the polynomial expansions of the feature vectors in the training set. When the correlation matrix $\mathbf{S}$ is a diagonal matrix, the GLDS kernel is given as

$$K^{\text{GLDS}}(\mathbf{X},\mathbf{Y}) = \left(\frac{1}{M}\sum_{m=1}^{M}\mathbf{S}^{-1/2}\boldsymbol{\Psi}(\mathbf{x}_m)\right)^t \left(\frac{1}{N}\sum_{n=1}^{N}\mathbf{S}^{-1/2}\boldsymbol{\Psi}(\mathbf{y}_n)\right) \tag{14.54}$$

When the identity matrix is considered for $\mathbf{S}$, the GLDS kernel in (14.52) turns out to be

$$K^{\text{GLDS}}(\mathbf{X},\mathbf{Y}) = \frac{1}{M}\frac{1}{N}\sum_{m=1}^{M}\sum_{n=1}^{N}\boldsymbol{\Psi}(\mathbf{x}_m)^t\boldsymbol{\Psi}(\mathbf{y}_n) = \frac{1}{M}\frac{1}{N}\sum_{m=1}^{M}\sum_{n=1}^{N}k(\mathbf{x}_m, \mathbf{y}_n) \tag{14.55}$$

where $k(\mathbf{x}_m, \mathbf{y}_n)$ is the polynomial kernel function of degree p between $\mathbf{x}_m$ and $\mathbf{y}_n$.

14.4.2 *Probabilistic Sequence Kernel*

Probabilistic sequence kernel (PSK) [6] maps a set of feature vectors onto a probabilistic feature vector obtained using generative models. It is known that the Gaussian components of a GMM trained for a speaker correspond to the underlying broad phonetic classes for that speaker [34] and the different phonetic classes have unequal discrimination power between the speakers [39]. This is the motivation for using speaker-dependent weighing of the Gaussian probabilities in the design of PSK to enhance separation of that speaker from the others. The PSK uses the universal background model (UBM) with Q mixtures [34] and the class-specific GMM obtained by adapting UBM. The likelihood of a feature vector $\mathbf{x}$ being generated by the $2Q$-mixture GMM that includes the UBM and class-specific GMM is given as

$$p(\mathbf{x}) = \sum_{q=1}^{2Q} p(\mathbf{x}|q)P(q) \tag{14.56}$$

where $P(q)$ denotes the mixture weight and $p(\mathbf{x}|q) = \mathcal{N}(\mathbf{x}|\boldsymbol{\mu}_q, \boldsymbol{\Sigma}_q)$. The normalized Gaussian basis function for the qth component is defined as

$$\psi_q(\mathbf{x}) = \frac{p(\mathbf{x}\,|q\,)P(q)}{\sum_{q'=1}^{2Q} p(\mathbf{x}\,|q'\,)P(q')} \tag{14.57}$$

A feature vector $\mathbf{x}$ is represented in a higher dimensional feature space as a vector of normalized Gaussian basis functions, $\boldsymbol{\Psi}(\mathbf{x})=[\psi_1(\mathbf{x}), \psi_2(\mathbf{x}), \ldots, \psi_{2Q}(\mathbf{x})]^t$. Since the element $\psi_q(\mathbf{x})$ indicates the probabilistic alignment of $\mathbf{x}$ to the qth component, $\boldsymbol{\Psi}(\mathbf{x})$ is called as the probabilistic alignment vector. A set of feature vectors $\mathbf{X}=\{\mathbf{x}_1, \mathbf{x}_2, \ldots, \mathbf{x}_M\}$ is represented as a fixed dimensional vector $\boldsymbol{\Phi}(\mathbf{X})$ in the higher dimensional space, as given by

$$\boldsymbol{\Phi}^{\mathrm{PSK}}(\mathbf{X}) = \frac{1}{M}\sum_{m=1}^{M} \boldsymbol{\Psi}(\mathbf{x}_m) \tag{14.58}$$

Then, the PSK between two examples $\mathbf{X}=\{\mathbf{x}_1, \mathbf{x}_2, \ldots, \mathbf{x}_M\}$ and $\mathbf{Y}=\{\mathbf{y}_1, \mathbf{y}_2, \ldots, \mathbf{y}_N\}$ is given as

$$K^{\mathrm{PSK}}(\mathbf{X},\mathbf{Y}) = \left(\frac{1}{M}\sum_{m=1}^{M} \boldsymbol{\Psi}(\mathbf{x}_m)\right)^t \mathbf{S}^{-1} \left(\frac{1}{N}\sum_{n=1}^{N} \boldsymbol{\Psi}(\mathbf{y}_n)\right) \tag{14.59}$$

where $\mathbf{S}$ is the correlation matrix as in (14.53), except that it is obtained using the probabilistic alignment vectors.

14.4.3 GMM Supervector Kernel

The GMM supervector (GMMSV) kernel [2, 7] performs a mapping of a set of feature vectors onto a higher dimensional feature vector corresponding to a GMM supervector. An UBM is built using the training examples of all the classes. An example-specific GMM is built for each example by adapting the UBM using the data of that example. An example is represented by a supervector obtained by stacking the mean vectors of the components of the example-specific GMM. A GMM supervector kernel is designed using a distance measure between the supervectors of two examples. In [7], Kullback-Leibler (KL) divergence is used as the distance measure between supervectors.

Let $p(\mathbf{x}) = \sum_{q=1}^{Q} \pi_q \mathcal{N}(\mathbf{x}|\boldsymbol{\mu}_q \boldsymbol{\Sigma}_q)$ be the probability density function represented by the UBM with Q components. Let $p_{\mathbf{X}}(\mathbf{x}) = \sum_{q-1}^{Q} \pi_q^{(\mathbf{X})} \mathcal{N}\left(\mathbf{x}\,|\boldsymbol{\mu}_q^{(\mathbf{X})} \boldsymbol{\Sigma}_q^{(\mathbf{X})}\right)$ and $p_{\mathbf{Y}}(\mathbf{x}) = \sum_{q=1}^{Q} \pi_q^{(\mathbf{Y})} \mathcal{N}(\mathbf{x}|\boldsymbol{\mu}_q^{(\mathbf{Y})} \Sigma_q^{(\mathbf{Y})})$ be the example-specific GMMs obtained by adapting the UBM to the examples $\mathbf{X}=\{\mathbf{x}_1, \mathbf{x}_2, \ldots, \mathbf{x}_M\}$ and $\mathbf{Y}=\{\mathbf{y}_1, \mathbf{y}_2, \ldots, \mathbf{y}_N\}$ respectively. When only mean vector adaptation is considered, an approximation for the distance between two GMMs is considered by bounding the KL divergence with the log-sum inequality as follows:

$$KL(p_{\mathbf{X}}(\mathbf{x})||p_{\mathbf{Y}}(\mathbf{x})) \leq \sum_{q=1}^{Q} \pi_q KL\left(\mathcal{N}(\mathbf{x}|\boldsymbol{\mu}_q^{(\mathbf{X})}\boldsymbol{\Sigma}_q)||\mathcal{N}(\mathbf{x}|\boldsymbol{\mu}_q^{(\mathbf{Y})}\boldsymbol{\Sigma}_q)\right) \quad (14.60)$$

For diagonal covariance matrices, the closed form expression for the distance between two example-specific GMMs is given by

$$D\left(\boldsymbol{\mu}_q^{(\mathbf{X})}, \boldsymbol{\mu}_q^{(\mathbf{Y})}\right) = \frac{1}{2}\sum_{q=1}^{Q} \pi_q \left(\boldsymbol{\mu}_q^{(\mathbf{X})} - \boldsymbol{\mu}_q^{(\mathbf{Y})}\right)^t \Sigma_q^{-1} \left(\boldsymbol{\mu}_q^{(\mathbf{X})} - \boldsymbol{\mu}_q^{(\mathbf{Y})}\right) \quad (14.61)$$

The distance in (14.60) is symmetric and can be used for kernel computation. The resulting GMMSV kernel is given as

$$K^{\text{GMMSV}}(\mathbf{X},\mathbf{Y}) = \sum_{q=1}^{Q} \left(\sqrt{\pi_q}\boldsymbol{\Sigma}_q^{-1/2}\boldsymbol{\mu}_q^{(\mathbf{X})}\right)^t \left(\sqrt{\pi_q}\boldsymbol{\Sigma}_q^{-1/2}\boldsymbol{\mu}_q^{(\mathbf{Y})}\right) \quad (14.62)$$

It is seen that the GMMSV kernel is a linear kernel on to the GMM supervector representations of two examples. The feature space of the GMMSV kernel represents a diagonal scaling using $\sqrt{\pi_q}\boldsymbol{\Sigma}_q$ of the GMM supervector, i.e., $\boldsymbol{\Phi}(\mathbf{X}) = \sum_{q=1}^{Q} \sqrt{\pi_q}\boldsymbol{\Sigma}_q^{-1/2}\boldsymbol{\mu}_q^{(\mathbf{X})}$. Hence the resulting kernel satisfies the Mercer property.

In [7], only the adapted means are considered in forming a supervector. However, significant information is present in the covariance terms. In [40], the covariance terms are also considered to compute the kernel. Here the symmetric KL divergence is used as the distance measure to compare two GMMs. The supervector kernel for two sets of feature vectors **X** and **Y** is given by

$$\begin{aligned} K^{\text{COVSV}}(\mathbf{X},\mathbf{Y}) = &\sum_{q=1}^{Q} \left(\sqrt{\pi_q}\boldsymbol{\Sigma}_q^{-1/2}\boldsymbol{\mu}_q^{(\mathbf{X})}\right)^t \left(\sqrt{\pi_q}\boldsymbol{\Sigma}_q^{-1/2}\boldsymbol{\mu}_q^{(\mathbf{Y})}\right) \\ &+\sum_{q=1}^{Q} \frac{\pi_q}{2}\text{tr}\left(\boldsymbol{\Sigma}_q^{(\mathbf{X})}\boldsymbol{\Sigma}_q^{-2}\boldsymbol{\Sigma}_q^{(\mathbf{Y})}\right) \end{aligned} \quad (14.63)$$

where $\boldsymbol{\Sigma}_q$ is the diagonal covariance matrix of qth component of UBM, $\boldsymbol{\Sigma}_q^{(\mathbf{X})}$ and $\boldsymbol{\Sigma}_q^{(\mathbf{Y})}$ are the diagonal covariance matrices of qth adapted components corresponding to **X** and **Y**.

One way of obtaining the kernel function is exponentiating a distance metric [21]. In [41], a nonlinear GMM supervector kernel is introduced. It is seen that the distance $D(\boldsymbol{\mu}_q^{(\mathbf{X})}, \boldsymbol{\mu}_q^{(\mathbf{Y})})$ in (14.60) is symmetric and satisfies the Mercer property. The nonlinear GMM supervector (NLGMMSV) kernel is obtained as

$$K^{\text{NLGMMSV}}(\mathbf{X},\mathbf{Y}) = e^{-\delta D\left(\boldsymbol{\mu}_q^{(\mathbf{X})},\boldsymbol{\mu}_q^{(\mathbf{Y})}\right)} \tag{14.64}$$

where δ is a constant used for numerical stability.

14.4.4 Bhattacharyya Distance Based Kernel

An alternative measure of similarity between two distributions is the Bhattacharyya affinity measure [42, 43]. The Bhattacharyya distance between two probability distributions $p(\mathbf{x})$ and $g(\mathbf{x})$ defined over $\mathbf{x}$ is given by

$$\mathcal{B}(p(\mathbf{x})\,\|g(\mathbf{x})) = \int_{-\infty}^{\infty} \sqrt{p(\mathbf{x})}\sqrt{g(\mathbf{x})}d\mathbf{x} \tag{14.65}$$

Let $p(\mathbf{x}) = \mathcal{N}(\mathbf{x}\,|\boldsymbol{\mu}^{(p)}, \boldsymbol{\Sigma}^{(p)})$ and $g(\mathbf{x}) = \mathcal{N}(\mathbf{x}\,|\boldsymbol{\mu}^{(g)}, \boldsymbol{\Sigma}^{(g)})$ be two Gaussian distributions. The closed form expression for Bhattacharyya distance [44] between $p(\mathbf{x})$ and $g(\mathbf{x})$ is given by

$$\mathcal{B}(p(\mathbf{x})\,\|g(\mathbf{x})) = \frac{1}{8}(\boldsymbol{\mu}^{(p)} - \boldsymbol{\mu}^{(g)})^t \left(\frac{\boldsymbol{\Sigma}^{(p)} + \boldsymbol{\Sigma}^{(g)}}{2}\right)^{-1} (\boldsymbol{\mu}^{(p)} - \boldsymbol{\mu}^{(g)}) + \frac{1}{2}\ln\frac{\left|\frac{\boldsymbol{\Sigma}^{(p)} + \boldsymbol{\Sigma}^{(g)}}{2}\right|}{\sqrt{|\boldsymbol{\Sigma}^{(p)}||\boldsymbol{\Sigma}^{(g)}|}} \tag{14.66}$$

This can be extended to compare two distributions represented by GMMs. Let $p_{\mathbf{X}}(\mathbf{x}) = \sum_{q=1}^{Q} \pi_q^{(\mathbf{X})}\mathbf{N}\left(\mathbf{x}\,|\boldsymbol{\mu}_q^{(\mathbf{X})},\boldsymbol{\Sigma}_q^{(\mathbf{X})}\right)$ and $p_{\mathbf{Y}}(\mathbf{x}) = \sum_{q=1}^{Q} \pi_q^{(\mathbf{Y})}\mathbf{N}\left(\mathbf{x}\,|\boldsymbol{\mu}_q^{(\mathbf{Y})},\boldsymbol{\Sigma}_q^{(\mathbf{Y})}\right)$ be the GMMs for the examples $\mathbf{X}=\{\mathbf{x}_1, \mathbf{x}_2, \ldots, \mathbf{x}_M\}$ and $\mathbf{Y}=\{\mathbf{y}_1, \mathbf{y}_2, \ldots, \mathbf{y}_N\}$ respectively. The closed form expression for Bhattacharyya distance between $p_{\mathbf{X}}(\mathbf{x})$ and $p_{\mathbf{Y}}(\mathbf{x})$ is given using the log-sum inequality as

$$\mathcal{B}\,(p_{\mathbf{X}}(\mathbf{x})\,\|\,p_{\mathbf{Y}}(\mathbf{x})) = \frac{1}{8}\sum_{q=1}^{Q}\left[\left(\boldsymbol{\mu}_q^{(\mathbf{X})} - \boldsymbol{\mu}_q^{(\mathbf{Y})}\right)^t \left(\frac{\boldsymbol{\Sigma}_q^{(\mathbf{X})} + \boldsymbol{\Sigma}_q^{(\mathbf{Y})}}{2}\right)^{-1} \left(\boldsymbol{\mu}_q^{(\mathbf{X})} - \boldsymbol{\mu}_q^{(\mathbf{Y})}\right)\right] + \frac{1}{2}\sum_{q=1}^{Q}\left[\ln\frac{\left|\frac{\boldsymbol{\Sigma}_q^{(\mathbf{X})} + \boldsymbol{\Sigma}_q^{(\mathbf{Y})}}{2}\right|}{\sqrt{|\boldsymbol{\Sigma}_q^{(\mathbf{X})}||\boldsymbol{\Sigma}_q^{(\mathbf{Y})}|}}\right] - \frac{1}{2}\sum_{q=1}^{Q}\ln\left(\pi_q^{(\mathbf{X})}\pi_q^{(\mathbf{Y})}\right) \tag{14.67}$$

The Bhattacharyya distance measure is symmetric and the corresponding kernel gram matrix is shown to be positive semidefinite in [44]. Hence it can be used as a kernel function.

In [45], the Bhattacharyya mean distance is used to represent the similarity between two GMMs. Let $p(\mathbf{x}) = \sum_{q=1}^{Q} \pi_q \mathcal{N}(\mathbf{x} | \boldsymbol{\mu}_q, \boldsymbol{\Sigma}_q)$ be the UBM with Q components and $p_{\mathbf{X}}(\mathbf{x}) = \sum_{q=1}^{Q} \pi_q^{(\mathbf{X})} \mathcal{N}\left(\mathbf{x} \middle| \boldsymbol{\mu}_q^{(\mathbf{X})}, \boldsymbol{\Sigma}_q^{(\mathbf{X})}\right)$ be the GMM obtained by adapting the UBM to the example $\mathbf{X}$. The GMM-UBM mean interval (GUMI) vector $\boldsymbol{\Phi}_q(\mathbf{X})$ is obtained from the approximation of Bhattacharyya mean distance between the qth component of the adapted GMM and the corresponding qth component of the UBM as follows:

$$\boldsymbol{\Phi}_q(\mathbf{X}) = \left(\frac{\boldsymbol{\Sigma}_q^{(\mathbf{X})} + \boldsymbol{\Sigma}_q}{2}\right)^{-1/2} \left(\boldsymbol{\mu}_q^{(\mathbf{X})} - \boldsymbol{\mu}_q\right) \tag{14.68}$$

The GUMI supervector is obtained by concatenating the GUMI vectors of different components as

$$\boldsymbol{\Phi}^{\text{GUMI}}(\mathbf{X}) = \left[\boldsymbol{\Phi}_1(\mathbf{X})^t, \boldsymbol{\Phi}_2(\mathbf{X})^t, \ldots, \boldsymbol{\Phi}_Q(\mathbf{X})^t\right]^t \tag{14.69}$$

The GUMI kernel is defined as the innerproduct of the GUMI supervectors of a pair of examples, and is given by

$$K^{\text{GUMI}}(\mathbf{X},\mathbf{Y}) = \left(\boldsymbol{\Phi}^{\text{GUMI}}(\mathbf{X})\right)^t \left(\boldsymbol{\Phi}^{\text{GUMI}}(\mathbf{Y})\right) \tag{14.70}$$

In the GUMI kernel, the supervector is obtained from the Bhattacharyya mean distance between a GMM and an UBM. However, significant information is present in covariance terms. In [8], the covariance terms are also considered to obtain the GUMI supervector. It is shown in [8] that a GUMI vector is obtained by concatenating the mean vector and the variance vector. The GUMI vector using the covariance terms for the qth component is given by

$$\boldsymbol{\Phi}_q^{\text{COV}}(\mathbf{X}) = \begin{bmatrix} \left(\frac{\boldsymbol{\Sigma}_q^{(\mathbf{X})} + \boldsymbol{\Sigma}_q}{2}\right)^{-1/2} \left(\boldsymbol{\mu}_q^{(\mathbf{X})} - \boldsymbol{\mu}_q\right) \\ \text{diag}\left(\left(\frac{\boldsymbol{\Sigma}_q^{(\mathbf{X})}+\boldsymbol{\Sigma}_q}{2}\right)^{-1/2} \left(\boldsymbol{\mu}_q^{(\mathbf{X})}\right)^{-1/2}\right) \end{bmatrix} \tag{14.71}$$

The supervector is obtained by concatenating the GUMI vectors using the covariance terms of the different components as

$$\boldsymbol{\Phi}^{\text{COVGUMI}}(\mathbf{X}) = \left[\boldsymbol{\Phi}_1^{\text{COV}}(\mathbf{X})^t, \boldsymbol{\Phi}_2^{\text{COV}}(\mathbf{X})^t, \ldots, \boldsymbol{\Phi}_Q^{\text{COV}}(\mathbf{X})^t\right]^t \tag{14.72}$$

Now the modified Bhattacharyya distance based kernel is obtained as

$$K^{\mathrm{COVGUMI}}(\mathbf{X},\mathbf{Y}) = \left(\boldsymbol{\Phi}^{\mathrm{COVGUMI}}(\mathbf{X})\right)^t \left(\boldsymbol{\Phi}^{\mathrm{COVGUMI}}(\mathbf{Y})\right) \tag{14.73}$$

The dynamic kernels presented in this section are computationally intensive. The GLDS kernel uses the kernel feature space of a polynomial kernel function. The dimensionality for the feature space increases as the degree of polynomial kernel increases. This makes the computation of $\mathbf{S}^{-1}$ in (14.53) complex. In PSK, dimensionality of the feature space is defined by the dimension of probabilistic feature vectors obtained using generative models. The construction of PSK needs to build a probabilistic model for each example and the dimensionality of the feature space is twice the number of components in the UBM. Though the PSK is computed similar to GLDS, it is computationally less intensive as the dimensionality of the feature space in PSK is small compared to that of the GLDS kernel space. However, building the probabilistic model for each example is difficult when the example does not have sufficient number of feature vectors in training set. The GMM supervector kernel and Bhattacharyya distance based kernel need the example-specific GMMs built by adapting the UBM.

In this work we propose to use the intermediate matching kernel (IMK) for the speaker identification task. It does not require any models to be built for the examples. In the next Section, we describe the IMK based SVM system.

14.5 Intermediate Matching Kernel Based SVM

In this section, we present the methods used to construct kernels suitable for image and speech processing tasks where an example is represented by a set of local feature vectors. In these methods, the kernel for a pair of examples is constructed by matching the two sets of local feature vectors representing the two examples. Let $\mathbf{X}_i = \{\mathbf{x}_{i1}, \mathbf{x}_{i2}, \ldots, \mathbf{x}_{iMi}\}$ and $\mathbf{X}_j = \{\mathbf{x}_{j1}, \mathbf{x}_{j2}, \ldots, \mathbf{x}_{jMj}\}$ be the sets of local feature vectors for the examples $\mathbf{X}_i$ and $\mathbf{X}_j$ respectively. The summation kernel [37] is computed by matching every local feature vector in $\mathbf{X}_i$ with every local feature vector in $\mathbf{X}_j$ as follows:

$$K^{\mathrm{S}}(\mathbf{X}_i, \mathbf{X}_j) = \sum_{m=1}^{M_i} \sum_{n=1}^{M_j} K(\mathbf{x}_{im}, \mathbf{x}_{jn}) \tag{14.74}$$

where $k(.,.)$ is a basic Mercer kernel for two local feature vectors that gives a measure of similarity between them. The matching kernel [36] is constructed by considering the closest local feature vector of an example for each local feature vector in the other example as follows:

$$K^{\mathrm{MK}}(\mathbf{X}_i,\ \mathbf{X}_j) = \sum_{m=1}^{M_i} \max_n K(\mathbf{x}_{im},\ \mathbf{x}_{jn}) + \sum_{n=1}^{M_j} \max_m K(\mathbf{x}_{im},\ \mathbf{x}_{jn}) \tag{14.75}$$

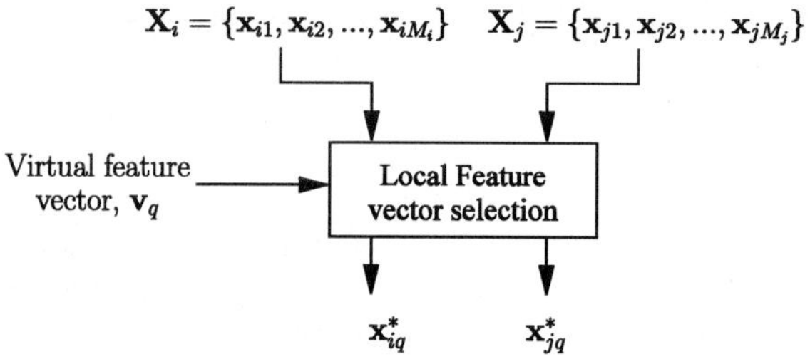

Fig. 14.5 Selection of a local feature vector from each of the examples, based on the closeness to a virtual feature vector $\mathbf{v}_q$

The summation kernel is a Mercer kernel. However, the matching kernel is not proven to be a Mercer kernel [30, 37]. Construction of the summation kernel or the matching kernel is computationally intensive because each local feature vector of an example is compared with every local feature vector of the other example. The number of basic kernel computations is $M_i * M_j$ for the summation kernel and $2 * M_i * M_j$ for the matching kernel. The summation kernel and the matching kernel give a measure of global similarity between a pair of examples.

Next we present the intermediate matching kernel (IMK) that is constructed by matching the sets of local feature vectors using a set of virtual feature vectors. The construction of IMK uses a set of virtual feature vectors obtained from the training data of all the classes. The IMK for a pair of examples, with each example represented as a set of local feature vectors, is constructed by matching the local feature vectors of the examples with each of the virtual feature vectors. Consider a pair of examples $\mathbf{X}_i = \{\mathbf{x}_{i1}, \mathbf{x}_{i2}, ..., \mathbf{x}_{iMi}\}$ and $\mathbf{X}_j = \{\mathbf{x}_{j1}, \mathbf{x}_{j2}, ..., \mathbf{x}_{jMj}\}$ that need to be matched. Let $\mathbf{V} = \{\mathbf{v}_1, \mathbf{v}_2, ..., \mathbf{v}_Q\}$ be the set of virtual feature vectors extracted from the training data of all the classes. The feature vectors $\mathbf{x}^*_{iq}$ and $\mathbf{x}^*_{jq}$ in $\mathbf{X}_i$ and $\mathbf{X}_j$ that are closest to $\mathbf{v}_q$ are determined as follows:

$$\mathbf{x}^*_{iq} = \arg\min_{\mathbf{x} \in \mathbf{X}_i} \mathcal{D}(\mathbf{x}, \mathbf{v}_q) \quad \text{and} \quad \mathbf{x}^*_{jq} = \arg\min_{\mathbf{x} \in \mathbf{X}_j} \mathcal{D}(\mathbf{x}, \mathbf{v}_q) \tag{14.76}$$

where $\mathcal{D}(.,.)$ is a distance function that measures the distance of a local feature vector to a virtual feature vector. The process of selection of local feature vectors that are closest to the virtual feature vector is shown in Fig. 14.5. Similarly, a pair of feature vectors from $\mathbf{X}_i$ and $\mathbf{X}_j$ is selected for each of the virtual feature vectors in the set. The selection of the closest local feature vectors for each virtual feature vectors involves computation of $M_i + M_j$ distance functions.

A basic kernel is computed for each of the Q pairs of selected local feature vectors. The intermediate matching kernel (IMK) is computed as the sum of all the Q kernel values and is given as

$$K^{\mathrm{IMK}}(\mathbf{X}_i, \mathbf{X}_j) = \sum_{q=1}^{Q} K(\mathbf{x}^*_{iq}, \mathbf{x}^*_{jq}) \tag{14.77}$$

The computation of IMK involves a total of $Q * (M_i + M_j)$ computations of distance function $\mathcal{D}$ and Q computations of the basic kernel. When Q is significantly smaller than M_i and M_j, the construction of IMK is computationally less intensive than constructing the GLDS kernel in (14.55) or the summation kernel in (14.74). When Q is greater than M_i and M_j, the construction of IMK is computationally more intensive than the computation of GLDS kernel in (14.55) or the summation kernel in (14.74). However, it is desirable that Q is smaller than the typical size of the set of local feature vectors of examples. Otherwise, a local feature vector of an example may be associated with more than one virtual feature vector.

In [30], the set of the centers of clusters formed from the training data of all classes is considered as the set of virtual feature vectors. Fuzzy k-means clustering method is used for clustering the local feature vectors extracted from training data of all the classes. The method used to form the clusters is similar to the LBG method for vector quantization. The local feature vectors $\mathbf{x}_{iq}^*$ and $\mathbf{x}_{jq}^*$ in $\mathbf{X}_i$ and $\mathbf{X}_j$ that are closest to the qth center $\mathbf{v}_q$ are determined as follows:

$$\mathbf{x}_{iq}^* = \arg\min_{\mathbf{x}\in\mathbf{X}_i} \left\|\mathbf{x}-\mathbf{v}_q\right\| \quad \text{and} \quad \mathbf{x}_{jq}^* = \arg\min_{\mathbf{x}\in\mathbf{X}_j} \left\|\mathbf{x}-\mathbf{v}_q\right\| \tag{14.78}$$

A basic kernel $K(\mathbf{x}_{iq}^*, \mathbf{x}_{jq}^*) = \exp\left(-\delta\left\|\mathbf{x}_{iq}^* - \mathbf{x}_{jq}^*\right\|^2\right)$ is computed for each of the Q pairs of selected feature vectors, where δ is a constant scaling term required for numerical stability. An IMK is computed as in (14.77). Then the SVM classifier is built using the IMK. This method for construction of IMK uses only the information about the centers of clusters for representing the set of virtual feature vectors. The key issue in IMK is the selection of the set of virtual feature vectors. A better representation for the set of virtual feature vectors can be provided by considering additional information.

In this work we propose to use components of the UBM built using the training data of all the classes for obtaining a better representation for the set of virtual feature vectors. This representation for the set of virtual feature vectors makes use of information about means of components, covariance of components and mixture coefficients. The additional information, in the form of covariance and mixture coefficients, is expected to give a better representation for the set of virtual feature vectors as compared to the cluster centers as the set of virtual feature vectors in [30]. The UBM is a large GMM of Q components built using the training data of all the classes. The local feature vectors from the pair of examples $\mathbf{X}_i$ and $\mathbf{X}_j$ that are closest to the component q are selected for matching. Since the Gaussian components are used as the virtual feature vectors, the responsibility term is considered as a measure of closeness of the local feature vector to the component q. The responsibility of the component q of UBM for the feature vector $\mathbf{x}$, $\gamma_q(\mathbf{x})$, is given by

$$\gamma_q(\mathbf{x}) = \frac{\pi_q \mathcal{N}(\mathbf{x}\,|\,\boldsymbol{\mu}_q, \boldsymbol{\Sigma}_q)}{\sum_{j=1}^{Q} \pi_j \mathcal{N}(\mathbf{x}\,|\,\boldsymbol{\mu}_j, \boldsymbol{\Sigma}_j)} \tag{14.79}$$

where π_q is the mixture coefficient of the component q, and $\mathcal{N}(\mathbf{x}|\boldsymbol{\mu}_q, \boldsymbol{\Sigma}_q)$ is the normal density for the component q with mean $\boldsymbol{\mu}_q$ and covariance $\boldsymbol{\Sigma}_q$. The feature

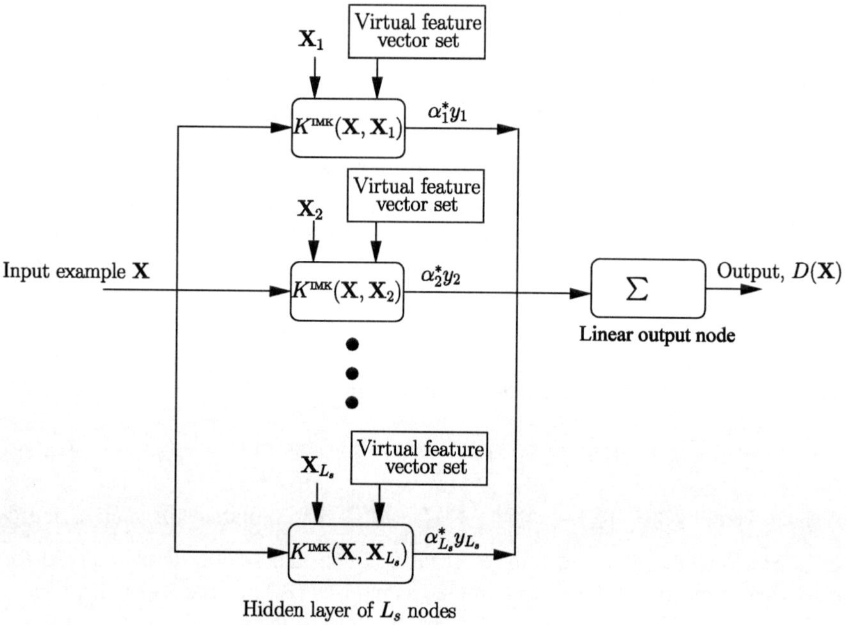

Fig. 14.6 Block diagram of architecture of SVM using IMK

vectors $\mathbf{x}_{iq}^*$ and $\mathbf{x}_{jq}^*$ in $\mathbf{X}_i$ and $\mathbf{X}_j$ that are closest to the component q of UBM are given by

$$\mathbf{x}_{iq}^* = \underset{\mathbf{x} \in \mathbf{X}_i}{\arg\max}\ \gamma_q(\mathbf{x}) \text{ and } \mathbf{x}_{jq}^* = \underset{\mathbf{x} \in \mathbf{X}_j}{\arg\max}\ \gamma_q(\mathbf{x}) \tag{14.80}$$

A pair of local feature vectors from $\mathbf{X}_i$ and $\mathbf{X}_j$ is selected for each of the components in the UBM. The IMK is computed using a basic kernel for each of the Q pairs of selected local feature vectors as in (14.77). The SVM classifier is built using the IMK. An SVM is a two-class classifier. Let the set of L training examples be, $\{(\mathbf{X}_l, y_l)\}_{l=1}^{L}$. Each of the examples $\mathbf{X}_l = \{\mathbf{x}_{lm}\}_{m=1}^{M_l}$, is a set of feature vectors and y_l is the corresponding class label. Let L_s be the number of support vectors and $\{\alpha_l^*\}_{l=1}^{L_s}$ be the optimal values of Lagrangian coefficients obtained after training the SVM. The architecture of the SVM using IMK for classifying an input example $\mathbf{X}$ is as shown in Fig. 14.6. The number of nodes in the hidden layer is decided by the number of support vectors. The output discriminant function for a given input example $\mathbf{X}$ is given as

$$D(\mathbf{X}) = \sum_{l=1}^{L_s} \alpha_l^* y_l K^{\text{IMK}}(\mathbf{X}, \mathbf{X}_l) \tag{14.81}$$

The sign of value of $D(\mathbf{X})$ is used to determine the class of $\mathbf{X}$. The approaches presented in Sect. 14.2.4 can be used for building a multiclass pattern classification system using the IMK based SVMs for sequential patterns represented using sets of local feature vectors.

The advantage of the proposed IMK based SVM is that the construction of IMK involves only identification of the set of virtual feature vectors. There is no need for building the example specific probabilistic models or adapted models as in the case of PSK, GMM supervector kernel or Bhattacharyya distance based kernel. In the next section we present our studies on speaker recognition using the IMK based SVM. We also compare its performance with that of the GMM-based system, the adapted GMM-based system and the GLDS kernel SVM-based system.

14.6 Studies on Speaker Identification

In this section, we present our studies on speaker identification. We first describe the implementation details of studies on speaker identification. The implementation details includes the details of dataset used and features considered for the study. Then we present the speaker identification accuracy obtained using GMM-based classifiers and SVM-based classifiers.

14.6.1 Dataset and Features Used

We performed experiments on the 2002 and 2003 NIST speaker recognition (SRE) corpora [46, 47]. We considered the 122 male speakers that are common in 2002 and 2003 NIST SRE corpora. Training data for a speaker includes a total of about 3 min of speech from the single conversations in the training set of 2002 and 2003 NIST SRE corpora. The test data from the 2003 NIST SRE corpus is used for testing the speaker recognition systems. Each test utterance is around 30 s long. After removing the silence portions in the speech utterance, we divide each utterance in the training and test sets into segments of around 5 s. Each speech segment is considered as an example. This leads to a total of 3,617 training examples with each speaker class having about 30 examples. The test set includes a total of 3,044 examples. A frame size of 20 ms and a shift of 10 ms are used for feature extraction from the speech signal of an example. Every frame is represented using a 39-dimensional feature vector consisting of 12 Mel frequency cepstral coefficients (MFCC), log energy, and their delta and acceleration coefficients. Each of the training and test examples is represented by a set of about 500 local feature vectors. The speaker identification accuracy presented in this section is the classification accuracy obtained for 3,044 test examples. The classification accuracy gives the percentage of test examples that are correctly classified by the classifier. In the context of speaker identification, the classification accuracy indicates the speaker identification rate. In order to ascertain the statistical importance of the result, classification accuracy is presented along with 95% confidence interval. A simple asymptotic method (Wald method) [48]

Table 14.1 Comparison of classification accuracy (in %), estimated at 95% confidence intervals, given by the GMM-based system and the adapted GMM-based system for speaker identification task

Model	Number of components (Q)	Classification accuracy (in %)
GMM	32	75.81±1.52
	64	76.50±1.51
	128	71.26±1.61
Adapted GMM	1,024	83.08±1.33

is employed to estimate 95% confidence intervals of classification accuracy. The confidence interval (*CI*), of classification accuracy is computed as

$$CI = z\sqrt{\frac{a(1-a)}{L_{test}}} \tag{14.82}$$

where a is the accuracy in decimals, and L_{test} is the number of test examples. The z is the $(1-\alpha/2)$ point of the standard Normal distribution associated with a two-tailed probability α. For 95% confidence interval, z takes the value of 1.96.

In this study, we compare the accuracy of speaker recognition systems built using GMMs, adapted GMMs, and dynamic kernel based SVMs.

14.6.2 Studies on Speaker Identification Using GMM and SVM-Based Classifiers

For the GMM-based system, we consider diagonal covariance matrices. We perform a line search to select the number of components to be used in GMMs and estimate the parameters of the model using the ML method. The best performance of the GMM-based system is observed for 64 components in the GMM for each speaker. The adapted GMM-based system uses an UBM with 1,024 components. The adapted GMMs are built by adapting the means, variances, and mixture coefficients with the value of relevance factor as 16 [34]. The classification accuracies on the test data, estimated at 95% confidence intervals, obtained for the GMM-based system and the adapted GMM-based system is given in Table 14.1. It is seen that the adapted GMM-based system gives a better performance than the GMM-based system. The better performance of the adapted GMM-based system is mainly due to robust estimation of parameters using the limited amount of training data available from a speaker, as explained in Sect. 14.3.2.

Next we study the performance of the SVM-based approaches to speaker recognition. We consider the GLDS based SVM, PSK based SVM, GMM supervector kernel based SVM and IMK based SVM for building the speaker recognition systems. LIBSVM [49] is used for training the SVM classifier. The one-against-rest approach is used to build the 122-class speaker recognition systems. The value of trade-off parameter, C, is chosen empirically as 10. The GLDS kernel for a pair of

Table 14.2 Classification accuracy (in %), estimated at 95% confidence intervals, and average number of support vectors of the PSK based SVM classifiers for speaker identification

Number of components (Q)	Classification accuracy (in %)	Average number of support vectors
512	58.32±1.75	182
1,024	64.68±1.70	176

Table 14.3 Classification accuracy (in %), estimated at 95% confidence intervals, and average number of support vectors of GMMSV and GUMI kernel based SVM classifiers for speaker identification

Number of components (Q)	Classification accuracy (in %)		Average number of support vectors	
	GMMSV	GUMI	GMMSV	GUMI
512	87.93±1.16	90.31±1.05	396	387
1,024	86.23±1.22	89.91±1.07	405	396

examples is constructed using the value of degree as 2 and 3 for the polynomial kernel in (14.55). The classification accuracy, estimated at 95% confidence intervals, is 76.77±1.5% for degree 2 and 78.62±1.46% for degree 3. This performance is better than that of the GMM-based system using 64 components. For the PSK based SVM, we considered UBMs with 512, and 1,024 components. We adapt the UBM with the data of each speaker to get the speaker dependent GMM. Table 14.2 presents the classification accuracy, estimated at 95% confidence intervals, for the PSK based SVM. It is seen that the PSK constructed using the 1,024 components gives the best performance. However, this performance is significantly less compared to that of the GMM-based systems given in Table 14.1.

For the GMM supervector (GMMSV) kernel based SVM and the Bhattacharyya distance based GMM-UBM mean interval (GUMI) kernel based SVM, we considered UBMs with 512 and 1,024 components. Table 14.3 presents the classification accuracy, estimated at 95% confidence intervals, for GMMSV kernel and GUMI kernel based SVMs. It is seen that the GMMSV kernel and GUMI kernel constructed using the 512 components of example-specific GMMs gives the best performance. This performance is significantly better than that of the PSK based SVM (Table 14.2). The performance of GUMI kernel based SVM is significantly better compared to that of the GMMSV kernel based SVM and the adapted GMM-based system using a UBM with 1,024 components (Table 14.1).

The IMK based SVMs are built using a value of 128 or 256 or 512 for Q corresponding to the size of the set of virtual feature vectors. The classification accuracy, estimated at 95% confidence intervals, for the IMK constructed using the components of UBM and the IMK constructed using the set of centers of clusters is given in Table 14.4, for different values of Q. It is seen that the IMK constructed using the UBM with 512 components as the set of virtual feature vectors gives the best performance of 88.12±1.15%. This performance is better than that of the adapted

Table 14.4 Classification accuracy (in %), estimated at 95% confidence intervals, and average number of support vectors of the IMK based SVM classifiers for speaker identification

Set of virtual feature vectors	Number of virtual feature vectors (Q)	Classification accuracy (in %)	Average number of support vectors
Components of UBM	128	84.30±1.29	279
	256	86.21±1.22	226
	512	88.12±1.15	254
Center of clusters	128	76.52±1.51	263
	256	78.54±1.46	290
	512	79.38±1.44	287

Table 14.5 Comparison of classification accuracy (in %), estimated at 95% confidence intervals, of the GMM-based classifiers and dynamic kernel based SVM classifiers for speaker identification

Classifier	Kernel	Number of mixtures (Q)	Classification accuracy (in %)
GMM	–	64	76.50±1.51
Adapted GMM	–	1,024	83.08±1.33
SVM	GLDS kernel	–	78.62±1.46
	PSK	1,024	64.68±1.70
	GMM supervector kernel	512	87.93±1.16
	GUMI kernel	512	90.31±1.05
	IMK with center of clusters	512	79.38±1.44
	IMK with components of UBM	512	88.12±1.15

GMM-based system using a UBM with 1,024 components (Table 14.1). This is mainly because the GMM-based classifier is trained using a non-discriminative learning based technique, where as the IMK based SVM classifier is built using a discriminative learning based technique. It is seen that the average number of support vectors for the SVMs using IMK is in the range of 225–300.

Table 14.5 compares the speaker identification accuracies obtained using GMM-based classifiers, SVM-based classifiers using GLDS kernel, PSK, GMM supervector kernel, GUMI kernel, IMK with centers of clusters as virtual feature vectors (IMK-CB) and IMK with components of UBM as virtual feature vectors (IMK-UBM). It is seen that the SVM classifiers using the dynamic kernels such as GMMSV kernel, GUMI kernel and IMK-UBM kernel give a better performance than the adapted GMM-based classifier. Though the GUMI kernel gives a marginally better performance than the IMK-UBM kernel, the average number of support vectors for SVMs using the GUMI kernel is significantly higher than the number of support vectors for the IMK-UBM based SVM. Additionally, the representation of an example in the GUMI kernel based method is of a large dimension that is proportional to the number of components, Q. Therefore, the storage requirements and the computational complexity during the recognition phase of the classifier are significantly lower for the IMK-UBM based SVM compared to the GUMI kernel based SVM. In comparison to the other dynamic kernels, the limitation of the IMK is that it is necessary to use a set of virtual feature vectors and the performance of the IMK-based SVM is dependent on the choice of the set of virtual feature vectors.

14.7 Summary and Conclusions

In this chapter, we presented the GMM-based approaches and the SVM-based approaches to speaker recognition. The discriminative training based large margin method for estimation of GMM parameters is expected to give a better performance than the maximum likelihood method. However, when the number of classes is large, the optimization problem solving in the large margin GMMs is computationally highly intensive. Development of the discriminative training based SVM classifiers used for speaker recognition requires the design of a suitable dynamic kernel for varying length patterns represented by sets of feature vectors. The dynamic kernels such as the generalized linear discriminant sequence (GLDS) kernel are computationally intensive. The construction of dynamic kernels such as probabilistic sequence kernel, GMM supervector kernel and Bhattacharyya distance based GUMI kernel involves building a probabilistic model for each example. The intermediate matching kernel (IMK) is computationally less intensive than the GLDS kernel. The construction of IMK does not involve building a probabilistic model for each example. The main issue in the construction of IMK is the choice of the set of virtual feature vectors used for intermediate matching. We proposed to use the components of the universal background model (UBM) as the set of virtual feature vectors. Our studies on the 122-class speaker recognition task show that the IMK based SVM using the UBM gives a better performance than the adapted GMMs. The size of UBM used in construction of the IMK is significantly smaller than the size of UBM used in building the adapted GMMs. The better performance of SVM-based classifiers using dynamic kernels (except PSK) compared to the GMM-based classifiers is mainly due to the discriminative learning based techniques used to build the SVM-based classifiers. Differences in the performance of different dynamic kernels are mainly due to different criteria considered in designing the kernel function used to compute a measure of similarity between the varying length patterns represented as sets of local feature vectors.

The SVM-based classifier for multi-class pattern classification is built using the one-against-the-rest approach. In this approach, an SVM is built for each class to separate the examples of that class from the examples of all the other classes. The IMK used in the SVM for a class can be constructed using the adapted GMM of the class obtained by adapting the UBM with the data of that class. This method leads to construction of the class-specific IMKs used in SVMs. The performance of the class-specific IMK-based SVM classifier is expected to be better than the performance for the IMK constructed using the UBM. As a result, we propose this novel IMK-SVM based technique to capture inter and intra speaker variability via training and testing feature vectors, as compared to state-of-the-art GMM based techniques, for forensic speaker recognition systems.

References

1. Reynolds DA (1995) Speaker identification and verification using Gaussian mixture speaker models. Speech Commun 17:91–108
2. Kinnunen T, Li H (2010) An overview of text-independent speaker recognition: from features to supervectors. Speech Commun 52(1):12–40
3. Sha F, Saul L (2006) Large margin Gaussian mixture modeling for phonetic classification and recognition. Proc. of IEEE International Conference on Acoustics, Speech and Signal Processing, (ICASSP 2006), Toulouse, France, pp 265–268
4. Burges CJC (1998) A tutorial on support vector machines for pattern recognition. Data Min Knowl Discov 2(2):121–167
5. Campbell WM, Campbell JP, Reynolds DA, Singer E, Torres-Carrasquillo PA (2006) Support vector machines for speaker and language recognition. Comput Speech Lang 20(2–3):210–229
6. Lee K-A, You C, Li H, Kinnunen T (2007) A GMM-based probabilistic sequence kernel for speaker verification. Proc. of INTERSPEECH, Antwerp, Belgium, pp 294–297
7. Campbell WM, Sturim DE, Reynolds DA (2006) Support vector machines using GMM supervectors for speaker verification. IEEE Signal Process Lett 13(5):308–311
8. You CH, Lee KA, Li H (2009) A GMM supervector kernel with the Bhattacharyya distance for SVM based speaker recognition. Proc. IEEE International Conference on Acoustics, Speech and Signal Processing (ICASSP 2009), Taipei, Taiwan, pp 4221–4224
9. Alexander A, Drygajlo A (2004) Scoring and direct methods for the interpretation of evidence in forensic speaker recognition. Proc. of INTERSPEECH, Jeju, Korea, pp 2397–2400
10. Campbell JP, Nakasone H, Cieri C, Miller D, Walker K, Martin AF, Przybocki MA (2004) The MMSR bilingual and cross channel corpora for speaker recognition research and evaluation. Proc. of the Speaker and Language Recognition Workshop, Odyssey'04, Toledo, Spain, pp 29–32
11. Drygajlo A, Meuwly D, Alexander A (2003) Statistical methods and Bayesian interpretation of evidence in forensic automatic speaker recognition. Proc. of Eurospeech, Geneva, Switzerland, pp 689–692
12. Campbell WM, Brady KJ, Campbell JP, Granville R, Reynolds DA,(2006) Understanding scores in forensic speaker recognition, Speaker and Language Recognition Workshop, The IEEE Odyssey 2006, pp 1–8
13. Gonzalez-Rodriguez J, Drygajlo A, Ramos-Castro D, Garcia-Gomar M, Ortega-Garcia J (2006) Robust estimation, interpretation and assessment of likelihood ratios in forensic speaker recognition. Comput Speech Lang 20:331–355
14. Thiruvaran T, Ambikairajah E, Epps J (2008) FM features for automatic forensic speaker recognition. Proc. of INTERSPEECH 2008 special session: forensic speaker recognition—traditional and automatic approach, Brisbane, Queensland, Australia
15. Cristianini N, Shawe-Taylor J (2000) An introduction to support vector machines and other kernel-based learning methods. Cambridge University Press, Cambridge
16. Sekhar CC, Takeda K, Itakura F (2003) Recognition of subword units of speech using support vector machines. Proc. recent research developments in electronics and communication. Trivandrum, Kerala, India: Transworld Research Network, pp 101–136
17. Haykin S (1999) Neural networks: a comprehensive foundation, 2nd edn. Prentice Hall, New Jersey
18. Kaufman L (1999) Solving the quadratic programming problem arising in support vector classification. In: Scholkopf B, Burges C, Smola A (eds) Advances in kernel methods: support vector learning. MIT Press, Cambridge, pp 147–167
19. Cover TM (1965) Geometrical and statistical properties of systems of linear inequalities with applications in pattern recognition. IEEE Trans Electron Comput EC-14(3):326–334
20. Scholkopf B, Mika S, Burges C, Knirsch P, Muller K-R, Ratsch G, Smola A (1999) Input space versus feature space in kernel-based methods. IEEE Trans Neural Netw 10(5):1000–1017

21. Borgwardt KM (2007) Graph kernels. Ph.D Thesis, Faculty of Mathematics, Computer Science and Statistics, LudwigMaximilians Universität, Munich
22. Shawe-Taylor J, Cristianini N (2004) Kernel methods for pattern analysis. Cambridge University Press, Cambridge
23. Leslie C, Eskin E, Noble WS (2002) The spectrum kernel: a string kernel for svm protein classification. Proc. the pacific symposium on biocomputing, River Edge, NJ, pp 564–575
24. Leslie C, Eskin E, Weston J, Noble WS (2003) Mismatch string kernels for SVM protein classification. In: Becker S, Thrun S, Obermayer K (eds) Advances in neural information processing. MIT Press, Cambridge, pp 1417–1424
25. Leslie C, Eskin E, Cohen A, Weston J, Noble WS (2004) Mismatch string kernels for discriminative protein classification. Bioinformatics 20:467–476
26. Lodhi H, Saunders C, Shawe-Taylor J, Christianini N, Watkins C (2002) Text classification using string kernels. J Mach Learn Res 2:419–444
27. Tsuda K, Kin T, Asai K (2002) Mariginalized kernels for biological sequences. Bioinformatics 18:S268–S275
28. Allwein EL, Schapire RE, Singer Y (2001) Reducing multiclass to binary: a unifying approach for margin classifiers. J Mach Learn Res 1:113–141
29. Kressel UH-G (1999) Pairwise classification and support vector machines. In: Scholkopf B, Burges C, Smola A (eds) Advances in kernel methods: support vector learning. MIT Press, Cambridge, pp 255–268
30. Boughorsbel S, Tarel JP, Boujemaa N (2005) The intermediate matching kernel for image local features. Proc. international joint conference on neural networks, Montreal, Canada, pp 889–894
31. Jayaraman A (2008) Modular approach to online handwritten character recognition of Telugu script. Master's thesis, Department of CSE, IIT Madras, Chennai, India
32. Hu H, Xu M-X, Wu W (2007) GMM supervector based SVM with spectral features for speech emotion recognition. Proc. IEEE International Conference on Acoustics, Speech and Signal Processing, 2007, ICASSP 2007, 4, Honolulu, Hawaii, USA, pp 413–416
33. Veena T, Dileep AD, Sekhar CC (2010) Scene categorization using large margin Gaussian mixture models. Proc. 2010 International Conference on Image Processing, Computer Vision, & Pattern Recognition, (IPCV 2010), 1, Las Vegas, Navada, USA, pp 395–401
34. Reynolds DA, Quatieri TF, Dunn RB (2000) Speaker verification using adapted Gaussian mixture models. Digit Signal Process 10:19–41
35. Wan V, Renals S (2002) Evaluation of kernel methods for speaker verification and identification. Proc. of IEEE international conference on acoustics, speech and signal processing, Orlando, Florida, US, pp 669–672
36. Wallraven C, Caputo B, Graf A (2003) Recognition with local features: the kernel recipe. Proc. Ninth IEEE International Conference on Computer Vision (ICCV 2003), pp 257–264
37. Boughorbel S, Tarel J-P, Fleuret F (2004) Non-Mercer kernels for SVM object recognition. Proc. British Machine Vision Conference (BMVC 2004), pp 137–146
38. Campbell W, Assaleh K, Broun C (2002) Speaker recognition with polynomial classifiers. IEEE Trans Speech Audio Process 10(4):205–212
39. Auckenthaler R, Parris ES, Carey MJ (1999) Improving a GMM speaker verification system by phonetic weighting. Proc. IEEE International Conference on Acoustics, Speech and Signal Processing (ICASSP1999), 1, Phoenix, Arizona, USA, pp 313–316
40. Campbell W (2008) A covariance kernel for SVM language recognition. Proc. IEEE International Conference on Acoustics, Speech and Signal Processing, (ICASSP 2008), Las Vegas, Nevada, USA, pp 4141–4144
41. Dehak R, Dehak N, Kenny P, Dumouchel P (2007) Linear and nonlinear kernel GMM supervector machines for speaker verification. Proc. INTERSPEECH, Antwerp, Belgium, pp 302–305
42. Bhattacharyya A (1943) On a measure of divergence between two statistical populations defined by their probability distributions. Bull Calcutta Math Soc 35:99–109
43. Kailath T (1967) The divergence and Bhattacharyya distance measures in signal selection. IEEE Trans Commun Technol 15(1):52–60

44. Kondor R, Jebara T (2003) A kernel between sets of vectors. Proc. International Conference on Machine Learning, (ICML 2003), Washington DC, USA
45. You CH, Lee KA, Li H (2009) An SVM kernel with GMM-supervector based on the Bhattacharyya distance for speaker recognition. IEEE Signal Process Lett 16(1):49–52
46. The NIST year 2002 speaker recognition evaluation plan. http://www.itl.nist.gov/iad/mig/tests/spk/2002/, 2002
47. The NIST year 2003 speaker recognition evaluation plan. http://www.itl.nist.gov/iad/mig/tests/sre/2003/, 2003
48. Newcombe RG. (1998) Two-sided confidence intervals for the single proportion: comparison of seven methods. Stat Med 17:857–872
49. Chang C-C, Lin C-J (2001) LIBSVM: a library for support vector machines. Software available at http://www.csie.ntu.edu.tw/cjlin/libsvm

Part IV
Applications to Law Enforcement and Counter-Terrorism

Chapter 15
Speaker Spotting: Automatic Telephony Surveillance for Homeland Security

V. Ramasubramanian

Abstract Automating telephony surveillance is an appealing and appropriate technology from the view point of being able to detect/spot if a person from a specific watch-list is on line. Such an automatic solution is of considerable interest in the context of homeland security where a potentially large number of wire tapped conversations may have to be processed in parallel, in different deployment scenarios and demographic conditions, and with typically large watch-lists, all of which make manual lawful interception unmanageable, tedious and perhaps even impossible. In this chapter, we first introduce this problem domain starting with a sketch of a glamorous fictitious example, followed by an outline of lawful interception and wire-tapping; we then take a brief look at similar watch-list based negative recognition application using the now very successful Iris biometrics and consider equivalent scenarios in the context of speaker-spotting based on voice as a biometric. Further, in the main body of this chapter, we first provide the basic framework for watch-list based speaker-spotting, namely, open-set speaker identification, subsequently refined into a 'multi-target detection' framework. We then examine in some detail the main theoretical analysis available within the framework of multi-target identification, leading to performance predictions of such systems with respect to the watch-list size as the critical factor. In a related note, we also briefly touch on the prioritization mode of operation which also lends itself to interesting theoretical analysis and performance predictions. Speaker-spotting systems face unique challenges, in a way combining the difficulties inherent in conventional speaker authentication applications as well as forensic speaker recognition applications; we consider these, while using the NIST SRE evaluation results to gain insights on the performances achievable presently and the latent performance limitations which seem to warrant a cautionary approach before widespread use of speaker recognition technology for surveillance applications becomes possible. In the later part of the chapter, we outline related topics such as speaker change detection, speaker segmentation and speaker diarization, followed by a summary of product level solutions currently available in the context of surveillance and homeland security applications, finally concluding with discussions highlighting the state-of-the-art and potential future research directions.

V. Ramasubramanian (✉)
Siemens Corporate Research & Technologies—India, Bangalore 560100, India
e-mail: V.Ramasubramanian@siemens.com

A. Neustein, H. A. Patil (eds.), *Forensic Speaker Recognition*,
DOI 10.1007/978-1-4614-0263-3_15, © Springer Science+Business Media, LLC 2012

15.1 Introduction

15.1.1 Speaker Spotting for Automatic Telephony Surveillance

Most of us, at one time or another, have heard or read about telephone wire-tapping, typically by investigative agencies in the form of 'eavesdropping' on telephonic conversations between suspect persons in order to acquire evidence of illegal activities. Conventionally, wire-tapping is conducted 'manually', including the main steps of tapping into a specific telephone number, recording and listening to the ensuing conversations to generate a transcript of who spoke and what is spoken. Alternately, one can look at 'automatic' solutions, wherein the process of generating a transcript is done by a machine, i.e., a system that can identify who spoke as well as recognize the words spoken. While the former part of identifying the speaker(s) in the conversation belongs to the topic of 'speaker recognition', the latter part of recognizing the spoken words belongs to the topic of 'speech recognition'. In this chapter, we will exclusively focus on the speaker recognition aspects, which acquires a more specialized name of 'speaker spotting' or 'speaker detection' in the context of surveillance applications.

More specifically, we will define the framework of speaker-spotting in discussion here as follows: A telephone conversation is tapped and the resulting audio-stream is processed and analyzed by a system to perform primarily the task of identifying the speaker(s) in the conversation from among a set of 'target' (or 'suspect') speakers. Here, the set of suspect speakers is typically called the 'watch-list', and it is of interest for the system to arrive at a decision if the speaker in the conversation is a speaker in the watch-list. The purpose of such a watch-list based speaker spotting is more or less obvious: the investigative agency already possesses a list of persons considered as potential perpetrators of a crime and needs to verify if a wire-tapped conversation originates from such persons. Such a purpose could be in various contexts: the watch-list could be a list of persons known to commit on-line frauds in banking or e-commerce scenarios typically in a call-center application, or known to be involved in illegal transactions such as corruption, bribery etc., or a list of terrorists suspected to maintain covert communications threatening the security of the nation. The last part specifically comes within the jurisdiction of homeland security, although almost all the other watch-list applications would necessarily also qualify as posing risk to the general welfare of a nation to varying degrees. In any of these scenarios, the detection (or spotting) of a speaker as belonging to a watch-list could trigger one or more of possible actions: triggering a real-time alarm so as to bring in a human listener to listen to the on-going conversation, trigger a recording of the conversation, locate the origin of the telephonic conversations, activate emergency response systems or teams to physically intercept the callers etc.

Note that a speaker-spotting task (based on a list of speakers of interest) has other potential applications too, such as searching an audio database of recorded meetings, broadcast news, audio documents etc.

15.1.2 Hollywood Examples

As with most advanced technologies, and even futuristic science-fiction concepts and ideas, art—in the form of fiction, cinema etc.—has always been steps ahead of actual reality either in rendering these ideas in an exaggerated form of poetic liberty or paving the way for the shape of things to come, in a manner far more appealing than might actually be possible by the more mundane state-of-the-art of the real technologies in question. Technology based surveillance, in the form of big-brother-is-watching-you scenarios, has been no exception and has fostered rich imagination towards what such technologies have in store and/or can accomplish in the current and conceivable future. Examples date back to the HAL computer of the '2001: A Space Odyssey' [1] (being able to think, converse and plot at par with the human astronauts on board, being able to lip-read the astronauts' secret conversations etc.).

Of specific interest to this chapter is the particular example of telephony surveillance portrayed in a more realistic manner in the context of matching a suspect speaker's voice in two separate telephone recordings [2]. The flow of events is more or less as follows: the investigative agency has a recording of a suspect's voice in one telephone conversation (on an answering machine), a murder is committed by the suspect, the agency has a recording of the suspect's voice in another telephone conversation (at a possibly much later time than the first recording), and the homeland security analyst requests for a match between these two recordings and a name to be associated with it. While the speech-expert runs the two recordings through the 'machine' to get a 'match', the movie clip shows the spectrogram of one of the recordings moving back and forth with respect to the other recording on a large wall-screen, even while displaying a 'number' indicating the percentage of match, which in a dramatic manner is ever increasing and reaches a 90.1% mark, much to the satisfaction of the speech expert. Of course, the identity of the recorded speaker is yet unknown at this stage and the analyst demands a 'name' on it, to which the speech-expert is aghast that there are simply too many names (hundreds of thousands) to match it against (meaning a potential 'watch-list').

This short, but very effective episode in the movie assumes (or rather, brings out) several serious technologies in question, all of which seem so easy in the cinematic context. To name a few, these are: (1) the fact that two distinct recordings could be brought into question for a match is remarkable (from among a host of such possible recordings), (2) obtaining a 'match' on such short recordings, particularly done via very disparate communication channels and background conditions, (3) quantifying the match by a single, simple-minded number (in a sense of closer-to-100%-means-better kind of intuitive and layman's assumption) and (4) various remarks by the speech-expert bringing out subtle underlying extra-speaker information—such as '...Cuban...', '...age 35–45...', '...educated in United States...' followed by an added emphasis on "...'eastern' United States...", "...it is the voice..., same voice..." etc.—all of which is actually inferred by the speech expert by listening alone; these snippet comments, being very cleverly interspersed even while the automatic matching is in progress, manages to create an awe-inspiring, but possibly equally misleading, impression to the audience that the 'matching system' might actually be giving out such extra-speaker cues as easy-to-obtain 'incidental' by-products of its own 'matching process'.

While all the items (1)–(4) are serious and difficult from the standpoint of voice-biometrics, and which are as yet to reach the level of maturity and reliability as portrayed in the movie, the underlying mechanism by which such a lawful interception (such as required for acquiring the telephony recordings in item (1) in the first place) becomes possible is also no less challenging in its own right.

Here again, in order to set the tone for discussing the kind of infrastructural aspects underlying such lawful interceptions, it is interesting to refer to yet another Hollywood movie [67] whose theme is set in a contemporary surveillance context replete with a curious and seamless mix of current technology capabilities (involing highly coordinated telephony and camera based surveillance and satellite based imaging and tracking) and a somewhat liberal dose of imaginative extrapolations into the not-too-distant future. Specifically, it is illustrative to note the spectrum of plausible surveillance practice alluded to by one of the protagonists (a former communication analyst with the National Security Agency (NSA) in the story) such as: 'they (referring to investigative agencies) can get into your bank statements, your computer files, e-mail, listen to your phone calls', 'Fort Meade (a metonym for NSA) has acres of mainframe computers underground. You're talking on the phone and you use the word, "bomb", "president", ... any of a hundred key words, the computer recognizes it, automatically records it, red flags it for analysis; that was years ago' and 'They've got over a hundred spy satellites looking down at us... In the old days, we actually had to tap a wire into your phone line. Now calls bouncing around on satellite, they snatch right out of the air'.

We briefly discuss in the following section, technologies underlying such lawful interception and wire-tapping, minimally necessary and currently possible in practice, in a more realistic setting and in a narrower context of telephony surveillance.

15.1.3 Lawful Interception and Wire-Tapping

The specific problem of speaker-spotting for telephony surveillance is best viewed in the broader context of lawful interception, which forms an integral part of homeland security and whose goal is to obtain communications network data pursuant to lawful authority for the purpose of analysis or evidence. Such data in general includes network management information as well as content of the communications, as is of interest in a telephony surveillance application which involves interception of telecommunications by law enforcement agencies, regulatory or administrative agencies, and intelligence services, in accordance with local law. Classically, this comes under the broad notion of 'wire-tapping', a name with an historic origin, being literally derived from the early methods where the monitoring connection was an actual electrical tap on the telephone line. In the case of PSTN systems, such taps were performed by accessing the mechanical or digital switches supporting the targets' calls. The more recent emergence of wireless (mobile), packet switched networks (VoIP), softswitch and server-based technologies have however altered the form and function of lawful interception via wire-tapping significantly. Presently, wire-tapping is to be understood to broadly refer to listening in on electronic communications on telephones, computers and other devices. Thus, there are several variants of a wiretapping operation, ranging from concealing electronic devices in a phone to tapping into a telecommunications line somewhere along its travel from the device to a routing or exchange center.

Wire-tapping is, by definition, a covert operation (where the targeted party is not aware that their lines are tapped) and is hence usually tightly controlled by laws which are designed to protect privacy rights, which in turn depends on the country of implementation. In many countries, governments have agreements with telecommunications companies which ensure easy access to lines of communication, which could typically be in the form of providing for remote wiretapping ports into the central exchange of digital switches which allows for "point-and-click" wiretapping, so that an investigative agency can listen and monitor a phone call from remote locations, without any longer requiring physical access to the specific phone lines, routers, or exchanges of interest.

More specifically, the following illustrates a traditional wire-tap of a fixed-line telephone (landline) and the now pervasive mobile telephony to essentially highlight the vast differences in the technologies involved and to set the ground for a latter discussion on the challenges these pose to a speaker-spotting system working on such tapped speech data.

Fixed-line Phone: Wire-tapping here involves attaching a 'load' (to tap the voice information that is mapped on to electrical impulses) in the wall socket, on the handset or anywhere along the phone line—in a manner analogous to plugging an extra phone line into a jack in a house. In order to record the tapped conversation, a 'bug' is usually installed, which typically receives audio information and broadcasts it usually via radio waves such that complex recording equipment can be kept away from the phone lines, in a concealed location (for example, a traditional receiving spot is a van parked outside the subject's home).

Mobile Phone: The second generation mobile phones (1978 through 1990) essentially used an analog transmission system—like an ordinary radio transmitter and therefore could be easily monitored by a scanning all-band receiver. However, the third generation digital phones make it much harder to intercept, due to the underlying digital compression and encryption techniques used in the GSM/CDMA protocols. However, law enforcement agencies can tap a phone without seeking the necessary 'key' (code) from the telecom companies to decrypt private conversations. This is done as follows: While it is necessary for the mobile phone to authenticate itself to the mobile telephony network (by sending an unique 15-digit IMSI (International Mobile Subscriber Identity) embedded in every SIM card), the network does not have to authenticate itself. A device called IMSI-catcher can now pose as a base station to the mobile phone and deactivates GSM encryption using a special flag. Subsequently, all calls made from the tapped mobile phone go through the IMSI-catcher and are then passed on to the mobile network. This is a viable technique, as there is no defense against IMSI-catcher, except by end-to-end encryption, such as offered by secure telephones; by nature of being incompatible to each other, such secure phones are not yet proliferate and leave the general second and third generation phones amenable for tapping and surveillance.

In this context, it is noteworthy that interception and intelligence gathering mechanisms—with a much broader scope (than the above instances of wire-tapping)—exist and operate under the name of Communications intelligence (Comint), loosely defined as 'technical and intelligence information derived from foreign communications by other than their intended recipient' [3] and constituting a major component of Sigint (Signals intelligence), which also includes the collection of non-communications signals, such as radar emissions. Among the many targets of Comint operations (including military messages, diplomatic communications, economic intelligence, scientific and technical information) are those concerned with homeland security in different ways, such as narcotics trafficking, money laundering, terrorism and organised crime. It is to be noted that agencies involved in Comint operations within the scope of targeting voice communications traffic (among a host of other possible information traffic and associated communciations infrastructures) have consistently strived to use speech technologies in varied ways, such as using speech activity detection, speech recognition, key-word spotting, speaker-spotting etc., to detect and select conversational data of critical interest allowing such systems to scale beyond what is possible now (i.e., without such content based automatic interception mechanisms). However, it has been difficult to assess the extent and success of these attempts, since they are neither reported in conventional speech research literature nor available as public domain information.

15.1.4 Negative Recognition Using Watch-Lists—Iris Example and Equivalent Voice Scenarios

Biometrics are usually associated with applications such as access control (using various biometric modalities such as finger print, voice, Iris, hand-geometry etc.) which typically work in a 'positive recognition' mode—to prevent multiple people from using the same identity. However, there are several applications requiring 'negative recognition'—to prevent a single person from using multiple identities. The specific application of determining whether an individual is part of a blacklist or watch-list of cheats, criminals and terrorists falls under this category. Enterprises maintaining such a blacklist would like to avoid doing business with such undesirable individuals, but without inconveniencing legitimate customers. To illustrate, examples could include face recognition in casinos and retail stores in conjunction with a watch-list of blacklisted card counters and shoplifters; businesses can even share their watch-list databases with each other to deny entry or access of their casinos or stores to undesirable individuals. Such negative recognition biometric systems are being used extensively and very successfully in government applications (ex: criminal identification, back ground checks, welfare disbursement, border control etc.) [4].

Before dwelling into the mechanisms of such 'watch-list' applications using voice as a biometric modality (i.e., speaker-spotting), it is pertinent to briefly look at a very successful watch-list biometric application, namely, the application of Iris biometric recognition in the context of deportation tracking currently deployed at a large scale [5–7].

Iris has been established as the foremost biometric in terms of the ROC performance characteristic (False-alarm, False-reject); a study by the U.K. government showed the superiority of Iris over other biometrics (fingerprint, voice, face, etc.), with the lowest FA of 0% and FR of 0.2% [38]. Based on this Iris biometric, an Iris Deportation Tracking System (IDTS) is currently in operation at the UAE, since 2001, to prevent re-entry of deportees into the country after they have been expelled. The IDTS is a successful field deployment in UAE, functioning across 35 air, land and sea ports in real-time. As of late 2006 [7], the system uses a watch-list of 1,050,000 expellees (IrisCodes of persons expelled from the UAE for various violations). The system uses this as a 'negative watch-list' where it is of interest to prevent these former expellees from entering the country using fraudulent travel documents. (On a related note, it is worth noting that such Iris biometric based systems are also deployed in airports in UK, The Netherlands, Canada and Frankfurt in a 'positive recognition' mode, wherein the goal is to provide enhanced convenience, speed and efficiency of border crossing formalities for legitimate customers). The UAE watch-list of size 1,050,000 is perhaps the largest in any real-world biometric application so far, representing more than 180 nationalities. The system functions on a daily basis, screening thousands of passengers entering the UAE via the 35 ports and matches each passenger's Iris pattern with the enrolled watch-list database requiring less than one second to yield a decision for one passenger, made possible by an extensive 'IrisFarm' of networked distributed server and communication architecture. The system's performance is to considered very successful, considering that no False Acceptances have been reported yet, while over 9,800,000 searches have been performed and over 115,000 attempts at re-entry prevented (all representing successful detections since these were subsequently confirmed by other records as attempting to enter with fraudulent travel documents).

In comparison to such an Iris based deployed real-world scenario of 'watch-list' based 'negative recognition', it can be said that there are as yet no similar large scale practical deployment of voice-biometric based system. While Iris based 'watch-list' application as above is in the border-crossing setting, voice-based 'watch-list' applications will have their unique characteristics. In fact, this raises the question of what the equivalent scenarios are in the context of voice as a biometric for watch-list applications. While Sects. 15.1.1, 15.1.2 and 15.1.3 outlined a potential application in the form of wire-tapping and subsequent speaker-spotting to detect a potential watch-list speaker on-line, we can consider two practical variations of such a setting.

Surveillance Speaker-Spotting: Ideally, a telephony surveillance scenario is assumed to have a watch-list of speakers; but more primarily, the phones that are wire-tapped are themselves associated with unique phone numbers and in turn to specific speakers who would call from these numbers. Therefore, detecting a call originating from a watch-list of phone numbers in itself serves the purpose of being able to keep track of speech conversations of interest. However, the need for speaker-spotting from a watch-list goes beyond such a simple phone number based detection; for example, while a specific phone number could be of interest in a watch-list, there could be several potential users of the phone (say, from a family or an organization) and more importantly, the incoming calls to a particular phone leaves the identity of the caller wide open and represents the case which warrants identification of the calling speaker from a watch-list.

A closely related issue of high practical importance and necessity is the expansion of a watch-list (such as by a process called 'link analysis' [8]), wherein individuals can be added to watch-lists in a kind of multiplier effect of adding a list of phone numbers or speakers that are called from (or calling) an existing watch-list of phone numbers or speakers. In the context of speaker watch-list expansion by this process, this raises the important and interesting problem of using an unknown speaker's (detected by the link-analysis process) voice for enrolment to the existing watch-list (and thereby create speaker-model for this unknown speaker), perhaps under a tag-name, and further in being able to detect this unknown speaker in future conversations (by the given tag-name); of interest in such applications, is usually, the ability to track conversations and critical contents as happening between speakers of interest (as defined by the watch-list) without necessarily knowing precisely the real identity of the speakers involved.

Negative Recognition: While the above relates to surveillance domain in a lawful interception context, with interesting espionage patterns of operation, a more mundane application would be the scenario dictated from very practical requirements of an organization (say, a bank or a financial institution) to maintain a watch-list of potential recidivists (those who are known to commit fraudulent transactions or have undesirable financial behavior) and deny access to new accounts or line of credit to those in the watch-list [4]. Note that such a blacklist is useful only for preventing repeat frauds and where the first instance of fraud is detected by other means. More specifically, this calls for enrollment by all customers being dealt with by the institution, followed by black-listing a customer whenever a fraud is committed and detected by other means. Alternately, institutions or enterprises can share their watch-list databases, so that an institution (say, a bank without prior enrollment data of potential undesirable individuals), can benefit from other institutions' information.

Here, a typical telephony dialog system (either with a human agent or an automated IVRS system) can listen-in to the incoming call (from the potential client or recidivists) and be able to make a 'negative recognition' decision of accept or reject based on the watch-list. This calls for a routine telephony infrastructure no more than a typical automated IVRS system to handle incoming calls, and which can have the speaker-spotting system running in a back-end server to listen-in to the incoming call to make the above decision on the calling speaker. Note that this in clear contrast to the more elaborate and covert wire-tapping kind of infrastructure and mechanisms arising in a surveillance speaker-spotting context as outlined above.

15.2 Watch-List Based Speaker Spotting

The speaker-spotting problem can be defined as being comprised of answering two questions with respect to the identity of the input speaker:

1. Does the input speaker belong to the watch-list (i.e., the set of speakers in the watch-list).
2. If so, what is the identity of the input speaker (from among the set of speakers in the watch-list).

As we will discuss later, in some constrained applications, it might be adequate to merely answer the first question, without having to assign an identity to the input speaker.

This problem of speaker-spotting based on a watch-list is essentially solved in the framework of "open-set speaker identification" (OSI). The watch-list consists of N speakers (representing the set of speakers to be detected or spotted in an on-going conversation in a telephone wire-tap or who represent a set of individuals who need to identified in the form of 'negative recognition' in a telephony transaction with a call-center or financial institution). It is assumed that the watch-list contains the speaker models of the speakers in question, typically from an enrollment session, where the speech of these speakers are made available for speaker-model training (note that this could be a voluntary/mandatory enrollment as in the example of negative recognition for financial institutions or an enrollment obtained by other means, such as through covert prior wire-tapping or speech data of the speakers in question from other sources or other contexts). The speaker models are typically Gaussian mixture models (GMMs) for a text-independent mode of operation. Note that speaker-spotting is almost always set in a text-independent mode of operation, since there is more or less no practical control on the text that the speaker can speak, being a covert operation in the case of surveillance speaker-spotting (except perhaps in the negative recognition scenario, which can work in a 'prompted' mode, where the input speech can correspond to known text and hence can use text-dependent mode of operation).

The earliest known work in speaker-spotting [9] is set in an audio (broadcast television news) annotation context, and used both closed-set speaker-identification and open-set speaker-identification frameworks for a watch-list (called 'target' set in [9]) of 5 speakers. For a work that represents perhaps the first of its kind, the thesis was quite comprehensive in its scope of defining the open-set framework for speaker-spotting, and reported 70–80% spotting accuracies for both target and non-target speakers.

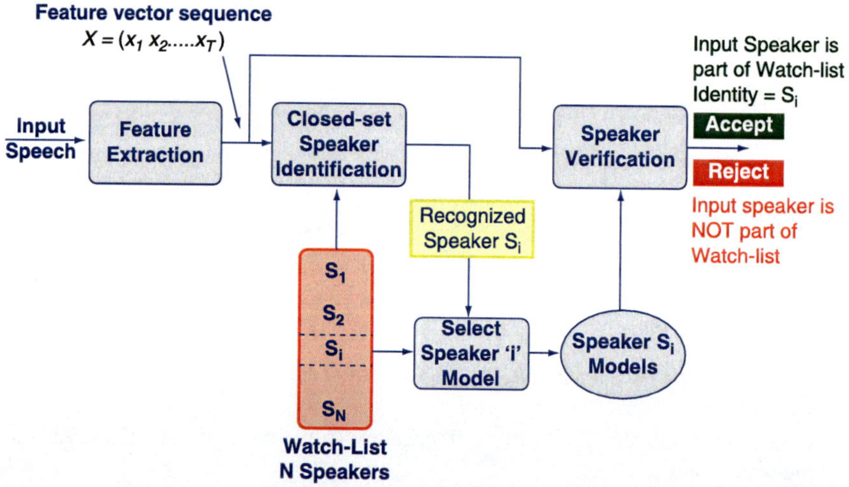

Fig. 15.1 Basic open-set speaker-identification framework for watch-list based speaker-spotting

Note that, in comparison to the other two types of speaker recognition, namely, the closed-set speaker-identification (CSI) and speaker-verification (SV) which are well studied in speaker-recognition literature [10–24], OSI has received somewhat less attention [25–31], despite its unique applications such as watch-list based detection. Here, we will examine the details of such an open-set speaker identification framework in its basic form, and further elaborate the specific variations brought in within this framework for the purpose of watch-list based detection and the associated performance measures and analysis available till date.

15.2.1 Open-Set Speaker Identification Framework

The open-set speaker identification task is illustrated in Fig. 15.1. This comprises two components:

1. Closed-set speaker-identification (CSI)
2. Speaker-verification (SV)

Given input speech in the form of a continuous audio stream (for example, from the telephony wire-tap), the speaker-spotting (or the OSI system here) processes short clips of speech (typically of 1–5 s duration) defined by a sliding window and is required to yield a decision result for each clip. Feature extraction provides a sequence of feature vectors $X = x_1, x_2, \ldots, x_T$ (such as the well known MFCCs) for each clip of audio, with one feature vector for every analysis frame (usually of duration 10 ms); the number of feature vectors 'T' per clip is determined by the frame-rate of the system, which is 100 for a frame-size of 10 ms (thereby yielding T=500 for a 5-s clip).

The OSI system first performs a closed-set speaker-identification task, where it recognizes the input speaker as one of the N speakers (watch-list). Let the recognized speaker-identity be 'S_i', i.e., the input speaker is classified as speaker S_i. Next, the speaker verification component verifies that the input speaker is indeed speaker S_i, i.e., it assumes an identity claim of S_i for the input speaker and accepts or rejects the input speaker as S_i. The final result is therefore as decided by the SV component: either an 'accept', i.e., the input speaker is part of the watch-list and his/her identity his S_i or 'reject', i.e., the input speaker is not part of the watch-list.

In the context of GMM based modeling (for text-independent mode of operation) and identification/verification, the process of OSI can be stated as

$$\max_{1 \le n \le N} \{p(X|\lambda_n)\} > \theta \rightarrow X \in W; \quad X \in \lambda_i, \quad i = \arg\max_{1 \le n \le N} \{p(X|\lambda_n)\}$$
$$< \theta \rightarrow X \notin W$$

where $\{\lambda_n\}, n = 1, \ldots, N$ are the GMMs of the N speakers $S_n, n = 1, \ldots, N$ in the watch-list and $p(X|\lambda_n)$ is the GMM likelihood of the input feature vector sequence X with respect to model of speaker S_n. To state this more briefly, the input utterance X is classified as the most likely speaker S_i (i.e., assigned to the speaker whose model yields the maximum likelihood over all speaker models in the watch-list W), followed by a verification whether X actually comes from the speaker S_i (accept) or not (reject), by comparing the corresponding maximum likelihood score to a threshold. In principle, the speaker-verification component can use a variable threshold, i.e., the threshold can be speaker-dependent (namely, θ_i) considering that the speaker-claim (S_i) against which the input utterance is being verified is known as S_i [31].

Figure 15.2 illustrates this OSI system in a schematic speaker-space. The squares (blue) correspond to the speakers of the watch-list, with the blob like pattern around each of the squares indicating the potential feature space from which the input utterance of that speaker can come from. This essentially represents the intra-speaker variability, having partial overlaps with other speakers in the watch-list, being the source of misclassification error in the CSI stage. The filled (green) and the unfilled (red) circles respectively show the two cases of an input speaker belonging to the watch-list and not belonging to the watch-list.

We now consider the types of errors that arise in a conventional OSI system as described here. The overall output of the OSI system can be the following:

1. Accept input speaker as belonging to the N speaker watch-list and the identity of the speaker is S_i.
2. Reject input speaker as not belonging to the N speaker watch-list.

In this process, various kinds of decision combinations occur at both the closed-set speaker-identification (CSI) and speaker verification (SV) taken together.

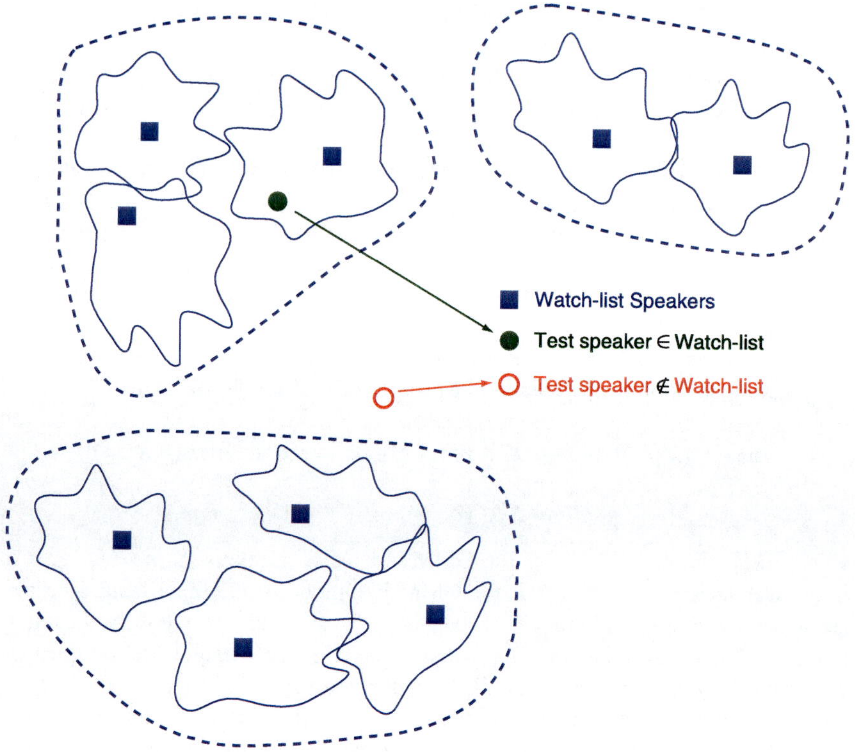

Fig. 15.2 Schematic of speaker-space for open-set speaker-identification framework for watch-list based speaker-spotting

Table 15.1 Errors in a basic open-set speaker-identification framework

Input Speaker Identity	Closed-set Speaker-identification decision	Speaker-verification decision	Errors	Overall Open-Set Speaker-identification Errors	Error Count
j (One of N) speakers)	j (No Error)	Accept as j	No Error	Correct Acceptance	
		Reject as j	Error	False Rejection	Nfr = Nfr + 1
	k (Error)	Accept as k	Error	Confusion; j accepted as k; hence counted as False Rejection (Miss)	Nfr = Nfr + 1
		Reject as k	Error	False Rejection	
m (Not one of N) speakers)	j (Any one of N) speakers)	Accept as j	Error	False acceptance	Nfa = Nfa + 1
		Reject as j	No Error	Correct rejection	

Table 15.1 gives all possible scenarios that arise. As seen from the table, the system makes the following types of errors:

1. **False-rejection:** The input speaker 'j' is part of the watch-list (one of N), but the system rejects him/her (after identifying him/her correctly as 'j'); this error is a 'false-rejection' arising from the speaker-verification component.

2. **Confusion:** The input speaker 'j' is part of the watch-list and the CSI makes an error in classifying this speaker as 'k'. The subsequent SV can either accept or reject, both leading to an error in the overall outcome. This CSI error is thus categorized as leading to an overall error termed 'confusion'. Regardless of an accept/reject decision by the speaker-verification system, this error is counted as a 'false-rejection' (or Miss) for the following reason: A accept (as 'k') corresponds to a Miss of the true-speaker 'j'; a reject (as 'k') is a reject of the true speaker 'j'. Thus, considering the overall decision with respect to the true input speaker identity, the CSI 'confusion' error is treated as 'false rejection' (or Miss), regardless of the speaker-verification's decision.
3. **False-acceptance:** The input speaker 'm' is not part of the watch-list and the CSI forcibly classifies this speaker as some speaker 'j' in the watch-list (CSI has no other possibility). The subsequent speaker-verification introduces an error if it 'accepts' this speaker. This again is a speaker-verification error and is referred to as 'false-acceptance'.

Thus, for a given threshold used in the speaker-verification system within the OSI system, we get two overall error measures:

1. The number of false rejections N_{fr} from a given set of target (within watch-list) speaker trials $N_{target\text{-}trials}$ and,
2. The number of false acceptances N_{fa} from a given set of non-target trials $N_{non\text{-}target\text{-}trials}$.

The overall performance of the OSI system is measured in terms of the probability of 'false rejections' (misses) and 'false acceptances' (false-alarms) as in the case of a speaker-verification (SV) system. The probability of false rejection (not detecting the speaker) $p_{fr}(\theta)$ for a given threshold θ, is given as

$$p_{fr}(\theta) = \frac{N_{fr}}{N_{target-trials}}$$

where, $N_{target-trials}$ and N_{fr} are respectively the number of target trials and the number of those trials for which the target speaker was not detected (falsely rejected).

The probability of false acceptance $p_{fa}(\theta)$ for a given threshold θ, is given by

$$p_{fa}(\theta) = \frac{N_{fa}}{N_{non\text{-}target-trials}}$$

where, $N_{non\text{-}target\text{-}trials}$ is the number of non-target trials and N_{fa} is the number of those trials for which the input speaker was falsely accepted.

These two measures $p_{fr}(\theta)$ and $p_{fa}(\theta)$ are combined in the 'Receiver Operating Characteristics' (ROC) curve by plotting $p_{fr}(\theta)$ against $p_{fa}(\theta)$ for various thresholds θ as in the case of a SV system.

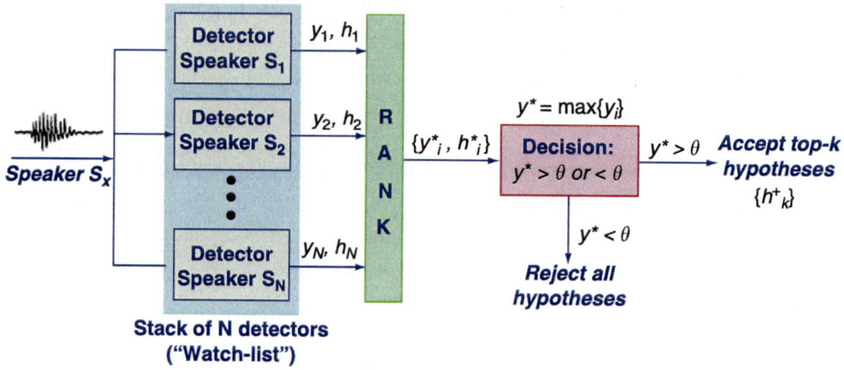

Fig. 15.3 Schematic of multi-target detection framework for watch-list based speaker-spotting [34]

15.2.2 Watch-List or Multi-Target Recognition

In the context of watch-list based speaker-spotting, the above OSI framework has lead to the 'multi-target' detection framework [32–35], derived as a generalization of the more conventional single target detection system. Figure 15.3 shows the schematic of such a multi-target detector framework [34]. For a watch-list of N speakers, the first stage detection is made of N single speaker detectors for speakers $S_1, S_2, \ldots, S_N$, also called the 'stack' or multi-target (stack) detectors, operating in parallel. A single class detector (of say, speaker S_n), computes a likelihood score y, such as $p(X|\lambda_n)$ given in Sect. 15.2.1, which is the GMM likelihood of the input feature vector sequence X with respect to model of speaker S_n. Thus the set of N parallel detectors yield a set of N scores $y_1, y_2, \ldots, y_N$ with associated hypotheses $h_1, h_2, \ldots, h_N$. Note that these N scores are essentially the N scores computed by the closed-set speaker-identification system in Sect. 15.2.1. These N scores are further ranked in order of decreasing score, with y^* being the maximum score, corresponding to the score of the speaker identified by the CSI module in Sect. 15.2.1. This score is now compared with a threshold 'θ' to arrive at a decision of accept or reject—where, 'accept' leads to acceptance of the top k hypotheses $\{h^*_k\}$ (hypotheses corresponding to the top k scores in the ranked score list), $1<k<N$; 'reject' leads to rejection of the input test speaker as not belonging to the watch-list.

Note that the above multi-target detector formulation is identical to the conventional OSI framework in Sect. 15.2.1, except for the acceptance of the top-k hypotheses in the multi-target framework, which are the k-best CSI speakers according to the rank ordered CSI likelihoods. The conventional OSI framework becomes a special case of the multi-target framework for $k=1$.

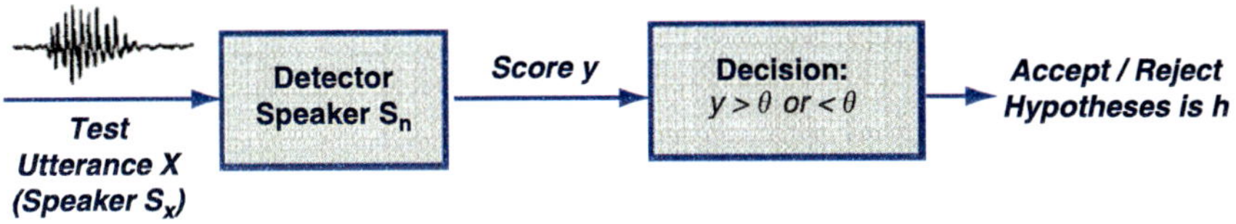

Fig. 15.4 Single-detector block diagram [34]

15.2.3 Performance Measures and Analysis

While the error types and performance measures of the conventional OSI system as given in Sect. 15.2.1 characterizes the performance of a watch-list based application, the multi-target detector formulation of Sect. 15.2.2 [34] was specifically tailored to yield a more rigorous analysis of the overall Miss and False-alarm probabilities of the multi-target detector in terms of the constituent single detectors' Miss and False-alarm probabilities (the single detectors being assumed to operate independently and to have the same Miss and False-alarm probabilities).

It is to be noted that, as with the less studied OSI framework, the multi-target detection framework is studied even less in the speaker-recognition literature. While there has been some detailed development of the watch-list framework in the context of face recognition [32] and biometrics [33], we discuss in this section the analysis developed in [34] in the context of multi-target speaker and language recognition and further extended in [35]. Specifically, [34] analyzed the multi-target detector framework both theoretically and empirically to characterize the performance of the multi-target detectors in terms of the constituent individual (single speaker) target detectors.

In order to be able to derive the multi-target detector performance in terms of the single detector performances, we specify the operation and errors of a single detector. Figure 15.4 shows the single detector for a speaker S_n (called the 'target' speaker). Let the true speaker identity of the input utterance X be S_x. The detector of speaker S_n produces a score y, as in the GMM likelihood $p(X|\lambda_n)$, where λ_n is the GMM of speaker S_n. The score y is compared with a threshold θ to arrive at a decision of accept or reject—the decision to accept $(y > \theta)$ has an associated hypothesis (h) that the input speaker is S_n.

The two types of errors that occur here are shown in Table 15.2 (first row). These are Miss and False alarm (FA), defined respectively as,

- **Miss:** Input speaker is target speaker, but score $y < \theta$
- **False alarm:** Input speaker in non-target speaker, but score $y > \theta$

The corresponding definitions of the miss and false alarm probabilities are also given in row one of Table 15.2.

We now define the types of errors in the multi-target formulation. Figure 15.5 (adopted from [34]) shows a tree diagram of the types of errors considering the various possible scenarios in the different stages of such a system:

Table 15.2 Miss and False-alarm error definitions and corresponding probabilities for single-detector and multi-target detector

Detector	Miss	False-alarm (FA)
Single detectzor errors	$Miss$: $y < \theta \mid S_x = S_n$ $P_{miss}(\theta) = \Pr[y < \theta \mid S_x = S_n]$	FA: $y > \theta \mid S_x \neq S_n$ $P_{fa}(\theta) = \Pr[y > \theta \mid S_x \neq S_n]$
Multi-target detector errors	$Miss1$: $y^* < \theta \mid S_x \in \{S_1, \ldots, S_N\}$ $Miss2$: $y^* > \theta, S_x \notin \{h_k^*\} \mid S_x \in \{S_1, \ldots, S_N\}$ $Miss' = Miss1 \cup Miss2$ $[(y^* < \theta) \cup (y^* > \theta, S_x \notin \{h_k^*\})] \mid S_x \in \{S_1, \ldots, S_N\}$	FA': $y^* > \theta \mid S_x \notin \{S_1, \ldots, S_N\}$
Top-N stack detector	$P'_{miss}(\theta, N) = P_{miss}(\theta) \cdot (1 - P_{fa}(\theta))^{N-1}$	$P'_{fa}(\theta) = 1 - (1 - P_{fa}(\theta))^N$
Top-1 stack detector	$P'_{miss}(\theta, 1) = P_{miss}(\theta) \cdot (1 - P_{fa}(\theta))^{N-1} +$ $(1 - P_{miss}(\theta) \cdot (1 - P_{fa}(\theta))^{N-1}) \cdot$ $\left(1 - \int_{-\infty}^{\infty} (1 - P_{fa}(\tau))^{N-1} p_{tar}(\tau) d\tau\right)$	$P'_{fa}(\theta) = 1 - (1 - P_{fa}(\theta))^N$
Top-k stack detector	$P'_{miss}(\theta, k) = P_{miss}(\theta) \cdot (1 - P_{fa}(\theta))^{N-1} +$ $\left(1 - P_{miss}(\theta) \cdot (1 - P_{fa}(\theta))^{N-1}\right) \cdot$ $\left(1 - \sum_{j=1}^{k} \int_{-\infty}^{\infty} \binom{N-1}{j-1} (P_{fa}(\tau))^{j-1} (1 - P_{fa}(\tau))^{N-j} p_{tar}(\tau) d\tau\right)$	$P'_{fa}(\theta) = 1 - (1 - P_{fa}(\theta))^N$

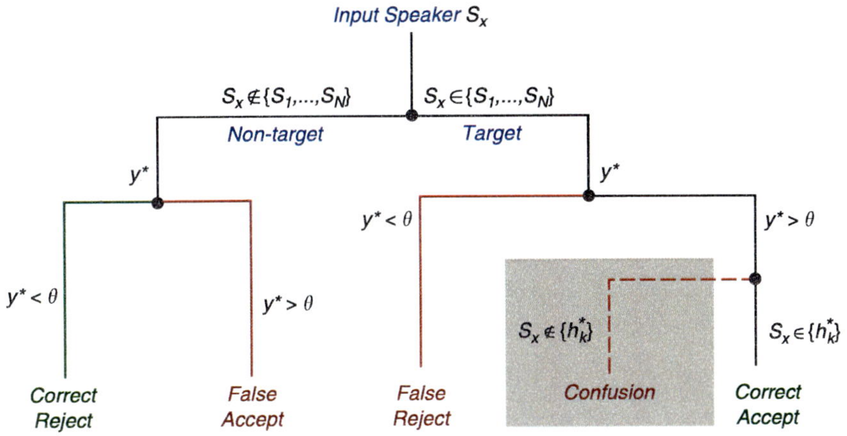

Fig. 15.5 Multi-target detector decisions and errors [34]

1. The input speaker S_x is part of the watch-list or not.
2. The top-rank score y^* (resulting after the N detectors and ranking) is less than or greater than the decision threshold 'θ'.
3. The case when the input speaker is part of the watch-list, but is part or not part of the top-k hypotheses list produced after $y^* > \theta$.

The various decisions and errors that arise in this multi-target speaker formulation are as follows:

1. For non-target input speaker (a non-target trial refers to the case when the input speaker is not one of the speakers in the watch-list or the stack)
   a. **Correct reject:** When a non-target speaker is correctly rejected $(y^* < \theta)$.
   b. **False accept (or false alarm):** When a non-target speaker is accepted $(y^* > \theta)$ and some top-k list is generated (which naturally will not have the non-target speaker, by definition). This occurs when one or more of the individual detectors produce a false alarm.
2. For target input speaker (a target trial refers to the case when the input speaker is one of the N watch-list speakers or the stack)
   a. **False reject:** When a target speaker is rejected $(y^* < \theta)$.
   b. **Correct accept:** When a target speaker is accepted $(y^* > \theta)$ and input speaker S_x is part of the top-k hypotheses (regardless of whether the associated score is the highest).
   c. **Confusion error:** When the input is a target speaker, $y^* > \theta$, but the input target speaker is not present in the top-k hypotheses generated. This represents the error shown in shaded region as 'confusion'. In this case, since the input speaker is eventually 'missed' (not present in the top-k hypotheses), this confusion error is counted as a 'false rejection' or 'miss'.

The above error definitions are given in the second row of Table 15.2. Note that these errors correspond to the 'overall errors' shown in Table 15.1 for the conventional OSI framework, when k=1. Once the Miss (False-rejection) and False-alarm counts are accumulated by these error definitions, the detection performance is provided by plotting miss rate as a function of the false alarm rate for various possible thresholds, as given in Sect. 15.2.1. In the case of this multi-target formulation, the detection error trade-off (DET) becomes a function of 'k' (in the top-k hypotheses) and a set of N DET plots for $k = 1, \ldots, N$ is needed to characterize the overall performance in its entirety. In the performance analysis provided by [34], the focus was restricted to top-1 and top-N conditions, as these provide the bounding performances. We discuss this briefly here.

Table 15.2 provides the performance of the multi-target formulation for three cases: top-N, top-1 and top-k hypotheses, in terms of Miss and False-alarm probabilities which are derived from the miss and false alarm probabilities of the single detector, (referred to as $p_{miss}(\theta)$ and $p_{fa}(\theta)$ of the 'prototype' single target detectors); the single detectors are assumed to operate independently and to have the same miss and false alarm probabilities. We consider the bounding cases of top-N and top-1 which further yields a generalization to the top-k stack detector.

15.2.3.1 Top-N Stack Detector

False-Alarm Probability: As defined as FA' in second row of Table 15.2, a false alarm occurs if one or more individual detectors false alarm. The probability of this event is the complement of the probability that none of the detectors false alarm. This is given as $p'_{fa}(\theta)$ in the third row of Table 15.2.

Miss Probability: For a multi-target detector and top-N hypotheses, a miss occurs as defined as Miss1 in second row of Table 15.2. This occurs when the detector of actual input speaker S_x misses and none of the other N-1 detectors false-alarm. The corresponding $p'_{miss}(\theta, N)$ is given in the third row of Table 15.2. Note that Miss2 does not occur for the top-N detector, since the correct speaker is bound to be in the top-N hypotheses, thus causing no confusion error. Note also that, the top-N detector essentially answers the question whether the input speaker belongs to the watch-list or not, without being able to yield the exact identity of the speaker. This is also referred to as the 'group detector', where it is of interest to merely know if the input speaker is a member of a given group of speakers (watch-list).

In a related work, [35] derives the DET curves of the predicted Miss and False-alarm of the top-N multi-target detector (as given by the equations in the third row of Table 15.2) from a simulation test, by assuming Gaussian distributions for the prototype (single detector) target and non-target score distribution. This is shown in Fig. 15.6 (in a stylized form as would be derived from the equations in Table 15.2 using the same Gaussian parameters as in [35] for the score distributions). Here, the DET curves are shown for different stack sizes (watch-list sizes) N=1, 2, 10, 100 and 1000. It can be noted that while N=1 yields an EER of 5%, the EERs for larger watch-lists are 12.8% for N=10 and 23.9% for N=100. This clearly reveals that in

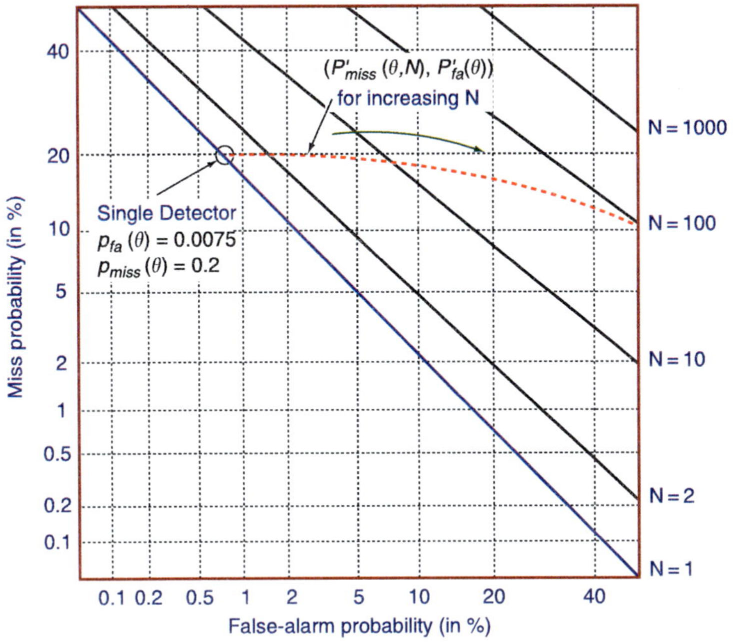

Fig. 15.6 DET family of curves for multi-target detector Top-N case for various watch-list size (stack) N (1, 2, 10, 100, 1000)

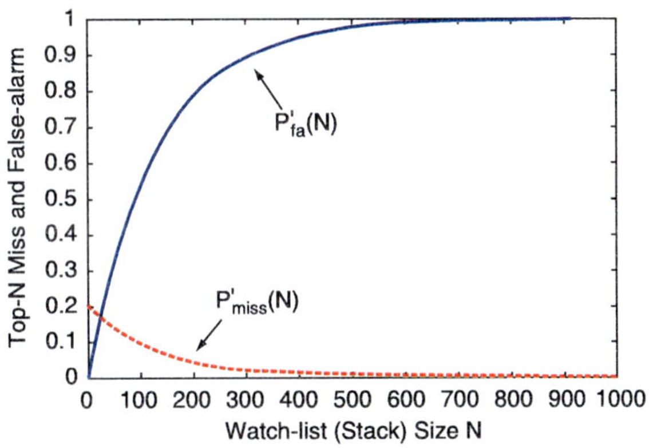

Fig. 15.7 Multi-target detector (Top-N case) Miss and False-alarm probabilities as a function of watch-list size N

applications where it might be necessary to work with very large watch-list sizes (say > 100), these EERs are very high to be of much practical use.

A similar set of DET curves were obtained by [34], but for N = 10 and N = 100, with additional 'actual' DETs obtained from experiments with NIST speaker recognition evaluation databases and Switchboard-II databases. It was shown that the performance predictions based on the multi-target detection framework (as briefly outlined above and in Table 15.2) is accurate and matches with the actual measured DET characteristics.

By assuming a prototype operating point of $p_{miss}(\theta) = 0.2$ and $p_{fa}(\theta) = 0.0075$, the corresponding multi-target $p'_{miss}(\theta, N)$ and $p'_{fa}(\theta)$ (from third row equations in Table 15.2) are shown in Fig. 15.7 as a function of the watch-list size (N=1, ..., 1000). It can be seen that the false-alarm rate increases dramatically with increasing watch-list size (N). The variation of these $p'_{miss}(\theta, N)$ and $p'_{fa}(\theta)$ points (as a function of N) in also shown as a dashed (red) line across the DET curves in Fig. 15.6. This behavior is also referred to as the 'false alarm rate problem' in [35] and has lead to the natural observation [33–35] that the single detectors have to operate at very low false alarm rates for a multi-target stack detector with large stack (watch-list) sizes to have a practically acceptable and useful false-alarm rate. Note also that the $p'_{miss}(\theta, N)$ shows a decrease with increase in stack size N, tending to zero for very large stack sizes. This follows from the fact that, for increasing N and with the top-N decision rule, Miss is made up of only the Miss1 component since Miss2 is non-existent in this case (since it is required to detect the input target speaker as any one of the N speakers for a correct detection, leading to no Miss2 errors), leading to increasing probability of accepting an input target speaker for increasing N, which in turn leads to decreasing probability of miss (i.e., Miss1 in this case).

15.2.3.2 Top-1 Stack Detector

False-Alarm Probability: This is same as defined as FA' in second row of Table 15.2 (the probability being given in the third row for top-N case), since the top-k decision rule does not effect the false-acceptance errors.

Miss Probability: A miss occurs in the top-1 detector due to both the miss cases (Miss1 and Miss2) shown in second row of Table 15.2 (combined as a single Miss). The probability of Miss1 is as given in the third row for Top-N case, with N=1. For the top-1 case, Miss2 corresponds to the condition that the input is a target trial but for which the maximum score $y^* > \theta$, and the resultant top-1 hypotheses h^* is not the target speaker. The probability of Miss2 is therefore given as the 'complement of the probability of Miss1' multiplied by 'the probability that the score of detector of speaker S_x is not the maximum y^* score'. [34] derives the latter probability using a moment model as in [32]. The overall probability of miss $p'_{miss}(\theta, 1)$ for this top-1 case is then as given in the fourth row (Top-1) under Miss. In this equation, $p_{tar}(\tau)$ is the target score distribution (probability density function of target input speakers' score with respect to single detector of the same target speaker).

In order to gain an understanding of the probability $p'_{miss}(\theta, 1)$, [34] considers it as a function of θ for N=10 and for top-N and top-1 conditions using the equations for $p'_{miss}(\theta, N)$ and $p'_{miss}(\theta, 1)$ respectively. Some of the observations to be noted in this regard are:

1. Top-1 miss rates are higher (than both the conventional speaker verification system (single detector) and the top-10 case) over all θ values, since this represents

a stringent condition for the target speaker to be found in the Top-1 hypotheses list under potential closed-set speaker identification errors.

2. The closed-set identification error sets the best miss rate achievable, and Top-1 miss rate asymptotically converges to this closed-set ID error as θ gets smaller.
3. The closed-set identification error serves as the asymptote value (also called the 'top-1 confusion rate'), since this represents the confusion cases when the maximum score is not that of the target speaker due to closed-set identification errors, and which therefore constitute the 'misses' that cannot be corrected by lowering the threshold θ.

In order to understand how the closed-set identification error becomes the asymptotic lower-bound of $p'_{miss}(\theta, 1)$ as θ decreases, consider the definition of the closed-set id error: For the top-1 case, it represents the condition when the score corresponding to the target speaker model is not the maximum y^*, and is thus the complement of the 'probability that the target detector score is the maximum'. The closed-set speaker-identification error is then given by

$$p'_{conf}(N) = \left(1 - \int_{-\infty}^{\infty} (1 - P_{fa}(\tau))^{N-1} p_{tar}(\tau) d\tau\right)$$

which is part of the Miss2 probability in the equation in row 4 of Table 15.2.

For θ at its lowest possible extreme, (say, $-\infty$), $p_{miss}(\theta) = 0$. Accordingly, $p'_{miss}(\theta, 1)$ (equation for Miss in row 4 of Table 15.2) reduces to $p'_{conf}(N)$ (Miss1 becomes zero, $y^* > \theta$ with probability 1 and Miss2 becomes the closed-set identification error), which is the closed-set id error, thus validating the above observation that the closed-set identification error serves as the asymptotic lower bound for $p'_{miss}(\theta, 1)$ as θ gets smaller—representing the 'misses' due to the confusion error which cannot be corrected by lowering θ.

The above behavior is further strengthened in the actual 'measured' DET performance for N=10 in [34]. In this case, the top-1 DET is observed to converge asymptotically to the top-1 confusion rate. Moreover, it was noted that there was a discrepancy between the predicted performance and the measured performance (DET curve) for top-1, attributed as possibly due to incorrect assumptions about the independence of single detectors and inaccuracies in numerical integration techniques (used in the evaluation of the miss probabilities of the multi-target detector in terms of single target detector performance, as in Table 15.2).

More importantly, [34] also observed that for N=10, the top-1 performance is poorer (higher DET curve and corresponding EER) than top-10, showing how an increase in k improves the performance, by lowering the overall miss rate, by taking advantage of the top-k list which provides an increase in 'correct acceptance', as is also discussed further in Sect. 15.2.4.

15.2.3.3 Top-k Stack Detector

The above top-1 case can be generalized to top-k case; this was done by the moment method in [32, 34]. Here, for the case when $y^* > \theta$ and yielding a top-k hypotheses list, a correct detection requires that the input target speaker must be in the top-k list. The probability of Miss2 (as in second row of Table 15.2 for the top-k case) is given as 'the complement of the probability of Miss1' multiplied by 'the probability that the score of detector of speaker S_x is not in the top-k scores'. Accordingly, the probability of the latter condition is computed as the complement of the condition that the target speaker score is among the top k scores—which in turn is computed by enumerating all possibilities by which this can occur [34]. As derived in [34], this finally yields the probability of miss (for top-k stack detector) $p'_{miss}(\theta, k)$ as shown in the last row of Table 15.2.

15.2.4 *Qualitative Summary of Performance*

While Sect. 15.2.3 provided a close look into the performance of the multi-target framework, mainly as formulated in [32] and [34], a qualitative view of the performance will offer broad insights into the speaker-spotting framework, that otherwise might not be available from the theoretical analysis alone.

First, we consider the open-set identification framework (OSI) of Sect. 15.2.1. The performance is best characterized in terms of the behavior of misses (false rejections) and false-alarms (false acceptances) of target inputs and non-target inputs respectively, for increasing watch-list sizes (N). We refer to Fig. 15.1 and Table 15.1 for this discussion.

- **Target input:** For target trials, as N increases, the closed-set identification first stage incurs higher classification errors, and recognizes input 'j' as 'k', causing increase in 'confusion errors'. As defined, the speaker verification system makes an accept/reject decision on this recognized speaker identity, and both decisions lead to false-rejection (miss), as eventually speaker 'j' is missed. Thus the Miss error increases with increase in watch-list size. However, if we were to relax the constraint as one of deciding if the input target speaker is one of the N speakers in the watch-list, as in a 'group-detector', defined in Sect. 15.2.2, it is quite evident that the Miss errors should actually decrease for increasing N. This is obtained by relaxing the constraint on the speaker-verification decision, by considering an accept decision as 'k' (for any 'k') to be counted as a correct acceptance (rather than as a false-rejection as is indicated in Table 15.1). This then becomes an exact equivalence of the top-N multi-target detector in Sect. 15.2.2—which also yields the same behavior of decreasing 'Miss' for increasing N for the top-N case.
- **Non-target input:** For non-target trials, the CSI system invariably classifies the input speaker as one of the N speakers in the watch-list (incorrectly, but unavoid-

ably, as this is a forced decision) followed by an accept or reject decision by the speaker verification system. However, for increasing N, the score on which the decision is made is higher (owing to larger N and presence of watch-list speaker models that match the input non-target speaker better), which in turn leads to increased accept decisions by the speaker verification system. This, therefore, leads to increased false alarm for increasing N.

Thus, it is intuitively pleasing to see that for increasing N, the speaker-spotting system as an OSI system yields increasing False alarm for non-target trials and decreasing Miss for target trials, if the problem is defined as a 'group detector', where it is required to answer only whether the input speaker is one among the N watch-list speakers or not—larger N increases makes the watch-list speaker models occupy large extents of the feature space, thereby facilitating a non-target speaker to be increasingly confused as one of the N speakers (increasing false alarm), and enabling a target speaker to be favorably decided as among the N speakers (decreasing miss).

Next, we consider the multi-target (stack) detector framework of Sects. 15.2.2 and 15.2.3. The following can be noted here.

- **Target input:** For increasing N, the first stage rank-ordering is prone to error (as in the CSI system of the OSI framework), where the maximum score y^* is not associated with the target speaker. However, considering that y^* tends to increase with increase in N, the second stage yields a decision $y^* > \theta$ more often, with the final result dependent only on the size of the top-k hypotheses (i.e., value of k). For k=1, this system is identical to the OSI framework, and results in high false-rejections when the target speaker is not in the top-1 hypotheses (confusion errors). Thus, in this case, increasing N actually causes increasing Misses. However, as k increases (i.e., $1 < k < N$), the chances of the input target speaker to be part of the top-k list increases, thereby increasing 'correct acceptances' or conversely decreasing 'false rejections' (under confusion error). The top-k hypotheses therefore serves as a compensation mechanism (as in k-best strategies) to counter the increased CSI errors with increase in N, thereby acting to 'decrease' the Misses with increase in 'k'. For the limiting case of k=N (i.e., the top-N stack detector), the Misses are the lowest possible, since the system behaves as a 'group detector' where it is required to determine only whether the input (target) speaker is among the full watch-list of N speakers or not (which it will always be); therefore, Miss2 (Table 15.2) is non-existent and Miss1 (Table 15.2) alone manifests as the overall Miss. Miss1 by itself is a decreasing function of N, since the condition $y^* < \theta$ is satisfied less and less due to higher y^* with increasing N for the same input utterances; in other words, the input utterance has higher likelihoods from a larger set of watch-list models, even if this highest score y^* is with an incorrect speaker in the watch-list; i.e., y^* is to be viewed simply as the best score that the watch-list, taken as a whole, can yield for an input target speaker, constituting the overall target score distribution with respect to watch-list target speaker models; i.e., the target score distribution shifts rightward (for a given theta) for increasing N, thereby decreasing Miss1, which is the reducing area under the curve of the target score distribution to the left of θ.

- **Non-target input:** For non-target trials, and for increasing N, the maximum score y^* tends to increase (rightward shift of the non-target score distribution for the given set of input non-target utterances and given θ), causing the condition of $y^* > \theta$ to be more likely, resulting in increasing false acceptances. The top-k hypotheses do not play any role here, and the only consideration is that a large watch-list is conducive for a non-target speaker to be confused with a watch-list speaker in a qualitative sense.

Thus, this also yields the intuitively pleasing behavior of the performance of 'increasing False-alarm for non-target speakers and decreasing Miss for target speakers' with increasing watch-list size N; here, the top-k hypotheses serves as a mechanism to allow increased likelihoods of the target speaker to be (correctly) 'accepted' even in the presence of first stage classification errors where the target speaker does not have the maximum score y^* in the rank ordered score, but could be a little down the ranked list (which is then picked up by the top-k hypotheses). Thus, for a given N, increasing k tends to offer lower miss-rates. Moreover, an increase in N will require an increase in k to maintain the miss-rate at the same level (i.e., to continue to detect the target speaker among the top-k list and thus maintain the same 'correct acceptances').

In short, the above give a qualitative insight into the behavior of the performance (Miss, False-alarm) of the speaker-spotting system, both in the OSI framework and the multi-target stack detector framework, and shows the primary effect of increasing false alarm with increasing watch-list size, decreasing overall Miss with increasing watch-list size (for top-k conditions, with $1 \ll k < N$), and top-N serving as the 'group detector' bound of the performance.

15.2.5 Related Work

15.2.5.1 Top-Norm

Following the multi-target detector formulation and framework in [34], [35] extended the framework further by proposing a new score-normalization technique, called Top-norm, and a method for adapting the normalization parameters. The Top-norm was motivated by the so called 'false alarm error rate problem' of the multi-target framework, where the false-alarm rate increases dramatically with increasing watch-list size (N). More specifically, the Top-norm is motivated from the observations that (1) the problem of increase in false-alarm rate in multi-target stack formulation exists even after conventional score normalizations such as world model normalization (WMN) or Z-norm [21], (2) the normalization parameters used in the Z-norm, namely, mean and variance of the non-target score distribution assume that the score distribution of the non-target scores is Gaussian and (3) this is not accurate for the entire distribution, but possibly true only for the right-tail of the score distribution. Accordingly, [35] proposes the top-norm which calculates the normalization parameters (mean and variance) only from 'top non-target scores' which contribute to the false-alarm errors. These parameters are thus calculated from a pre-defined percent of the top-scores, but used as in Z-norm. [35] compares this method with other score normalization methods such as WMN, Z-norm, T-norm [21] and unconstrained cohort normalization (UCN), with experiments using the NIST99 speaker evaluation database and Switchboard databases with up to N=183 watch-list sizes, and show that the Top-norm is superior to the other methods.

15.2.5.2 Normalization Parameter Adaptation

It is interesting to note that in watch-list applications (in the context of surveillance speaker-spotting or negative recognition applications such as fraudster detection), the watch-list consists of the speakers in question (potential fraudsters), and the majority of the test speakers in the surveillance audio (obtained via wire-tapping or through incoming customer calls) are non-fraudsters, i.e., non-target speakers (not part of the watch-list). Taking advantage of this fact and towards realizing lower false-alarm rates in multi-target recognition, [34] proposed the use of an adaptation scheme for continuously adapting the Top-norm parameters (as above) from the test data of the non-target speakers collected during test. The adaptation essentially consists of estimating the mean and variance of the top p% of the non-target scores using the previous values of the mean and variance and the current score value, after using a threshold correction to decide whether a current score (non-target) qualifies to update the parameters reliably. With experiments done using up to 183 watch-list speakers from NIST99 database, [35] showed that the Top-norm with parameter adaptation offers better performance in the low FA rate regions.

15.2.5.3 Prioritization Mode of Operation

In what can be considered as a related but mildly alternate framework for surveillance applications and associated analysis formulation, [36] proposed and studied the so called 'prioritization mode' of operation. Here, in a variant of watch-list of multi-target speakers of interest to be detected in surveillance audio, the problem of surveillance is posed as that of locating a given speaker of interest (watch-list size=1) in a large set of speech samples (input audio data), such as arising in law enforcement or intelligence activities or even news indexing, but with the focus shifting to the analysis of a 'queue' of ordered list of samples in decreasing order of likelihood (of being the target speaker of interest). Such a surveillance system is seen more as a tool to 'prioritize' samples for further examination, assessment and action by human experts, who can inspect the samples in the given order, to determine whether the target speaker is present or not in the large set of samples in the surveillance audio. It should be noted that such a mode of operation is however restricted to ‘batch’ mode of processing stored data (such as for example, a system that gathers 24 hours data followed by off-line processing and presentation of the queued result to the human experts), rather than on sample-by-sample basis, where a decision might be expected for each sample, as and when they occur.

To motivate the proposed approach, [36] considers the conventional verification mode of operation as having a baseline performance of $\frac{N_f p(f)}{1-p(m)}$ false alarms per hit, while missing $p(m)$ of the targets; here, $p(f)$ and $p(m)$ are the probability of false-alarm and miss of the verification system, and N_f false trials are attempted for every true trial, yielding $N_f p(f)$ false alarms for $1-p(m)$ successful hits. [36] gives an interesting argument forwarded by [37], by considering an example where a surveillance system has to search 10^{11} samples looking for 1 target speaker, being verified by a system with $p(m) = 0.001$ and $p(f) = 0.001$. Based on the false-alarm per hit performance given above, [37] concludes that such an “unrealistically-accurate system will generate one billion false alarms for every real terrorist plot it uncovers”. The primary basis of [36] is that while such a reasoning is valid for a system operating in verification mode (where a hard accept/reject decision is made for each sample), the performance can be enhanced significantly in some surveillance applications by the ‘prioritization mode’ of operation. In such a mode of operation, the system need not make such hard decisions, but yields only the queue of samples in decreasing order of likelihood, which the human expert examines in the same order (from top of the queue downwards); the usefulness of such a system rests on the possibility that the target speaker appears as close as possible to the top of the queue.

The effectiveness of the queue to provide such a successful search near the top of the queue can therefore be characterized by the position of the target speaker in the queue or equivalently, in terms of the proportion of non-target speakers who are expected to score higher than the target speaker of interest. [36] develops a theoretical formulation to determine statistics related to this relative position of target speaker, as a figure of merit of the queue in terms of the target- and non-target score distribution parameters (mean and variance, assuming these to be univariate Gaussian). Specifically, [36] derives the distribution of the target speaker position, the mean target position, median target position, and the truncated queue length (defined as L_q, the length of the truncated queue which contains the target speaker with probability q), in order to further analyze the model with simulated score distributions, as well as real experiments with the data of the NIST SRE 2005 tasks.

With respect to simulated score distributions, [36] considers three distributions (three different mean and variance for the target score distribution) and show that while the conventional speaker verification would exhibit the same performance, their proposed prioritization mode of operation allows for very favorable mean and median target positions. Specifically, considering a total sample size of $N=10^{11}$ (where N is the number of non-target speaker samples plus one target speaker sample), they show an improvement of approximately 5 times, 40 times and 150 times for all these three cases over the system operating in conventional verification mode. With reference to the median position, they show that in order to be successful half the time, an expert analyst need to look only within the top 16 entries for one of the systems, and only the top entry for the other two systems, as obtained using Monte Carlo simulations on queues of 10,000 entries repeated 100,000 times. These results serve to strengthen the case in [36] that surveillance systems are more usefully operated in prioritization mode instead of the verification mode.

Experiments with the NIST SRE 2005 task data also strengthened the queuing approach. Specifically, [36] showed that the DET curve for prioritization has dramatically better false alarm rates for the same miss rates, when compared to a verification system and that the simulated DET from Gaussian data matches well with the real data, though with some performance difference, attributed to deviation of the real data from the Gaussian assumption.

15.3 Challenges in Speaker-Spotting

Voice as a biometric signal lends itself to a wide variety of applications, and with a degree of success perhaps next only to the IRIS or fingerprint biometric [38–40] identifies three major types of applications which arise from the biometric character of speech, namely, voice authentication (access control, either for physical locations or over telephone channels), speaker detection (blacklisting detection, also known as speaker-spotting, as is the focus in this chapter) and forensic speaker recognition (where voice is used as evidence in courts of law and police investigations). Accordingly, the technologies inherent in these three applications share similar properties and constraints, the most important of which includes the vulnerability of the speech signal to a host of sources of variability; while some of the variability arises from the inherent personal and socio-linguistic characteristics, a major component of the variability come from environmental aspects.

This variability poses the biggest challenge to all the three modes of applications defined above. More specifically, the variability manifests primarily as the mismatch between training and testing sessions and can severely degrade the performance of the speaker recognition system underlying any of the three applications above. The mismatch comes from the various variability factors between any two recordings: environment (background noise), microphone or handset, transmission channel, psychological and pathological state of the speaker, linguistic content, voice aging etc. Thus any attempts to enhance the robustness of an automatic speaker recognition system to these sources of variability will impact all of the three modes of applications. Thus a speaker-spotting system or application faces more or less the same challenges as a voice-authentication system (say, in telephony access control kind of scenario) and the forensic speaker recognition application (where a voice sample or a trace has to be matched against a host of suspect recordings) does, in addition to having highly unique challenges idiosyncratic to the speaker-spotting application. More specifically, it can be said that speaker-spotting shares more of its challenging conditions with forensic speaker recognition, which can be categorized as 'realistic conditions'—implying no control, assumption or forecast can be made on the conditions under which the signal is acquired [21]. More importantly, speaker-spotting (as a negative recognition), shares a peculiar characteristic with forensic application in that the perpetrator is not a collaborative partner (as could be in access control or similar positive recognition application), but is someone trying to defeat any mechanism which attempts to derive a decision from such recordings (wire-tapped or incoming call) and which has the potential to convict him.

In a generic manner, the speaker-spotting system is essentially an open-set identification system (as examined in Sect. 15.2.1) or a set of detectors followed by speaker verification thresholding (as examined in Sect. 15.2.2) and has an underlying component of likelihood calculation (the likelihood that a given sample of speech is from a speaker model in the watch-list), as well as a speaker-verification type of decision making involving comparing the (possibly 'normalized') target-speaker score against a threshold. In such a context, the performance of the system is affected in a manner similar to open-set speaker-identification or speaker-verification. Thus, all the challenges facing such an overall system can be viewed as those factors impacting the performance of the underlying score calculating component or the speaker verification component, in addition to the unique challenges such a system could face in terms of large watch-lists, difficult access to enrollment data, test data that could be short, of varied linguistic content, noisy, from different handsets, telephony channels, background conditions, over large session differences, voice aging etc.

Specifically, we consider in the following, the main challenges that a speaker-spotting system would face in a practical deployment scenario:

1. **Large watch-lists:** As dealt with in detail in Sect. 15.2, the performance of the multi-target recognition system is critically affected by the size of the watch-list (or the stack). Various performance characteristics were considered in terms of top-1, top-k and top-N stack detectors, all of which share the 'false alarm error rate' problem of dramatically increasing false-alarm with watch-list size. Clearly, a practical system may have to work with sizeable watch-lists (as demonstrated in the Iris application in Sect. 15.1.4, and which is likely to be the case in equivalent voice scenarios too), and a severe degradation in performance with watch-list size can be detrimental to widespread use, success and acceptance of such a system. As noted in Sect. 15.2.3.1, for the multi-target stack detector with large stack (watch-list) sizes to have a practically acceptable and useful false-alarm rate, it then becomes necessary that the single detectors have to operate at very low false alarm rates, among other considerations.
2. **Linguistic content:** A speaker-spotting system has to necessarily operate in a text-independent mode (except perhaps in prompted negative recognition scenarios to detect a black-listed person wishing to enroll into a bank or an enterprise), and hence needs to be robust to linguistic differences between enrollment and testing. Intra-speaker variability arising from phonetic content, duration etc. can play a vital role in accentuating the training/testing mismatch, which nevertheless is handled in conventional text-independent approaches. However, as in

forensic applications, there could be severe mismatches, sometimes purposely induced by the speaker (in such a non-cooperative application such as speaker-spotting), such as when the enrollment data could be loud and the test data is feeble (causing severe difference in degree and nature of articulation of underlying phonetic content), or of very short durations and mismatching phonetic content and in distinctly different language, accent, dialect etc.

3. **Handset-variability:** It has been long established that the mismatch in handset type (for example, carbon button vs electret) between training and test can cause significant loss of performance in a speaker recognition system [41]. This is clearly a very pertinent problem in the context of speaker-spotting, where the test data of the speaker to be spotted can come from diverse and unknown hand-sets, something that can be deliberately introduced. However, score normalization techniques based on Hnorm (variant of the Z-norm) and HTnorm (a variant of the Tnorm, based on Hnorm) have been proposed to compensate for this mismatch [21]. For instance, in Hnorm, handset-dependent normalization parameters are estimated by scoring handset-dependent speech of non-target speakers against target speaker models and the appropriate set of parameters are used for score normalization during testing, after determining the type of handset of the test speech. However, the relevance and success of such normalization in a speaker-spotting application does not seem to have received particular attention in the literature; in an idiosyncratic manner, it should be noted that given the nature of speaker-spotting or negative recognition applications (where adequate training data of target speakers, i.e., watch-list speakers, might not be available), which of these normalization methods is most applicable remains a question.
4. **Channel-variability:** This variability is more or less as encountered by voice authentication application for telephony access control. Considering the variability inherent in the telephony channels, where the training and test data could originate across a variety of telecommunication networks (such as landline (PSTN), cellular (mobile), VoIP etc.), channel variability does pose a serious issue in the context of speaker spotting. While basic techniques for channel compensation such as cepstral mean normalization at the feature extraction level can impart some degree of robustness [21], the applicability of such techniques to real and large-scale speaker-spotting is yet to be undertaken and reported.
5. **Session-variability:** This is perhaps a significant problem in all voice biometric based applications. In the context of speaker-spotting, there could be significant session variability between training and test sessions, as well as over test sessions. Recent advances in session variability modeling techniques include latent factor analysis (FA) or nuisance attribute projection (NAP) [42, 43]. However, studies in the context of speaker-spotting incorporating such techniques are yet to be reported.
6. **Background noise:** Robustness to background noise is a perennial problem in all practical applications; in the context of speaker-spotting in telephony application, the background conditions are as could be expected for access control or

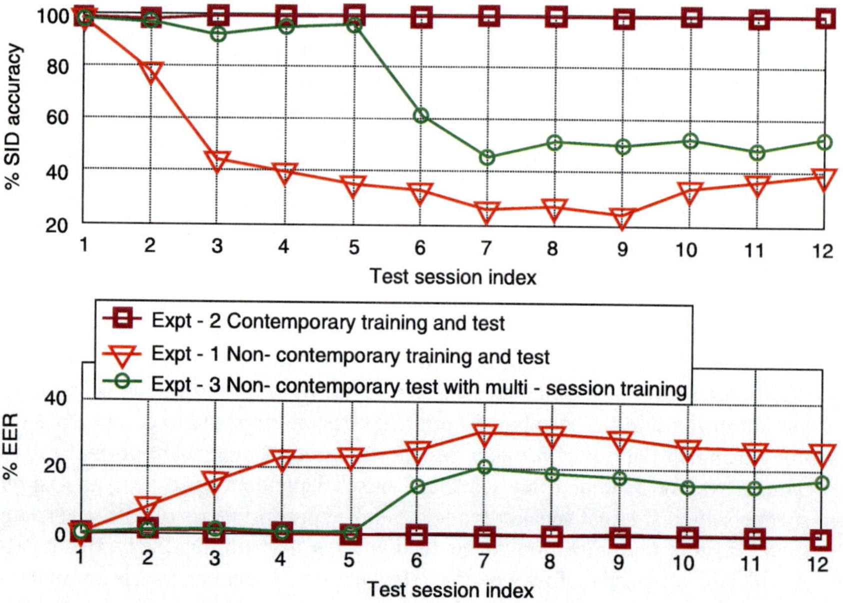

Fig. 15.8 Contemporary and non-contemporary performances over 12 test sessions (2 years); *top:* closed-set speaker-identification; *bottom:* speaker verification EER

forensic application—that of uncontrolled background environment. This could be combated with varying degrees of effectiveness in the signal processing stage (speech enhancement), feature extraction (robust features), or modeling stage, each with varying degree of effectiveness.

7. **Voice aging:** An important and unavoidable component of intra-speaker variability that manifests within session variability is the phenomenon of voice-aging, where an enrolled speaker's voice changes over time, and produces significant mismatch between the models (trained during enrollment session) and later test sessions. Longer the elapsed time difference between the training and test sessions, the more pronounced is the aging effect, and can cause considerable degradation in performance, as has been observed consistently in most studies. Also called 'non-contemporary testing', such test sessions can have very poor performance when compared to contemporary data (where the elapsed time is very less). For instance, we show in Fig. 15.8, the effect of voice aging on closed set speaker-identification accuracies and speaker-verification EER, which we obtained [44] using the multi-session "Speaker-recognition corpus version 1.1" database of the Center for Spoken Language Understanding (CSLP) from LDC. Here, we used a set of 70 speakers from this database with each speaker providing speech utterances over the telephone, spanning a 2 year period in 12 sessions. This database thus represents a very long period

over which voice-aging can manifest. It can be seen from the top plot, that the closed-set accuracy (%SID accuracy) is very good for contemporary testing (Expt 2, square (purple) line at nearly 100%), while it drops severely to as low as 20% for non-contemporary testing (Expt 1, triangle (red) line). Likewise, for speaker verification, the EER is very low (~0%) for contemporary testing, and increases to beyond 30% for non-contemporary testing. Creating 'non-contemporary robustness', i.e., robustness of a system to such voice-aging is still a serious challenge and is very relevant to speaker-spotting applications where the test sessions are invariably distributed over arbitrarily long elapsed-times after the enrollment session (of target speakers in the watch-list). While multi-session training data can help realize improved performances (as shown as Expt. 3, circle (green) line), more complex session adaptation strategies are required to handle this problem.

8. **Voice disguise:** In the context of speaker-spotting to detect a watch-list of speakers, and given the legal context of such a requirement, it is not surprising that the speakers involved might attempt any means of ensuring that their voice is not detected. Voice disguise is easily a commonplace strategy a speaker could adopt to conceal or morph his/her voice to avoid detection. Speaker-spotting shares this problem with forensic application and it is not clear if there are effective methods to undo such a disguise mode of operation. This clearly can be the most challenging of problems, considering it is a conscious act on the part of the speaker to not appear as his natural voice, particularly in the context of negative recognition.
9. **Data length:** As in forensic applications, the data available for training (enrollment) and test can be seriously compromised in speaker-spotting application also. While the enrollment of potential watch-list speakers poses no difficulty in the negative recognition applications (such as a financial institution requiring an enrollment of every customer, to be followed only later by a shifting of a customer committing fraud to a watch-list), this could be a serious issue in speaker-spotting in wire-tapping applications, where either adequate prior training data is not available from other modalities or telephony conversations are not available in sufficient length for training of speaker models. Test data could also pose severe constraints, unless they are of the nature of long conversations, which is an unpredictable and uncontrolled condition in speaker-spotting.
10. **Real-time requirements:** The fact that speaker-spotting in its surveillance context is expected to either trigger an alarm (to bring an expert listener online) or recording into a database, calls for real-time recognition; this becomes a considerable demand if watch-list sizes are very large for a given telephone channel under surveillance or if a large number of parallel channels are being monitored. In contrast, the 'prioritization mode' of operation [36], which yields a rank ordered list of samples for further expert examination can work in off-line mode on stored data, as pointed out in Sect. 15.2.5, but is not viable for decisions required on a sample by sample basis, as in real-time operation.

15.4 NIST SRE Results and Implications on Speaker-Spotting

As indicated above, the watch-list based speaker-spotting or negative recognition system faces formidable challenges arising from the many sources of variability, some common to voice authentication scenarios and some others more in line with forensic speaker recognition settings. However, the underlying mechanism that gets affected by such variety of variability is the score calculation giving the likelihood of an input sample from an unknown speaker (test utterance) with respect to a speaker model in the watch-list, and at the system level, the verification decision that an input utterance is from a target speaker or not.

In this context, it is important to take into consideration that the NIST Speaker Recognition Evaluation (SRE) campaigns (organized since 1996 annually with a few exception years) [45–47] indeed address the very same issue of evaluating the approaches and systems under various challenging scenarios notably those such as variability arising from environment, microphone or handset, transmission channel, session variability, training data size, voice aging etc. Thus, in order to be able to assess the performance levels of a speaker-spotting system, it is appropriate to turn to the results obtained from the NIST SRE campaigns and understand their implications on the potential and limitations governing such speaker-spotting systems. This is particularly pertinent, considering that NIST SRE has been concerned mainly with text-independent speaker recognition tasks based on telephonic conversational speech, and have to answer the question "did speaker X produce the speech recording Y and to what degree" [48]—which is precisely the setting and quantification underlying a speaker-spotting system for each speaker in the watch-list; a degradation in performance of such a scoring and decision due to any of the sources of variability impacts the overall performance of the speaker-spotting system either as a open set identification system or a multi-target detector system as outlined earlier in Sect. 15.2—thereby serving as an indicator of the performances that can be expected from a speaker-spotting system under such challenging conditions of variability.

The following represents some of the observations from the NIST SRE results during the period 2004–2008 [48]:

1. With respect to training duration, it was clearly shown that the use of three times more data for training a speaker model reduced the EER drastically from 2.96 to 1.94% (NIST SRE 08, short2-short3 condition versus 3conv-short3 condition). This has immediate implications on speaker-spotting, in that non-cooperative conditions of not being able to acquire adequate training (enrollment) data for watch-list models can seriously degrade performance.
2. In the case of speaker-spotting, as noted earlier, the target speakers (in the watch-list), being the black-listed speakers, occur rarely, and the non-target speakers (representing the normal population not in the watch-list) occur more frequently. This leads to the question whether conventional notions of unsupervised adaptation of the target/non-target models using test session data can play a role. Clearly, considering the natural abundance of the non-target speaker data, it

appears that only the non-watch-list speakers can perhaps be modeled (possibly in the GMM-UBM background or cohort models) in such an adaptive manner, while the target speakers may not benefit from such an adaptation. In general, NIST SRE results do show that such an adaptation can be significantly useful—with an EER reduction from 4.55 to 2.36% in unsupervised adaptation mode (NIST SRE 06, 1conv-1conv, English only, male set) for the GMM-Factor-Analysis system, which compares favorably to the achievable baseline of oracle performance of 1.62%, clearly indicating that the effective training data used is a key factor in speaker recognition performance.

3. With respect to voice aging, it was shown (NIST-SRE '05) that elapsed time between training and test can seriously degrade speaker verification performance by a factor of two when the elapsed time exceeds one month (a result with a caveat that other factors, such as bias in corpus collection, were correlated with elapsed time in these experiments).

In general, despite significant improvements in error rates, under various controlled variability factors as above in the NIST SRE campaigns over the years, it is concluded that this may not be a sufficient criterion for evaluating performance of speaker recognition, particularly for forensic applications [48]. Specifically, attention is drawn to the fact that in forensic applications, only short pieces of speech are available, and hence the performances are bound to be poor, considering the requirement of large training and test data for high performances. In addition, [48] points to the fact that there are significant differences between commercial applications and forensic field, the former (such as access control) involving a clear scenario with well defined environment and variability factors, while the latter has factors affecting performance varying significantly. It can be noted that this applies readily to the speaker-spotting scenario which has aspects of both commercial application and forensic field combined in a curious manner (telephonic application like access control, but difficult enrollment, test conditions such as in forensic applications).

Moreover, [48] points out that evaluations need to include more variability factors, more heterogeneous factors, and scaled up test conditions to include 'thousands of speakers, hundred of thousands of tests and hundreds of mixed conditions' for better generalization of the results. This is again very pertinent for evaluating and assessing speaker-spotting kind of applications, where watch-list sizes could be large, and surveillance data being processed for speaker-spotting could be proportionately even larger (refer to Sect. 15.1.4 for the more mature Iris watch-list application in this context).

Taking into account the various performance characteristics that NIST SRE experiments and systems point to, along with the limitations that these still pose with respect to 'realistic conditions', [48] conclude that 'forensic applications of speaker recognition should still be taken under a necessary need for caution', in a way reinforcing a similar note of caution made much earlier in 2003 by [49] that 'currently, it is not possible to completely determine whether the similarity between two recordings is due to the speaker or to other factors…', though there has been significant progress in speaker recognition since then (2003) until now.

Given the fact that speaker-spotting and watch-list based negative recognition pose a primarily more difficult dimension in terms of a large watch-list and proportionately large test data (samples), even while considering all other conditions being same as in an access control application (say, telephony) or forensic application, it becomes clear that the same note of caution as above needs to be exercised on the use of speaker recognition technologies in the context of surveillance or negative recognition.

This however, should not deter further progress in this very difficult task, which perhaps ought to be viewed as a uniquely challenging speaker-recognition application with its own idiosyncratic challenges and solutions. Some of the potential research directions are evident from the challenges listed above, and it remains to be seen how successful future solutions would emerge.

15.5 Related Topics: Speaker Change Detection, Speaker Segmentation, Speaker Diarization

Speaker-spotting as defined and discussed in this chapter is set in a telephony surveillance and negative recognition context and essentially involves processing continuous stream of speech data consisting of at least two speakers in conversation (in the case of surveillance) and one speaker speaking into the system (in the case of negative recognition). In either case, the system performs a multi-target decision using a watch-list on short clips (also called sample or trace, in forensic parlance) of speech (typically of 1–5 s duration making up a sequence of 100 to 500 feature vectors). The speaker-spotting system has to provide an accept or reject decision to each clip, along with a speaker identity (from the watch-list) accompanying an accept decision. Here, the decision for each clip is done using speaker models, i.e., models of speakers in the watch-list.

In this section, we briefly point to a related class of problem which shares a similar, but not exactly the same, objective of processing a stream of speech spoken by two or more speakers in turns (such as in a conversation or a meeting or broadcast news or multi-media data or movie content) towards detecting speaker changes or segmentation of the speech stream in terms of speakers and assigning speaker identities to each segment where possible. This class of problem has applications such as speaker tracking, speaker diarization, facilitating speaker normalization and adaptation to improve speech recognition systems, indexing audio recordings, meeting capture data (live or telephony meetings), and providing cues for scene-, topic- or program-changes in multi-media applications (broadcast news, television, movie content etc.).

The primary problem of 'speaker change detection' (SCD) underlying these applications has received considerable attention in speech literature [50–57]. Note that unlike speaker-spotting with watch-lists (i.e., knowledge of speakers of interest and their models obtained a priori by some means), SCD is defined as detecting speaker changes without prior acoustic information on the speakers. Thus, these algorithms aim to segment the input audio stream into acoustically homogenous segments, where the measure of acoustic homogeneity is meant to yield 'speaker-pure' segments, i.e., each segment corresponds to one speaker.

A specific task where SCD is applied is speaker diarization [58], which is typically applied for meeting capture, where it is required to index the meeting audio data to yield a 'who spoke when' information, for later retrieval. Here, SCD is followed by speaker clustering which clusters the 'speaker-pure' segments, typically by agglomerative clustering methods, in the process determining the optimal number of (unknown speakers) and also yielding clusters of speaker data which can be used to train the speaker models. These models can further be associated with information on speaker-location in smart meeting rooms (using microphone arrays for location information or simply by using TDOA as a feature in combination with the speech features), thereby yielding a consistent speaker indexing of the audio stream in terms of the individual speaker-models thus derived [59].

In the context of speaker-spotting, it is clear that similar 'who spoke when' indexing of the surveillance audio is obtained, but in a stronger framework of using the watch-list of speaker models. A speaker-diarization process, being model-independent seems redundant in such a scenario. However, it is likely that important advantages can be derived by the use of SCD principles in a speaker-spotting system for the case when the watch-list sizes are large and speaker spotting system can result in high errors. We consider this now.

When the watch-list size (N) is large, speaker spotting is bound to make two kinds of errors: (1) high misses (Miss2) for top-1 or top-k (small k) for target speakers (due to increased confusion errors in the first stage CSI or score ranking) and (2) high false-alarms for non-target speakers. Note that the Miss errors are bound to be quite low for large N only for the top-N (or group detector scenario) which might not be appropriate in applications that also require the identity of the 'accepted' speaker accurately. Both of these errors result in what is called 'over-segmentation' errors, where a target segment or non-target segment is segmented into several sub-segments with arbitrary speaker labels. In this context, neither the knowledge that there are only two speakers in the conversation is effectively utilized, nor is the fact that a target or non-target segment (occurring in turns in a telephone conversation) is fairly homogenous (in terms of having been spoken by a single speaker) and can be processed in some way to yield the result that it is one 'speaker-pure' segment (thereby overriding the kind of over-segmentation errors made by speaker-spotting under large watch-list sizes). It is in this context, that a SCD algorithm can provide a parallel segmentation of the input audio stream, which can be used to reinforce and/or correct the speaker-spotting decisions, thereby taking advantage of the acoustic homogeneity considerations underlying the SCD principle. This can be utilized in several ways within the conventional speaker-spotting framework:

1. The SCD corrected segmentation yields more consistent speaker-wise segments which are now longer test-data than was originally available to the first pass speaker-spotting, and can therefore be subjected to a second-pass of speaker-spotting or simply re-classification with watch-list models, to yield corrected speaker labels, but now under large test data conditions and hence yielding more accurate decisions.
2. The longer (and more 'speaker-pure') test data segments of target speakers (in watch-list) can be reused to adapt the target speaker models, providing the advantage of cumulatively larger training data and session adaptation to handle session variability, voice-aging etc.
3. The longer (and more 'speaker-pure') non-target segments (non watch-list speaker data) can be used as enrollment data for creating the non-target speaker's model for inclusion in the watch-list under 'link analysis' based expansion of the watch-list (as noted in Sect. 15.1.4).

In this context, it is interesting to ask whether speaker-spotting itself can be re-formulated as a 'SCD and speaker-diarization' process (at least in a first-pass or parallel-pass), wherein a model-free segmentation, clustering and modeling (utilizing the fact that there are only two speakers in the conversation and as in a meeting capture and indexing setting) can itself yield a first-pass information for further segment-wise classification with watch-list speaker models, for increased performance, particularly under large watch-list scenarios. This essentially stems from the possibility that a model-free SCR and diarization could perform better (at least for the first-pass segmentation into speaker-pure segments, to be followed by a watch-list based classification on long test segments) than a solely model-based (watch-list models) multi-target speaker-spotting framework (on short isolated test clips), under the condition of large watch-list sizes, where the large number of models to be used in itself becomes a liability by being a causative factor for high miss (top-1 and top-k cases) and false-alarm errors.

15.6 Speaker Spotting Solutions and Product Spaces

As noted earlier in Sect. 15.1.3, surveillance and intelligence gathering operations in the context of voice communications provide conisderable scope for use of speech technologies in varied ways, such as speech activity detection, speech recognition, key-word spotting, speaker-spotting etc., to detect and select conversational data of critical interest.

In the more broader and critical context of homeland security, [60] refers to the National Strategy for Homeland Security's [61] concern on the need for effective performance of varied sectors such as border security, critical infrastructure etc. In this context, [60] identifies a host of critical homeland security applications including 'identification and tracking of individuals using biometrics (speech, face, iris etc.), ability to detect and track criminal or adversary communications, searching conversational speech' etc. Specifically, among various speech technologies that can have a bearing on such applications, [60] highlights 'audio hot spotting' [62] as an all-encompassing audio technology for surveillance, 'aiming to support natural querying of audio and video, meetings, news broadcasts, telephone conversations and tactical communications/surveillance', which, in the process, calls for an integration of a variety of technologies such as 'speaker identification, language identification, nonspeech audio detection, keyword spotting, transcription, prosodic feature and speech rate detection (for speaker emotional detection) and cross language search'.

Such vital homeland security applications have in turn facilitated the emergence of a product-space that specifically addresses the underlying critical requirements and in encouraging commercial-off-the-shelf (COTS) solutions. Alongside such surveillance and homeland security applications also lie more commercial speech-analytics requirements that are geared to provide business-intelligence or actionable-intelligence typically arising in the context of contact-center and call-center transactions and conversations cutting across a wide cross-section of services and customer bases.

In the larger context of speech technologies mentioned above, we list below a few notable solutions, in the context of automated surveillance and which includes speaker recognition in a form that might have bearing on 'speaker-spotting'. More mainstream telephony speaker and speech technologies such as telephony voice-authentication systems, telephony IVRS/dialog systems etc. are not mentioned here.

- **Mitre:** Audio hot spotting (process of finding "hot spots" within audio files, such as words of interest, speakers, or key sound effects) [62]
- **Virage:** Speaker change and speaker recognition, word spotting and phrase recognition [63]
- **Agnitio:** [64], ASIS: The Next Generation Biometrics Database [65], BS^3—Biometric Speaker Spotting System Family: Voice Surveillance for Military and Intelligence Organizations [66]

Despite the existence of the above mentioned solutions and product spaces, it should be noted that it is not clear whether there are, as yet, speaker-spotting surveillance deployments matching the Iris based deportation tracking system (Sect. 15.1.4) in terms of full-fledged automation, scale, routine operation, accuracy and robustness. Any short fall in such solutions towards a full-fledged automated and large scale deployment can at best be attributed to the caveats and cautions inherent in the field as of now, as pointed out in Sect. 15.4, which therefore consigns the field to research, particularly as one still in its infancy requiring several challenges to be solved, rather than as a proven technology, before it can be considered robust and ready to meet the real-world conditions.

15.7 Discussion and Conclusions

In this chapter, we have introduced and described the problem of speaker-spotting in the context of covert surveillance as well as negative recognition in the context of detecting a speaker in a watch-list from gaining access to some services. Specifically, speaker-spotting was treated in the frameworks of open-set speaker-identification and multi-target detection or recognition. Given that the field seems to be still in its early stages of research, the known theoretical formulations were provided, with accompanying performance analysis. These formulations available till date allows prediction of performances of the basic framework alongside measured performances from actual experiments on real data. There do seem instances where predicted performances and measured performances do not match exactly, and there seem additional factors that need to be considered before more precise modeling can emerge. Such theoretical formulations and models of the performance of the system allow understanding of the parameters underlying the framework and arrive at ways of determining optimal operating conditions and methods to enhance the performance.

Notable among such considerations are the interplay between θ, k (in top-k hypotheses) and watch-list size N and their effect on the overall Miss and False-alarm probability of the multi-target detector. The performance of a multi-target detector is noted to degrade seriously with increase in watch-list size—specifically, high false-alarm rates for increasing N, independent of k (and hence even in a top-N or group-detector mode of operation) and high miss rates for increasing N for small k. Thus, enhancing the performance of the framework for top-1 or top-k mode of operation, for large N, is crucial to the eventual success of such systems under scaled-up conditions. In very general terms, it seems important to ask how to make the performance of the underlying CSI and SV components in the OSI framework (Sect. 15.2.1) and the multi-detector framework (Sect. 15.2.2) more robust to large watch-list sizes.

In addition to such performance considerations based on the inherent parameters of the system, it was noted that the multitude of different sources of variability that such a system has to face makes the system vulnerable to realistic conditions. These sources of variability manifest as variability in the score distributions which in turn degrade the performance for unseen test data. The question of whether conventional score-normalization methods [21] will suffice to handle this or whether it is required to explore new score normalization techniques (such as the Top-norm in Sect. 15.2.5) remains an important issue.

Fundamental to the entire concept of speaker-spotting is the underlying notion of 'uniqueness' of a bio-metric signal (such as voice) in identifying a person. While much effort has been placed towards establishing uniqueness of Iris as a biometric or finger-print as a biometric across large populations, this is still something yet to be established for voice. This leads to the question of whether it is even correct to expect an accurate identification of a speaker from among large watch-lists. In this regard, it seems easy to resort to the basic question of what performance can be expected from human listeners. While this in itself is also yet to be established, the next question that naturally arises is whether a machine performance can be poorer or as good as or even surpass human performance. In keeping with the spirit of automation, it is appealing to hope that machines can outperform humans in very well defined contexts, in this case, that of identifying a person from a large watch-list, excelling in some dimensions such as the size of the list, size of the test population, speed and accuracy. However, in order to reach there, it is not clear whether the underlying tenets of the framework employed (namely, the features used, models used, score calculations, decision mechanisms taken individually and together) have the requisite discriminatory power to ensure unique identification of a speaker in a large watch-list (or rejection of a speaker from an arbitrarily larger and unseen population as not in the watch-list). Note that this has been done to some extent for the Iris based system (Sect. 15.1.4). Therefore, it appears that it is only a question of time and effort before similar biometric properties can be established for voice as a biometric. In this context, it is to be noted that the absence of a 'voice-print' (equivalent of the finger-print or an IRIScode) that uniquely represents and identifies a person is perhaps the underlying difficulty in voice biometrics. Despite popular, but non-existent (and hence misconceived), notions of a 'voice-print', practical systems work only with features (that carry information about the linguistic content in addition to speaker-specific information) and models (which try to capture the speaker characteristics in terms of the feature distribution of speech spoken by the person), and are not yet geared towards extraction or modeling of very fine and accurate physiological or behavioral aspects of the voice of a person (devoid of any associated linguistic characteristics), to the extent required for unique mapping between such model parameters and the person—such as will be required for large watch-list operations—in an equivalence to the Iris or fingerprint biometrics.

In conclusion, the following can be stated about speaker-spotting:

1. Speaker-spotting is viable, in that there exist clear need, formulation, analysis and system-level details towards realizing a practical solution.
2. The field is still in its infancy and draws from speaker-recognition techniques available till date, but has several idiosyncratic problems and challenges to be addressed before truly large scale deployable solutions are realized.
3. The Iris watch-list based detection will serve as a reference to compare such an emerging solution based on voice as a biometric.
4. Speaker-spotting, due to its unique positioning in the intersection of voice-authentication (typically telephony based access control) and speaker forensics, can draw from their solutions as much as it is challenged by the difficulties of both these domains. This makes it an attractive research field, even while holding the promise to solve several fundamental problems in voice-biometrics, speaker-forensics, and practical speaker-recognition technologies with far-reaching practical impact.

Acknowledgements The author wishes to thank the editors of this volume for their continued support and understanding throughout the long period over which this chapter was prepared. Special thanks are due to the author's colleagues S. Thiyagarajan and R. Karthik for their help with specific experiments and results included in this chapter.

References

1. "2001: A Space Odyssey" (Movie), based on Arthur C. Clarke (1968), "2001: A Space Odyssey" (novel), New American Library (also Signet, 2005)
2. "Clear and Present Danger" (Movie), based on Tom Clancy (1993), "Clear and Present Danger" (novel), HarperCollingPublishers
3. Duncan Campbell, IPTV Ltd (1999) Interception capabilities 2000. e-prints, STOA report, 1999 (e-prints, Federation of American Scientists (FAS) Intelligence Research Program. http://www.fas.org/irp/eprint/ic2000/ic2000.htm)
4. Prabhakar S, Bjorn V (2008) Biometrics in the commercial sector. In: Jain AK, Fynn P, Ross AA (eds) Handbook of biometrics. Springer, New York, pp 479–507 (Chap 23)
5. Daugman J, Malhas I (2004) Iris recognition border-crossing system in the UAE. Int Airport Rev 2:49–53
6. IrisGuard Incorporated. www.irisguard.com
7. Lazarick R, Cambier JL (2008) Biometrics in the government sector. In: Jain AK, Fynn P, Ross AA (eds) Handbook of biometrics. Springer, New York, pp 461–478 (Chap 22)
8. Link analysis workbench (2004) SRI International, Air Force Research Laboratory, final technical report, AFRL-IF-RS-TR-2004-247 (Sept 2004)
9. Kwon P (1998) Speaker spotting: automatic annotation of audio data with speaker identity. ME thesis, Electrical Engineering and Computer Science, MIT, Boston
10. Atal BS (1976) Automatic recognition of speakers from their voices. Proc IEEE 64:460–475
11. Rosenberg AE (1976) Automatic speaker verification: a review. Proc IEEE 64:475–487
12. Doddington GR (1985) Speaker recognition—identifying people by their voices. Proc IEEE 73:1651–1664
13. O'Shaugnessy D (1986) Speaker recognition. IEEE ASSP Mag 3:4–17
14. Naik JM Speaker verification: a tutorial. IEEE Commun Mag 28:42–48 (Jan 1990)
15. Rosenberg AK, Soong FK (1991) Recent research in automatic speaker recognition. In: Furui S, Sondhi MM (eds) Advances in speech signal processing. Marcel and Dekker, New York, pp 701–738 (Chap 22)
16. Furui S (1994) An overview of speaker recognition technology. In ESCA workshop on automatic speaker recognition, identification and verification, pp 1–9
17. Furui S (1996) An overview of speaker recognition technology. In: Lee CH, Soong FK, Paliwal KK (eds) Automatic speech and speaker recognition—advanced topics. Kluwer, Boston, pp 31–54 (Chap 2)
18. Gish H, Schmidt M (1994) Text-independent speaker identification. IEEE Signal Process Mag 11:18–32 (Oct 1994)
19. Campbell JP (1997) Speaker recognition: a tutorial. Proc IEEE 85(9):1437–1462 (Sept 1997)
20. Quatieri TF (2002) Discrete-time speech signal processing—principles and practice. Pearson Education, Delhi, pp 709–766 (Chap 14, Speaker recognition)
21. Bimbot F et al (2004) A tutorial on text-independent speaker verification. EURASIP J Appl Signal Process 4:430–451
22. Rosenberg AE, Bimbot F, Parthasarathy S (2008) Overview of speaker recognition. In: Benesty J, Sondhi MM, Huang Y (eds) Handbook of speech processing. Springer, Berlin, pp 725–741 (Chap 36)
23. Hebert M (2008) Text-dependent speaker recognition. In: Benesty J, Sondhi MM, Huang Y (eds) Handbook of speech processing. Springer, Berlin, pp 743–762 (Chap 37)

24. Reynolds DA, Campbell WM (2008) Text-independent speaker recognition. In: Benesty J, Sondhi MM, Huang Y (eds) Handbook of speech processing. Springer, Berlin, pp 763–781 (Chap 38)
25. Gong Y (2002) Noise-robust open-set speaker-recognition using noise dependent Gaussian mixture classifier. Proc. ICASSP, I:133–I:136, Orlando, FL
26. Deng J, Hiu Q (2003) Open-set text-independent speaker recognition based on set-score pattern classification. Proc. ICASSP, II:73–II.76, Hong Kong
27. Sivakumaran P, Fortuna J, Ariyaeeinia AM (2003) Score normalization applied to open-set, text-independent speaker identification. Proc. Eurospeech/Interspeech, 2669–2672, Geneva
28. Fortuna J, Sivakumaran P, Ariyaeeinia A, Malegaonkar A (2004) Relative effectiveness of score normalization methods in open-set speaker identification. Proc. Odyssey 2004 The Speaker and Language Recognition Workshop, Toledo, pp 369–376
29. Fortuna J, Sivakumaran P, Ariyaeeinia A, Malegaonkar A (2005) Open-set speaker identification using adapted Gaussian mixture models. Proc. Interspeech 2005, 1997–2000, Lisbon, Portugal
30. Angkititrakul P, Hansen JHL (2006) Discriminative in-set/out-of-set speaker recognition. IEEE Trans Audio, Speech and Lang Process 15(2):498–508
31. Ramasubramanian V et al (2006) Text-dependent speaker-recognition systems based on one-pass dynamics programming algorithm. Proc. Odyssey 2006 The Speaker and Language Recognition Workshop, San Juan
32. Philips PJ, Grother P, Michaels RJ, Blackburn DM, Tabassi E, Bone JM (2003) FRVT 2002: evaluation report. http://frvt.org/FRVT2002/documents.htm (Mar 2003)
33. Daugman J (2000) Biometric decision landscapes. Technical report No. TR482, University of Cambridge Computer Laboratory. http://www.cl.cam.ac.uk/users/jgd1000
34. Singer E, Reynolds D (2004) Analysis of multi-target detection for speaker and language recognition. Proc. Odyssey 2004 The Speaker and Language Recognition Workshop, Toledo
35. Zigel Y, Wasserblat M (2006) How to deal with multiple-targets in speaker identification systems? Proc. Odyssey 2006 The Speaker and Language Recognition Workshop, San Juan
36. Barger PJ, Sridharan S (2006) On the performance and use of speaker recognition systems for surveillance. Proc. IEEE International conference on Video and Signal based Surveillance (AVSS' 06)
37. Schneier B (2006) Data mining for terrorists. Crypto-Gram (15 Mar 2006)
38. UK Communications-Electronics Security Group (CESG) http://www.cesg.gov.uk/policy_technologies/biometrics/media/biometrictestreportpt1.pdf
39. Jain AK, Fynn P, Ross AA (2008) Handbook of biometrics. Springer, New York
40. Gonzalez-Rodriguez J, Toledano DT, Ortega-Garcia J (2008) Voice Biometrics. In: Jain AK, Fynn P, Ross AA (eds) Handbook of biometrics. Springer, New York, pp 151–170 (Chap 8)
41. Martin A, Przybocki M (1999) The NIST 1999 speaker recognition evaluation—an overview. National Institute of Standards and Technology (NIST)
42. Kenny P, Demouchel P (2005) Eigenvoices modeling with sparse training data. IEEE Trans Speech and Audio Proc 13(3):345–354
43. Campbell WM, Sturim D, Reynolds DA (2006) Support vector machines using GMM supervectors for speaker verification. IEEE Signal Process Lett 13:308–311
44. Ramasubramanian V, Thiyagarajan S (2008) Handling inter-session intra-speaker variability: unsupervised on-line session-adaptation for text-independent speaker-recognition. Technical report, Siemens Corporate Research & Technologies—India, Bangalore
45. Martin A, Przybocki M The NIST speaker recognition evaluation series. National Institute of Standards and Technology's web site (Online). http://www.nist.gov/speech/test/sre
46. Przybocki MA, Martin AF, Le AN (2006) NIST speaker recognition evaluation chronicles, Part 2. Proc. IEEE Odyssey, ISCA speaker recognition workshop, pp 1–6
47. Przybocki MA, Martin AF, Le AN (2007) NIST speaker recognition evaluations utilizing mixer corpora—2004, 2005, 2006. IEEE Trans Audio, Speech and Lang Proc 15(7):1951–1959

48. Campbell JP, Shun W, Campbell WM, Schwartz R, Bonastre J-F, Matrouf D (2009) Forensic speaker recognition: A need for caution. IEEE Signal Process Mag 26(2):95–103
49. Bonastre J-F, Bimbot F, Boe L-J, Campbell JP, Reynolds DA, Magrin-Chagnolleau I (2003) Person authentication by voice: A need for caution. In Proc. Eurospeech, ISCA, Geneva, Switzerland, pp 33–36
50. Gish H et al (1991) Segregation of speakers for speech recognition and speaker identification. Proc ICASSP 2:873–876
51. Delacourt P, Wellekens CJ (2000) DISTBIC: a speaker-based segmentation for audio data indexing. Speech Commun 32(1–2):111–126
52. Johnson S (1997) Speaker tracking. MPhil thesis, CUED, University of Cambridge, Cambridge, UK
53. Chen S, Gopalakrishnan P (1998) Speaker, environment, and channel change detection and clustering via the Bayesian information criterion. In: Proc. DARPA speech recognition workshop, pp 127–132
54. Tritschler A, Gopinath R (1999) Improved speaker segmentation and segments clustering using the Bayesian information criterion. In Proc. Eurospeech, vol 2, pp 679–682
55. Malegaonkar A, Ariyaeeinia V, Sivakumaran P, Fortuna J (2006) Unsupervised speaker change detection using probabilistic pattern matching. IEEE Signal Process Lett 13(8):509–512
56. Vuorinen O, Peltola J, Mäkelä S-M (2007) Unsupervised speaker change detection for mobile device recorded speech. Proc. ICASSP, II-757:760
57. Malegaonkar AS, Ariyaeeinia AM, Sivakumaran P (2007) Efficient speaker change detection using adapted Gaussian mixture models. IEEE Trans Audio, Speech and Lang Proc 15(6):1859–1869
58. Vijayasenan D (2010) An information theoretic approach to speaker diarization of meeting recordings, PhD thesis, Ecole Polytechnique Federale de Lausanne (EPFL)
59. Ajmera J (2004) Robust audio segmentation, PhD thesis, Ecole Polytechnique Federale de Lausanne (EPFL)
60. Maybury M (2009) Speech and video processing for homeland security. In: Voeller JG (ed) Wiley handbook of science and technology for homeland security. Wiley, Hoboken, pp 1–17
61. Office of Homeland Security (2002) National Strategy for Homeland Security. http://www.whitehouse.gov/homeland/book
62. Mitre. http://www.mitre.org/news/digest/advanced_research/02_10/audio.html
63. Autonomy Virage. http://www.virage.com/
64. Agnitio. http://www.agnitio.es/index.php
65. Agnitio, ASIS. http://www.agnitio.es/producto.php?id_producto=3
66. Agnitio, BS^3. http://www.agnitio.es/producto.php?id_producto=4
67. Enemy of the state (1998) (Movie)

Chapter 16
Helping the Forensic Research Institute of the French *Gendarmerie* to Identify a Suspect in the Presence of Voice Disguise or Voice Forgery

Patrick Perrot and Gérard Chollet

Abstract In the field of forensic speaker recognition, the question of voice disguise presents a specific interest. Most criminals try to disguise their voice before making a malefic call or a terrorist threat. Their aim is to change the register of their voice quality in order to falsify their identity (voice disguise) or to mimic the voice of another person (voice forgery). This chapter proposes to analyse two different kinds of disguise: The first is the transformation of the voice by non-electronic and deliberate means; the second is the conversion of the voice by electronic and deliberate means. By considering both kinds of disguise (electronic and non-electronic) our analyses of voice transformation are based on an acoustic approach, which we use to measure specific changes in speech, and on an automatic approach to detect voice disguise. Four kinds of disguises which are considered the most common are studied: high pitched voice, low pitched voice, a hand over the mouth and pinched nostrils. A constraint of audibility and intelligibility has been imposed on the speakers who have recorded the database. The acoustic analysis of specific features reveals some differences according to the form of disguise, while in the automatic experiment we found the best way to detect a voice disguise is to use Support Vector Machines (SVM) technique. The level of performance is an AUC (area under curve) at 0.79. Voice conversion techniques are also proposed and applied in two forensic scenarios: first, the imitation of a politician from an Internet recording; and second, the application of voice disguise reversibility. Different kinds of tests are proposed to evaluate the relevance of the results, which are based on objective and subjective measurements. The best conversion is obtained from a GMM-ALISP voice conversion.

P. Perrot · G. Chollet (✉)
Gendarmerie Operational Unit, Gendarmerie Nationale, 3 rue de Dampierre,
17400 Saint Jean d'Angely, France
e-mail: patrick.perrot@gendarmerie.interieur.gouv.fr

G. Chollet
CNRS-LTCI, Telecom ParisTech, 46 rue Barrault, Paris 75013, France
e-mail:gerard.chollet@telecom-paristech.fr

A. Neustein, H. A. Patil (eds.), *Forensic Speaker Recognition*,
DOI 10.1007/978-1-4614-0263-3_16,

16.1 Introduction

Identification and detection of voice disguise must often serve as a prerequisite task to performing speaker recognition so that one may determine if the suspect's voice truly matches the speech sample collected during the commission of a crime or terrorist act. In fact, in criminal investigations where prosecutors focus on deliberate actions (wilful violations of criminal statutes) the disguise of one's voice is especially crucial to explore, as it underscores the deliberate actions of the offender to conceal his/her identity. Prosecutors face both the possibility of finding a false negative in their investigative work (where they fail to recognize a suspect's voice because it is disguised)—or a false positive where they confuse the suspect's voice with that of someone else because the disguise was truly effective. For this reason, it is imperative that robust testing methods be designed and implemented to better detect disguised voice.

Our research focuses on specific voice characteristics to evaluate the recognition of a suspect's voice in the presence of disguise. In so doing, we use an automatic classification method to detect the possibility of disguise. We propose two experiments: (1) the automatic imitation of a politician voice and (2) the analysis of reversibility of a disguise.

16.2 Background

In a criminal context, offenders, whatever the seriousness of their crime, tend more and more to disguise their voices. Their goal is simply to hide their actual identity, and sometimes impersonate someone else. A disguised voice is generally used in telephone threats, malicious calls, extortion and blackmail or terrorist demands. True, there are those cases when there are involuntary voice changes, as when there are alterations in voice characteristics due to poor transmission of telephonic communication, a strong overriding emotion that overrides the clarity of the communication, or even pathologies (both acute and chronic) that morph speech production. Nevertheless, we limit this discussion to disguise which consists of a person who *deliberately* conceals his identity. In such cases, the suspect sees as their objective in disguising their voice a means to mislead the human ear and the automatic speaker recognition system.

In recordings of notable international terrorists or calls of less well known individuals but still of a malefic nature, determining the identity of an individual from his voice can be crucial. Yet, we must ask ourselves whether someday it might be possible to use voice with the same confidence that we use fingerprints or DNA in criminal prosecutions. While one can say the jury is still out on this issue, advances in signal processing combined with a better understanding of the speech mechanisms have no doubt significantly increased the performance of automatic speaker recognition systems. Still, there are scientists opposed to the use of speech in applications other than non-criminal, commercial settings because of reliability factors that present themselves with using voice to make a case beyond reasonable doubt which is the accepted evidentiary standard in criminal law. And while forensic scientists look toward speaker recognition as a way of characterizing an individual from his voice, the voice in contrast to a fingerprint can be easily (and quite consciously) modified for deception purposes. The objective of the person who disguises his voice is to deliver an understandable message without being recognized. Changing his voice to conceal his identity is often used in criminal cases and the techniques that are used are often varied.

For several years now the scientific community, working with law enforcement, has shown that they are not only concerned with assessing the impact of deliberate disguise on speaker recognition performance in general, but with helping to identify the individual who hides his voice. These studies have led to much progress in the field of research into forensic speaker recognition, but unfortunately in the area of actual forensic applications to real cases there are few concrete results. This should not be cause for despair. Instead, it gives us an idea where we stand today with this particular science and how we can come closer to achieving our goal of making forensic speaker recognition a robust method akin to the other biometrics regularly used by law enforcement and counter-terrorism units. One way of embarking on this challenge is to assiduously examine the work of a forensic expert in the field of speaker recognition and in particular his grappling with voice disguise.

We propose the following example excerpt from a real situation: a victim of a blackmail attempt made over the phone brought a complaint to the police services. The victim had been placed under wiretaps because of the seriousness of the facts at hand. The thief, a relative of the victim, made serious threats of death, but no doubt did not wish to be recognized under any conditions. He therefore tried to disguise his voice in the different calls to the victim, giving the forensic speech expert a set of recorded calls with which to work.

The aim of investigators is to compare these speech samples of the person making the threats with known samples of voice recording of the suspect's voice. But before considering any speaker identification matter, by comparing the voice on the telephone recordings with the voice of an individual regarded as suspect, one must consider the presence or absence of disguise. The risk of making a false positive identification of the voice sample as that of the wrong criminal suspect exists when the disguised voice is not accurately identified as a disguised voice, but is considered by mistake as a normal voice within a large pool of voice recordings. Disguised voices take various forms: from simple means, as pinched nose or by putting a handkerchief in front of mouth, to the most advanced, as in the use of automatic imitation of a target voice. Many studies break down disguise into different categories: the use of a whispered voice [35]; a falsetto voice [25]; or a foreign accent [56].

The studied features of voice are generally suprasegmental parameters: evolution of the fundamental frequency [25]; variation of speech rate [35]; distribution of energy per frequency band; and changing formants. Impersonation studies relate to both techniques of imitation: phonetics and conversion techniques based on spectral transformation. These imitations have a real impact on speaker recognition [27].

Because the disguise of voice is a real challenge in the field of automatic speaker recognition in forensic sciences, we wish to address the issue of disguised voice as broadly as possible. We include under the name of deliberate disguise two different techniques:

- alteration of one's own voice in order to conceal one's identity,
- automatic imitation in order to impersonate another voice.

The work developed in this chapter covers the following areas: descriptive analysis of data and automatic voice conversion. We focus first on a closed set of disguises among the most commonly used in criminal cases by the simplicity of their implementation, and second on techniques based on automatic voice conversion that involve more complex methods, aiming to traverse the spectrum of vocal techniques of forgery.

16.2.1 Domain of Voice Disguise

Nowadays forensic techniques appear in more and more TV shows as *Crime Scene Investigation (CSI), Naval Criminal Investigative Service (NCIS),* and so on. The interest in these TV shows provide a new way to consider forensic sciences for attorneys, judges, scientists, policemen, journalists, and criminals. This is the "*CSI effect*". This effect is purported to skew public perceptions of real-world forensic science, as well as the behavior of criminal justice system actors; this is of particular concern in the courtroom setting, where many prosecutors feel pressured to deliver more and more forensic evidence [31, 46]. It is not surprising to notice that the most popular courtroom dramas focus on the use of new science and technology in solving crimes. *CSI* has been called the most popular television show in the world. Not only is *CSI* so popular that it has spawned other versions that dominate the traditional television ratings, it has also prompted similar forensic dramas, such as *Cold Case*, *Bones*, and *Numb3rs*. One of the consequences of this *CSI effect* is criminals' use of different disguises to falsify their own identity, especially with the alteration of their voices. This section aims to present not only the different domains where a disguise is used, but also the techniques developed in the literature to identify cases of disguised voices. There are currently several different applications where voice is deliberately disguised. The framework of entertainment, where we hear voice imitators, is the most common, but the disguise may also be a marketing issue and sometimes criminal.

16.2.1.1 Entertainment

The use of imitation is actually very common in our daily lives. Many professional imitators parody politicians or business stars. The purpose of an imitation in an entertainment context is to impersonate another person through caricatural features.

Indeed, the imitator seeks first to imitate gestural, facial expressions, and nonverbal characteristics. For voice, imitation also focuses on caricatural features, as vocal mannerisms or accents which characterize a particular known person. In case of imitation, it is essential that the person is well known to listeners.

The object of imitation is not always to impersonate the identity of another person; it can also be to impersonate an accent of a region or even of another country. As we perceive it, imitation in entertainment is generally synonymous of caricature. The imitator exaggerates the features of an individual and while not really trying to imitate him. However, the imitators sometimes use their talent without exaggerated caricature. We can cite the case of Gerald Dahan, a French imitator, who phoned over a radio antenna to the national coach of French soccer team Raymond Domenech and to the team captain Zinedine Zidane, in imitating the French President Jacques Chirac.

Everyone is not a professional in imitation. Therefore, the automatic voice imitation is an issue of interest to exploit. Many companies offer possibilities to customize their welcome by using avatars or virtual agents. It is even possible to create a virtual agent very close to reality. Born in video games, the virtual agent is gradually invading multimedia and becomes progressively our partner in a job search or in a request to a courier. The purpose of the avatar is to humanize the virtual reality. This enthusiasm for virtual worlds is illustrated by the rise of social worlds on the Internet:

- Second Life allows one to create a second life in a virtual world [38],
- MyLife3D is a virtual meeting space and real-time communication and 3D in a playful environment,
- IMVU (www.imvu.com) is a 3D instant messaging service

The similarity between an avatar and a person is visual but it can also be complemented by a customized voice. Therefore, this effort requires a similar imitation of the voice. In such applications techniques, transformed voices are not based on a professional impersonator but on the use of automatic voice conversion.

16.2.1.2 Games

Commercial applications of voice disguise, and especially in the field of games, are also booming. A quick Internet search reveals the attractiveness of voice disguise and we can find different proposals to promote tools dedicated to voice transformation: "*We already knew the devices that connect phones to change the voice. With this phone, no need to buy a separate box, you can become the greatest impostor by tricking your friends easily?*"

Or, "*Voice Changer Software now has Voice Comparator to compare your changed voice with another's voice. It tells you how successful you are done and what need further adjustment to imitate or simulate that target voice.*"

Or "*So you speak normally over phone, your correspondent will hear you as if you had inhaled a balloon filled with helium or as if you had a Darth Vader voice.*"

Or, "*The voice processor of the mark 'X' is a funny accessory that will amaze children" and without any doubt police forces and forensic experts, etc.*

These popular slogans, given above, reveal many different aims of the proposed systems. Unfortunately, misuse and abuse of these objects are always possible and perhaps even totally predictable.

16.2.1.3 Crimes

Besides the commercial and recreational activities, voice disguise is also used in a criminal context. It is an especially relevant means to conceal one's identity. Generally an individual seeks to conceal his identity when he imagines himself to be tapped, or is likely to be recognized by a relative. The most common criminal situations where a voice disguise is used are:

- Terrorists threats
- Malicious calls
- Ransom demands
- Blackmail extortions

It is appropriate to ask whether specific means of disguise are more popular than others under varying conditions. Unfortunately it is difficult to establish meaningful statistics because it is not easy to ensure that a voice is disguised. Studies illustrate that point by showing differences in their results. Masthoff [32], and Gfroerer (BKA: BunderKriminalamt) evaluated between the years 1989 and 1994, the number of documented cases where a disguise was assumed. It appears that an offender disguises his voice in 52% of cases. This percentage reaches a level of 62% in case of malicious calls. In general, the caller knows the victim. These figures are revised to a reduction by Künzel [24] which estimates the number of disguise cases between 15 and 25% from his expertise in speaker recognition. He explains this noticeable difference with the results proposed by Masthoff and Gfroerer by the difference between the type of expertises assigned to the University of Trier and these to Bunder-Kriminalamt (BKA). In Brazil, according to Molina and Figueiró Souzza Britto [8], disguise is widely used in kidnapping. The most common technique is based on the use of an object placed in the mouth.

However, the authors of those works, cited above, do not propose quantitative information. In the JP French Association in Great Britain, the number of cases is estimated to a level of 2.5%. Within L'_Institut de recherche criminelle de la gendarmerie nationale (IRCGN), otherwise known as the Forensic Research Institute of the French Gendarmerie, the number of identified voice disguise cases is between 5 and 10% of speaker recognition expertise. These figures are derived from real cases of disguise and do not take into account cases of disguise suspicion. We therefore notice a real heterogeneity in the percentage of case where the voice disguise is highlighted. Indeed, this ratio varies between 2.5 and 52%, revealing the difficulty for experts to detect a disguise.

Hirson and Duckworth reported a specific case of a voice disguise at a ransom demand made to a bank manager, following the kidnapping of his wife. The thief used a "creaky voice" combined with a slow speech rate.

In [20], authors present an experimental comparison of the performance of phoneticians and non-phoneticians working on speaker recognition involving such disguises. The results of the perceptual experience reveal the importance of expert witnesses phonetically trained in the analysis of recording controversial. Hirson and Lindsey have worked in 1999 on the case of a bomb threat, where disguise was the use of a foreign accent. In [28], authors analyse read sentences from the /r/ using their original accent and using an imitation accent. The analysis revealed that the /r/ is considered as non-standard if it has a higher F3 than 2000 Hz for men and 2350 Hz for women. Koester and Wagner [59] in 1999 highlighted the use of a falsetto voice in 40 malicious calls by the same author.

Another technique to disguise the voice also used in criminal applications is imitation by a professional impersonator [45], but we can also include in this approach techniques for automatic voice conversion. Imitating an individual in a criminal context is not for everybody. It is indeed necessary to use an imitator or possess oneself an ability to imitate. We can very easily consider the ability of a terrorist organization to imitate voice of its most notable terrorist in charge of the organization. This hypothesis was made by Samy Bengio in an analysis of the Bin Laden voice[1]. Moreover, an individual wishing to impersonate someone else knows him and therefore prefers to avoid the risk of being found out.

16.3 State of the Art of Voice Disguise

Disguised voices have been studied in the literature and Rodman [43, 44], proposes a very well defined classification of them. There are several techniques that can significantly alter voice parameters. Rodman distinguishes them as two dual criteria: electronic/non-electronic—deliberate/non deliberate. From these two criteria, four different combinations are possible:

- Electronic–Deliberate
- Electronic–Non Deliberate
- Non-electronic–Deliberate
- Non-electronic–Non Deliberate

For example, an Electronic–Deliberate disguise is found in deliberate manipulation made by journalists to hide the identity of an interviewee. An Electronic–Non deliberate disguise can result from the transmission channel on a fixed line or GSM. This transformation of the signal is the only one that could be considered as a non disguised voice. A Deliberate–Non electronic disguise is what we find most commonly

[1] http://tempsreel.nouvelobs.com/actualite/monde/20021129.OBS3376/la-voix-de-ben-ladenimitee-par-un-imposteur.html.

in the criminal field. It is a voice that is falsely high, low or falsely pronounced with a handkerchief in front of the mouth. A Non electronic–Non Deliberate disguise may be caused by a disturbed state of health, such as the common cold.

As previously, we distinguish two different techniques of disguise: transformation and conversion (or imitation).

16.3.1 Voice Transformation Analysis

With regard to voice transformation analysis, there are few studies to begin with, most of which are often dedicated to a specific category of voice disguise. Techniques are also generally very focused on phonetic peculiarities. In truth, changing one's voice does not require special ability. Just go one octave higher or lower voice or take a foreign accent and that will be enough to disturb the recognition level.

The impact of voice disguise on speaker recognition is present in the literature [7, 22, 24, 26, 41, 42, 46, 53]. The study proposed by [47] focuses on the use of different dialects as disguise technique and the analysis of the power of dialect in the identification of voice from hearing tests. The results show that if a criminal uses a native dialect at the time of the commission of an offence and uses a new dialect during recording of comparison, the level of performance degradation on recognition is particularly important. The witness would not only be able to identify the perpetrator, but may even mistake him for another suspect.

Clark et al. [7] present a study of electronic disguise impact on speaker identification. Pitch was chosen as the parameter for investigation because it can be manipulated in almost all commercially-available voice changing devices. SoundForge software was used to change voices. To control for listeners' degree of familiarity with the voices they trained listeners to identify a group of four male speakers. Training data consisted of samples of approximately 90 s, extracted from readings of the Cinderella story to control the spoken material. 36 listeners were recruited to participate in the experiments. They were then asked to listen to 8 experimental stimuli, which consisted of 10 s samples extracted from other sections of the story. They were given a closed test in which the names of the four potential speakers were listed. The results of the disguised conditions showed, as predicted, that identification rates fall when listeners hear disguised voices. It appears that when pitch was raised recognition performances were better than when pitch was lowered. The speaker identification results are legitimately the worst obtained from the more extreme disguises, ranging from ± 8 semitons.

According to [53], each person has a unique voice characteristics because of the unique physical characteristics such as the vocal tract, vocal cords, the form of oral and nasal cavities, and the particular shape of the articulators (lips, tongue, …). Taseer considers that the glottal pulse and the value of the associated energy facilitate the identification of the speaker in the presence of disguised voices.

After considering specifically the impact of disguise on the fundamental frequency [25], Kunzel joins Gonzalez-Rodriguez and Ortega-Garcia [26] to study the effects of common types of voice disguise, including increased voice pitch (even falsetto speech), lowered voice pitch and pinching the nose while speaking, on forensic speaker recognition (FSR) techniques. Natural and disguised speech data from 100 German speakers recorded 5 times over a period of 7–9 months were used in a series of speaker recognition experiments. This work is limited to estimate the performance degradation when the suspect is known to be the author of the disguised test speech (no impostor trials are reported). Results indicate that the three types of voice disguise selected affect only marginally the performance of the system if reference populations contain speech data which are based on the same type of disguise. If, however, the reference population is based on normal speech only, effects are generally more severe and also different for the three kinds of disguise.

Meuwly [33] also discussed the impact of disguise on the automatic recognition of the speaker. He demonstrates that the presence of a disguise in the query recording significantly impacts the level of system performance, and confirms the lack of robustness of the automatic recognition of the speaker in front of disguises. In fact, they generate too much intra speaker variability.

The study proposed by [42] measures the impact of disguise on speaker identification from perceptual tests. The auditors have to spot whether two heard phrases were uttered by the same speaker or by different speakers, and give an order of magnitude of the confidence of their response. The speakers pronounce a sentence in a normal voice and a sentence in one of five selected disguises. The listeners were divided into two groups: one group of naïve listeners and a group of three listeners familiar with the area of speech processing. Recognition often falls sharply when speakers attempt to disguise their voices e.g., 59–81% accuracy rate depending on the disguise vs. 92% accuracy rate for normal voices.

Whispered voice is also a technique of disguise which has been a subject of study. In fact, whisper inhibits voiced part and thus causes the loss of information on the perception of the fundamental frequency. Orchard et al. [35] highlights the difficulty in recognizing a voice whispered, particularly if the query samples are compared to samples of non whispered voice. Fan et al. [14] propose an acoustic analysis for speaker identification in front of whispered voice.

Thus the different techniques of voice disguise directly impact the performance of speaker recognition. However, it is difficult to have a reference level of degradation because of the multiplicity of tests.

16.3.2 Voice Conversion Technique

Human voices can be disguised by means of human impersonation, but also by means of voice conversion. In both cases, disguise appears as a relevant and real question for forensic considerations, since the aim is to hide or falsify one's own identity. This section describes these two ways to disguise one's own voice.

16.3.2.1 Voice Impersonation

Voice imitation is one of the potential threats to security systems that use automatic speaker recognition [4] and is likewise a serious challenge for forensic experts [61]. The question of identification of professional impersonators has also been raised by Patil [36, 37]. Since prosodic features have been considered for state-of-the-art recognition systems in recent years, the question arises as to how vulnerable these features are to voice mimicking.

Eriksson and Wretling [13] studies professional mimics and shows that professionals match speaking rate and mean fundamental frequency. It was found that a professional impersonator is able to mimic global speech rate very closely, but timing at the segmental level showed little or no change in the direction of the targets. Mean fundamental frequency and variation matched the targets very closely. Target formant frequencies were attained with varying success. For two of the three target voices the vowel space of the imitation was intermediate between that of the artist's own voice and the target.

In [16], two experiments are conducted for 12 individual features in order to determine how a prosodic speaker identification system would perform against professionally imitated voices. The results show that the identification error rate increases for all the features except F0 range when the impersonators' modified voices are used instead of the impersonators natural voices. Moreover, it seems easier to copy prosody on the basis of a whole sentence than for a specific word.

Zetterholm [62] works on which aspects of the human voice need to be altered to successfully mislead the listener through voice imitation. This suggests that voice and speech imitation can be exploited as a methodological tool to find out which features a voice impersonator picks out in the target voice and which features in the human voice are not changed, thereby making it possible to identify the impersonator instead of the target voice. Her work examines whether three impersonators, two professionals and one amateur, selected the same features and speaker characteristics when imitating the same target speakers and whether they achieved similar degrees of success. The acoustic-auditory results give an insight into how difficult it is to focus on only one or two features when trying to identify one speaker from his voice.

16.3.2.2 Voice Conversion

An automatic voice conversion is the process of transforming the characteristics of speech uttered by a source speaker, such that a listener would believe the speech was produced by the target speaker: the person who is the subject/target of the impersonation.

Different kinds of information are included in the speech signal: environmental noise, speech message, and speaker identity. The question of voice conversion is first, to establish the most characteristic features of a source individual in order to

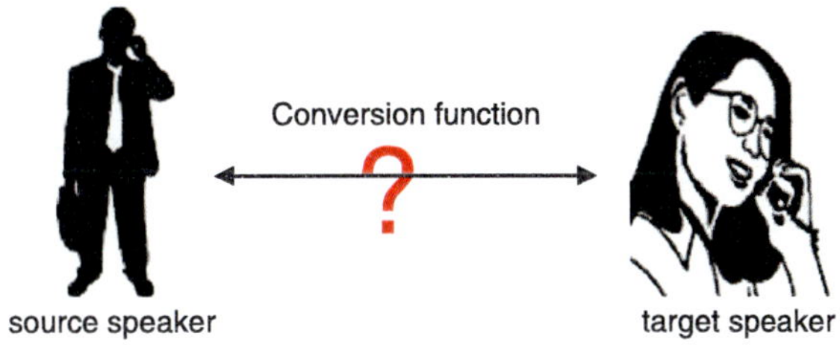

Fig. 16.1 Voice conversion principle

transform the voice to their target counterpart. The analysis part of a voice conversion algorithm focuses on the extraction of speaker identity. Second, to calculate the transformation function required in transforming the voice to its target counterpart. Both operations must be performed independent of the environment and of the voice message. Finally, a synthesis step will be achieved to replace the source speaker characteristics with the target speaker characteristics.

Consider a sequence of spectral vectors articulated by the source speaker:

$$X_s = [x_1, x_2, \ldots x_n]$$

And a sequence articulated by the target speaker composed of the same words:

$$Y_t = [y_1, y_2, \ldots y_n]$$

Voice conversion is based on the calculation of a conversion function F that minimizes the mean square error:

$$\varepsilon_{mse} = E\left\lfloor \|y - F(x)\|^2 \right\rfloor, \text{ where E is the expectation}$$

Two steps are useful to build a conversion system: a training step and a conversion step. In the training phase speech samples obtained from the source and the target speaker are analysed to extract the main features. Then these features are time aligned and a conversion function is estimated to map the source and the target characteristics.

The aim of the conversion step is to apply the estimated transformation rule to an original speech produced by the source speaker. The new utterance sounds like the same speech articulated by the target speaker, that is to say, produced by replacing the source characteristic with those of the target voice.

The last step is the re-synthesis of the signal in order to reconstruct the speech segment of the source voice after the conversion.

The most representative techniques on voice conversion, developed in the past decades, are based on vector quantization [1, 34] on Linear Multivariate Regression [57], on vocal tract modification [50], and on Gaussian Mixture Models and derived, [11, 12, 21, 50–52, 54, 55].

Through this state of the art we understand the possibilities for offenders to alter their voices in criminal cases in order to trick forensic experts. Several experiments were conducted on the question of voice forgery that are described in the following section.

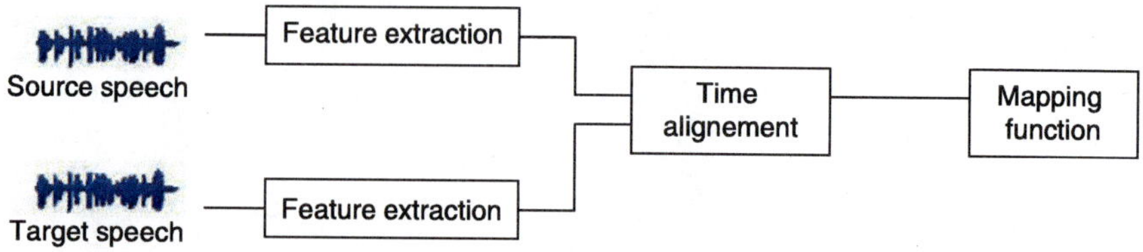

Fig. 16.2 Training step

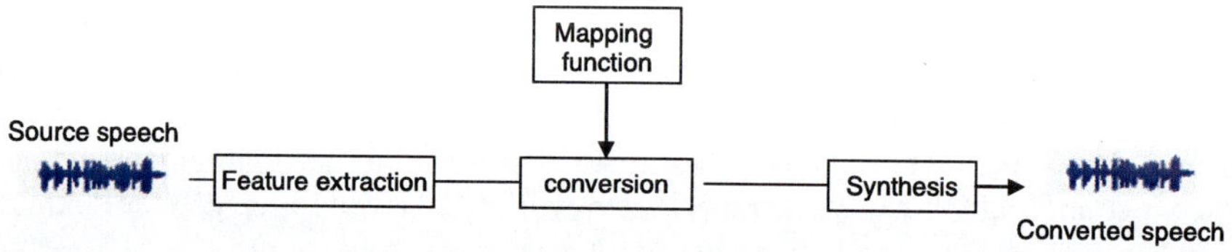

Fig. 16.3 Conversion step

16.4 Acoustic Analysis of Voice Disguise

During a 60 month period, starting in 2005, one of the authors in pursuit of his doctoral work, conducted a study of four different voice disguises among the most commonly used form of voice disguise; they are included in the category non-electronic and deliberate. The choices of voice disguise that were used in our experiment were guided by the findings of the IRCGN (L'_Institut de recherche criminelle de la gendarmerie nationale), or the Forensic Research Institute of the French Gendarmerie, derived from a questionnaire presented to 100 people who were asked what disguise they would most likely use over the phone (Table 16.1).

From the nine forms of disguise used in the IRCGN questionnaire, we chose for our own study those forms of disguise which were most popular according to the IRCGN study participants. Four forms of disguise were isolated. Here is a list of the forms of disguise we used in our study of transformed voices from the following processes:

1. pinched nostrils
2. the fundamental frequency increasing
3. the fundamental frequency decreasing
4. placing ones hand over the mouth

Characterizing specific disguised voices requires attention to parameters that change from a normal voice. The most useful descriptors for the disguised voice characterization are intended to model the acoustic signal changes related to physiological changes. These changes while they occur at the source level itself are also involved at the level of the filter: the vocal tract; mouth; nasal resonators; and lips. For instance, if one hides his mouth this influences the level radiation on the lips, or if one holds his nose that causes a change in the nasal resonator as well as a change

Table 16.1 Choice of disguise

Type of disguise	Percentage
Pinched nose	17
Hand over the mouth	34
Foreign accent	8
High voice	16
Low voice	11
Whispered voice	3
Electronic device	4
Helium absorption	2
Others	5

in the way of breathing. The parameters that we study can be fitted into two categories: parameters related to voice quality; and parameters related to spectral domain. The extraction of these elements is done using Matlab programs and PRAAT scripts [5]. The parameters that we considered are:

- speech rate
- Harmonics to Noise Ratio (HNR)
- jitter
- shimmer
- voiced part
- fundamental frequency
- formants

The analysis of voice disguised requires before any study the development of a database. Our end goal was to characterize types of disguised voices so that they can be readily detected and identified. The choice of disguise was guided by the use of what can be found in criminal cases. Since there is currently no database of disguised voices in France, we had to build a database ourselves. Our database, built upon referrals of subjects from IRCGN, consisted of fifty (50) male voices: each of the subjects was recorded in a normal voice and in four different disguises: pinched nose; hand over mouth; elevated voice; and lowered voice. We note that this database is structured around the study of the following disguises:

- Pinched nose,
- Hand over mouth
- Higher voice,
- Lower voice

All research subjects used the same text in testing the disguise across the four categories. The text consisted of the following:

- A series of 10 phonetically balanced sentences
- French vowels
- Phonetically balanced text of 7 lines (from the well-known "North Wind and the Sun" text)

Table 16.2 Speech rate

Disguise	ph/s	Std
Normal voice	11.6	0.8
High voice	11.3	1.5
Low voice	10.9	0.7
Pinched nose	11.7	1.1
Hand over the mouth	11.5	1.2

The recording sessions took place in a quiet room, using a digital recorder brand and a microphone headset, both professional grade. The data were stored on a Compact Flash card, and subsequently on a CD-ROM. Waveforms were digitized without compression (PCM) using a 48 kHz sampling frequency and 16-bit uniform linear quantization. The interest of such a sampling frequency is to accurately capture the different changes of disguised voice, even if this sampling frequency is not common in a forensic context.

16.4.1 Speech Rate

Speech rate is an important paralinguistic parameter of our means of expression. We consider that this flow can be different depending on the chosen disguise. Indeed, some disguises require longer breaks to take one's breath, or at times an increased speed to get to the end of the sentence more quickly. We focused the study on phoneme rate from the 10 phonetically balanced sentences. The results are not significant in terms of discrimination.

The average values for all ten sentences ranged up to 11.7 ph/s (in the pinched nostrils case) and down to 10.9 ph/s (using lowered voice). The observed standard deviation span was between 0.7 and 1.7. Average speech rate for normal voice is 11.6 ph/s, with a standard deviation equal to 0.8 (Table 16.2).

It is important to notice that these values are measured using very short sentences, and that it would not be surprising to notice more significant changes if the sentences were longer. Nevertheless, short sentences are more realistic in a forensic context.

16.4.2 Harmonics to Noise Ratio

Harmonics to Noise Ratio: HNR [9, 60] is considered as a voice quality parameter. It is based on calculating the ratio of the energy of the harmonics related to the noise energy present in the voice (both measured in dB). This is a measurement of the voice pureness. We worked on the oral vowel [a] and the nasal vowel [ã]. These vowels are sustained three seconds long. Table 16.3 shows average HNRs and associated standard deviations considering the vowels appearing above.

Table 16.3 HNR value

	[a]		[ã]	
Disguise	Mean	Std	Mean	Std
Normal voice	21	1.4	25	1.2
High voice	22	1.1	28	1.8
Low voice	20	1	24	1.1
Pinched nose	14	0.9	34	1.6
Hand over the mouth	12	1.6	24	1.2

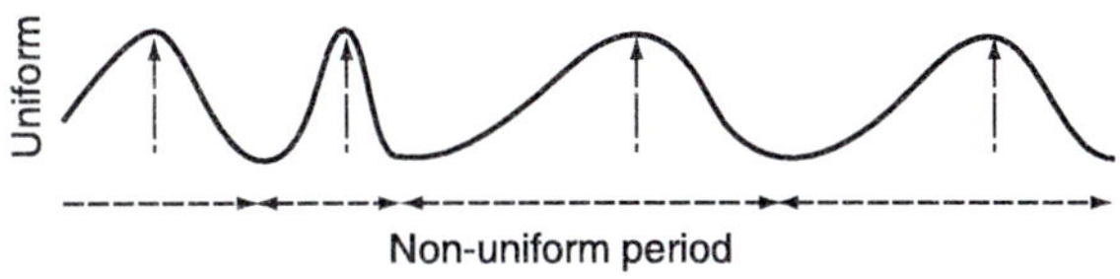

Fig. 16.4 Graphical interpretation of Jitter

It appears that HNR analysis distinguished the hand over the mouth and the pinched nostrils disguise from the normal voice on the vowel [a], and only the pinched nostrils voice for the nasal vowel [ã].

16.4.3 *Jitter*

Jitter is also a qualitative parameter of the voice. It corresponds to a short-term, cycle to cycle, and can thus be considered as aperiodicity measurement. The mathematical expression of the jitter is:

$$\text{Jitter Ratio} = \frac{\sum_{i=1}^{n-1} \frac{|T_i - T_{i+1}|}{n-1}}{\sum_{i=1}^{n} \frac{T_i}{n}}$$

Graphically, the jitter could be represented by Fig. 16.4.

Jitter is studied on the oral vowel [a] and the nasal vowel [ã] by asking speakers to sustain the sound over a three-second period. This parameter is often taken into account in the search for pathological voices. In such a context, a 1.04% value [5] is considered as a maximum for a normal voice. Beyond this value, the voice is considered pathological. It is reasonable to expect disguise to affect voice quality, moving it closer to the pathological threshold. Given the results below, it appears that hand over the mouth and pinched nose voices have values above the pathological threshold (Table 16.4). This result is consistent with the nature of these disguises that generate a mechanical modification of the vocal tract.

Table 16.4 Jitter value

	[a]		[ã]	
Disguise	Mean	Std	Mean	Std
Normal voice	0.6	0.03	0.4	0.03
High voice	0.7	0.03	0.8	0.04
Low voice	0.5	0.06	0.7	0.05
Pinched nose	1.2	0.06	1.4	0.1
Hand over the mouth	1.3	0.04	1.3	0.08

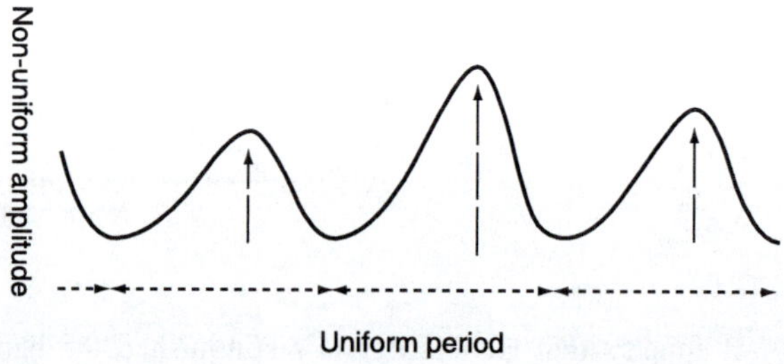

Fig. 16.5 Graphical interpretation of Shimmer

16.4.4 Shimmer

The shimmer is also considered a voice quality parameter. It stands for a short-term, cycle to cycle, perturbation in the amplitude of the voice. Mathematically the shimmer is expressed by:

$$\text{Shimmer} = \frac{\sum_{i=1}^{n-1} \frac{|A_i - A_{i+1}|}{n-1}}{\sum_{i=1}^{n} \frac{A_i}{n}}$$

Graphically the shimmer could be represented by Fig. 16.5.

A value above 3.81 [5] characterizes a pathological voice. The observed values (Table 16.5) show that, in the case of hand over the mouth disguise, the voice is considered pathological. Other disguises remain in the normal voice range.

16.4.5 Unvoiced Frames

As in the case of pathological voice detection, unvoiced frames are interesting to investigate when it comes to studying voice disguise. Generally, some frames are unvoiced because of the impossibility of carrying out a periodic glottal closure. Table 16.6 shows the average unvoiced frames ratio for all the balanced sentences.

Table 16.5 Shimmer value

Disguise	[a]	Std	[ã]	Std
Normal voice	2.1	0.06	1.9	0.1
High voice	3.5	0.1	3.6	0.2
Low voice	3.4	0.4	3.9	0.3
Pinched nose	3	0.07	2.8	0.2
Hand over the mouth	4.2	0.1	5.4	0.08

Table 16.6 Unvoiced frames ratio

	Unvoiced frames ratio	
	Avg (%)	Std
Normal voice	28	0.7
High voice	23	1.1
Low voice	25	1.1
Pinched nose	27	0.8
Hand over the mouth	23	1.2

This table reveals that disguise causes a slight decrease of the unvoiced part, especially for high pitched and hand over the mouth voices.

16.4.6 Fundamental Frequency

Raw analysis of the fundamental frequency mean (see Table 16.7) provides a three-class clustering: E1=(low voice), E2=(normal voice, voice with hand over the mouth, pinched nose voice), E3=(High voice). Nevertheless, the measure of the fundamental frequency mean cannot be considered as truly discriminative, because it is highly dependent on the context in which the voice is produced or emotional state of the speaker. In addition, the measured values for normal, low, pinched nose, and hand over the mouth voice sets, are still in the range of the male normal voice [80–200 Hz]. The standard deviation values confirm the underlying confusion in discriminating these sets. Therefore, the value of the fundamental frequency mean can be determined without taking into consideration a range of uncertainty. To increase the relevance of this criterion, the range of F0 variation is interesting to look at. This range is particularly important not only for high voice (298 Hz), but also (to a lesser extent) for the low voice with 140 Hz span. For these two disguises, the control of fundamental frequency stability seems to be more difficult (see Table 16.7).

16.4.7 Formant analysis

Formants F1 and F2 are also interesting features to focus on. Under normal voice production, F1 can vary from 300 to 1000 Hz depending on the vowel. The lower F1, the closer the tongue is to the top of the mouth. The F2 value is proportional to the anterior/

Table 16.7 Fundamental frequency analysis

Disguise	Normal voice	High	Low	Pinched nose	Hand over the mouth
Mean	127	255	114	128	131
Std	14	28	16	16	18
Min	75	110	75	75	75
Max	190	408	215	194	202
Min–max difference	115	298	140	119	127

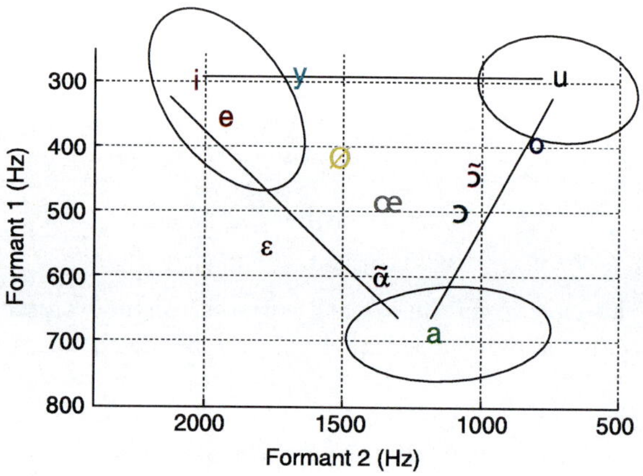

Fig. 16.6 French vocalic triangle under normal condition

posterior position of the highest part of the tongue while producing the vowel. In addition, lip rounding causes a lower F2 than in the case where there is no lip rounding.

Formant analysis is computed using a LPC analysis of the signal. The linear prediction coefficients are calculated from the Burg method [18]. The formants, even more than the fundamental frequency, are subject to the influence of phonetic context during vowels production (co-articulation). That is the reason why we conducted our experiments on vowels having the same articulation context (consonantal neighbourhood). The measurements were done at the time-central position of the vowel realization. We then studied changes in the vocalic triangle F1–F2 considering disguise as a parameter. Figure 16.6 shows the vocalic triangle vowels under normal conditions.

Figures 16.7, 16.8, 16.9, and 16.10 show the vocalic triangles under given disguises.

At first sight, formant analysis seems relevant in order to discriminate the normal voice from disguised voices, and even to distinguish disguised voices one from another. The differences can be explained by the modification of the various resonators and articulators during the use of disguise like hand over the mouth voice and pinched nose voice. These immediately affect the timbre of the voice. What is really significant is a narrowing vocalic triangle in the case of hand over the mouth voice and an enlargement of the triangle in the case of pinched nose voice. Considering hand over the mouth the results are a little bit surprising according to the theory of wave propagation. Cavities closed at one end (or followed by a smaller cavity

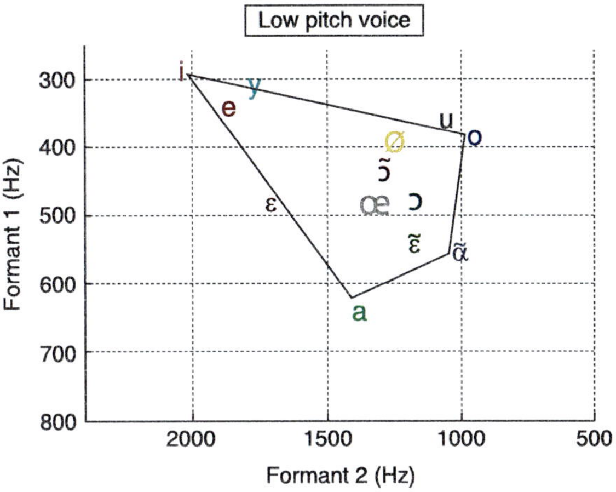

Fig. 16.7 Vocalic triangle: low voice

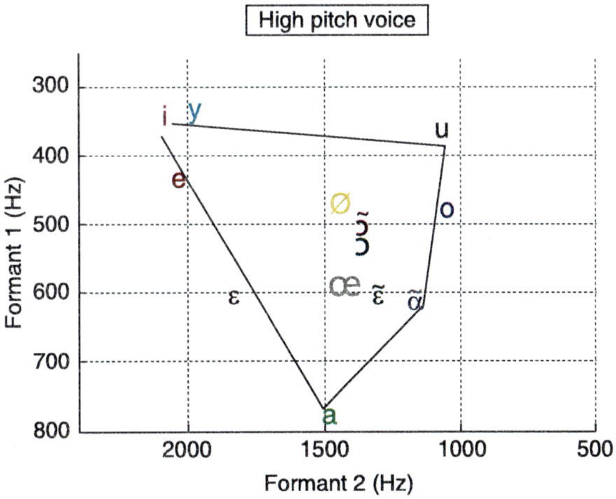

Fig. 16.8 Vocalic triangle: high voice

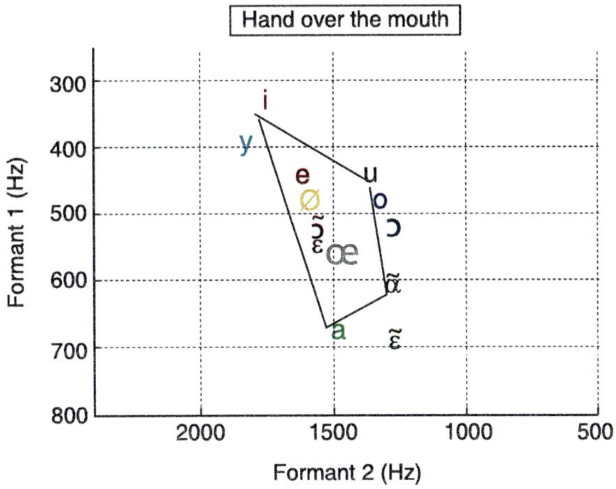

Fig. 16.9 Vocalic triangle: hand over the mouth

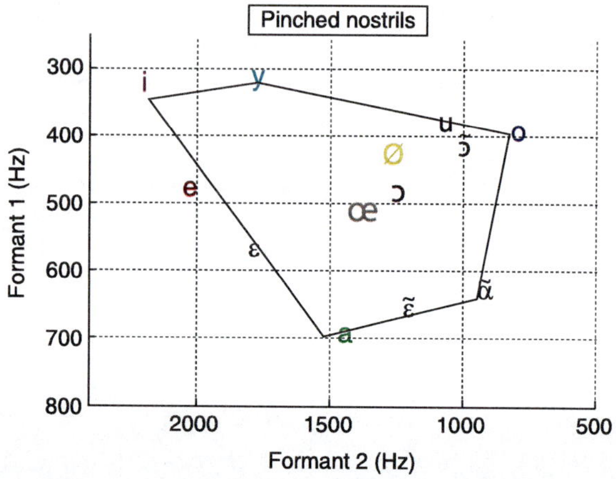

Fig. 16.10 Vocalic triangle: pinched nose voice

volume) and open to another (or followed by a cavity of greater volume), can be modelled by cascading uniform tubes with different sections [15, 29, 30]. In the open tube case, wave propagation frequency modes are given by:

$$f = \{nv/2L\}$$

where n is the resonant frequency number (1,2,3...), L is the tube length and v is the sound celerity.

In a closed tube, wave propagation frequencies correspond to:

$$f = \{nv/4L\}\,.$$

So, we expect half reduction for F1 in the hand over the mouth disguise case. Measurements show this is not the case. We explain this situation by a compensation phenomenon that consists in exaggerating the mouth movement to increase the intelligibility of the sound.

Concerning the pinched nostrils case, nasal vowels seem less robust than for other disguises. It can be explained by the alteration of the transfer function of the speech signal. We can assume that this disguise affects the position of zeros in this function.

The descriptive analysis of acoustic parameters allowed us to evaluate the impact of voice disguises on (auditory-instrumental or acoustic) descriptors. We focused on spectral and voice quality parameters. The study of different features is important because it allows, by performing a comparison, one to better understand the changes involved during disguises and avoid confusion in identification.

16.5 Automatic Detection of Voice Disguise

In this section, we propose different supervised learning methods to detect disguises. The automatic detection of voice disguise is addressed by means of k-nearest neighbours and Support Vector Machines (SVM). Input features are non-parametric

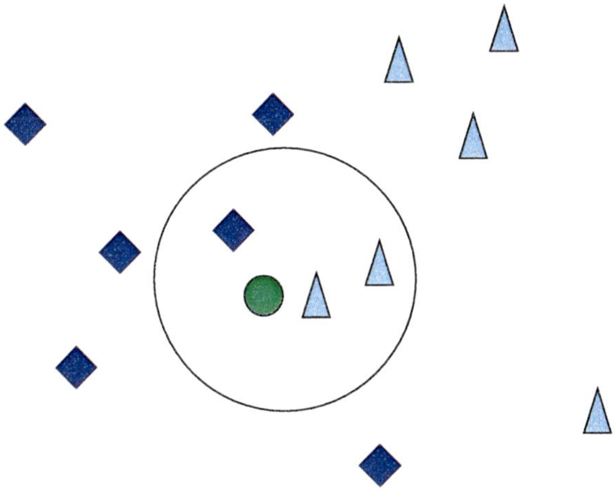

Fig. 16.11 2-NN classifier

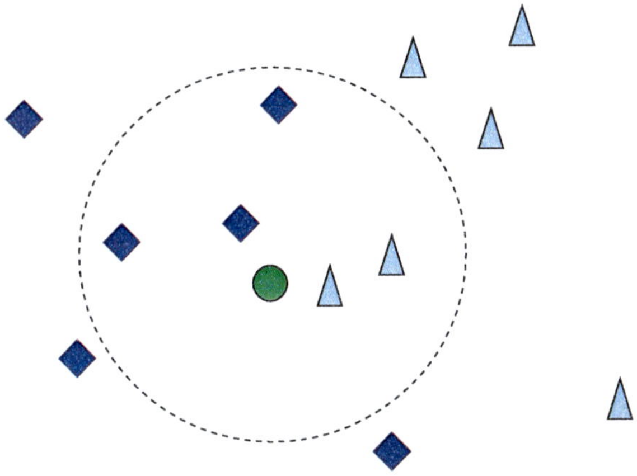

Fig. 16.12 3-NN classifier

short-term Mel Frequency Cepstral Coefficients (MFCC) [11], and their first derivatives, complemented with acoustic parameters (excepted HNR, jitter and shimmer) as previously described.

The k-nearest neighbours were considered at first. It consists in creating and using a learning database including known objects, for which we already know what the correct class is. Then, when the system is given a query (i.e., an unknown object to classify), the system simply finds its k-nearest neighbours in the learning database (i.e., the database objects that are the most similar to the query). Then, the query is given the same class as the most represented among the nearest neighbours. The relevance of the classification depends on k (the size of the neighbourhood) as described in Figs. 16.11 and 16.12.

Another technique used to separate normal voice from disguised voice is the Support Vector Machines (SVMs) approach. SVMs have become a popular tool

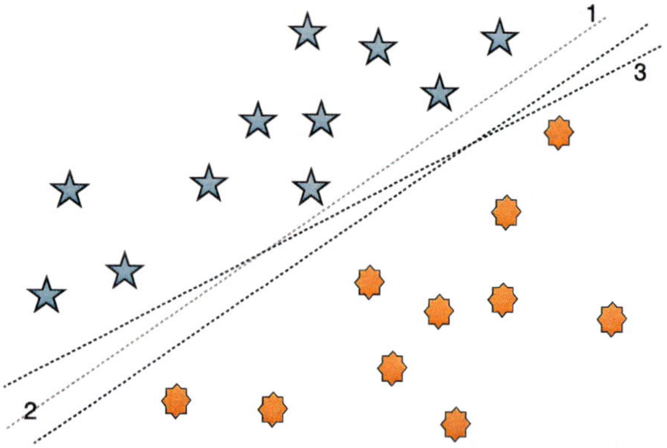

Fig. 16.13 SVM classifier: boundaries computation

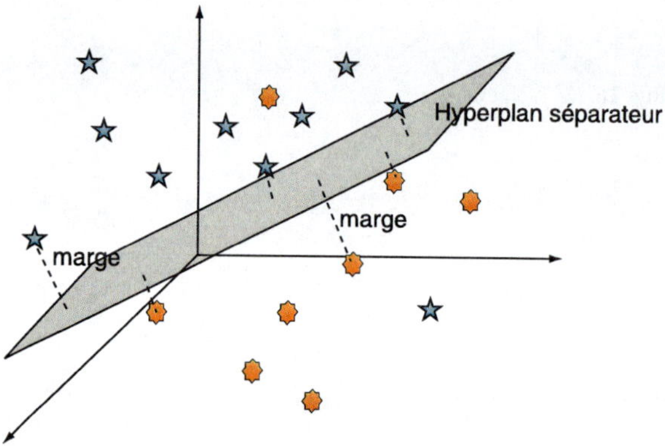

Fig. 16.14 SVM classifier: hyperplane optimization

for discriminative classification and were originally proposed by Vapnik [58]. SVMs are effective discriminant classifiers, capable of maximizing the error margin. Thanks to the use of kernel functions, SVMs can make decisions in non-linearly separable contexts. Choosing an appropriate kernel function allows SVMs to mimic other classifiers that are based on linear discrimination. The algorithm creates a maximum margin hyper-plane to discriminate two classes aside by linear separation in a higher dimensional space. SVM non-linearly map their original n-dimensional input space into a higher dimensional feature space. In this high dimensional feature space, a linear classifier is constructed. The Figs. 16.13 and 16.14 describe the SVM principles. The first step consists in separating data in two different classes by finding different hyperplanes and the second step computes the best hyperplane in the sense of support vectors to hyperplane distance maximization.

Classifier fusion [19, 23] appears as a good way to improve overall performance. Because different fusion methods are designed for different problems and different results, we propose two architectures as described by Figs. 16.15 and 16.16.

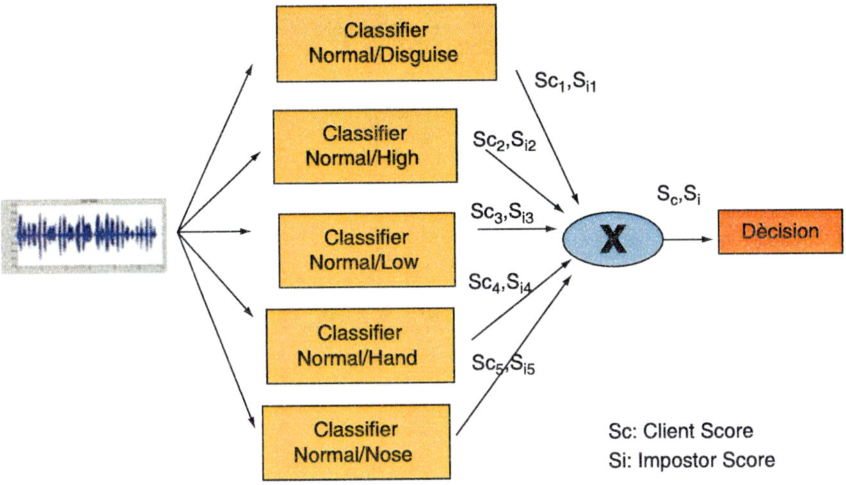

Fig. 16.15 Fusion classifier: parallel architecture

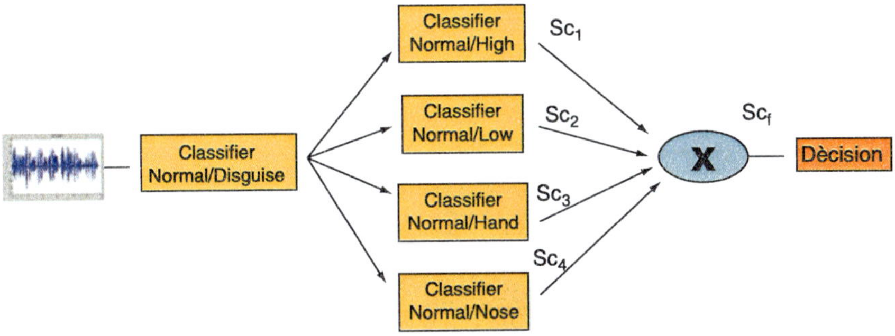

Fig. 16.16 Fusion classifier: mixed architecture

The performance evaluation of the automatic detection are based on Receiver Operating Characteristic curves (or ROC curves) [17]. These curves are drawn from two measurements resulting from the confusion matrix:

- *Sensitivity:* probability that a test result will be positive when the disguise is present (true positive rate, expressed as a percentage).
- *Specificity:* probability that a test result will be negative when the disguise is not present (true negative rate, expressed as a percentage).

In a Receiver Operating Characteristic (ROC) curve the true positive rate (Sensitivity) is plotted in function of the false positive rate (100-Specificity) for different cut-off points. Each point on the ROC plot represents a sensitivity/specificity pair corresponding to a particular decision threshold. A test with perfect discrimination

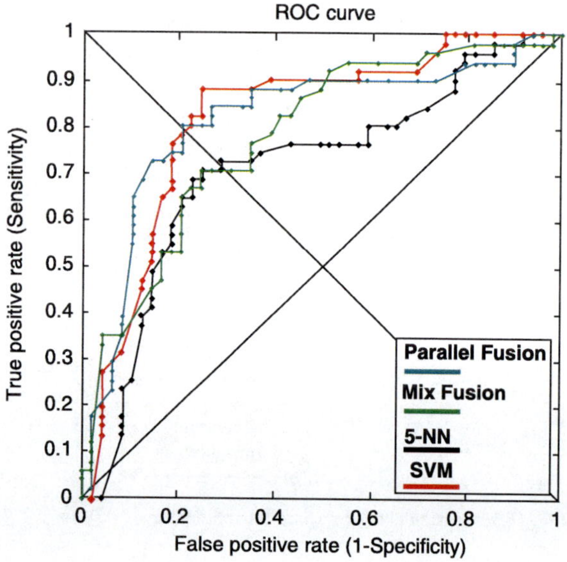

Fig. 16.17 ROC curve: high/normal

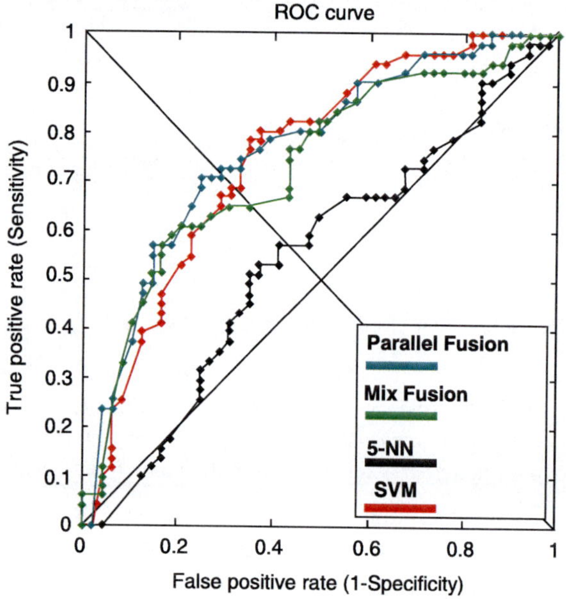

Fig. 16.18 ROC curve: low/normal

(no overlap in the two distributions) has a ROC plot that passes through the upper left corner (100% sensitivity, 100% specificity).

Figures 16.17, 16.18, 16.19 and 16.20 show results of discrimination between normal and each disguise case. Figure 16.21 shows results of discrimination considering all disguises as belonging to the same class (normal/overall disguised discrimination).

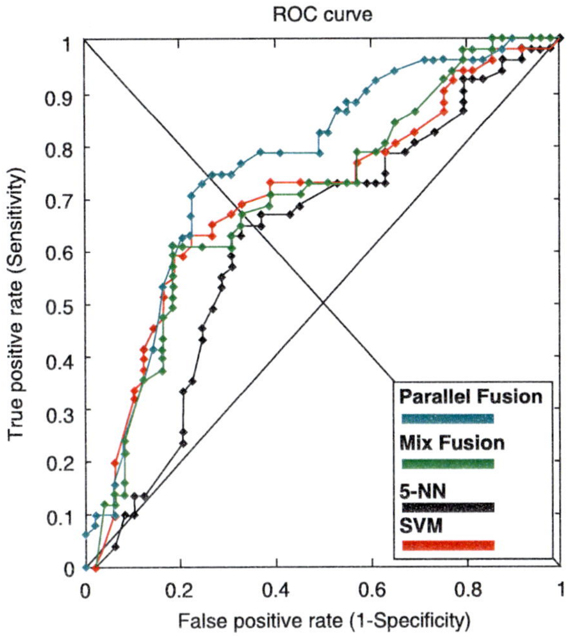

Fig. 16.19 ROC curve: pinched nose/normal

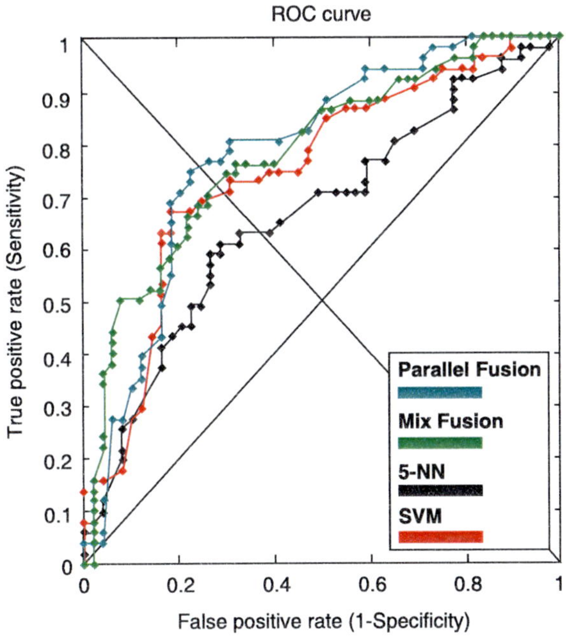

Fig. 16.20 ROC curve: hand over the mouth/normal

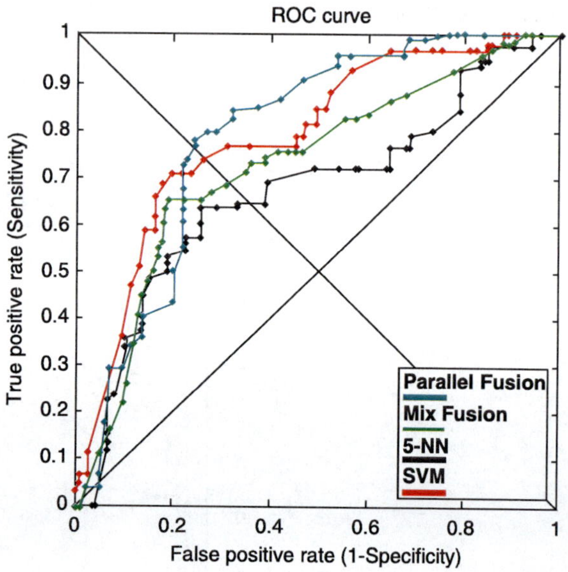

Fig. 16.21 ROC curve: disguise/normal

Table 16.8 AUC criterion

	High voices	Low voices	Hand over the mouth	Pinched nose	Disguised voices
5-NN	0.63/0.71/0.78	0.45/0.54/0.63	0.58/0.66/0.73	0.53/0.62/0.71	0.63/0.67/0.71
SVM	0.75/0.82/0.89	0.67/0.75/0.83	0.66/0.74/0.82	0.61/0.69/0.77	0.75/0.78/0.82
Parallel fusion	0.74/0.81/0.88	0.70/0.77/0.84	0.70/0.78/0.86	0.69/0.76/0.84	0.75/0.79/0.83
Mix fusion	0.71/0.78/0.85	0.65/0.73/0.81	0.69/0.77/0.85	0.61/0.69/0.77	0.69/0.72/0.77

What we notice is that a quite similar performance level is reached between three techniques: SVM, parallel fusion and mix fusion. The k-nearest neighbours (with k=5) approach exhibits the worst performance level. Because of the difficulties in separating the following classifiers (parallel fusion, SVM and mix fusion), it is better to use another performance criterion. This criterion, linked to the ROC curve, is the AUC (Area Under Curve).

AUC values, presented along with associated confidence intervals in Table 16.8, confirm previous results and the best classifiers to discriminate each disguise against normal voices appear to be the parallel fusion and the SVM classifier.

16.6 Voice Conversion as a Tool for Voice Disguise

Two original techniques are proposed, which apply to two different forensic scenarios. The first technique is described in [39]. This conversion is based on a technique of voice forgery using ALISP (Automatic Language Independent Speech Processing)

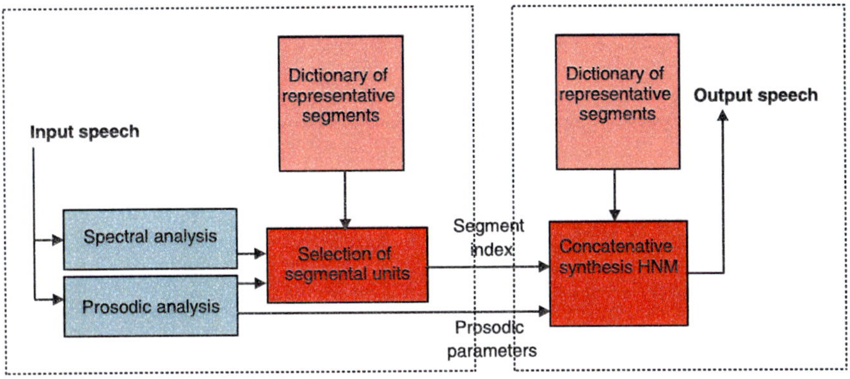

Fig. 16.22 ALISP encoder/decoder

[2, 6]. Features extraction is performed using the temporal decomposition algorithm [3] on MFCC (Mel Frequency Cepstral Coefficient) features. The different segments from the temporal decomposition are clustered using vector quantization into 64 classes. The training data (only from the target speaker) are labelled. A set of HMM (Hidden Markov Models) are then trained on these data and a new segmentation of the data results from the iterative re-estimation of the model parameters. The training step provides an inventory of client speech segment divided into 64 classes according to a 64-symbol codebook. The different speech segments are represented by their Harmonics plus Noise model parameters (HNM) [10] in order to get a smooth concatenative synthesis of the new segments.

The second step of this technique consists in encoding the impostor's voice using the above ALISP codebook, and then performing decoding, using synthesis units taken from the segment inventory obtained from client's voice. A collection of speech segments is created by segmenting a set of training sentences pronounced by the target speaker.

This technique of conversion provides interesting results as presented in the following figures (Fig. 16.23a and b), where a significant displacement of the impostors score distribution towards the client score distribution can be seen. These tests have been realized on the NIST 2004 corpus [10]. 1729 imposters, 1320 clients are used and the languages used by speakers were English, Russian, Chinese, Spanish and Arab. The DET curve indicates the decrease of the recognition rate.

The second original technique of voice conversion consists in combining GMM and ALISP approaches to create a new system that outperforms the previous ones. The idea is that to make ALISP recognize the good segments in the inventory, the signal to encode must have the good spectral characteristics. To do so, after building the ALISP inventory and training the GMM conversion function, as done previously, we apply the GMM conversion function to the source signal. The advantage of this technique is to move spectral characteristics of the source towards those of the target. This transformed signal is then encoded by ALISP. The transformed speech signal produced by this method is free of artefacts, thanks to the smooth concatena-

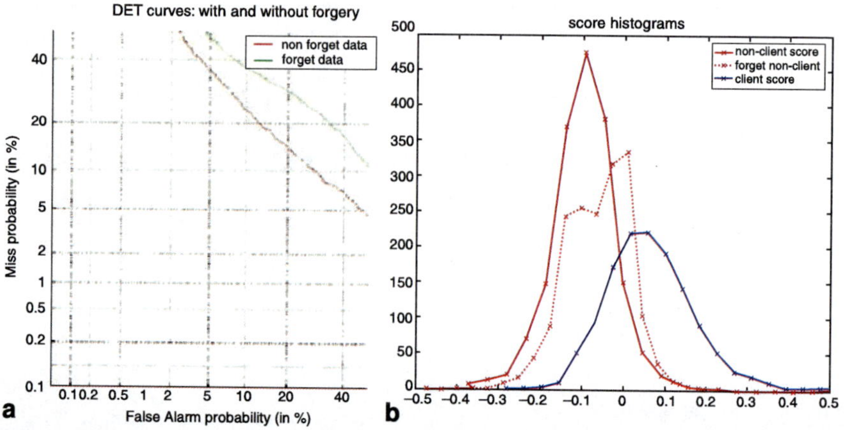

Fig. 16.23 **a** DET curve. **b** Score distribution

Fig. 16.24 Training phase

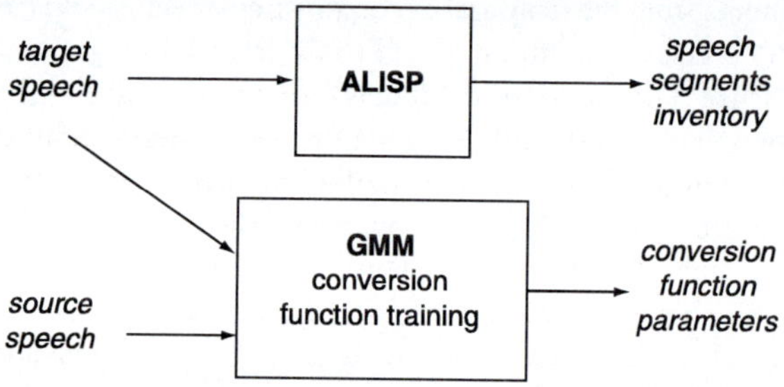

tion of the ALISP acoustical units. Figures 16.24 and 16.25 give an overview of this new conversion system.

To be close to a realistic scenario in which an impostor would use a voice conversion system to forge his identity, our experiments were carried out on an authentic database We used a recording of France President Nicolas Sarkozy as the target speaker voice: this recording was a 40-min discourse addressed to the parliamentarians of the majority, on July 20th, 2007.

Two males and one female were recorded, articulating the 10 first minutes of Sarkozy's discourse. In our experiments, 8 min of each source speaker were used for the training step of the GMM conversion system, along with the corresponding 8 min of Sarkozy's speech signal. To build the ALISP inventory, we used 35 min of the French president discourse. Two minutes of both the source speakers' and Sarkozy's speech signal (used in neither the GMM or the ALISP training) were kept for the tests. For the conversion using GMM, we used a model with 8 Gaussians, which gave us the best results with our database. In order to compare the performance of the three systems, the spectral distortion measure between the converted and the target signal and the log-likelihood of the converted signal against a GMM

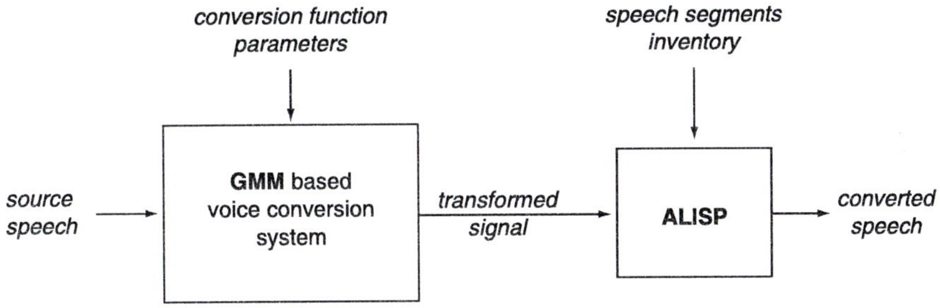

Fig. 16.25 Transformation phase

Table 16.9 Average normalized spectral distortion for the three conversion methods

Conversion system	GMM	ALISP	GMM+ALISP
Spectral distortion	0.77	0.78	0.73
Confidence interval at 95%	±0.02	±0.02	±0.02

modelling the target speaker were computed. Obtained results are presented in the following section.

A frequently used estimator for a voice conversion system is the logspectral distortion. After the temporal alignment of the converted and the target signal, this estimator is calculated using the quadratic distance between the MFCC's features:

$$d_{t,c}(t_i) = \sum_{k=1}^{p} \left|c_k^c - c_k^t\right|^2$$

where $\left[c_0^c, c_1^c, \ldots, c_p^c\right]$, $\left[c_0^t, c_1^t, \ldots, c_p^t\right]$ are respectively MFCC's coefficient at time t, of respectively the converted and target signal.

On all the converted signal, the spectral distortion is expressed by:

$$D_{t,c} = \sum_{i=1}^{N} d_{t,c}(t_i)$$

Moreover, the use of the MFCC's implies that the distortion is measured using a mel scale, which is perceptually more relevant. This distortion was normalized by the initial distortion between the source and the target speaker.

$$DSN = \frac{D_{t,c}}{D_{s,c}}$$

Then, a value of the normalized spectral distortion under 1 indicates that the converted speech is closer to the target speech than the source speech is. A value of 0 indicates that the converted speech perfectly matches the target speech. Table 16.9 presents the average normalized log-spectral distortion as measured on the test corpus, for our three voice conversion methods.

Table 16.10 Log-likelihood against the GMM modelling the target speaker for the three conversion methods

System	GMM	ALISP	GMM+ALISP	Target
Score	0.32	0.34	0.60	1.02
Confidence	±0.05	±0.05	±0.05	±0.05

In order to check the reliability of the results, the variance of the spectral distortion over one frame was estimated for each conversion system. Then, given the number of frames over which the mean spectral distortion is calculated, a 95% confidence interval was computed.

These results show that the combined system outperforms each method taken separately. The difference between the GMM alone and the ALISP alone systems do not seem to be significant. One can be surprised that the ALISP alone system does not behave better than the GMM system, as it produces a signal which is the concatenation of speech segments of the target speaker. Though, it has to be kept in mind that these segments have been selected for being the closest to the source speech signal. On the contrary, GMM regression aims precisely at minimizing the distance between the converted MFCC's and the target ones. Another possible way of estimating the performance of a voice conversion system is the following. From a 2-min speech signal of the target speaker (independent of the tests sentences), MFCC's features are extracted every 20 ms, and a 16-GMM is fitted to these cepstral feature-vectors thanks to the EM algorithm (omitting the first cepstral coefficient, as previously). Note that a silence removal has been applied to the signal prior estimation of GMM parameters. Another GMM is built similarly for the source speaker. The converted speech signals can then be gauged, computing the probability that the MFCC's of these signals have been produced by the GMM modelling the target speaker. In this work, we used the following log-likelihood difference:

$$P = \log(P(X|Mt)) - \log(P(X|Ms))$$

where P(X|Mt) and P(X|Ms) are the probability that the transformed signal X has been produced respectively by the GMM modelling the target or the source speaker. This method based on the use of GMM to model the speakers is the root of many automatic speaker recognition systems.

Table 16.10 shows the relative performances of the three conversion methods according to that estimator. The last column is the score obtained when the test sentences are pronounced by the target speaker himself. Again, a 95% confidence interval was determined from the variance of the score over one frame for each system.

As expected, in the three cases, the figures obtained are positive, which means that the "recognized" speaker is the target speaker, Sarkozy. The difference in log-probability is much higher (nearly twice) for the GMM+ALISP system, which means the conversion fits better the acoustic space of the target in that case. More importantly, the score obtained by the sentences pronounced by the target speaker himself shows that the combined system significantly moves the converted speech closer to the target model. Once more, the difference of performance between the two other systems is not significant.

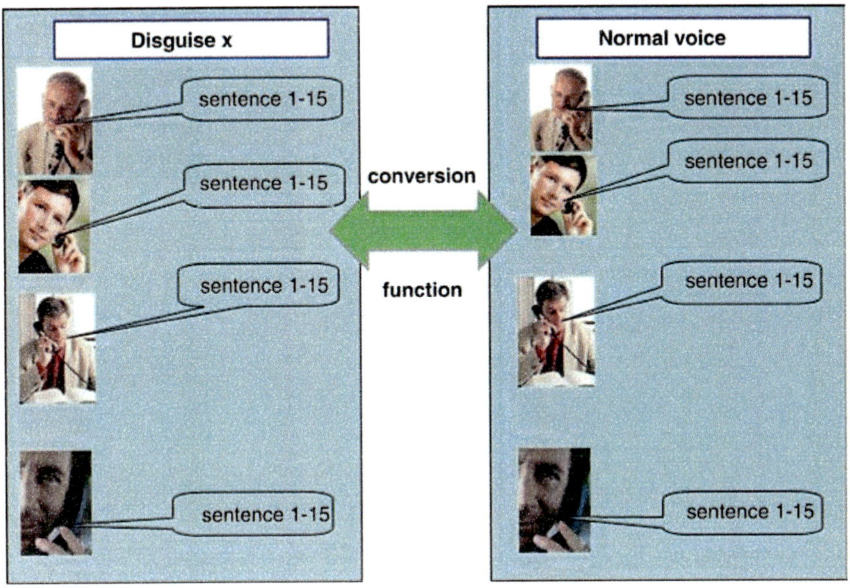

Fig. 16.26 Training phase

Voice conversion has numerous applications for security, entertainment or speech synthesis. It constitutes a real threat to public safety and an effective means to trip up automatic speaker recognition systems that are used in the field of forensic sciences. The proposed analysis of two automatic voice conversions provides interesting results in the field of forgery. The influence of voice conversion significantly decreases the performance of an automatic speaker recognition system even if the synthesis is not perfect.

16.7 Reversing Disguised Voice

Another original application that combines forensic interests with voice conversion is the question of the reversibility of voice disguise itself. The aim is to be able to get back to a normal speaking voice from a disguised voice. We work on four different kinds of reversibility:

- high to normal voice
- low to normal voice
- hand over the mouth to normal voice
- pinched nose to normal voice

The training part is defined as proposed on Fig. 16.26. 15 similar sentences pronounced by 20 speakers in disguised mode and in normal mode are used to create the conversion function.

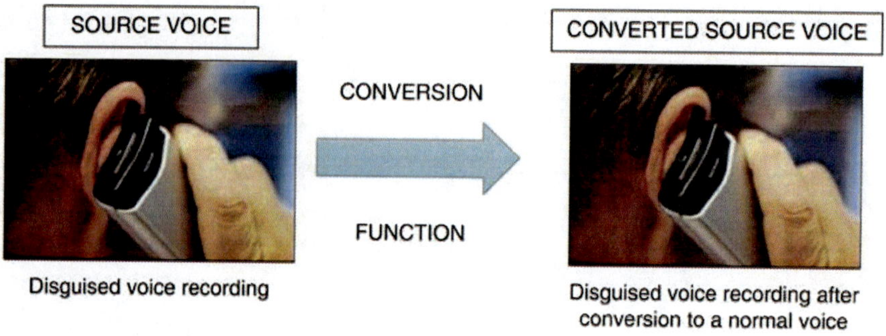

Fig. 16.27 Transformation phase

Table 16.11 Results of disguise reversibility

Conversion technique	High-normal	Low-normal	Nose-normal	Hand-normal
Spectral distortion	0.65	0.76	0.72	0.76
Confidence interval at 95%	±0.02	±0.02	±0.02	±0.02
LLR (Log likelihood Ratio) P=log (P(Xc|Mt))−log(P(Xc|Ms))	0.9	0.2	1.1	0.4
Confidence interval at 95%	±0.06	±0.08	±0.09	±0.04

The conversion function based on the GMM regression is applied on 5 sentences pronounced by 10 speakers. In the transformation step speakers and sentences are different from the training set. This step is described by the Fig. 16.27. GMM regression is well described in [21, 40, 51].

The analysis of the reversibility reveals satisfactory results, proposed in Table 16.11, in view of the problem to solve. The best performance is obtained for the most significantly distinguished disguise from the normal voice, including the conversion between the high voice and normal voice. The disguise reversibility is more effective if the disguise is accentuated.

16.8 Conclusion

Voice forgery presents forensic science with an enormous challenge. Today, more and more possibilities are offered to criminals to transform their own distinct voice into that of another so that they can elude detection by law enforcement and counter terrorism agencies. This chapter proposes two different methods to analyse voice disguise. The first method is based on an acoustic analysis of different features that become seem relevant according to the form of disguise. Our study showed that the effect of the disguise on voice characteristics is dependent upon the kind of disguise that is used. The second one proposes an automatic analysis to detect the most common disguises. We found that parallel fusion and SVM classifier provide the best results with a good level of discrimination.

Two different applications of voice conversion in the forensic setting were examined and tested. First, an automatic impersonation of the French President was realized with success. Second, voice disguise was reversed. While it is not possible today to fully reverse a voice disguise in such a way that that the resulting waveform would sound completely natural to a listener (mainly due to limitations with the quality of converted voice synthesis) our study demonstrates, nevertheless, that a disguised voice could be reversed to a relatively 'normal' voice as evaluated by current state of the art speaker verification systems. And to that end, the analysis of voice disguise reversibility serves as a fertile topic for future research. The main interest in a forensic context is to be able to perform speaker recognition on a reverse disguised voice (a voice that has already been converted back from its disguised form to normal speech) and to evaluate the performance of speech applications in such contexts.

References

1. Abe M, Nakamura S, Shikano K, Kuwabara H (1988) Voice conversion through vector quantization. In: Proceedings IEEE Int conf on acoustics, speech and signal processing, pp 655–658
2. Baverel C, Chollet G, Gournay P (2001) Amélioration d'un codeur de parole à très bas débit par indexation d'unités de taille variable. In: GRETSI
3. Bimbot F, Chollet G, Deleglise G, Montacie C (1988) Temporal decomposition and acoustic-phonetic decoding of speech. In Proceedings of the international conference of acoustics, speech, and signal processing, ICASSP, pp 445–448
4. Blomberg M, Elenius D, Zetterholm E (2004) Relating acoustic features of a professional impersonator with the score of a speaker verification system. In: Proceedings of Fonetik
5. Boersma P, Weenink D (2008) Praat: doing phonetics by computer. http://www.praat.org/
6. Chollet G, Cernocky J, Constantinescu A, Deligne S, Bimbot F (1999) Towards ALISP: a proposal for automatic language independent speech processing. In: Computational models of speech pattern processing, NATO ASI Series, Series F: computer and system sciences, vol 169. Springer, pp 375–387
7. Clark J, Foulkes P (2007) Identification of voices in electronically disguised speech. Int J Speech Lang Law 14:2
8. de Figueiredo RM, Souza Britto H (1996) A report on the acoustic effects of one type of disguise. Forensic Linguist 3(1):168–175
9. de Krom G (1993) A cepstrum-based technique for determining a harmonics-to-noise ratio in speech signals. J Speech Hear Res 36:254–266
10. Doddington DR, Przybocki M, Martin AF, Reynolds DA (2000) The NIST speaker recognition evaluation—overview, methodolgy, systems, results, perspectives. Speech Commun 31:225–254
11. Duxans H, Erro D, Pérez J, Diego F, Bonafonte A, Moreno A (2006) Voice conversion of non-aligned data using unit selection. In: TC-STAR workshop on speech to speech translation
12. En-najjary T (2005) Conversion de voix pour la synthèse de la parole. In Rapport de Thèse, Université de Rennes I
13. Eriksson A, Wretling P (1997) How flexible is the human voice? A case study of mimicry. In: Proceedings in European conference speech technology, Rhodes

14. Fan X, Hansen JHL (2010) Acoustic analysis for speaker identification of whispered speech. In: Proceedings of ICASSP
15. Fant G (1960) Acoustic theory of speech production. Mouton & Co, The Hague
16. Farrus M, Wagner M, Anguita J, Hernando J (2008) Robustness of prosodic features to voice imitation. In: Proceedings of Interspeech, Brisbane, Australia, Sep 2008, pp 613–616
17. Fawcett T (2005) An introduction to ROC analysis. Pattern Recog Lett 27:861–874 (special issue on ROC analysis)
18. Gray AH, Wong DY (1980) The burg algorithm for LPC speech analysis/synthesis. IEEE Trans Acoust Speech Signal Process 28:609–615
19. Hatef M, Kitter J, Duin R(1996) Combining classifiers. In: Proceedings of ICPR, pp 897–901
20. Hirson A, Duckworth M (1995) Forensic implications of vocal creak as voicedisguise. Beitr Phonetik Linguist 64:67–76
21. Kain A, Maccon MW (1998) Spectral voice conversion for text to speech synthesis. In: Proceedings of the ICASSP
22. Kajarekar S, Bratt H, Shriberg E, de Leon R (2006) A study of intentional voice modifications for evading automatic speaker recognition. In: Proceedings Odyssey
23. Kittler J, Hatef M, Duin R, Matas J (1998) On combining classifiers. IEEE Trans Pattern Anal Machine Intell 20;226–239
24. Künzel HJ (1994) Current approach to forensic speaker recognition. In: Proceedings ESCA workshop on automatic speaker recognition, identification, and verification, pp 135–141
25. Künzel HJ (2000) Effects of voice disguise on speaking fundamental frequency. Forensic Linguist 7(2):149–179
26. Künzel H, Gonzalez-Rodriguez J, Ortega-Garcia J (2004) Effect of voice disguise on the performance of a forensic automatic speaker recognition system. In: Proceedings of Odyssey
27. Lau YW, Wagner M, Tran D (2004) Vulnerability of speaker verification to voice mimicking. In: Proceedings of international symposium on intelligent multimedia, video and speech processing
28. Lindsey G, Hirson A (1999) Variable robustness of nonstandard /r/ in English: evidence from accent disguise. Forensic Linguist 6(2):278–288
29. Maeda S (1990) Compensatory articulation during speech: evidence from the analysis and synthesis of vocal-tract shapes using an articulatory model. In: Hardcastle WJ, Marchal A (eds) Speech production and speech modelling. Kluwer, Amsterdam, pp 131–149
30. Maeda S (1992) Modélisation articulatoire du conduit vocal. J Phys IV(2):191–198
31. Mann MD (2006) The 'CSI Effect': better jurors through television and science? Buff Pub Int Law J 211:215–218
32. Masthoff H (1996) A report on voice disguise experiment. Forensic Linguist 3:160–167
33. Meuwly D (2001) Reconnaissance de locuteur en sciences forensiques: l'apport d'une approche automatique. PhD thesis
34. Nakamura S, Shikano K (1989) Spectrogram normalization using fuzzy vector quantization. Int J Acoust Soc Jpn 45:107–109
35. Orchard TL, Yarmey AD (1995) The effects of whispers, voice sample duration, and voice distinctiveness on criminal speaker identification. J Appl Cognit Psychol 9(3):249–260
36. Patil HA, Basu TK (2008) LP spectra vs. Mel spectra for identification of professional mimics in Indian languages. Int J Speech Tech IJST, Springer 11(1):1–16
37. Patil HA, Dutta PK, Basu TK (2006) Effectiveness of LP based features for identification of professional mimics in Indian languages. In: International workshop on multimodal user authentication, MMUA06, Toulouse, France, May 11–12, 2006
38. Perrot P, Chollet G (2009) Les mondes virtuels: un nouvel espace ouvert à la criminalité. In: Proceedings WISG workshop interdisciplinaire sur la Sécurité globale
39. Perrot P, Aversano G, Blouet R, Charbit M, Chollet G (2005) Voice forgery using ALISP: indexation in a client memory. In: ICASSP
40. Perrot P, Razik J, Chollet G (2009) Vocal forgery in forensic sciences. In: Proceedings of E Forensics, Adelaïde

41. Reich AR (1977) Speaker identification: effects of vocal disguise upon listener performance. J Acoust Soc Am 62(S1):S4
42. Reich AR, Duke JE (1979) Effect of selective vocal disguise upon speaker identification by listening. J Acoust Soc Am 66:1023–1028
43. Rodman RD (1988) Speaker recognition of disguised voices. In: Proceedings of the consortium on speech technology Conference on speaker recognition by man and machine: directions for forensic applications COST250, pp 9–22
44. Rodman RD, Powell MS (2000) Computer recognition of speakers who disguise their voice. In: Proceedings of the international conference on signal processing applications and technology, ICSPAT
45. Schlichting FF, Sullivan KPH (1997) The imitated voice—a problem for voice lineups? Forensic Linguist 4(1):148–166
46. Schweitzer NJ, Michael JS (2007) The CSI effect: popular fiction about forensic science affects public expectations about real forensic science. Jurimetrics, Spring
47. Sjötröm M, Eriksson J, Zetterholm E, Sullivan KPH (2006) A switch of dialect as disguise. Lund University, centre for languages and literature, Department of Linguistics and phonetics working papers
48. Solewicz YA, Sofer MK (2004) A robust framework for forensic speaker erification. In: SPECOM 2004: 9th Conference Speech and Computer
49. Sreenivasa Rao K, Yegnanarayana B (2006) Voice conversion by prosody and vocal tract modification. Proceedings of ninth international conference on information technology, Bhubaneswar, Orissa, pp 111–116
50. Stylianou Y (1996) Harmonics plus noise models for speech, combined with statistical methods for speech and speaker modifications. Phd thesis, Telecom Paris
51. Stylianou Y, Cappe O (1995) Statistical methods for voice quality transformation. In: EUROSPEECH
52. Sündermann D, Bonafonte A, Höge H, Ney H (2004) Voice conversion using exclusively unaligned training data. In: Proceedings Spanish society for natural language processing conference
53. Taseer SK (2005) Speaker identification for speakers with deliberately disguised voices using glottal pulse information. In: Proceedings of the 3rd international workshop on frontiers of information technology
54. Toda T, Black A, Tokuda K (2005) Spectral conversion based on maximum likelihood estimation considering global variance of converted parameter. In: ICASSP, pp 9–12
55. Toda T, Black A, Tokuda K (2007) Voice conversion based on maximum-likelihood estimation of spectral parameter trajectory. IEEE TASLP 15(8):2222
56. Torstensson N, Kirk P, Sullivan H, Erik JE (2004) Mimicked accents: do speakers have similar cognitive prototype? In: Proceedings of SST2004: the 10th Australian international conference on speech science and technology
57. Valbret H, Moulines E, Tubach JP (1992) Voice transformation using TDPSOLA technique. In: ICASSP
58. Vapnik VN (1998) Statistical learning theory. Wiley, New York
59. Wagner I, Köster O (1999) Perceptual recognition of familiar voices using falsetto as a type of voice disguise. In: Proceedings of the XIVth international congress of phonetic sciences, USA, pp 1381–1385
60. Yumoto E (1982) Harmonics to noise ratio as a degree of hoarseness. J Acoust Soc Am 71(6):1544–1549
61. Zetterholm E (2003) Voice imitation. A phonetic study of perceptual illusions and acoustic success. PhD dissertation
62. Zetterholm E (2007) Detection of speaker characteristics using voice imitation. In: Müller C (ed), Speaker classification II. Lecture Notes in Computer Science, vol 4441. Springer, Berlin, pp 192–205

Chapter 17
Applying Lessons Learned from Commercial Voice Biometric Deployments to Forensic Investigations

Chuck Buffum

Abstract Commercial deployments of voice biometrics have predictably focused primarily on automating the correct acceptance of true users for telephony self-service. However, over the past few years, a trend has developed within the financial institutions to begin using voice biometric technology to look for duplicate enrollments or to investigate suspicious transaction activity. This trend opens the discussion of bringing relevant techniques and experiences from commercial voice biometric deployments into the forensic voice biometric space. This chapter explores those techniques that show promise for crossing over from commercial to forensic use.

17.1 The Objective of Commercial Solutions

Commercial voice biometric applications are primarily focused on identification and verification (ID&V) applications. These are characterized by the user making a specific identity claim (user id, account number, etc) and one or more authentication credentials (voiceprint, secret question, device id, etc) being used to verify that the user is who they claim to be.

ID&V processes are often the first step in the majority of traditional transactions, ranging from internet transactions (user id + password) to phone transactions (account number and PIN or one or more knowledge questions). Voice biometrics are now being applied to the traditional ID&V process to streamline and simplify the user experience (no more password to remember, no more password to forget) while providing stronger, multi-factor authentication.

These enhanced ID&V solutions are being implemented first on phone based transactions in customer contact centers, where the benefits of automated authentication provide a compelling return on investment. However, we are now seeing a number of prototypes and pilots using voice biometric ID&V solutions for out of

C. Buffum (✉)
Nuance Communications, 1198 E. Arques Avenue, Sunnyvale, CA 94085, USA
e-mail: Chuck.Buffum@nuance.com

A. Neustein, H. A. Patil (eds.), *Forensic Speaker Recognition*,
DOI 10.1007/978-1-4614-0263-3_17, © Springer Science+Business Media, LLC 2012

band confirmation of internet transactions (voice signatures for e-commerce) and enhanced security on mobile devices (secure mobile access for m-commerce).

While enhanced security is a common thread for all these commercial applications, the primary objectives tend to be an increased automation rate with a simpler and more streamlined user experience for the ID&V process. As a result, the key performance indicators (KPI) for these solutions tend to be the following:

- Enrollment Success Rate—the measure of the ease and effectiveness of the voiceprint enrollment process
- Automated Authentication Rate—the measure of ease and effectiveness of the ID&V process for true users
- Security Rate—the measure of the effectiveness of the ID&V process at rejecting imposter access attempts

17.2 Text Dependent Voice Prints

In order to achieve these high levels of automation and convenience, most commercial deployments use text dependent voice biometrics. A user is typically asked to speak a pass phrase three times to enroll the voiceprint. Pass phrases may be a text string ("in Canada, my voice is my password") or a digit string (phone number, account number, etc). After enrollment, the text dependent voiceprint can be verified by a single utterance of the pass phrase. Thus, the text dependent voiceprint can typically be enrolled in less than 30 s and verified in less than 5 s.

In addition, when the pass phrase is designed to be the same as the identity claim (phone number or account number), the ID&V solution can deliver multi-factor authentication in a two step process which is faster and easier for the user and more secure for the enterprise. For example, the user speaks their phone number or account number which identifies the user (via speech recognition) and provides the first factor of authentication (via voice biometrics). Then the user enters their secret knowledge (date, password, etc), providing the second factor of authentication.

In a well designed, commercially deployed ID&V solution, it is reasonable to achieve enrollment success rates greater than 90%, automated authentication rates as high as 95%, and security rates greater than 98%—all significant improvements over traditional ID&V processes.

17.3 The Use of Text Independent Voice Prints in Commercial Applications

Although the use of text dependent voice biometrics in an enhanced ID&V application can deliver impressive commercial results, there remains an important role for text independent verification. That role is to be an effective deterrent to recorded audio attacks.

Text dependent voice biometric solutions are inherently vulnerable to recorded audio attack. In low risk (non financial) transactions, this risk is often mitigated by the use of a second factor (knowledge). However, in high risk or high privacy transactions (financial, health) aggressive imposters can be expected to record the specific utterance of the targeted victim and play that audio to the system. Since the voice biometrics system is comparing the presented audio sample (from the true user) with the enrolled user's voiceprint, it is expected that a positive match would be the likely outcome. This is exactly the place where a text independent voice biometric can be applied to overcome such an attack.

The text independent voice biometric is applied by prompting the user to speak a specific set of words or numbers, often randomly selected from a large list. This approach is often referred to as text prompted voice biometrics and burdens the imposter with the task of acquiring audio for all potentially prompted words or numbers from the victim and then assembling them in the proper sequence in response to the prompt.

The first generation of this approach used random digits, often a 4 digit string repeated twice (4-3-9-7 4-3-9-7). This approach was a reasonable first effort, but was too easy for the imposters to crack. The imposters quickly realized that they only needed to acquire the audio for the 10 digits and it was rather easy to assemble a PC tool to arrange the audio clips in the appropriate order in response to the prompt.

A second generation of text prompted voice biometrics is now in use and proving to be much more effective. It is based on the notion of prompting the user to speak two or three words or names randomly selected from a pre-defined list (Kevin Ivan Bates). This approach leaves the imposter with the ill defined task of acquiring audio from the victim for all words or names that might be prompted. Since that list is of unknown size and content, the audio acquisition task for the imposter is daunting. As a result, this approach is considered significantly more secure and effective in the use of mitigating the risk of recorded audio attack.

Text prompted voice biometric accuracy is lower than text dependent voice biometrics. The specific accuracy is a function of the enrollment approach used in creating the text independent voiceprint as well as the acoustic modeling techniques used to implement the performance of the text prompted approach. When these elements are well optimized, it is reasonable to achieve an EER in the 10% range. This level of performance is not acceptable as a primary authentication credential, thus it is not used in the primary ID&V process. However, it can be very effective to mitigate the risk of recorded audio attack in high risk transactions.

One note of caution: It is important in text prompted verification to make sure that the system is capable of understanding what the speaker says, not just score the biometric data. For this reason, most solutions implement automated speech recognition (ASR) as a quality control measure in systems using text prompted verification. This allows a well designed system to confirm that the user spoke the correct words or names in response to the prompt prior to scoring the voice sample. This approach prevents the imposter from simply playing the "closest audio samples" he has in response to the prompt. This is very important to the effectiveness of a text prompted system, which depends inherently on text independent voice biometric technology.

17.4 First Signs of Forensic Implications

Although the primary objective of commercial ID&V solutions is to correctly allow true users simple and secure access to self service applications, a secondary objective emerged as systems were implemented. That secondary objective was to prevent imposters from gaining access to those self service applications, and should they attempt access, then collect and retain as much evidence as possible to be used to investigate and potentially prosecute them. This second objective begins to bridge the commercial use of voice biometrics with related forensic uses of the technology

This forensic-related objective is now causing a number of new use cases and companion application solutions of voice biometrics as described below.

17.5 Analysis of Newly Enrolled Voiceprints, Looking for Duplicates

One source of potential fraud is the takeover of a customer's phone self service account. This might be accomplished by enrolling an imposter's voiceprint as a credential related to the intended victim's account. The proactive mitigation of this risk involves verification of the user with an appropriately strong set of authentication credentials prior to voiceprint enrollment.

In situations where such pre-enrollment authentication credentials are unavailable, or as post enrollment validation check, some commercial deployments are now creating a utility to compare new voiceprint enrollments with existing voiceprints in the database, looking for possible duplicates. Should a matching voiceprint be found with different identity information, at least one of them should be considered a potential imposter.

Depending on the nature of the voiceprint token, the duplicate identification process will vary. When the voiceprint token is text dependent and the same for all users ("my voice is my password") potential duplicates can be accurately identified using text dependent verification processes. However, when the voiceprint tokens are user specific (account number, phone number, etc) text independent voiceprint analysis must be used, resulting in a less accurate, but still effective process.

Another unique aspect of this utility is its user identification characteristics. Unlike an ID&V application, where a single user identity is claimed allowing the audio sample to be compared to a single voiceprint, the duplicate analysis utility needs to compare the specific voiceprint or audio sample to all available voiceprints, looking for potential duplicates. As the database of voiceprints grows, this task can become rather compute-intensive, leading to an offline process.

These user identification characteristics, text independent matching against a large database of potential target voiceprints, closely approximate the characteristics of many forensics voice biometric applications.

17.6 The Creation of Negative Voiceprint Files

While the duplicate voiceprint enrollment utility lends itself to an offline processing mode, some commercial customers are also interested in creating a real time solution for catching imposters in the act of trying to access self service systems or transactions. This requirement leads to the development of a "black list" or "negative voiceprint database".

The negative voiceprint database contains the voiceprint of known fraudsters and may also contain the voiceprints of suspected fraudsters. These voiceprints may be either text dependent or text independent, depending on the voiceprint token used. The typical usage scenario shows a negative voiceprint database being used in real time alongside a commercial ID&V solution. The system sends the user audio to both the ID&V solution so that it can look for a positive match for the claimed user, and to the negative voiceprint database respectively, so that it can search for a negative match in the fraudster database.

A low to moderate score on the positive voiceprint match raises some suspicion of the user's identity. A moderate to high score on the negative match raises significant suspicion of the user's identity. However, since the negative voiceprint comparison is one audio sample against all enrollment voiceprints (one of many), if the negative database becomes too large (1000 voiceprints or more) the testing result may not be returned quickly enough to be effectively used in real time applications. Further optimization of this technology is required to make this capability commercially viable for real time applications, but will also optimize its use in forensic (1:n) applications.

17.7 Implications for Fraud Investigation

Financial institutions are often instrumenting their ID&V applications such that when confidence scores are low to marginal in ID&V applications, all ID&V information presented is logged for potential fraud/imposter investigation. When properly implemented, such a system will record the audio sample from an imposter for post transaction analysis by fraud investigations personnel.

In the event of an imposter attempt, the imposter's audio is captured and saved, not only for comparison against currently enrolled voiceprint database (to look for duplicates), but also against the negative voiceprint file as an attempt to identify the imposter as a repeat offender. In such systems, the imposter always leaves his or her voiceprint behind as potential evidence of their attempt to fraudulently access the system.

17.8 Voiceprints as Evidence

This captured audio, freely volunteered by the imposter during the attempted transaction, can then be used as biometric evidence against the imposter once identified. This is easily accomplished by having the identified imposter create a voiceprint (text dependent will be the most accurate and convincing) after being apprehended and using the voiceprint matching score as evidence against him or her.

In this way, the commercial ID&V application is used to capture the primary evidence that is used in the forensic investigation to prosecute the identity thief.

17.9 Conclusion

Although the commercial uses for voice biometrics have strongly commercial objectives, there is a common thread with the forensics market. Both are keenly interested in identifying and constraining identity thieves and fraudsters. The commercial systems will be the ones that are most likely to capture audio evidence from these nefarious users. The forensic applications will use that evidence to help identify and prosecute the criminals. By leveraging common technology and optimizing it for both use cases, voice biometrics can be effectively deployed to address both markets.

Chapter 18
Designing Better Speaker Verification Systems: Bridging the Gap Between Creators and Implementers of Investigatory Voice Biometric Technologies

Avery Glasser

Abstract Though automated and semi-automated speech analysis and identification technologies have massive potential within law enforcement, forensics labs and intelligence communities, adoption has been slow and sporadic. This is partly due to poor experience with the previous generations of voice biometric technologies combined with a cultural mis-perception of voice biometrics being considered easily "spoofable" due to television and movies. However, the voice biometric technology vendors also have contributed to this challenge by producing products that fail to address critical implementer challenges. There are critical problems that only voice biometrics can solve, but getting the solutions well positioned requires a deep understanding of the nature of government implementations that seems to escape the grasp of too many vendors. This chapter will explore a number of critical use cases and provide perspective on how technology creators can position their solutions to meet those needs.

18.1 Introduction

The voice biometric industry faces a challenge that is unique to voice when compared with all other biometric methods. Fingerprint and iris analysis (as well as most of the physical biometrics) were studied primarily with the goal of being able to identify individuals. As automated biometric technologies were developed, the process of analysis moved from human observable details to minutiae that required specialized equipment. Even biometric technologies such as vein scanners and DNA analysis work on minutiae that, with the right equipment, essentially become observable. However, voice biometrics faces the special challenge that it's not visually observable. There are visual representations, but you can't look at a person's actual voice—even with the most powerful electron microscope.

A. Glasser (✉)
Loquendo S.p.A., a Telecom Italia Group Company, Via Arrigo Olivetti, 6, 10148 Torino, Italy
e-mail: averyglasser@loquendo.com

A. Neustein, H. A. Patil (eds.), *Forensic Speaker Recognition*,
DOI 10.1007/978-1-4614-0263-3_18,

At this point, fingerprint and DNA biometrics are accepted for use in forensic labs because there is the belief that with a good enough (magnifying), you can observe the exact physical biometric, and with enough time, an expert could manually view a set of samples and see what matches and what doesn't. This level of perceived certainty grants these technologies usage in courts under the Daubert standards. However, as voice can never be "seen", meeting the Daubert standard will require a significant effort in order to become admissible in court. However, even outside of court proceedings, there are significant uses for voice biometrics in the investigatory and forensic communities. However, if we're going to get these groups to accept voice biometrics, we need to better understand how the technologies will ultimately be used.

18.2 A Common Disconnect Between Implementers and Researchers

The First FBI-CJIS sponsored Investigatory Voice Biometrics Symposium was held at the NIST campus in March, 2009. For the first time, a group of voice biometric technology vendors, academic researchers, system integrators, members of the standards bodies, federal law enforcement, intelligence and forensic scientists were brought together by the FBI to discuss the challenges surrounding the use of voice biometrics for investigatory purposes, and I was fortunate enough to be in attendance. Now, "investigatory purposes" is an extremely wide field to address, it includes major use cases such as:

- forensic scientists building court testimony
- forensic investigators who are trying to determine if an arrest warrant for a suspect should be issued
- intelligence officers attempting to identify a voice from an intercepted phone call from a pool of previously recorded "individuals of interest"

We broke into smaller groups to talk about standards, use cases, interoperability and other challenges that investigators would face—most everybody was excited as we finally saw a path towards wider acceptance of the technology within domestic law enforcement. As we came back together, to present our working group findings, an academic researcher attached to one of the most prestigious university laboratories simply declared that voice biometrics was not usable for the purposes of an investigator.

To those in the industry, it is widely understood that voice biometrics have been used for investigatory purposes throughout the world since the mid 2000s. Deployments have been put in place by intelligence and law enforcement in three of the 5-Eye nations, implemented as a national program by a number of NATO countries and is used extensively and publically in Mexico, Latin America, South America and Asia. In Germany, tools have been approved for use by forensic scientists to perform deep analyses of voice samples for use in court. In fact, the most surprising country that doesn't use voice biometrics for investigatory purposes is the United States of America.

The academic researcher's position was factually correct, but demonstrated the naïveté that many researchers fall prey to. He stated that even if one could have a false match rate of less than 0.5%, that comparing a voice sample against a list of 100,000 previously recorded voices for the purposes of identification would result in a list of 500 possible matches. He argued that 500 voices are too many for an investigator to make use of—and many academics supported his position.

Without voice biometrics, if voice is the sole or primary piece of evidence that one has, finding a match is a combination of deduction and blind luck. You can use other intelligence to try and create a list of individuals who were known to be active in the area where the voice was intercepted or try to figure out nationality by accent. But aside from passing the voice to an investigator who can say "Hey, that's Bob's voice. We went to college together—I'd recognize that voice anywhere," without a rigorous method of deciphering the suspect's voice what do really have?

When investigators hit a brick wall when trying to find a suspect, many times, they call upon the public for help—they give suspect descriptions, show photos and sketches. However, when voice is the sole piece of identifying evidence, this isn't always possible to expose, as the suspect's voice sample contains content that may not be suitable for public broadcast as it may pertain to some future crime. Voice is a behavioral representation of a physical biometric: the physical biometric is the vocal tract but it is only observable when an individual is speaking.

So, without voice biometrics, having a voice sample is essentially only useful to confirm/deny the identity of a suspect that was identified using some other technique.

The researcher was challenged back by a few in the crowd: with voice biometrics, we go from not being able to identify an individual solely by voice, to having a chance at identifying the individual using his or her voice. From an infinite set of possible matches, we end up with a list of a few hundred or a few thousand suspects. Once that list is made, it can be culled using other investigatory techniques, for example, excluding people who have alibis. This list can then be double-ranked, once based on the biometric score returned by the technology (biometric likelihood) and once based on other investigatory methods. The list may still run into the hundreds or thousands, but at least there should be a clear ordering of the priority for investigators—a list of who to consider as a suspect and in what order to follow up with people.

Investigators in the room agreed—positioned this way, as a tool to help create and order a likely suspect list, regardless of the length of the list produced, would be extremely helpful. More important, this is exactly how fingerprint identification works. Unlike what we see in shows like CSI and NCIS, investigators rarely get back a single fingerprint that's a 100% match—instead, they get back a list of fingerprints with associated likelihood ratios, which are then compared to other intelligence to determine if one of them is the likely suspect. The researcher was technically correct, but at the same time, the facts led him to an incorrect assumption.

Simply changing the conversation from Identification *by* Voice to Identification *utilizing* Voice changed the tone for the rest of the session.

18.3 Understanding the Use Cases

Through this narrative it starts to become clear. Though the technology can always improve, it's not the technology that is holding back implementation—it's the technology vendors' inability to understand the use cases of the intelligence and law enforcement sector that has held back the implementation of voice biometrics. If the technology vendors can't "speak the language" of their customers in the same way that fingerprint vendors can, investigatory voice biometrics will fail to be a reality in the United States.

The first task that we, as an industry, must take is to better understand the nature of how the technology is going to be used. In a broad sense, we can set a number of categories in which voice biometrics can be useful: Access Control, Surveillance, Target Identification and Forensics.

18.3.1 Access Control

Access Control is the most common category of biometrics usage in general, and of this broad category, identity verification is the most typical application. Identity verification typically starts with an individual providing some sort of identifying credential that is then validated biometrically. For example, Federal Information Processing Standard 201 (FIPS-201) describes a Personal Identify Verification process compliant with Homeland Security Presidential Directive #12 (HSPD-12).It combines a smartcard (an identity credential) that can be managed using a biometric method, principally a fingerprint. One can insert the smartcard into a reader on a computer and then provide a fingerprint as a method to confirm the identity of the user who presented the smart card.

Of course, access control can be much simpler than the smartcard based systems as well: it can operate a time card system in a factory, confirm an identity at a bank, or provide access to a laptop. In each of these cases, a user somehow identifies his or her self before presenting the biometric: the time card system is activated with a swipe card or PIN, at a bank an ID card is presented or ATM card is swiped, and on a laptop, the user clicks on their claimed logon (if there are multiple logons) before presenting a fingerprint.

18.3.2 Surveillance

One of the most common applications of voice biometrics in the intelligence world focuses on surveillance. As it pertains to voice biometrics, surveillance attempts to intercept conversations performed by a specific target speakers or set of target speakers. To properly conduct surveillance, two things must happen: first, you must have an idea of where the target will be (either physically or virtually) and second, you must have an appropriate probe that can intercept the audio communication. There is a third aspect that also needs to be considered: making sure that these intercepts fall within the proper legal framework.

In the old days, before cell phones were common, attempting to capture a target's communication was relatively straightforward. A court order was requested under the Electronic Communications Privacy Act of 1986 (commonly known as a Title 18 intercept) that would allow a law enforcement agency to place probes (aka "bugs") in the target's home, work or other location where he or she was known to regularly frequent. Some probes were tailored to specifically intercept telephone communications and others were general microphones used to intercept conversations. Officers would then physically place the devices and then sit in a monitoring van to pick up the transmissions from these probes. Requests could also be made to the phone operators to allow monitoring to occur from within the phone company's facilities. These phone company probes were true "wire taps" as they were remotely initiated.

However, in the modern age, voice conversations can come across many channels: instant messenger, IP telephony, satellite phones, walkie-talkie, cell phone and landline phones are the most common methods for communication. To perform monitoring, there are three common categories of probes now. Local probes have to be placed near the source of the communications. These can be the "bugs" and telephone probes previously mentioned or even advanced computer applications that are surreptitiously placed on a target's computer to intercept computer based voice communications. Remote probes can be handled at a phone base station (cellular) or central office as can IP probes from within the internet provider's facilities. If allowed by law, there is also the option of air probes, which allow law enforcement and intelligence agencies to passively or actively intercept (and potentially decrypt) GSM and CDMA communication.

What is legally permissable is typically based on what type of agency performs the request. In the USA, where law enforcement typically works under the Electronic Communications Privacy Act, intelligence agencies fall under the Foreign Intelligence Surveillance Act—which has a much wider scope regarding what is permissible. Other countries also have similar designations to what can be performed by law enforcement versus intelligence agencies.

For an intelligence agency operating overseas, they may have very broad capabilities. For example, military intelligence may have the right to monitor all cell phone transactions in a given city, or monitor the footprint of an entire cell tower searching generically for possible terrorist acts, individuals fitting a specific profile or hunting for a specific individual.

Conversely, in many jurisdictions, call intercepts performed by law enforcement can only be recorded if it has been determined that the target named in the warrant is on the phone and that the call has some sort of salient purpose. Only specific phone numbers or physical locations are probed to minimize any violation of individual civil liberties. In this example, if searching for a fugitive, law enforcement may get a warrant to tap the fugitive's family's phones as well as their known close associates. However, for each phone line, there needs to be an officer listening to the call to determine if the call can or cannot be recorded. Fugitives commonly attempt to outwit live listeners having a third party make these calls, talk about something innocuous and then quickly take the phone, pass a message, and then return it to the third party, hoping that the live listener has stopped listening, assuming the call is not of importance.

18.3.3 Target Identification

Target identification is essentially the inverse of surveillance. Where in surveillance we know who we are looking for but don't know where that individual is, target identification is when we have an individual's voice already but don't know who he or she is.

Classically, law enforcement agencies have held large databases of fingerprints and photographs of individuals. Depending on the country and the agency's charter, it can range from convicted criminals to anyone arrested under suspicion of committing a crime. Intelligence agencies also collect biographical and biometric information on people of interest. In addition, law enforcement and intelligence agencies can get access to databases from the military that list all active and retired soldiers, some Department of Motor Vehicle departments, individuals working in controlled sectors (financial service, law enforcement) and even commercial databases of biometric and biographical information.

Biometric Identification can be performed when you have an individual present or non-present. If an individual is present, biometric verification is also possible. If, for example, you have arrested a suspect who does not have any identification on him or herself, you can take a controlled set of fingerprints (where a trained officer collects the fingerprints) and run them through a fingerprint identification system to see if the individual appears in the available databases. If not, then these professionally collected fingerprints can be added to the database, even if the individual's identity is not fully confirmed. If the suspect also provides his or her identity, if a fingerprint record is on file, the suspect's fingerprint can be used to provide biometric verification of the claimed identity.

If a latent fingerprint is collected (for example, from a crime scene), then it would be considered a non-present acquisition of the biometric data. In this case, the officer is working from a sample that was not provided under a controlled environment. It could be a full print, or a partial print. It can be smudged or otherwise distorted due to environmental factors. This sample would be compared against the entire applicable database to see if there is a match. If there is no match, it may still be entered into the system to see if other future matches occur. In the case of a latent fingerprint, the police may use this to request a warrant from a district attorney so further surveillance, searches or arrests can be performed.

Another specific type of target identification is called link analysis. Link analysis is performed, at various levels, by both intelligence and law enforcement. Link analysis refers to the process of attempting to determine who an individual is interacting with. Let's say, for example, that an accountant is arrested for falsifying tax returns. In the process of the investigation, they check his phone records and find a number of calls back and forth to a phone number. Using link analysis software, they also see that a number of drug dealers call this number regularly as well. By following the links between individuals, they can determine that somehow the accountant and the drug dealers may somehow be linked and investigate who this common contact is. This type of analysis is commonly used in counterterrorism, drug enforcement and anti-mafia/organized crime investigation.

18.3.4 Forensics

One of the more significant challenges for technologists is understanding the role and scope of forensics and biometrics, especially in the United States. What has historically been lumped together into one major application use case needs to be cut in half. The primary use case that biometrics companies have focused on is that of the evidentiary preparation for trial. For the purposes of this section, we will call this the "forensic" use case.

In the forensic use case, a forensic scientist is retained to analyze large amounts of evidence in order to produce a scholarly, learned opinion that can then be entered into court as testimony. Therefore, in its most basic analysis, the pure forensic use case typically falls to a judiciary action. The nature of the forensic use case is that technology can (and typically will) be used by a forensic scientist who can be considered an "expert" in court. The output from these automated biometric analysis systems is not as much admissible—where a full report from the forensic scientist is admissible.

One of the most significant elements of the forensic use case is that it is not extremely time sensitive. Forensic scientists can spend weeks or months analyzing a single case.

On the other hand, we have a category that is just being better understood: pre-forensics, also known as forensic investigations. Where pure forensics are typically handled by scientists in labs as part of the overall prosecution (or defense) of an individual during court proceedings, pre-forensics is the act of performing basic forensic analysis in order to determine if a crime has been committed and to create a list of suspects. If you think of it this way, pre-forensics is an investigatory function of Law Enforcement that may result in an arrest or search warrant, while forensics is a function of the judiciary process that may result in incarceration.

Because pre-forensics is a law enforcement process, the operators typically have less time to work on any individual case and have less theoretical training than forensics labs. However, because the result of the pre-forensic analysis can only result in an arrest or search, the standards for assurance are much lower.

There is a major overarching limitation for the use of biometrics in the forensic space (though not as much in the pre-forensic space). The issue is the Daubert standard. The Daubert standard sets the bar for the use of technology for creating expert testimony. To pass the Daubert Trilogy, technology must: be peer reviewed, have a theory of usage that is unquestionable, have a defined error rate and be accepted by the scientific community as acceptable in court. To this point, polygraph, DNA and fingerprint methods all have successfully been proven to meet the Daubert challenge, while voice biometrics has not. This is not to say that today's voice biometrics cannot pass the Daubert criteria, simply that to this date it hasn't been appropriately presented to determine what, if any issues, may still need to be addressed.

18.3.5 Voice Biometrics and the Use Cases

Now that the categories of possible use have been laid out, we need to understand how voice biometrics can be used to increase the effectiveness of each of these use cases.

18.3.6 Access Control

Access control is a very simple application and it is one of the most common use cases for voice biometrics in general. In the vast majority of applications we have a number of conditions that create a very high acceptance rate, with low false accept and false rejects. The primary reason is the psychology of the interaction. When someone is registering their voice for a telebanking application, he or she wants to provide a good voice print as it will make their authentications more successful when attempting to call in and get their bank balance. This means that the act of both enrolling on the system and verifying their identity is a very controlled action. Scripts can be put into place for the enrollment and verification segments. Many access control applications still use text-dependent technologies: Hidden Markov Model (HMM) based algorithms that perform very well as long as the channel doesn't change (it will always be a call for verification from the same channel that enrollment was performed on) and when the enrollment text does not change from what the verification text is. It is essentially very inflexible: the equivalent of the physical key to access a door: if you change the lock (in this case the authentication phrase), you need to cut a new key (voice enrollment). New access control applications are starting to be built using Gaussian Mixture Models (GMM)/Joint Factor Analysis (JFA)/i-vector based text independent technologies. Here, a universal voiceprint is built using a longer set of utterances, and then, by coupling speech recognition to the voice biometrics, the identity challenges can be randomized, where the text of what was said is evaluated for accuracy by the speech recognition engine and the actual voice can be evaluated by the voice biometric engine.

With access control applications, there's a very specific psychological desire to be understood by the system so the caller can quickly get what he or she wants.

18.3.7 Surveillance

Surveillance is one of the key use cases that voice biometrics can be extremely successful on. One needs to consider that voice biometrics simply becomes a plug in for existing surveillance solutions. This is a key point that needs to be fully understood—barring some radical technology change, most law enforcement agencies and intelligence agencies have already made their choice regarding what sort of probes they are going to use, what their biometric database structure is and what kind of administrative tools they are going to provide the users access to. To this point, the voice biometric engine needs to simply be a plug-in to these existing surveillance systems.

The benefit of using voice biometrics becomes two-fold. First, because the system is fully automated, where before one person needed to be dedicated for each phone line that was being monitored, now a system can watch multiple T-1/E-1s without the need for regular human intervention. This means that mass monitoring can be performed: instead of having to pick specific phone numbers for "wire taps", a cell tower, neighborhood or community can have their calls run through a voice biometric surveillance system, searching for calls where there are voice matches.

The second benefit is that by its nature, voice biometrics removes the need for humans to listen to the audio, which is a possible affront to civil liberties. Aside from the sheer manpower requirements when listening to an entire community's phone calls, civil libertarians can make clear civil (and possibly criminal) complaints against indiscriminate monitoring. The voice biometric system simply collects calls and determines if there is a suspected voice match. If this is the case, then some further action can be made—the call could be recorded, a live listener could be alerted or the phone number can simply be flagged for a further wiretap warrant.

Because surveillance is typically planned, the probe types can be well defined before surveillance begins, allowing a solution to be well calibrated before it starts processing live audio. Uncalibrated systems or working with unknown probe types is typically a very large contributor to unsuccessful voice biometric implementations.

18.3.8 Target Identification

For decades, police departments worldwide have been photographing suspected criminals and collecting fingerprints in the police station. This type of controlled collection has allowed police (local, federal/national and international) unparalleled access to biometric data on known and suspected criminals. In many countries, such as Mexico and Spain, criminals are not only photographed and fingerprinted, but controlled voice samples are also collected as part of the intake process.

By building large databases of voices, similar to DNA and Fingerprint databases, law enforcement and intelligence agencies can attempt to identify individuals primarily based on their voices, then use other investigatory methods to determine if the individual is a fit. This is extremely helpful when dealing with issues such as phoned-in bomb threats or terrorist communiqué videos posted on the web.

From a link analysis perspective, the use of voice biometrics can become key when working on issues ranging from organized crime to terrorism. When a suspect places a call, target identification can be run both to confirm that the suspect is the one placing the call and to attempt to determine who the called party is. Even if the called party cannot be identified, his or her voice can then be used to check other recorded calls to see if there is a match. So, known person A talks to unknown person B. Unknown person B talks to unknown person C. Unknown person C is linked to a violent crime. Now, investigators can go to person A, to get access to person B, who can link the investigator to person C.

18.3.9 Forensics

In the United States, working in forensic labs for the purpose of creating testimony that can be used in court is not possible as no single voice biometric technology has passed the Daubert challenges. However, in many countries around the world, there are public cases where the use of voice biometrics by a trained acoustic forensic scientist is permitted as court testimony.

In these cases, the voice biometric engine is not used in an automated manner. Instead, in its simplest form, the engine is run in a manual mode, where the scientists have the ability to perform inter-session variability checks as well as intra-session variability checks. This works by taking multiple voice samples of a suspect and determining how the voice changes within a recording and between different recordings. These voice samples are then compared to a population of individuals considered by the forensic scientist to be similar in nature to the suspect—typically based on nationality, gender, age, relative health, regional dialect, education level, financial capabilities, etc. The goal is to produce a conclusive likelihood ratio in the same way that fingerprints are used in court. The process of building the appropriate reference populations can take weeks or months.

When we watch television shows like Law and Order, CSI or NCIS, there's always the pivotal scene when the forensic scientist declares that the fingerprint is a "14 point match" or "100% match" between the print and the suspect. However, in reality, this is not what is used in court. Instead, likelihood ratios are used. A 100% match, when balanced by the size of the reference population as well as the size of the reference population versus the total possible population, may equal something to the point of a 240,000::1 likelihood that the person is the match to the fingerprint. No biometrics are ever absolute.

Outside of the world of court and Daubert sanctioned voice biometrics is the world of pre-forensics or forensic investigation. Their responsibilities fit in between those of the forensic scientists and the target investigations. Here, forensic technicians (and some forensic scientists) use forensic techniques in order to determine if someone is or is not a suspect. Because of this, time is of the essence. At the same time, where in the courts an incorrect analysis can mean the difference between life and death, here, the worst consequence is an erroneous arrest. It's still a major consideration, but not one of such a dire nature.

18.4 Solution Requirements

At the same time, speed is of the essence. Where forensic scientists in a court case may have months to prepare the evidence, a forensic investigator may have minutes to be able to take a voice sample and confirm it's the suspect. To this point, tools need to be better tailored for this use case. We can still require the need to have multiple samples of the suspect's voice so inter and intra session variability can be tested. However, we do not necessarily have to build a custom reference population. To this point, Loquendo has taken an innovative approach of building a series of standard targeted reference populations for their pre-forensic tool (Loquendo Voice Investigation System (LVIS)—Pre-Forensic). Their system comes pre loaded with over 60 reference populations broken down by language, gender, country/region and probe type. So, if a forensic investigator has the voice of a Mexican national, male, collected by the phone, speaking Spanish, instead of spending the time to find 50 similar individuals to build a reference population from the same age, economic background and region, the investigator can simply select the Mexican-Male-Spanish-Telephony reference population and produce a good likelihood ratio. So, where the forensic scientist can drive down to the "150000::1 likelihood that this is a match" as court testimony, the forensic investigator can say to a judge issuing a warrant "Given that this is a Mexican male, we are comfortable saying that there is a 120000::1 likelihood that this is the suspect." This can be done in minutes and can provide feedback that can easily be used to get a court order for an arrest.

Instead of discussing the specifics of any given voice biometric algorithm, it makes more sense to discuss what we, as an industry, need to start focusing on in order to better handle all of these use cases. Loquendo, as one of the few vendors solely focused on the government and intelligence space, has been actively working to improve both the core biometric engine as well as these other ancillary features:

18.4.1 Calibration Testing

Operators need to be able to take an audio sample (or sets of samples) collected using a new probe or a new method and the system needs to be able to determine if the system is well calibrated for this acquisition method. If not, it needs to prescribe a course of action regarding how to recalibrate this system. Loquendo has implemented this in the LVIS Pre-Forensic version of their software.

18.4.2 Channel Handling

One of the most significant challenges for voice biometrics rises out of channel mismatch handling. This is when a sample recorded on a microphone is compared to a sample recorded using an air probe or a Very High Frequency (VHF) probe or even a hands free speaker phone. For surveillance, target identification systems and forensic solutions to be widely adopted, this needs to be improved across the board. This can be addressed in the core voice biometric algorithms or with updated normalization strategies.

18.4.3 Speed

To better handle massive surveillance and mass identification projects, speed becomes a critical issue. If the system is slow, then it becomes functionally unusable in many critical use cases.

18.4.4 Language Independence

This is a critical function that is required when performing surveillance and target identification. When individuals are recorded at a police station, for example, they may speak one language. However when their voice is intercepted during surveillance, they may be speaking another language or even code words. Systems need to be truly language independent when analyzing voice samples.

18.4.5 Voice Change

For the purpose of surveillance, being able to determine when in a stream or an audio file the voices change is critical. Being able to determine if the phone has been passed to a third party who may be providing information is a critical trigger for a live listener to determine if the call needs to be listened to.

18.4.6 Key Word Spotting

For the same reason as the voice change, it is important to have key word spotting available to analysts. Just as a voice chance may be a trigger for a live listener, so may the speaker uttering a specific word or phrase. Of course, implementing key word spotting in a vacuum can be an issue. For example, “I will detonate this bomb” and "My girlfriend is the bomb" both include the word bomb, but the semantics are completely different. Semantic processing of key words, potentially combined with emotional analysis, becomes a key technology that needs to be developed for the surveillance market.

18.4.7 Language Identification

As voice change and key word spotting are issues during surveillance, so is language identification. One reason is that by identifying the language being spoken, the correct live listener can be brought into the call. It doesn't make sense to have a Pashto translator listening to a call that ends up being Urdu. It also can be used for surveillance profiling. For example, if you are monitoring Chinese drug gangs and the system detects Mexican Spanish being spoken, that may be a person of interest to start monitoring. In addition, Gender Identification can be similarly helpful.

18.4.8 Anti-Spoofing

This has traditionally been the function of the Access Control space—determining if a voice is live or if it is synthetic or somehow manipulated to sound like a person's voice. However, in the surveillance space, a synthetic voice or a digitally modified voice may also raise concern. If an automated surveillance systems notices two people are speaking using obvious digital manipulation, that could be a signal to start monitoring.

Different types of analog and digital manipulation can also carry signatures. Though the original voice pattern may not be able to be extracted, it may be possible to determine out what kind of manipulation method was being used.

18.4.9 Signature Detection

Every device that transforms or records a voice leaves a signature when used—this includes a microphone, headset or telephone handset. Determining what types of phone/microphone are being used can be helpful to forensic investigators in the course of identifying a suspect for a crime.

18.4.10 Better End-Pointing

Unless you're taking a call in a sterile room without any form of background noise, there are going to be aberrations in any audio sample. It could be a car horn, music in the background, gunfire, wind or any other non-speech noise that can be introduced. Vendors need to improve their capabilities to identify in a speech sample, what parts are speech related and what parts are environmental.

18.4.11 Real Time

For many use cases, audio cannot be buffered or recorded. Loquendo has implemented a sliding-window method that avoids the need to do any buffering on a live stream in order to perform voice biometric analysis.

18.4.12 Support Field Devices

We are now seeing that military, law enforcement and intelligence agencies want access to self-contained voice biometric systems that can be used on a handheld device without network connectivity for target identification and field enrollment. Algorithms and capabilities that were designed for high powered server environments need to be re-designed to work on Android, Windows CE and mobile Linux platforms to meet the needs of the 21st Century Warfighter.

18.4.13 Shorter Utterances

When systems are deployed, there are times when only a small amount of audio can be provided for analysis. Vendors need to look at how they can reduce the amount of audio needed to build a comprehensive voice print or perform an analysis. This can also help in very noisy environments where the Signal to Noise Ratio (SNR) may be bad on average, but there may be small snippets of audio that are clean enough to be processed.

18.4.14 Likelihood Ratios

Voice biometric vendors typically report back their results as a confidence score or percentage. These numbers are essentially only useful in a relative sense… is a 3.1 good or bad? It's better than a 2.0 and not as good as a 4.0. Vendors need to start using a Bayesian approach to Likelihood Ratios, reporting back values such as "there is a 20000:1 likelihood that the voice is a match for the target versus a random sampling of other voices from around the world" or "There is a 50000:1 likelihood that the voice is a match for the target versus a random sampling of other Mexican males."

18.5 Building the Short List

Though this is not a technology requirement, one of the key elements of success that needs to be understood is how to create a target list from a set of results. There are four strategies that are typically returned:

- Single Threshold: In a one threshold system, any response over that threshold is considered a possible match, and any response under that threshold is considered not to be a likely match
- Dual Threshold: In this case, two thresholds are set, where anything above the top threshold is a likely match, anything below the lower threshold is not a likely match and anything in the middle is a possible match
- N-Best: Here, an implementer simply states that they want the *N* closest matches to the target voice
- Full list: This is when all results are returned to an operator

Each of these strategies have their pros and their cons. The threshold based systems require live-testing calibration to determine where the calibration point should be. N-Best lists can typically provide short-lists that include the target speaker in a very high number of cases, however these can be thrown off if there is not a match in the system or if there is a cluster of many targets that score similarly (if *n* is set to the top 10 results but there are 30 results that all score very similarly). Full lists give a trained operator more granular control, but can be unwieldy if there are more than 20 possible targets to test against.

It is important to use logic when setting thresholds or N-Best lists. If a target is of very high importance, you may choose to favor a false match if it ensures that there are no false rejects. Conversely, for a low priority target, possibly failing to identify is more important than tracking down possible false matches.

For integrated solutions, there are other pieces of information that can be fed into a decision engine. For example, law enforcement and intelligence services typically have full dossiers on suspects that list items such as known whereabouts. If you get a voice in Chicago that matches the voice of someone known to be in federal lock-down in Miami, it's probably safe to exclude that as a match. In a full solution, it can be much more valuable to send all results to an application that can apply these types of rules instead of fixing a threshold in the biometric engine.

18.6 Project SPEC

We have spoken about the need to understand the implementer's use cases and to ensure that the technology is prepared to meet or exceed the real world requirements that will be placed upon it. However, if the strategy for rolling out the technology is flawed, even a technically successful deployment may be considered to be a failure. Though there have been books written about how to successfully manage complex technology deployments, it's worth keeping a simple acronym in mind: SPEC.

18.6.1 Scope

The first stem in ensuring a successful deployment is to take the time to fully understand how a solution will be used and what will make it successful. It's almost impossible to spend too much time scoping out what a project will truly entail. Will the voice biometric solution work independently, or will it be integrated into a multi-biometric searching system? What sort of probes will be used and what could cause new probes to be introduced? What is the overall charter of the project and what real life problem is it trying to solve? It's important to level-set the implementer regarding what can be achieved with the technology and to ensure that they understand what it will take to properly integrate and deploy a voice biometric solution.

Scoping a project is not passive. As much as vendors need to extract information from the implementers, it's just as important for the vendors to relay the best way for the technology to be implemented. Many times, what seems like an unreasonable technical goal may be a mis-statement of a reasonable operational goal.

18.6.2 Prototype

One should never deploy a system without first building a prototype to evaluate the technology's performance in a series of near-live environments. Depending on the complexity of the final integrated solution and if the solution will be deployed in a classified environment, sometimes two or more prototypes are necessary. To illustrate this point, let's discuss a voice biometric implementation for military checkpoint control.

During the scoping step, it was determined that the goal is to integrate voice biometrics into a multi-biometric acquisition and identification system that will be given to war fighters in an active theater of engagement. The system will be deployed using hardened laptops with proprietary, military use only microphones.

The first prototype would be to take the technology and build a simple voice biometric standalone application that could be run in a lab. The implementation partner would need to collect enough audio using that microphone—either in a live or simulated live environment—to see how the voice biometric solution works with that audio. It's possible that the biometric technology or the microphone may need to be adjusted, or that there are conditions where background noise may impact the performance. This is where proper scoping becomes critical as many technological crises can be mitigated with a programmatic/operational solution. Sometimes, just being able to return an error message saying "Have the target speak louder" satisfies an immediate need better than spending months to improve low signal to noise ratio performance.

The second prototype would be to integrate the solution into the overall system and re-test to see that the integration works properly. Having the first prototype to confirm that the technology works with the desired probe reduces finger pointing between vendor(s) and system integrators if there are problems when testing in a live environment.

18.6.3 Execute

Building a proper roll-out plan is critical for success. Does the system need to be calibrated using live audio? How many units should be deployed as a beta before certifying that everything is working properly? These are all critical questions, but the most important is this: will you have access to any live performance data to ensure that the system is working properly?

In many use cases, based on clearance levels, the vendor may never be able to gain access to live audio samples or get final specifications on the deployment hardware. In these cases, you need to determine if it is necessary to build an execution plan that includes a pre-deployment simulation for aspects of calibration or system training.

18.6.4 Control

When scoping a project, it is important to build the success criteria—not solely based on technology goals, but also on operational system goals. Once the system has been deployed, it is important to determine a point in time where the solution will be evaluated to determine if the goals have been met. This is the control point and the opportunity to proactively determine if the implementer is happy with the solution. If there are issues that need to be addressed, SPEC can be used for handling solution changes as well as implementations.

18.7 Conclusion

Worldwide, the case for using voice biometrics in investigatory, forensic and judicial processes has been made. As the international precedents have been set, what holds back implementations in the United States is the voice biometric community itself. We have a habit of speaking to our implementation customers with the guarded voice of a researcher, not the confident voice of a vendor. Though voice biometrics is not infallible, no statistical identification method is—fingerprints, DNA and iris scanning all have acceptable levels of tolerance for errors. These levels are set not by the technologists, but by the implementers. As an industry, we can no longer confuse voice biometric accuracy with speaker identification's utility.

About the Editors

Amy Neustein, Ph.D. is Editor-in-Chief of the *International Journal of Speech Technology*, Series Editor of *SpringerBriefs in Speech Technology* and editor of *Advances in Speech Recognition: Mobile Environments, Call Centers and Clinics* (Springer Verlag, 2010). Dr. Neustein is the Founder and CEO of *Linguistic Technology Systems*, a NJ-based think tank for intelligent design of advanced natural language understanding software to improve human response in monitoring recorded conversations of terror suspects and customers' calls into contact centers. She is a graduate of Boston University where she received her Ph.D. in sociology; her specialty area is Conversation Analysis. Dr. Neustein has published a number of scholarly articles, chapters and books and is the recipient of a pro Humanitate Literary Award, the ITNG (Information Technology Next Generation) Medical Informatics Award, and a Humanitarian Award. She serves on the visiting faculty of the National Judicial College and as a plenary speaker and moderator at academic and industry conferences in soft computing, information assurance and security, and speech processing and technology. She is a member of MIR (Machine Intelligence Research) Labs.

Hemant A. Patil, Ph.D. is assistant professor at Dhirubhai Ambani Institute of Information and Communication Technology (DA-IICT) Gandhinagar, India. Dr. Patil received his Ph.D. from IIT Kharagpur. Dr. Patil's research findings in speaker recognition and signal processing have been published in a number of conference proceedings, peer reviewed journals and edited books. His co-authored paper on variable length Teager energy operator was awarded ISCA grant during INTERSPEECH 2008, Brisbane, Australia. Dr. Patil was as a short-term scholar at the University of Minnesota in summer 2009. He is a member of editorial board for Engineering Letters, IAENG International Journal, Hong Kong.

A. Neustein, H. A. Patil (eds.), *Forensic Speaker Recognition*,
DOI 10.1007/978-1-4614-0263-3, © Springer Science+Business Media, LLC 2012

Index

A. Neustein, H. A. Patil (eds.), *Forensic Speaker Recognition*,
DOI 10.1007/978-1-4614-0263-3, © Springer Science+Business Media, LLC 2012